Mechanics of Microelectronics

SOLID MECHANICS AND ITS APPLICATIONS
Volume 141

Series Editor: **G.M.L. GLADWELL**
Department of Civil Engineering
University of Waterloo
Waterloo, Ontario, Canada N2L 3GI

Aims and Scope of the Series

The fundamental questions arising in mechanics are: *Why?, How?,* and *How much?* The aim of this series is to provide lucid accounts written by authoritative researchers giving vision and insight in answering these questions on the subject of mechanics as it relates to solids.

The scope of the series covers the entire spectrum of solid mechanics. Thus it includes the foundation of mechanics; variational formulations; computational mechanics; statics, kinematics and dynamics of rigid and elastic bodies: vibrations of solids and structures; dynamical systems and chaos; the theories of elasticity, plasticity and viscoelasticity; composite materials; rods, beams, shells and membranes; structural control and stability; soils, rocks and geomechanics; fracture; tribology; experimental mechanics; biomechanics and machine design.

The median level of presentation is the first year graduate student. Some texts are monographs defining the current state of the field; others are accessible to final year undergraduates; but essentially the emphasis is on readability and clarity.

For a list of related mechanics titles, see final pages.

Mechanics of Microelectronics

by

G.Q. ZHANG
*Philips Semiconductors and Delft University of Technology,
The Netherlands*

W.D. VAN DRIEL
*Philips Semiconductors and Delft University of Technology,
The Netherlands*

and

X.J. FAN
Intel Corporation, U.S.A.

A C.I.P. Catalogue record for this book is available from the Library of Congress.

ISBN-10 1-4020-4934-X (HB)
ISBN-13 978-1-4020-4934-7 (HB)
ISBN-10 1-4020-4935-8 (e-book)
ISBN-13 978-1-4020-4935-4 (e-book)

Published by Springer,
P.O. Box 17, 3300 AA Dordrecht, The Netherlands.

www.springer.com

Printed on acid-free paper

All Rights Reserved
© 2006 Springer
No part of this work may be reproduced, stored in a retrieval system, or transmitted
in any form or by any means, electronic, mechanical, photocopying, microfilming, recording
or otherwise, without written permission from the Publisher, with the exception
of any material supplied specifically for the purpose of being entered
and executed on a computer system, for exclusive use by the purchaser of the work.

Printed in the Netherlands.

CONTENTS

Preface xi
1. Microelectronics Technology 1
    1. *Introduction* 1
        1.1 A Heart of Silicon 3
        1.2 In a Little Black Box 7
    2. *Baseline CMOS* 8
        2.1 Diffusion 11
        2.2 Patterning 11
        2.3 Deposition 12
        2.4 Planarization 13
        2.5 Integration 14
        2.6 Interconnect 15
    3. *Non-CMOS Options* 17
        3.1 Memory 18
        3.2 Analog/RF 19
        3.3 Passive Integration 19
        3.4 High-Voltage/Power 21
        3.5 Sensors and Actuators 22
    4. *Packaging* 24
    5. *Systems* 28
    6. *Conclusions* 31
    7. *References* 33
    8. *Nomenclature* 34

2. Reliability Practice 35
    1. *Introduction* 35
    2. *Reliability Assessment* 37
        2.1 Burn In 38
        2.2 Biased Moisture Test 38
        2.3 Unbiased HAST and Steam Test 39
        2.4 Bake and Extended High Temperature Storage Test 39
        2.5 Electromigration Testing on Devices and Packages 40

2.6 Moisture Sensitivity Test 40
2.7 Temperature Cycling and Temperature Shock Tests 40
2.8 Power Cycle Testing 41
2.9 Mechanical Testing... 42
2.10 Design for Manufacturability, Reliability,
and Testability (DfMRT)... 43
3. *Reliability Statistics*.. 44
3.1 Life Distributions... 44
3.2 Confidence Level 51
4. *Acceleration Factor Models*... 52
4.1 Arrhenius Relationship 52
4.2 Peck's Model.. 53
4.3 Temperature-Voltage-Relative Humidity Model
(Eyring model)... 53
4.4 Coffin-Manson Model 54
4.5 Norris-Landzberg Model.. 55
5. *Failure Mechanisms*... 55
5.1 General 55
5.2 Examples of Failure Mechanisms 56
6. *Conclusions* 61
7. *References* 62
8. *Exercises* 63

3. Thermal Management... 65
1. *Introduction* 65
2. *Heat Transfer Basics* 68
2.1 Conduction 68
2.2 Convection 72
2.3 Radiation... 76
2.4 Remarks on Thermal Resistance 80
2.5 Typical Thermal Properties....................................... 81
3. *Thermal Design of Assemblies* 82
4. *Thermal Design for a SQFP* 87
5. *Heatsink Design Choices* 89
6. *Conclusions/Final Remarks* 91
7. *References* 92
8. *Exercises* 93

4. Introduction to Advanced Mechanics 95
1. *Introduction* 95
2. *Stress and Strain* 97
2.1 Analysis of Stress.. 97
2.2 Analysis of Strain 101
2.3 Thermal Strain and Thermal Stress 103

Contents   vii

  2.4 Thermoelasticity 104
  2.5 Geometric Nonlinearity 109
  2.6 Material Nonlinearity 110
  2.7 Contact Nonlinearity 120
 3. *Failure Criteria* 122
  3.1 Failure of Ductile Materials 122
  3.2 Failure of Brittle Materials – Maximum Normal Stress
   Theory... 124
  3.3 Fatigue Failure 124
 4. *Fracture Mechanics* 126
  4.1 Linear Elastic Fracture Mechanics 126
  4.2 Elasto-Plastic Fracture Mechanics...................... 132
  4.3 Fatigue Crack Propagation 134
  4.4 Mixed-Mode Fracture 135
  4.5 Crack Closure 138
  4.6 Singularity of Angular Corner of a Homogeneous
   Material.. 140
  4.7 Interface Fracture Mechanics 141
 5. *Finite Element Method* 147
  5.1 Introduction 147
  5.2 Treatment of Nonlinearity in Finite Element Analysis...... 148
  5.3 Finite Element Implementation in Fracture Mechanics..... 158
  5.4 Advanced Techniques in Finite Element Analysis 162
 6. *References* 165
 7. *Exercises* 166

5. Thermo-Mechanics of Integrated Circuits and Packages 169
 1. *Introduction* 169
 2. *Manufacturing Processes and Testing Methods* 171
  2.1 IC Backend Processes 171
  2.2 Packaging Processes 174
  2.3 Reliability Testing for IC Packages 177
 3. *Thermo-Mechanics of IC Backend Processes* 179
  3.1 IC Material Characterization.............................. 180
  3.2 Wafer Warpage as a Function of Temperature ... 183
  3.3 Nano-Indentation 185
  3.4 IC Interface Toughness Characterization 188
  3.5 Other IC Material Characterization Techniques 190
  3.6 Typical IC Material Properties 191
  3.7 Finite Element Modelling for Backend Processes 191
 4. *Thermo-Mechanics of Packaging Processes* 206
  4.1 Packaging Material Characterization 207
  4.2 Silicon Anisotropy 208

  4.3 Characterization of Thermosetting Resins 210
  4.4 Advanced Experimental Techniques for Packaging
    Stresses and Deformations.. 214
  4.5 Package Interface Toughness Characterization 216
  4.6 Typical Packaging Material Properties........................ 221
  4.7 Finite Element Modelling for Packaging 222
5. *Thermo-Mechanics of Coupled IC Backend and Packaging Processes*.. 245
  5.1 Effect of IC Metal Design on Passivation Crack
    and Pattern Shift Occurrence.. 246
  5.2 Effect of Package Structure on IC-Compound
    Interfacial Delamination... 253
6. *Case Studies* 259
  6.1 Reliability Predictions of Thermo-Mechanical Integrity
    of the Damascene Process 259
  6.2 Simulation-based Material Selection for a TBGA Package.. 263
7. *References* 271
8. *Exercises* 278

6. Characterization and Modelling of Moisture Behaviour............. 281
1. *Introduction* 282
2. *Moisture Diffusion Modelling* 285
  2.1 Diffusion in Multi-Material System 285
  2.2 Application to PBGA Package 289
  2.3 Moisture Desorption 291
3. *Characterization of Moisture Diffusivity and Saturation Concentration*... 292
  3.1 Diffusivity Measurement 292
  3.2 Saturated Moisture Concentration 296
4. *Vapour Pressure Modelling* 298
  4.1 Micromechanics-based Vapour Pressure Model 298
  4.2 Vapour Pressure as External Loading in Delaminated
    Areas... 305
  4.3 Vapour Pressure-Induced Expansion......................... 307
  4.4 Whole-Field Vapour Pressure Modelling 308
  4.5 Failure Mechanism... 309
  4.6 Underfill Selection for Flip Chip BGA Package
    for Moisture Performance 309
5. *Hygroscopic Swelling Characterization & Modelling* 311
  5.1 Hygroscopic Swelling Characterization 311
  5.2 Hygroscopic Swelling Modelling for FCBGA Package ... 327

Contents ix

    6. *Single Void Instability Behaviour Subjected to Vapour Pressure and Thermal Stress* 330
        6.1 Void Behaviour at Bulk 331
        6.2 Void Behaviour at Interface 336
    7. *Interface Strength Characterization and Modelling*.................. 338
        7.1 Mechanics of Interfacial Delamination 338
        7.2 Interfacial Fracture Toughness 341
        7.3 Interface Modelling using Cell Element 342
    8. *Case Studies* 345
        8.1 Integrated Stress Study for QFN Package 345
        8.2 BGA Moulding Compound Selection with Optimal Resistance to Moisture Induced Failures 361
    9. *References* 370
    10. *Exercises* 374

7. Characterization and Modelling of Solder Joint Reliability.......... 377
    1. *Introduction* 377
        1.1 Low Cycle Fatigue Loading 379
        1.2 Thermally Induced Solder Joint Reliability 381
    2. *Analytical-Empirical Prognosis of the Reliability* 385
    3. *Thermo-Mechanical Characteristics of Soft Solders* 389
        3.1 Eutectic Sn-Pb-(Ag) Solder 396
        3.2 Tin-Based Lead Free Solders 399
        3.3 Discussion on the Solder Creep Characteristics 405
        3.4 Primary Creep 408
    4. *Data Evaluation and Life-time Estimation* 410
    5. *Case Studies* 418
        5.1 Comparing Different Creep Laws for a Ceramic Capacitor and a PBGA on FR-4 Boards 418
        5.2 Comparison of Predicted and Test Results for Surface Mount Quartz Components 428
        5.3 Parametric Study on Chip Size Packages 438
        5.4 Flip Chip on Board Assemblies 449
    6. *References* 462
    7. *Exercises* 466

8. Virtual Thermo-Mechanical Prototyping 469
    1. *Introduction*... 469
    2. *Strategy, Methodology and Procedures of Virtual Prototyping*.... 473
        2.1 Strategy and Methodology..................................... 473
        2.2 Procedures 477

3. *Fundamentals of Simulation-based Optimisation* 480
      3.1 Design of Experiments (DOE) 480
      3.2 Response Surface Models (RSM) 482
      3.3 Design Optimisation 489
      3.4 Reliability and Robustness Analysis 507
   4. *Case Studies* 512
      4.1 Analytical Examples 512
      4.2 Industrial Application Cases................................... 517
   5. *Conclusions*.. 531
   6. *References* 532
   7. *Exercises* 534

9. Challenges and Future Perspectives 537
   1. *Introduction* 537
   2. *Mechanical Related Characteristics of Microelectronics* 538
   3. *Reliable Inputs* 542
      3.1 Design Inputs 542
      3.2 Failure Inputs 543
   4. *Tests and Experiments* 545
   5. *Material and Interface Behaviour* 546
      5.1 Material Behaviour 546
      5.2 Interface Strengths 547
   6. *Multi-scale Mechanics* 549
      6.1 Introduction 549
      6.2 Hierarchy of Methods for Mechanical Modelling 551
      6.3 Handshaking between Approaches 556
      6.4 Summary and Outlook 558
   7. *Multi-Physics Modelling* 560
   8. *Advanced Simulation Tools* 561
   9. *Conclusions*.. 561
   10. *References* 562

PREFACE

Microelectronics has pervaded our lives for the past fifty years, with massive penetration into health, mobility, security, communications, education, entertainment, and virtually every aspect of human lives. In the past decades, as the main stream, these progresses are powered by Moore's law, focusing on IC miniaturization down to nano dimensions and silicon-on-chip (SoC) based system integration. While microelectronics community continues to invent new solutions around the world to keep Moore's law alive, there are ever-increasing awareness, R&D effort, and business drivers to push the development and application of "More than Moore" (MtM) that are based upon or derived from silicon technologies but do not simply scale with Moore's law (with typical examples as RF, HV and power, sensors and actuators, MEMS/NEMS, system-in-package (SiP), heterogeneous integration, etc.). This emerging trend is partially triggered by the increasing social needs for high level microelectronic systems including non-digital functionalities, the necessity to speed up the innovative product creation and to broaden the product portfolio of existing wafer fabs, and the limiting cost and time factors of advanced SoC development. Along with the major technology development trends characterized by Moore's law and "More than Moore", the business trends are mainly characterized by cost reduction, shorter-time-to-market and outsourcing. The combination of these technology and business trends leads to increased design complexity, decreased design margins, increased chances and consequences of failures, decreased product development and qualification times, increased gap between technology advance and development of fundamental knowledge, and increased difficulties to meet quality, robustness and reliability requirements.

Based on the root cause analyses from observed failures of microelectronics during different life cycles, it is found that, among others, thermo-mechanical (thermal, mechanical and thermo-mechanical) related failures account for about 65% of total failures in microelectronics, and they originate mostly from the product/process design phase. Thermo-mechanical reliability is becoming one of the major bottlenecks for both current and future microelectronics technologies.

Due to the lack of available mechanics knowledge on one hand, and non-sufficient attention and R&D effort from both the academia and industry on the other hand, unfortunately, thermo-mechanical design and qualification of microelectronics are still largely depending on one's experience, supported by some qualitative numerical simulations. As a result, many design cycles are needed: from material development/pre-selection to concept design, to building and testing multiple physical prototypes. It is hard to achieve competitive designs with shorter-time-to-market, optimized performance, low costs, and guaranteed quality, robustness and reliability. Therefore, there is an urgent need to exploit and develop advanced knowledge of mechanics for microelectronics to enable the development of innovative thermo-mechanical design methods and tools.

Driven by our strong motivation and experience of leading and participating in many relevant research, development, and graduate education activities, ranging from microelectronics technologies to fundaments of mechanics, we present this book, as our obligation, to graduate students in universities, researchers, engineers and managers in industries. Our aims are to provide industry and academia with the cutting edge methods and solutions for various thermo-mechanical related problems of microelectronics in a systematic way, and also the development roadmap of mechanics for microelectronics. The book chapters are written by the worldwide leading experts with both profound theoretical achievement and rich industrial experience, containing essential and detailed information about the state-of-the-art theories, methodologies, way of working and real industrial case studies.

Acknowledgements

We would like to thank for their contributions to the book, A.J. van Roosmalen, J. Zhou, R. Dudek, E. Eggink, J.H.J. Janssen, F. G. Kuper, and N. Tzannetakis. We also would like to make acknowledgment to many of our colleagues who have contributed to this book in one way or another, among them, D. van Campen and M. Geers from Technical University of Eindhoven; L.J. Ernst, F. van Keulen, L.G. Wang, C. Yuan from Delft University of Technology; R. van Silfhout, M. van Gils, D.G. Yang, J. Beijer, O. van der Sluis, J. Bisschop, Y. Li, and many others from Philips.

G.Q. Zhang is particularly grateful to his wife Suping, his son Luke and his daughter Romy for their motivating, understanding and supporting. W.D. van Driel is grateful to his partner Ciel for her support and understanding on the many evenings at home he has spent on writing and editing this book. X.J. Fan is grateful to have the opportunity working together with his wife, Jenny, and the support from their son Bill.

G.Q. Zhang
W.D. van Driel
X.J. Fan

May of 2006

Chapter 1

MICROELECTRONICS TECHNOLOGY

A.J. van Roosmalen
Philips Semiconductors, HTC 60, 5656AG Eindhoven, The Netherlands

Abstract: A concise overview is given of the origin of microelectronics and its evolution into the nanotechnology era, from the principles of CMOS and other devices to the details of silicon wafer processing and packaging, and from units steps and integration to application. The system-on-chip and system-in-package concepts are explained and compared with each other, and brought into relation with Moore's Law as well as with the 'More than Moore' technologies not governed by digital circuit dimensions. Wherever applicable, an explorative view is given into future directions and roadmaps.

Key words: Microelectronics, nanotechnology, CMOS, silicon wafer processing, packaging, system-on-chip, system-in-package, Moore's Law, More than Moore, MEMS

1. INTRODUCTION

When 1965 Nobel Prize winner Richard Feynman gave his classic talk 'There's Plenty of Room at the Bottom' at the annual meeting of the American Physical Society on December 29th 1959 at the California Institute of Technology, few people will have imagined that most of his predictions were to turn into reality before the end of the century. Feynman's talk was on 'manipulating and controlling things on a small scale', covering technologies from printing, electron microscopy, biology, computing, evaporation, and micromachining to atomic scale arrangements [1]. The first question he asked his audience was 'Why cannot we write the entire 24 volumes of the Encyclopedia Britannica on the head of a pin?'. After which he quickly demonstrated that a conventional half-tone reproduction on this scale would allow for some 1000 atoms per dot, a level of detail that is readily visible through an electron microscope.

Of course, this argument does not say anything about practical execution, let alone about mass production, but that is not the point. Its strength is in the insight that many a seemingly impossible question on miniaturization can be answered using existing physical knowledge. Feynman ended his talk contemplating on how miniaturization could develop in future, stating 'They could have competition in high schools'. The Los Angeles high school could send a pin to the Venice high school on which it says, "How's this?" They get the pin back, and in the dot of the "i" it says, "Not so hot."

Now one needs to realize that back in 1959 none of the current day Microelectronics processes were available. To be more precise, Jack Kilby's original concept of an Integrated Circuit (IC) had just entered the patent application process, and the first real circuit existed only on the drawing board. But from this early start, microelectronics technology has raced forward at an enormous speed until the present day, driven by the same simple argument introduced by Feynman that the Laws of Physics do not exclude it. In stunning similarity to the pin story mentioned earlier, Figure 1 depicts a memory cell made in a 2002 technology lying on the bottom of a contact hole made in a 1978 technology. Note that the contact holes in the memory cell are not even visible at this magnification. The investments are huge, and consequently the competition on miniaturization is between global industries, not between high schools, but the spirit is still the same.

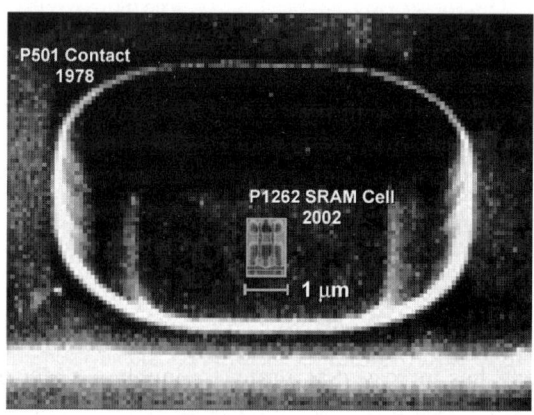

Figure 1. 2002 Memory cell in a 1978 contact hole (Source: Intel)

1.1 A Heart of Silicon

Microelectronics technology is commonly seen as synonymous with silicon technology. This underestimates the complexity of the value chain leading to the final product but the general idea is understandable. Because at the heart of any microelectronics system there is almost always silicon. Silicon (Si) combines excellent (semi) conductive and thermal properties with chemical versatility and mechanical stability, making it the most frequently used base substrate. Si frequently serves as a convenient carrier for many more exotic materials as well.

It has not always been like that, though. The very first commercial microelectronic devices were diodes and bipolar transistors made in germanium (Ge), a material having better mobility and a lower threshold voltage than Si. Silicon established its position only after the introduction of the planar technology that is fundamental for making ICs in general and for the field-effect transistor (FET) in particular, because Si grows a stable and protective thermal oxide, and Ge does not. Nevertheless, non-Si semiconductors continue to have a place at the table in specific domains. In fact, Ge has returned in the shape of SiGe alloys to reduce channel resistance in ultra-high speed bipolar transistors for radio-frequency (RF) applications, as well as in the newest generations of nanometer-scale metal-oxide-silicon (MOS) processing. The III-V compound semiconductor gallium arsenide (GaAs) is commonly used in RF power amplifiers, and indium phosphide (InP) is being investigated for use at even higher frequencies. Other and more complex compositions have laid the basis for light-emitting diode (LED) devices.

In 1959, incidentally the same year Feynman gave his talk on miniaturization, Jack Kilby, then at Texas Instruments, submitted a patent request on 'Miniaturized Electronic Circuits'. His invention demonstrated the feasibility of realizing resistors and capacitors based on semiconductor technology together with transistors in one and the same substrate. With that, the integrated circuit (IC) was born.

Just a few years later, in 1965, Fairchild engineer Gordon E. Moore postulated a bold theorem that predicted exponential growth in the semiconductors industry [2]. He stated 'The complexity for minimum component costs has increased at a rate of roughly a factor of two per year'. Certainly over the short term this rate can be expected to continue, if not to increase. Over the longer term, the rate of increase is a bit more uncertain, although there is no reason to believe it will not remain nearly constant for at least 10 years. That means by 1975, the number of components per integrated circuit for minimum cost will be 65,000. I believe that such a large circuit can be built on a single wafer'. When Moore co-founded Intel a few years later and became its first president and chief executive, the team used a business plan based on this same prediction.

Figure 2 shows Moore's original prediction together with the evolution of dynamic random-access memory (DRAM). DRAM stores its bits in cells of a capacitor and a transistor in series, so with two components. It has been the dominant stand-alone memory device for almost as long as microelectronics exists. Starting point, of course, is Kilby's invention marking an IC with one transistor, $T = 1$. The eventual result was not as dramatic as in Moore's forecast, but the growth turned out to be exponential indeed. At the time of writing, that is 2005, the industry has passed the line of $T = 10^9$, meaning that the largest circuits of today carry as much transistors as there are people on Planet Earth.

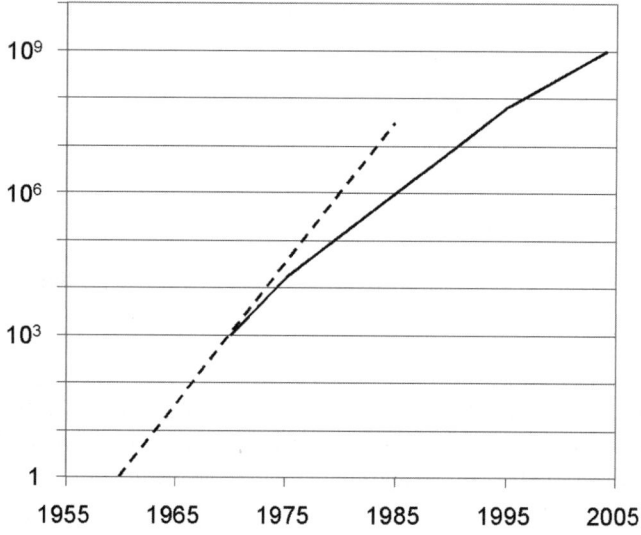

Figure 2. Evolution of number of transistors per circuit; dotted: Moore's original prediction, drawn: DRAM actual (Source: Moore [3])

Through the years, this spectacular growth in circuit complexity has become known as Moore's Law [3]. Frequently and erroneously quoted as being some force of nature that doubles almost anything in high-tech every 18 months or so, Moore's Law in reality notes that the number of components in a single product – like DRAM – has doubled every two years since 1970 [4], and suggests that the growth will continue. Note that Moore's original argument was strictly economic, implying that the cost per IC function – such as storing one bit of data – can be cut in half roughly every two years. It is this cost-down drive that has allowed the microelectronics industry to grow at double-digit figures for more than four decades.

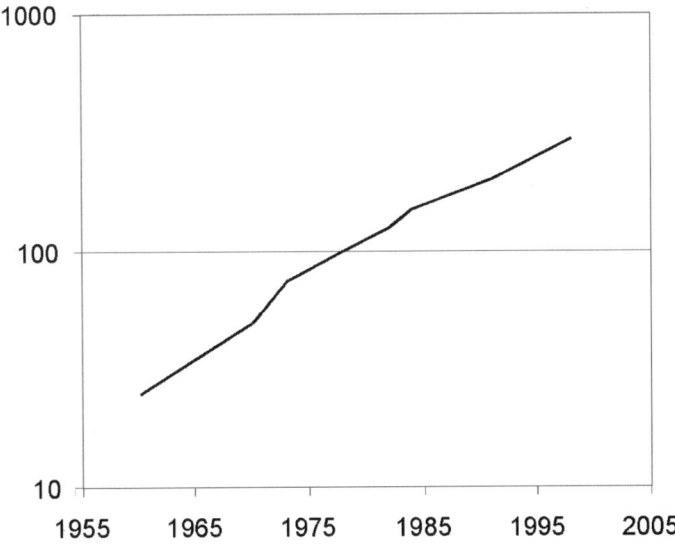

Figure 3. Evolution of Si wafer size [mm] (Source: ITRS [5], Sze [7])

The main technological developments that have made Moore's Law possible are wafer size (going up) and minimum pattern dimension (going down). Figure 3 shows that in 1965, IC's manufacturing relied on 1-inch Si wafers, while nowadays it's 300 mm. At the same time, dimensions of the smallest patterns on these wafers have been shrunk from the size of an amoeba (15 µm) to that of medium-size virus (100 nm), see Figure 4. From these graphs it can be seen that the time constant for doubling wafer size is 10 years, while the time constant for halving critical dimensions is 5 years. Net effect is that the number of IC's has remained almost the same in successive advanced process generations, that is, a few thousand per wafer for complex volume products such as stand-alone memories and microprocessors.

Tracking and predicting the key parameters and main technological challenges in the semiconductor industry has become a science in itself through the International Technology Roadmap for Semiconductors (ITRS) [5]. The ITRS has evolved from a regional initiative driven by a handful of companies into a global forum populated by IC makers, equipment and material suppliers, institutes and universities. The documents generated in its annual updates and symposia play a leading role in determining the world's microelectronics technology agenda.

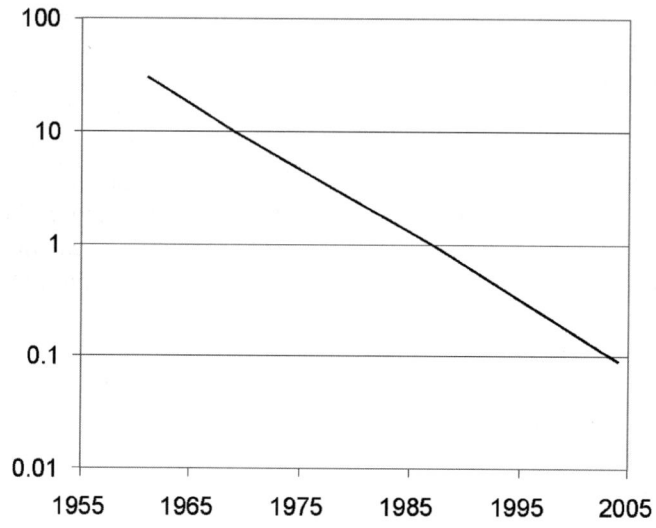

Figure 4. Evolution of Si technology critical dimension [μm] (Source: Moore [3])

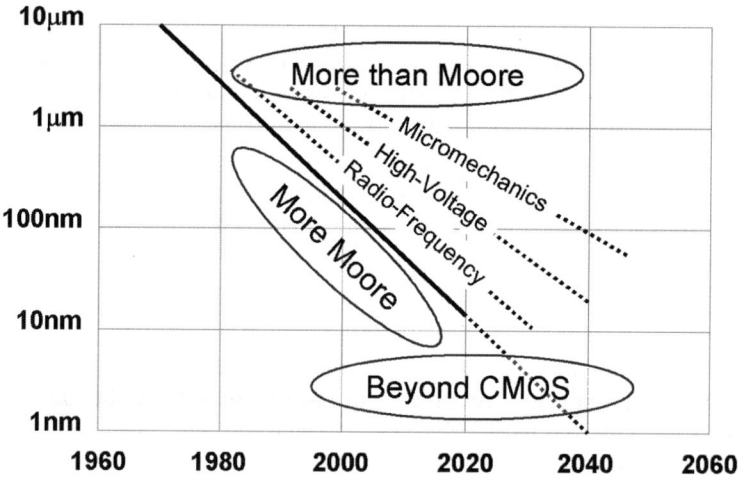

Figure 5. Wafer processing roadmap (Source: ENIAC)

1. Microelectronics Technology

Moore's Law is today's main mechanism driving the evolution of the 'brain' of microelectronic systems through increasing data storage capability and computing power. But the brain also needs intelligent interaction with its environment, which includes radio communication, mains power, and non-electrical interaction, such as the human interface. Microelectronics technology has made it possible to realize the necessary 'ears, eyes, arms and legs' on more or less conventional Si wafer processing, but the device structures associated with it usually do not scale with minimum pattern dimension. Radio communication, for example, needs inductors of which the dimension is determined by the transmission frequency and mains power circuits are limited by breakdown fields, while non-electrical interactions often relies on mechanical phenomena on mm-scale through micro-electromechanical switches (MEMS). A typical example of the latter is the motion sensor that is triggering car impact airbags. From the realization that these technologies evolve at right angles to the path of Moore's Law, the domain has become known as 'More than Moore', see Figure 5.

1.2 In a Little Black Box

ICs are commonly depicted as headless silver-legged black insects with a company stamp on their back. From which it would seem that packaging is straightforward, once the wafer processing is done. Just break up the finished wafers into individual product die, provide electrical contacts for soldering to a printed circuit board (PCB), then encase in plastic for ease of handling.

But real life is not that simple. Package specifications are driven by the 'silicon heart' (e.g., die dimension, number of connections, heat dissipation, operating frequency) but also by its eventual environment (e.g., size limitations, operating temperature, PCB process compatibility). And there is more. Functions for which it is technically difficult or commercially inconvenient to incorporate them in the same wafer process in order to built a system-on-chip (SoC) may still be combined in a system-in-package (SiP) holding multiple die in a single outline. One may say that SiP is to SoC what More than Moore is to Moore's Law, by providing the design community with an extra dimension to create overall system value.

Sze et al., expected ceramic dual-in-line to remain the dominant IC outline [7], and the 1998 advanced packaging handbook by Garrou and Turlik still states that multi-chip modules (MCM) are not likely be used in volume consumer markets any time soon [8]. Neither prediction turned out to be true, see the current packaging roadmap in Figure 6 showing that the little black box is evolving at the same high speed as its silicon heart.

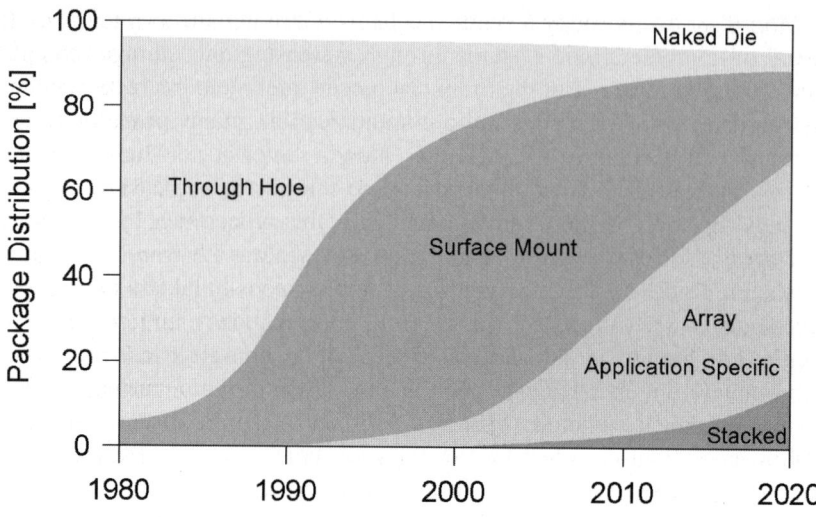

Figure 6. Packaging roadmap (Source: Prismark)

Though there is no unified guide in packaging equivalent to the critical dimension roadmap in Si technology, the ITRS publishes medium and long term predictions for a number of key parameters and difficult challenges in packaging along with those for wafer processing [5]. Common parameter in many of these challenges is the thermal and mechanical behaviour of the die-package combination, since that is the key-determining factor in predicting overall product reliability.

General observation today is that chip interconnect continues to shrink at a faster rate than package based interconnect, thereby limiting the benefit of chip scaling; longer term, the packaging challenges of evolving MEMS and future semiconductor device architectures will become critical. Both trends drive the integration of advanced interconnect, including passive components, in the package rather than in the IC, which in its turn causes a steady increase in the relative market share of multi-die as well as naked die solutions as shown in Figure 6.

2. BASELINE CMOS

Over 80% of today's microelectronic products depend on complementary MOS (CMOS) baseline technology, with Moore's Law as guiding light. A basic description of the CMOS device and how it can be made is given here; details about physical design in CMOS technology can be found elsewhere [6].

1. Microelectronics Technology

Pure Si has a resistivity of about $200 \cdot 10^3$ Ohm·cm at room temperature; compared to the $2 \cdot 10^{-6}$ Ohm·cm for copper it is an almost complete insulator. Through doping (controlled contamination with suitable elements) the resistance can be brought down to values ranging typically from 5 Ohm·cm in bulk to $5 \cdot 10^{-4}$ Ohm·cm for contact and source/drain areas, see below. The dopant phosphorus, for example, fits easily in the Si lattice, even though it has five electrons in its outer shell while Si has only four. When building one atom of phosphorus in the Si lattice, one electron is not needed for bonding and enters the conduction band. With enough dopant around, the conduction band fills up, electrons become the majority charge carriers, and the base material turns into a so-called n-type semiconductor. Other elements with these same properties are arsenic and nitrogen. Dopants holding three electrons in the outer shell, like boron and aluminium, miss one electron for complete bonding. This generates holes in the conduction band, and the crystal becomes a p-type semiconductor.

Where n-type and p-type domains meet, a charge-depleted non-conducting zone – the junction – forms at the interface. This junction originates from the space charge that forms after the initial annihilation of mobile electrons and holes to prevent further cross-border diffusion. Applying a positive voltage on the n side of the junction repels the mobile charge carriers even more, thereby widening the depleted zone, and lowering the cross-junction current to a level determined by thermal diffusion only. With a negative voltage on the n side, the energy of the mobile charge carriers is increased and allows them to cross the junction, eventually bringing the system in conduction. This is a rough description of the simplest semiconductor device, the diode. Using Si as a base material, the junction will function properly up to 150 deg C; beyond that, the leakage current spoils the overall performance.

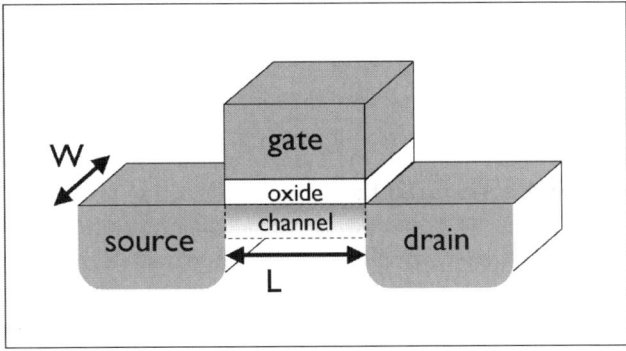

Figure 7. MOSFET schematic drawing

Figure 7 shows a basic MOSFET side view. Now consider this to n-type, or NMOS. In that case, the transistor is composed of two highly doped n+ domains (source and drain) in a lightly doped p- substrate, separated by a top electrode (gate) sitting on an isolator (oxide). Source and drain represent two diodes back-to-back, so no current can flow. At least, as long as the gate is left untouched. Making the gate positive drives the holes in the substrate away from the oxide interface and forms a local depleted zone in which a negative space charge builds up. With a sufficiently high voltage on the gate (threshold voltage) the electron energy in the depleted zone very close to the gate oxide will reach the conduction band of the Si crystal and an electron conduction path (channel) opens between source and drain. Current can now be drawn from the source by applying a positive voltage at the drain. The actual gate voltage and the transistor design control the level needed for that.

Overall performance of MOSFETs is described in terms of width (W) and length (L), which refers to the current path between source and drain. High W means high drain current, high L implies high breakdown voltage. Big transistors use a lot of power when switching, as charge has to be pumped to and from the gate in every cycle. This switching power can be reduced by shrinking the transistor, i.e., defining it with a shorter channel and a thinner oxide. A smaller transistor also gets faster, as it takes the electrons and holes less time to cross the channel. Side effect is that the drive voltage needs to come down as well to prevent electrical breakdown. A 2.5-V technology, e.g., will typically have an L of 0.25 µm and a 5-nm gate oxide. Dimensions in older, higher-voltage generations scale with drive voltage; in more recent generations dimensions drop faster than voltage.

PMOS is the mirror twin of NMOS, having a p+ source and drain in an n- overall environment, and negative drive voltages. The two are similar but not the same. In PMOS, charge is transported by holes, in NMOS by electrons. Since electron mobility is ~3 times larger then hole mobility, an NMOS device will carry ~3 times as much current as an PMOS for otherwise identical dimension and drive conditions. PMOS and NMOS can easily be combined in one wafer process flow, which offers considerable design advantages. This combination of two FET flavors has become known as CMOS.

CMOS devices are made in the substrate wafer in a group of processes commonly known as Front-end-of-Line (FEOL). Process temperatures are high, and great care is taken to prevent contamination as most foreign elements will interfere with primary device parameters. To realize ICs, transistors have to be built into cells, cells interconnected to blocks, and blocks arranged into systems, requiring multiple layers of crossing-over and mutually isolated conducting lines. These processes are on rather then in the wafer, together referred to as Back-end-of-Line (BEOL). Much lower

1. Microelectronics Technology

temperatures are involved to prevent recrystallization and melting of conductors like Al and the decomposition of the intermetal dielectrics.

In terms of CMOS unit processes, not much has changed really since Simon M. Sze edited and published the material from the Bell Labs internal training course on Very Large Scale Integration (VLSI) Technology [7]. Issued for the first time in 1983, the book quickly became – and still is – the standard reference in processing. Which comes down to diffusing dopants and growing oxides by thermal processing, transferring mask patterns by lithography and etching, and depositing conducting and isolating layers, helped by occasional planarization of excessive topography originating from previous process steps.

2.1 Diffusion

VLSI technology requires more then a dozen diffusions. Doping may be done in a furnace by feeding suitable volatile components in the gas flow, but most steps involve implantation. In the latter, dopant atoms are ionized and accelerated to the wafer in high vacuum, where they are injecting below the wafer surface at densities and depths determined by the particle beam current and voltage. Implanted atoms need to be thermally activated, which can be done in a furnace for smooth and deep profiles, and with lamps or lasers – rapid thermal annealing (RTA) – whenever very steep or very shallow profiles are needed.

High-temperature furnace steps are also used to grow the MOSFET gate oxide, through admitting a diluted stream of oxygen. Slightly different conditions lead to formation of the thick oxides used in the lateral isolation of transistors in a circuit.

2.2 Patterning

Patterning starts with a thin layer of photosensitive resist that is applied to the wafer by spinning or spraying, then hardened by baking or otherwise. After lithographic exposure and development to remove either the illuminated (positive resist) or the non-illuminated parts (negative resist) the resist pattern is used to selective etch of underlying layers or to provide local protection against doping by implantation.

Since the days of Rick Kilby, the exposure system has evolved from a common mercury lamp and lens plus a 1:1 chromium-on-glass mask in contact with or in close proximity to a family of tools with growing complexity and cost in order to keep pace with the continuous call for smaller pattern dimensions. The limitations in optical lithography can be visualized through the resolution formula.

$$CD = k_1 \cdot \lambda / NA \tag{1}$$

with CD the critical dimension, λ the wave length of the light source, NA the numerical aperture of the lens, and k_1 a factor <1 given by the exact process conditions. Production tools for VLSI are G-line (436 nm), I-line (356 nm), KrF laser (248 nm) and ArF laser (193 nm). These systems are 4-10x reduction steppers with an exposure field of 1-2 cm.

Not only λ, also k_1 and NA have improved over the years in order to keep pace with Moore's Law requirements. As a result, the smallest dimension that can be made at a given wavelength had shrunk from significantly larger than wavelength in the early days to about quarter wavelength in today's 45-nm CMOS development using 193-nm lithography. Admitted, at the cost of immersing the critical light path in water to push NA above the physical 1.0 limit and applying mask enhancements such as non-imaging phase shift (PSM) and optical proximity correction (OPC) patterns. Future developments include 13-nm extreme ultraviolet (EUV) and various alternatives in maskless lithography (ML2) and nanoimprint technology.

Making masks obviously also requires lithography. Here, the pattern is written directly from a data file by generating the pattern with a focused electron-beam or laser. Direct writing is also used in the development of prototype structures at the forefront of technology where lithographic production tools do not reach yet.

With the patterned resist as a local mask, the wafers may be patterned by immersion or spraying with suitable chemicals, like buffered HF for etching silicon oxide and KOH for silicon. Key disadvantage of wet etching is that material is removed in all directions, also underneath the mask, leading to loss in resolution. The original pattern dimensions can be preserved much better by dry etching with RF-exited reactive gas plasma's.

Dry etching is based on formation of volatile species from chemical reaction between the substrate and activated gas phase particles, usually halide and halocarbon chemistries. The process can be tuned such that this reaction takes place only where high-energy ions accelerating out of the plasma hit the wafer surface, while leaving the non-bombarded pattern sidewalls untouched. Tuning these processes is a delicate matter, especially with high aspect rations (trench isolation). Etching with ions alone (ion milling) is seldom used because the selectivity between the various materials in wafer processing is limited and etch rates are low.

2.3 Deposition

Thermal growth of silicon oxide is possible only if there is silicon to grow on, and the profiles of dopant already present in the wafer allow it. In

1. Microelectronics Technology

all other cases, the oxide needs to be deposited. This also holds for conductors, including silicon, and the other dielectric materials.

Chemical Vapour Deposition (CVD) is the dominant technology for dielectrics. Here, the layer precursors are formed in the gas phase by chemical reactor, either at high or at low temperature, at atmospheric or reduced pressure, and possibly enhanced by plasma excitation. The choice of technology depends on required composition, mechanical and electrical properties, allowable temperature budget, and step coverage needed. E.g., silicon nitride is deposited from a mixture of dichlorosilane and ammonia, at high temperature for thin and dense layers in FEOL, at low temperature with plasma enhancement for BEOL passivation. Other key steps based on CVD are the polycrystalline silicon (poly-Si) for gates and local interconnects in conventional MOSFET and, more recently, tungsten to fill deep contact holes.

A special variant of CVD is epitaxy. With high temperatures and starting from a monocrystalline silicon surface cleaned in the process chamber, monocrystalline layers are formed with the same lattice orientation as the substrate. The process allows Si-compatible components, like Ge and carbon, to be built into the crystal structure. Recent additions to the CVD portfolio are atomic layer deposition (ALD) and metal-organic CVD (MOCVD), technologies used to grow very thin multilayers and complicated formulations, respectively.

Physical Vapour Deposition (PVD) is used mostly for metals and metal alloys. This involves heating a slab of base material – evaporation – or exposing it to a noble gas plasma or ion beam – sputtering – which releases a steady stream of base material particles to be deposited on the wafer. Typical applications are aluminium for BEOL interconnect and the refractory metals Mo, Ti, Co, and Ni that are used to provide low-resistance shunting of poly-Si lines by deposition and subsequent thermal silicide formation.

Very different from the above processes is electroplating of copper, the metal used in interconnect structures for 0.18 μm CMOS and below. Electroplating starts with a seed layer uniformly deposited by conventional techniques after which copper is grown in an electrochemical cell much the same way as decorative chrome on steel or brass.

2.4 Planarization

When lithographic tools are pushed to their resolution limits, the practical depth of focus shrinks to values close to the critical pattern dimension of the system. Which for the more recent CMOS generations is of the same order of magnitude as the thickness of any single deposited layer, making planarization a necessity.

An early process used to this purpose is spin-on-glass, involving spinning an organosilicate solution on the wafer, then baking it to form a stable BEOL dielectric. A related technology is spinning a planarizing resist layer on a BEOL dielectric, then applying a uniform reactive etch plasma that removes resist and dielectric at the same rate.

From 0.25 µm CMOS down, Chemical Mechanical Polishing (CMP) is the dominant planarization process. The principle is simple, just grinding the wafer between rotating plates in the presence of slurry that both polishes and etches. CMP in practice is less straightforward, requiring a careful chemical and physical balance in the slurry between bulk and surface.

CMP is an essential element in copper damascene technology. This involves structuring of the deposited intermetal dielectric with line trenches and – for the second metal layer and up – contact holes (dual damascene), electroplating the wafer with copper, then removing excess material through CMP. Damascene processing is in fact the only practical way in use today to pattern copper. The metal has no volatile halides and can, therefore, not be patterned by dry (plasma) etching. Seed and barrier layers are needed to smoothly start the plating process, to protect the intermetal dielectric from deterioration by chemical reaction, and to prevent the copper from diffusing to transistor level.

2.5   Integration

Semiconductor processing invariable starts with ~0.5-mm slices cut from the <100> or <111> face of an ultra pure Si monocrystal and grouped in lots that are processed together in the wafer fab clean room. Typical lot size is 25, notable exceptions being traditional discrete device manufacturing where lot size is determined by how many wafers fit in a furnace tube, and the latest 300-mm generations where many unit steps are single-wafer anyway and the lot size may be 5 or less for reasons of cycle time reduction.

By far the simplest wafer process is that for making traditional power diodes, and requires diffusion only. Wafers are coated on one side with n-type dopant solution, on the other side with p-type, then stacked back-to-back and front-to-front in a SiC furnace pipe and heated to ~1000 deg C for a day or so. Doing so, the diffusion profiles come in from either side and meet in the middle to form a junction, turning each wafer into a single wafer-size diode that only need to be cut in pieces before assembly.

Unfortunately, VLSI processing requires more than that. A simple CMOS circuit may be built using the following recipe.
- Start with a p- wafer, thermally grow thin silicon oxide, apply thin silicon nitride by CVD; apply a resist mask on the future active NMOS and PMOS areas; plasma etch the nitride; strip the resist; through a second

1. Microelectronics Technology

high-temperature oxidation, selectively increase the thickness of the oxide not covered by nitride; strip the nitride to uncover the gate oxide.
- Mask the entire wafer except for the NMOS areas; implant with n-type dopant – it reaches the wafer only where no resist is present; strip the resist; thermally activate and drive in to form local n-environments.
- Apply doped poly-Si by CVD; mask the gate electrodes; plasma etch the poly-Si to define the gate; strip the resist.
- Mask the PMOS areas; implant the NMOS source and drain dopant which is now self-aligned with the gate; strip the resist; repeat the same procedure but now for the NMOS areas; thermally activate.
- Apply thick silicon oxide by plasma CVD; mask the entire wafer except for the future interconnect contacts with the source, drain, and gate areas; plasma etch the oxide; strip the resist.
- Apply aluminum by PVD; mask the interconnect; plasma etch the aluminum; strip the resist.
- Apply thick silicon nitride by low-temperature plasma CVD; masking the entire wafer except where the bond pads will connect the Si die with its package; plasma etch the nitride etch; strip the resist, thermally post-anneal.

The above recipe is a 8-mask flowchart suited for minimum dimensions down to about 1 μm. Passing the borderline into submicron technology, one step after the other from this simple sequence will fail, needing replacement by more elaborate processing.

2.6   Interconnect

Mainstream 90-nm CMOS needs about 35 mask steps and comes with at least 6 layers of interconnect. The net result is that most of the physical volume of advanced CMOS is in BEOL, as can be seen in Figure 8, showing the cross section of a conventional 90-nm process.

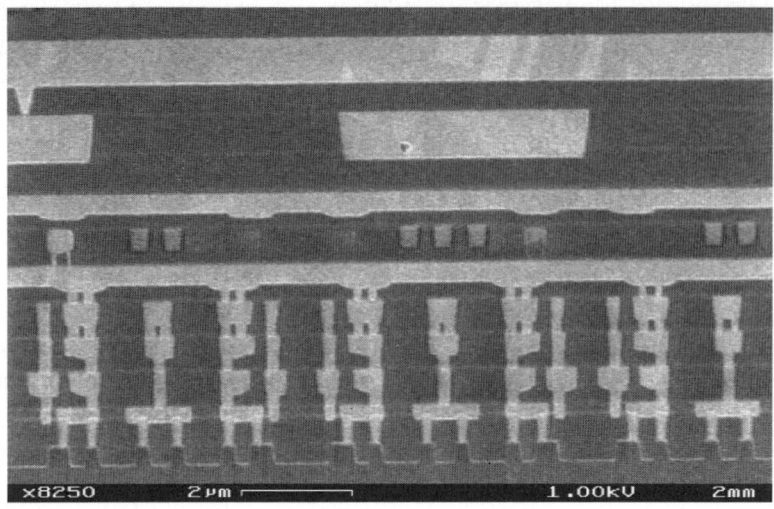

Figure 8. Cross section of a 9-metal- layer 90-nm CMOS process
(Source: Crolles2 Alliance)

Individual MOSFETs in advanced CMOS can be extremely fast, responding easily to microwave frequencies. In fact, the real speed challenge is in BEOL, not FEOL. The required high density of the interconnect structures leads to higher line resistance and higher mutual capacitance between the lines, and the inevitable combination of the two increases propagation delay. At 2 GHz, pulse raise and fall times should be less then 30 ps to perform critical computational tasks within the available 0.5-ns time frame. For conventional 0.25-µm CMOS, it can be calculated that the delay is already more than 0.5 ns on a 1-cm line [6]. It will be clear that a 2-GHz signal will not travel far, and certainly cannot cross a larger size die unattended. For this reason, larger circuits in advanced CMOS are broken up in regions running under a local clock regime only.

Having too many of these 'Islands of Synchronicity' has practical design limitations by laying a burden on design flexibility and design time. That is why copper was introduced to replace Al transiting from the 0.18 µm to the 130-nm node, and why there is a continuous drive to lower the dielectric k-factor of the intermetal isolation to as close as possible to vacuum level ($k = 1$). Typical low-k materials are based on silicon oxide with a significant percentage of fluorine or carbon grown either by CVD or spin-on techniques. To achieve a k-factor below ~2.5, dielectric materials have to be more or less porous, reducing their mechanical stability and making them susceptible for penetration of liquids and other unwanted intruders. These properties are serious complications when going into packaging.

1. Microelectronics Technology 17

Still other methods have been proposed to solve the interconnect delay problem. An approach receiving much attention nowadays is 3D Interconnect, where multiple very thin die (~10 μm) are stacked on top of each other, each die connected to the next by thru-wafer metallization. In this way, the number of metal layers in each die can be limited, and so the delay time. Whether the required combination of extreme wafer thinning, high-aspect ratio thru-hole etching, and accurate wafer-to-wafer bonding will result in cost-effective manufacturing still remains to be seen.

3. NON-CMOS OPTIONS

By design, CMOS VLSI is best suited for processing digital data. Adding process steps and device architectures to the CMOS baseline flowchart can create extra functionality created on top of standard features. A convenient way of categorizing these non-CMOS options is to group them by application domain to reflect the specific requirements from the overall system environment. The grouping that will be used here is memory, analog, radio-frequency (RF), passive integration, high-voltage/power, and sensors and actuators; the last group includes MEMS.

Each of these domains has specific requirements, which is reflected in specific non-CMOS technology solutions on wafer process and packaging level. Full CMOS functionality can almost always be maintained while implementing non-CMOS process features. This way of working is referred to as embedded process options; it comes at the price of adding mask steps and other complexity to the baseline process. If only limited CMOS capability is required, the flowchart can usually be simplified considerably. This results in a stand-alone process option meant to be combined with other off-chip functionalities and data processing, meaning elsewhere on a PCB or in a SiP.

By volume, the largest option domain is memory, most of it coming as stand-alone, a significant portion also as embedded in baseline CMOS. Storing and retrieving digital data requires a significant amount of peripheral circuit overhead to take of that also has to take care of clock, refresh and interface controls as well as redundancy management needed for error correction. Therefore, memory technology requirements stay close to CMOS and the evolution of the underlying processes closely follows Moore's Law. Requirements for the other domains are much less, up to not at all, determined by digital data considerations. These other options fall in the category 'More than Moore'; they either do not evolve in line with Moore's Law or follow with time delay of many years.

3.1 Memory

To store digital data one does not need special wafer processing. Six transistors in CMOS technology in a circuit known as an SRAM (Static Random Access Memory) cell are all that is needed [6]. An additional advantage of this type of memory, in fact a special version of a flip-flop, is that it needs very little periphery. Unfortunately, the SRAM cell is also the largest by far of all memories, though some efficiency can be gained in a stand-alone solution by trimming the technology down.

DRAM (dynamic random access memory) is much more compact than SRAM, its basic cell being just a capacitor and a transistor in series. As DRAM requires a lot of peripheral circuitry, it is mostly applied as stand alone feature. Stand-alone DRAM has the capacitor built into the substrate in the form of a very deep trench or pinhole, with the perimeter and a center plug functioning as electrodes. DRAM can also be embedded in a CMOS baseline, though not at the same high density as stand-alone. Reason for that is that the capacitor now has to be realized above transistor level to not interfere with the CMOS baseline transistors environment.

SRAM and DRAM are volatile, that is, they lose their information if disconnected from their supply voltage. Non-volatile memories in use today are EEPROM (electrically erasable programmable read-only memory), which is based on pushing charge to and from a floating gate through a semi-isolating (tunneling) oxide, and FRAM (ferro-electric) and MRAM (magnetoresistive), where the electrical resistance of ferromagnetic islands in different magnetic orientations determines the bit position. Also these memories require a fait amount of peripheral overhead, though not as much as DRAM. The EEPROM read-write cycle requires relatively high voltages to overcome the energy barrier for tunneling, FRAM and MRAM require a high to very high current to be able to switch the magnetic orientation of the bit cell. Flash memory refers to a special version of EEPROM in which the design allows for addressing complete blocks of data rather than individual cells, thereby significantly improving speed performance. EEPROM involves extra processing in FEOL; FRAM and MRAM are positioned high up in BEOL and involve foreign materials potentially harmful for proper functioning of the underlying CMOS.

There are large differences in performance between the various memories. At the risk of oversimplifying things, the below table can provide some guidance
- Speed: SRAM>MRAM>DRAM>FRAM>EEPROM
- Area efficiency: EEPROM>DRAM>MRAM>FRAM>SRAM
- Power efficiency: EEPROM>FRAM>MRAM>SRAM>DRAM

1. Microelectronics Technology

Memory options still under investigation include phase-change chalcogenide and large-area MEMS, both involving technologies remotely related to disk storage.

3.2 Analog/RF

Analog signal processing are found wherever the human interface is involved, which largely comes down to speech and vision, but also includes more general phenomena in the physical environment such as position, motion, temperature, and radiation. Signal accuracy and stability is crucial in this stage, and this is where bipolar transistors are fundamentally superior to MOSFETs, especially at frequencies over 100 MHz. The device architecture and operation of bipolar transistors is fundamentally different from that of MOSFETs, involving amongst others epitaxial growth in the switching area. A description of bipolar technology can be found elsewhere [6]. Bipolar transistors can be integrated with CMOS leading to a process option known as BiCMOS. Limiting the MOSFET part to one flavor only results in a much-simplified group of technologies called BiMOS.

BiCMOS is significantly more complex than baseline CMOS, but the eventual circuit can be relatively simple. Gate length has limited the maximum transmission frequency in older generations of CMOS, making BiCMOS a necessity for all RF applications above ~1 GHz. In more advanced CMOS generations it is intrinsic RF noise, not gate delay that spoils the MOSFET circuit performance. The transmission frequency and computing power of technology generations beyond 0.18 µm allows optimized CMOS-only circuits (RFCMOS) to compete with BiCMOS. Nevertheless, RFCMOS solutions come at a price, as a very large amount of circuit overhead is needed to compensate for the overall non-ideality of MOSFET, most notably the RF noise.

3.3 Passive Integration

Tuning and matching are invariably part of any RF circuit, implying the presence of relatively large inductors and capacitors on or in close proximity the active die. Placing these components nearby limits the negative effects of parasitics and cross talk that always complicates RF system design. Passive elements can be defined on many places in a regular baseline CMOS, but best performance requires extra process steps again. For example, tantalum oxide and tantalum oxynitride can be placed between BEOL interconnect layers to realize high-quality RF metal-insulator-metal (MIM) capacitors.

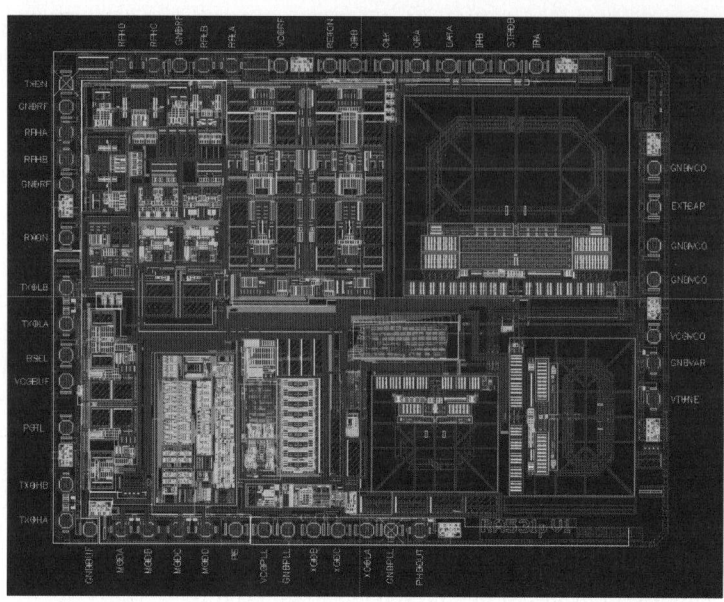

Figure 9. BiCMOS mask drawing, actual size 8 mm^2 (Source: Philips)

Figure 9 shows is the mask layout drawing of a typical BiCMOS design clearly showing large embedded inductors that occupy almost entirely the right half of the die. An alternative to embedding passives as described above is to separate the more bulky components from the active circuit and integrate them on a dedicated silicon die with very limited semiconductor functionality. The flowchart of process options for passive integration can be very simple and will be able, therefore, to effectively compete with passives on the active die and solutions where the passives are mounted elsewhere on the system PCB. Depending on overall system architecture, passive silicon dies may even serve as an interconnect substrate in advanced SiP. Obviously, RF is the dominant application, but passive integration can be used equally well in other environments where passives are important, such as in dc-dc voltage converters. Figure 10 shows an example of SiP passive integration; here, an active die is flip-chipped on a much larger but significantly less complicated passive die holding bulky high-quality inductors together with high-density trench capacitors.

1. Microelectronics Technology

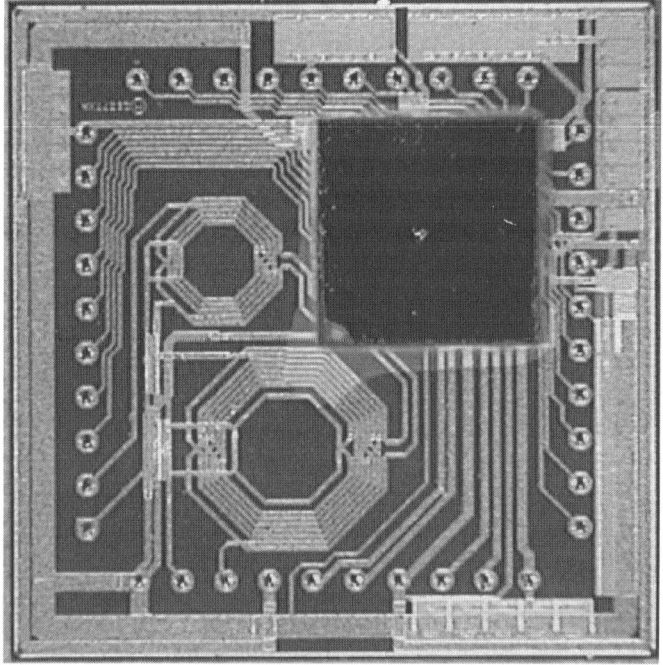

Figure 10. GSM transceiver with active die flip-chipped on passive die; actual size 27 mm² (Source: Philips)

3.4 High-Voltage/Power

For baseline CMOS, high voltage is a relative term. To be able to have a 1.2-V CMOS process interfacing with an external signal source of, e.g., 5 V, MOSFETs with a thicker gate oxide need to be embedded on the wafer to serve as input-output (I/O) devices, as the 1.2-V standard transistors will break down under this voltage stress. The conventional approach in all baseline CMOS is to 'borrow' the necessary transistors from an older CMOS generation. This allows reuse of existing and proven circuit designs. More is needed to reach the voltage levels typical for automotive (60 V) and mains (200-600 V). Of course, a large transistor with a double diffusion (DMOS) is needed to reduce the field in the source and drain areas, but also deep trench oxide isolation and buried implants to prevent lateral avalanche breakdown.

An elegant solution for high-voltage applications are to work from silicon-on-oxide wafers (SOI) having an oxide layer embedded very close to the top surface. Perfect lateral isolation can now be achieved by cutting trenches down to the buried oxide, allowing pockets of very different voltage ranges to be designed in very close proximity to each other. This approach also greatly simplifies the FEOL processing. Another application of SOI wafer technology not related to the high-voltage/power domain is in boosting the performance of baseline CMOS in microprocessor applications through reduction of substrate leakage and parasitic capacitances. A disadvantage of SOI is the reduced thermal conductivity caused by the embedded oxide, which puts higher demands on the package.

3.5 Sensors and Actuators

Micro-electromechanical switches (MEMS) is a term often used in relation to sensors and actuators. MEMS technology focuses on miniaturizing moving parts, usually involving Si wafers as substrate and CVD poly-Si as layered building blocks, with phosphorus-doped CVD silicon oxide as sacrificial layer to generate floating parts to allow horizontal and vertical displacement, and even rotation. Conventional chemicals such as buffered HF and KOH may be used to pattern the oxide and the silicon, respectively. From the relatively large dimensions, which may involve etching straight through a wafer, this mode of patterning is usually referred to as micromachining. More recently, high-rate dry etching with magnetic-field enhanced RF plasmas as well as direct processing of the Si wafer itself have found more application, thereby reducing the gap between the worlds of MEMS and baseline CMOS. MEMS have some specific issues when brought to practical use, most of them related with movement. This includes sticking by electrostatic charging or microadhesion ('stiction'), corrosion, debris, and particles in general.

Main fields of application for sensors are magnetic field, pressure, and acceleration; for actuators the fields are fluidics and optics. Virtually all devices listed here are MEMS-based, sensing magnetic field being the major exception. This device derives its signal from a special ferromagnetic layer of which the resistance depends on its orientation relative to an external field. The external field may be induced by a moving magnet in sensing position and speed, or it may be the earth magnetic field in which case the device is a solid-state compass.

One of the better-known MEMS sensors is the accelerometer used in many automotive airbag systems. Here, a freestanding comb-shaped strip is defined through a sacrificial layer technique. Forces generated in deceleration make the strip bend, an effect that can be detected by checking the capacitance between the strip and a stationary part of the structure. Airbag sensors are used in the 10-g range. More sensitive accelerometers use the earth gravity; by integrating multiple comb structures allowing in-plane (XY) as well as out-of-plane (Z) movement, these devices can be used as absolute 3D position detectors. Such products need to be relatively large

1. Microelectronics Technology

(>1 mm^2) to be able to have enough mass in the moving body. Pressure sensing involves a thin free-standing Si or piezoelectric material membrane. A special version of a pressure sensor is the solid-state microphone; in a modified design the microphone may double as loudspeaker.

The dominant MEMS actuators are the dies that form the ejection part of an inkjet printer head, and the digital mirror device that is the imaging part in many video projectors. Applications related to actuators are solid-state resonators, which includes surface and bulk acoustic wave devices, and RFMEMS, high-quality-factor mechanical RF switches operated by applying a DC voltage across the tongues. The latter is depicted in Figure 11.

Figure 11. RFMEMS (Source: Philips)

Solid-state biosensors are an emerging business. Most of the interest is going to the development of mechanisms to derive electrical signals from the biochemical interaction with the substrate in which the detection is done. The silicon plays a much less important role here, serving mainly as a convenient mechanical carrier and multi-field electrode for the active bio-layers.

Technically, it is usually very well possible to combine MEMS technology with baseline CMOS. Since all sensors and actuators need drive circuitry to be operated, and the signal levels are typically quite low, this may be a convenient solution. Whether it is also the best solution depends on the eventual overall cost level. MEMS devices tend to relatively large, much larger at least than the amount of silicon typically needed to realize the necessary analog signal processing. To be able to draw conclusions, the overall system including the packaging has to be taken into account.

4. PACKAGING

Finished wafers coming out of the wafer fab are electrically tested by contacting the exposed bond pads with a probe card. Purpose of this pretest is to early sort out circuits that will not likely result in functional products. The circuits not passing the test are marked with an ink dot or otherwise, the wafer thinned down to the required thickness by grinding, then singulated into individual dies by sawing or laser cutting. Typical height for many compact packages today is ~0.5 mm, almost the same as a standard wafer, implying that most of the wafer backside will need to be removed to fit the eventual outline. Connecting wafer level test with final product yield is relatively straightforward for digital devices, but much less so for RF products. This is caused by the intrinsic difficulty of doing RF tests on wafer level; parasitic capacitance and inductance values of regular probe cards make them useless for signals of 1 GHz and above. Alternative probe architectures and test strategies are being developed but not readily available today.

IC packages come in through-hole and surface-mount, through-hole subdivided in DIL (dual-in-line) and PGA (pin-grid array), surface mount classified as Dual (leads left and right), Quad (leads on all four sides), and BGA (ball-grid array). Then there are multi-chip modules (MCM) holding more then one IC or other device inside, the internal components mounted side by side, or stacked on top of each other, or both. Chip-size package (CSP) refers to a bundle of technologies having in common that the eventual package size is less than 1.2 times the size of the die. No real standardization exists; each package is identified by an acronym indicative of PCB mounting method and height, power handling capabilities, and technology used. HVQFN, for instance, means Heat-sink Very-thin Quad Flat-pack No-leads; it is a compact surface-mount package less than 1 mm high with lands and corresponding PCB landing patterns optimized to control the surface-tension forces that settle the package during solder reflow. Bottom part of the package can be either a metal heat-sink or an upside-down-mounted Si die.

The packaging process itself involves die attach and encapsulation, each cluster being a conglomerate of shapes and processes in itself. After packaging, the products are tested again, now attached to a real or simulated PCB and for full functionality. This may involve keeping them under high-endurance conditions for a prolonged time to sift out early device failure, also known as infant mortality. Such procedures are common practice for large stand-alone memories, especially DRAM.

To connect the die to its environment, the basic choice is between wire bonding and flip-chip. In the more mature of the two, wire bonding, the die is fixed into position with a metal-filled epoxy film or paste adhesive or, to allow operation at high die power dissipation, through soldering. A continuous wire, usually gold, is connected to the first bond pad by ultrasound-enhanced thermal compression, then stretched out and welded to the corresponding pin on the external connection structure. That structure will typically be a strip holding multiple patterns, meant to be encapsulated in the moulding process, the so-called leadframe. After the weld is made, the wire is cut with a flame of electric arc and brought to the next bond pad, restarting the sequence. In power applications, the thin gold wires may be replaced by thick copper wiring or by aluminium straps.

Wire bonding has RF issues similar to those in BEOL technology because of parasitic inductances. Also, the force exerted on the bond pad area during wire attach is significant, and may cause cracks and adhesion loss in the brittle low-density low-k materials in advanced copper interconnect.

Flip-chip is much more compact than wire bonding, but requires preprocessing of die and die landing area, the bumping process. Bumping involves building metal pillars or balls at the contact areas either before or after singulation. That can be done through electroplating, stencil printing, or stud bumping, the latter being an abbreviated version of the wire bonding process just described, but now with the wire sheared off just above the bond pad. Stud bumping does not require special die pretreatment. In most other cases, the bond pads have to be covered first with an underbump metallization to provide reliable contact. Alternatively, an extra interconnection, the redistribution layer, can be applied above bond pad level to create more space for the bumps and make it possible to use the entire die area for bonding. After the preprocessing is complete, the bumped dies are aligned to the substrate or leadframe and connected by thermo-compression, reflow soldering, or conductive glue.

Government-enforced environmental regulations ask for removal of lead from microelectronic systems. With the usual materials used in baseline CMOS, conventional PbSn eutectic soldering induces automatic alignment of misplaced or non-flat die by surface tension. This phenomenon, together with the quite limited temperature budget imposed by advanced Si wafer technologies, limits the range of reliable alternatives. The alloy SnAgCu is a frequently used lead-free solder replacement.

Figure 12. Stacked die mounting (Source: Amkor)

Next to conventional lead-frame bonding, the die may also be placed on ceramic or organic laminate structures, on glass, on a metal heat spreader for improved thermal management, or stacked onto one another as in Figure 12, the latter with or without an 'interposer' to accommodate mechanical stress.

Transfer moulding is the process most commonly used for die encapsulation. Epoxy resin liquefied by high temperature and pressure is forced through a mould chase over the die and leadframe and into the cavity on the frame where the die was placed earlier; the hardened epoxy eventually forms the body of the final package. The die may also have been attached to a laminate substrate holding the external interconnect wiring. In this case, the epoxy forms a mould cap that becomes the top half of a sandwich, with the laminate substrate forming the bottom half and the die in between. Less common today is ceramic packaging, where a cap of either ceramic or metal is welded or sealed over the die. Ceramic is relatively expensive, and used mostly for prototyping and application under severe conditions. Immediately before encapsulation, the die – leadframe combination is usually strengthened by a dedicated epoxy, the underfill, flowing in the capillary cavities between die and leadframe. Also the bond wires will then be covered with liquid epoxy and cured to prevent mechanical damage and shorting.

Wafer-scale packaging (WSP) is a growing group of widely varying techniques that all start with assembly steps on the wafer itself, and often involve singulation after moulding. WSP may include mounting die and caps on wafer, applying interconnect patterns, bonding, bumping, and wafer-to-wafer bonding. In the latter case, the second wafer usually has a capping function and may for instance be made of glass. As such, WSP is bridging the gap between die packaging and multilayer PCB processing. The resulting packages can be very compact and often fall in the CSP category.

WSP also offers solutions to protect sensitive non-electronic on-wafer functions from potential damage inflicted by conventional epoxy moulding. This protection is essential when working with sensors and actuators, as most of them are based on MEMS and, hence, require a cavity inside the final outline for proper operation. They may also need a transparent window or hole to communicate with the outside world. Very often, the cap or lid over the MEMS device needs to be hermetically sealed, and the remaining atmosphere in the cavity well controlled. Reliable lid sealing can be done through applying and subsequent sintering of glass frit, anodic processing, and direct fusion. Temperature budget tends to be the limiting factor here.

Naked dies are tested Si dies having undergone sufficient processing to be handled, which may include bumping and circuit protection. This 'outline' is applied as such in smartcards and on liquid-crystal display flexfoils, and as a component in more elaborate multi-die SiP solutions. It is also increasingly used for direct placement on PCB, especially when space is limited, as in many handheld consumer products. An example is the integrated FM radio shown in Figure 13.

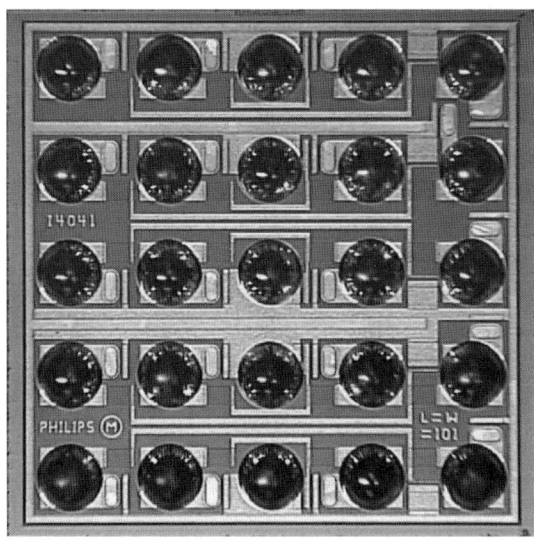

Figure 13. Bumped naked-die FM radio (Source: Philips)

An approach competing with stacking die in one package is stacking multiple packages. Standard surface-mount packages are thinned by grinding to reduce the overall height and to expose relevant portions of the supporting leadframe, and some further processing is done on the package to be able to make the necessary external soldered interconnects. Existing standard procedures for packaged products can now be used to test and qualify the individual components before final assembly. This can be major advantage compared to handling naked die, especially if one of the components is a stand-alone memory.

Outlines very different from what has been described so far can be found in discrete semiconductors, especially at the high-power side, that is, with ratings well above 1 A. Many of these high-volume products come in tin can style or with a passivating glass package.

5. SYSTEMS

Miniaturization in microelectronics is the ultimate step in the overall compaction process of electronic systems. In building the system, packages holding one or more Si devices are mounted on PCBs or other substrates, which are then assembled together and eventually stacked in an end-customer product. Through time, there has been a continuous drive to pull-in value and complexity from a higher level in the supply chain to lower levels, that is, from box to board to IC.

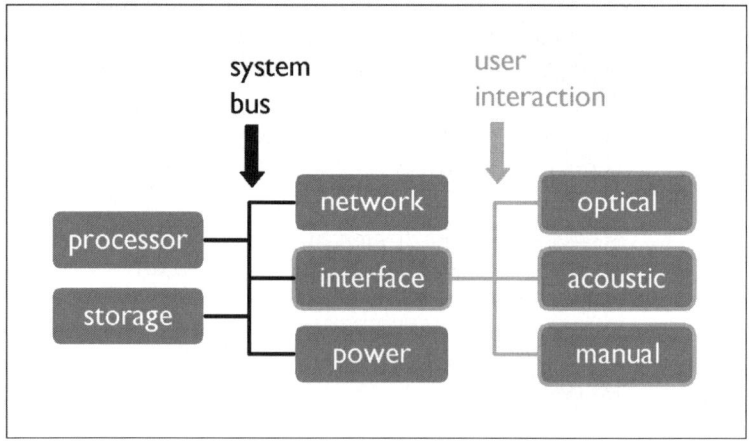

Figure 14. Microelectronic system schematic

1. Microelectronics Technology

Figure 14 shows the high-level schematic of a microelectronic system. This system may be a personal computer, in which case the block labelled storage contains the hard-disk, external disk drives, and plug-in (DRAM) memory boards, system bus is the motherboard, and user interface encompasses the usual set-up of screen, speakers, mouse and keyboard. A cell phone has the same basic structure, network now being the antenna and transmitter-receiver part, and a similar division can be made for very different other products such as a TV set, a digital camera, an inkjet printer, and even an airplane. But it also may be a single SoC or SiP.

Within one system, the technical requirements for each of the blocks in the above diagram can vary largely, from high-density software-dominated digital circuits for storage and processor following Moore's Law, to high-voltage for the power block, RF capability for wireless and web connections to MEMS for sensing the environment. This diversity of requirements and features can be combined on one wafer, SoC, or in one package, SiP, with multiple die each having its own dedicated technology, in which one or more may still be a more limited SoC.

The extra processing involved in embedding options in baseline CMOS has the unwanted side effect of raising the price of the rest of the die too, such as logic sitting next to an embedded block of DRAM. Also, developing embedding technology takes time, which may effectively result in the option being released at the time the CMOS node is already coming up. There are cases where all functions must be integrated in a single die, e.g., in the case of a smartcard, for reasons of security. In many other cases, the benefits of combining multiple technology requirements on one wafer need to be carefully balanced against the cost and time-to-market consequences.

With growing system complexity, it may be more cost effective and performance-wise equivalent or better to use dedicated wafer technologies and do the integration on package level. This approach also makes it possible to create additional value by including components that otherwise would need to be mounted separately. Figure 15 shows an example of extensive integration, including ICs as well as discrete passive components.

Figure 15. Bluetooth module with active and passive dies plus discrete elements on a laminate substrate (source: Philips)

A combination of more conventional wafer and packaging processes is depicted in Figure 16, showing a microcontroller attached to the face of a much larger power MOSFET. By doing so, the controller can instantaneously react on local heating caused by surge transients, and power down the MOSFET before real damage is done.

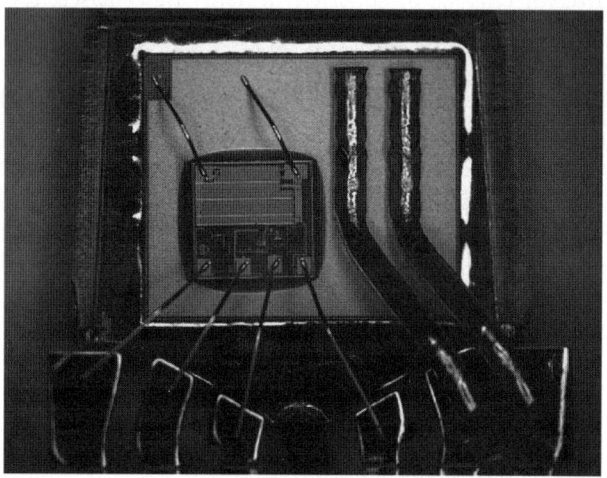

Figure 16. Power MOS with mounted CMOS controller (Source: Philips)

Sensors and actuators almost always need signal processing in close proximity of the active device to keep the noise level down. Also, the

1. Microelectronics Technology

MEMS technologies often involved ask for special packaging. Virtually all products in this domain are, therefore, a SiP of some sort, the outline often being an integral part of the overall system functionality. Figure 17 shows a solid-state microphone, a characteristic example for such an integrated solution.

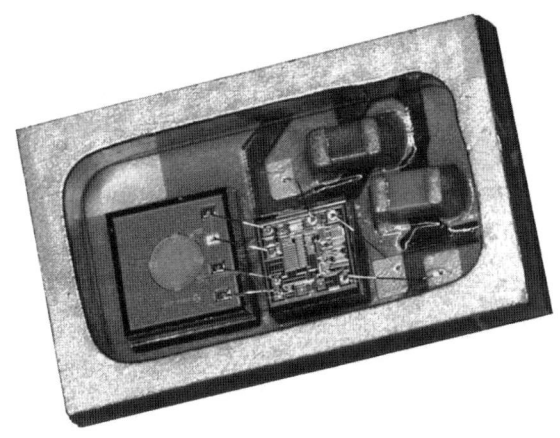

Figure 17. MEMS microphone with signal processor and noise-suppression capacitors (source: Knowle)

Overall trend in microelectronic systems is that the boundaries between die and package, and between package and PCB are fading. This is seen in the growing proportion of SiP solutions coming to the market, but also in the ongoing developments in advanced packaging [9]. Examples of innovative strategies being investigated are System-on-Package, that starts with a package and uses it as a substrate to build a system, and Chip-in-Board, where the PCB is given active functionality by embedding naked die in the substrate during lamination.

6. CONCLUSIONS

Virtually all functionality required for microelectronic products can be integrated on a single wafer, but is it also economically feasible? Innovative packaging solutions have become available allowing silicon system partitioning over multiple CMOS and non-CMOS solutions. The underlying technologies span at least 20 years of consecutive CMOS generation, with

1-μm still in production today while at the same time 65-nm is being released. Silicon wafer processing for the latest generations of 300-mm baseline CMOS is more than 10 times as expensive per mm^2 as the cost-optimised and stripped early generations. This is, of course, compensated by the fact that CMOS density improved by at least 100 times, but that principle applies to digital circuits only. Inductor and capacitor tuning elements do not shrink in size from one wafer technology generation to another, and neither do high-voltage transistors, while the scaling of MEMS is determined by mechanics rather than electronics.

To decide on how a given microelectronic product is to be designed and built, a good and reliable system partitioning and cost model is absolutely needed. Such a model needs to take into account the impact of multiple wafer and packaging technologies, the balancing between PCB assembly vs. SiP integration, and the cost of testing to guarantee that pre-tested components will be functional when combined into the final product ('known-good die'). And in all cases, volume manufacturing will be needed, not only to recover the initial costs of design and start-up, but also to be able to reach stable process and product yield. This is a concern for many sensors and actuators because of the extreme diversity of the underlying technologies.

Will there be an end to Moore's Law? Solutions to the technical problems that may occur at the 22-nm node of silicon wafer processing are being investigated already today, including the potential application of innovative nanotechnologies 'beyond CMOS'. The real question here is what will be size and price of the wafer fabs that are to build such products in a cost effective way, and what is their payback time. It should be always kept in mind that Moore's Law is an economic argument, not one limited to wafer technology. In other words, the value of the future microelectronic products may be as much in the package as in the die or dies inside. Financials rather than physics will decide on the eventual solution.

The growing complexity of microelectronic systems, the increasing proportion of packaging in the overall system cost, and the fading boundary between die and package make it necessary to bridge the gap between wafer processing and package assembly and develop transparent approaches that will allow designers and architects to be able to use the best of both worlds. This implies integral harmonization of design flows, tools, and libraries, as well as multi-technology multi-parameter modelling, simulation and verification across the entire supply chain, including electromagnetic as well as thermo-mechanical challenges.

7. REFERENCES

[1] Richard P. Feyman, 'There's Plenty of Room at the Bottom', Engineering & Science, Caltech, February 1960 (http://www.zyvex.com/nanotech/feynman.html)
[2] Gordon E. Moore, 'Cramming More Components onto Integrated Circuits', Electronics, Volume 38, April 1965
[3] Gordon E. Moore, 'No Exponential is Forever', ISSCC, February 2003
[4] Ilkka Tuomi, 'The Lives and Death of Moore's Law', First Monday, November 2002 (http://firstmonday.org/issues/issue7_11/tuomi)
[5] International Technology Roadmap for Semiconductors (http://public.itrs.net)
[6] Harry J.M. Veendrick, 'Deep-Submicron CMOS ICs', Kluwer, Deventer, 1998
[7] Simon M. Sze (editor), 'VLSI Technology (2nd edition)', McGraw-Hill, New York, 1988
[8] Garrou, P.E. Turlik I. (editors), 'Multichip Module Technology Handbook', McGraw-Hill, New York, 1998
[9] Bernd Michel, Rolf Aschenbrenner (editors), 'The World of Electronic Packaging and System Integration', DDP Goldenbogen, Dresden, 2004
[10] IEEE Transactions on Electron Devices
[11] IEEE Transactions on Electronics Packaging Manufacturing
[12] IEEE Transactions on Nanotechnology
[13] IEEE Transactions on Semiconductor Manufacturing
[14] Journal of Microelectromechanical Systems
[15] European Solid State Device Engineering Conference (ESSDERC)
[16] International Electron Device Meeting (IEDM)
[17] International Electronics Manufacturing Technology Symposium (IEMT)
[18] International Solid State Circuit Conference (ISSCC)
[19] Symposium on VLSI Technology (VLSI)
[20] Advanced Packaging (http://ap.pennnet.com)
[21] Electronic News (http://www.reed-electronics.com/electronicnews)
[22] Solid-State Technology (http://sst.pennnet.com)
[23] Electronic Components & Technology Conference (ECTC)
[24] Thermal, Mechanical, and Multiphysics Simulation and Experiments in Microelectronics and Microsystems (EuroSimE)
[25] International Conference on Electronic Packaging Technology (ICEPT)
[26] Electronics Packaging Technology Conference (EPTC)
[27] Reliability of Electron Devices, Failure Physics and Analysis (ESREF)
[28] Intersociety Conference on Thermal and Thermomechanical Phenomena in Electronic Systems (ITHERM)
[29] ASME International Mechanical Engineering Congress & Exhibition (IMECE)
[30] International Microelectronics and Packaging Society Conference (IMAPS)
[31] Journal of Microelectronics and Reliability
[32] IEEE Transactions on Advanced Packaging
[33] IEEE Transactions on Components and Packaging Technologies
[34] IEEE Transactions on Advanced Packaging
[35] ASME Journal of Electronic Packaging

8. NOMENCLATURE

BEOL:	back end of line (wafer process)
BGA:	ball grid array (outline)
BiCMOS:	bipolar technology combined with CMOS
CD:	critical pattern dimension
CMOS:	complementary MOS
CMP:	chemical mechanical polishing
CSP:	chip size package
CVD:	chemical vapor deposition
DIL:	dual in line (outline)
DRAM:	dynamic RAM
EEPROM:	electrical erasable programmable read-only memory
FEOL:	front end of line (wafer process)
FET:	field effect transistor
FRAM:	ferro-electric RAM
HVQFN:	heat-sink very-thin quad flat-pack no-leads (outline)
IC:	integrated circuit
k:	dielectric constant
k_1:	lithographic process factor
MCM:	multi chip module
MEMS:	micro electro mechanical switch
MIM:	metal insular metal (capacitor)
MOS:	metal oxide silicon (technology)
MRAM:	magneto resistive RAM
n:	donor-type semiconductor
NA:	numeric aperture
NMOS:	n-type MOS
p:	acceptor-type semiconductor
PCB:	printed circuit board
PGA:	pin grid array (outline)
PMOS:	p-type MOS
PVD:	physical vapor deposition
RAM:	random access memory
RF:	radio frequency
SiP:	system in package
SoC:	system on chip
SOI:	silicon on insulator (wafer)
SRAM:	static RAM
VLSI:	very large scale integration
WSP:	wafer scale packaging
λ:	wavelength (exposure system)

Chapter 2

RELIABILITY PRACTICE

F.G. Kuper[1], X.J. Fan[2]
[1]*Philips Semiconductors, P.O.Box 30008, 6503HK Nijmegen, The Netherlands*
[2]*Intel Corporation, CH5-263, 5000 W Chandler Blvd, Chandler, AZ 85226 USA*

Abstract: Reliability is one of key considerations in the development of new devices and packages that meet customer expectations. This chapter presents the basic concepts of reliability in microelectronics, and introduces the following: 1). reliability assessments by various environmental accelerated tests, 2). basic reliability statistics, 3). acceleration models, and 4). failure mechanisms.

Key words: Reliability, PPM, failure rate, probability, bathtub curve, accelerated test, HAST, highly accelerated stress test, JEDEC, thermal cycling, burn in, biased, unbiased, moisture, statistics, probability density function, cumulative probability function, MTTF, confidence level, failure mechanism, acceleration model.

1. INTRODUCTION

A common definition of reliability is the following:
"Reliability is the ability of a system or component to perform its required functions under stated conditions for a specified period of time [1]."
To quantify reliability, "ability" should be interpreted as a "probability". From this definition it is clear that all products always fail eventually. Indeed, a probability of zero failure during a certain amount of time is physically impossible, even for integrated circuit (IC).
 Nevertheless, in practice, zero is approached quite closely for ICs and discrete semiconductors. Therefore, the probability that an integrated circuit returns within the guarantee period is typically expressed in FPM's or PPM's ("failed parts per million" or simply "parts per million"), and the discrete elements reach numbers even expressed in PPB's ("parts per billion").

Next to PPM's and PPB's, there are many other indicators used to describe reliability. One of them is the failure rate. If one depicts the failure rate versus time, some interesting aspects of reliability can be explained. Such a curve appears to have a bathtub cross-section shape, and is therefore mostly referred to as a bathtub curve, as depicted in Figure 1.

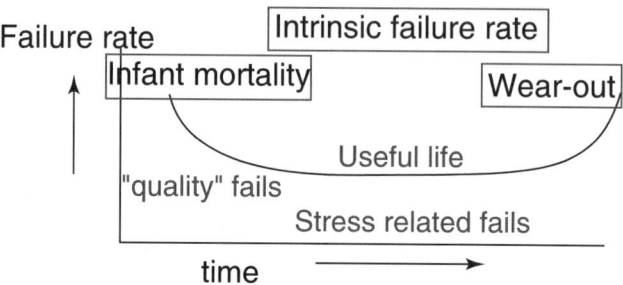

Figure 1. The bathtub curve: failure rate versus time. Note its three phases: infant mortality, useful life and wear-out

Three distinct periods of time exist: early failure, intrinsic failure, and wear-out. The early life of a product is also referred to as "infant mortality", another reference to reliability's actuarial origin. It is the period of time in which the product experiences failures also exclusively due to defects in the fabrication or assembly of the product. Because the defect represents weak or marginally functional parts, they quickly fail once turned on; hence they show up as a large rate of failure initially. However, once these few poorly manufactured parts are screened out and eliminated, the majority of the population is left that is otherwise robust. Thus, the rate of failure quickly diminishes. With the defects screened out of the population, a robust product will enjoy long and sustained period where intrinsic failures occur randomly. This stable region is identified through its near constant rate of failure. Finally, as the product ages, chemical, mechanical, or electrical stresses begin to weaken the product's performance to the point of failure. This is called the wear-out region. Sometimes the wear-out region doesn't appear, as shown in Figure 2. Apparently, the devices used in this analysis have been designed such that wear-out sets in long after 10 years of use. This is exactly what reliability engineering is all about: keep the failure rate in the early part of the bathtub low, and prevent any wear-out mechanism to occur before a required time.

2. Reliability Practice

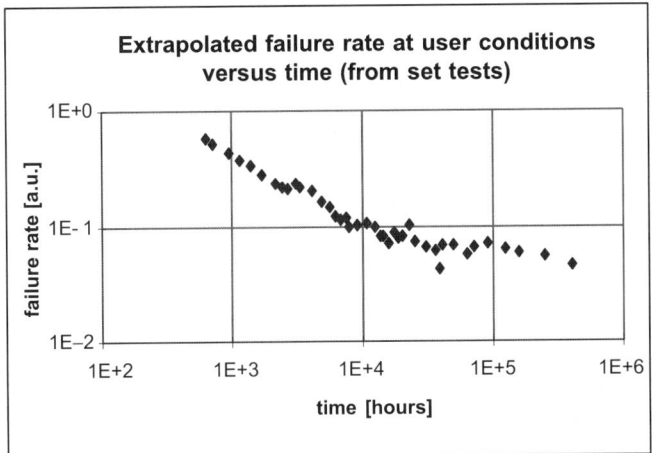

Figure 2. Failure rate versus time for various electronic components extrapolated to use conditions, as obtained from accelerated tests. Note the absence of the wear-out phase, even after 10 years (~1E5 hours) [2]

The reliability improvement over years is amazing. In the last 30 years, the number of transistors in an integrated circuit over a fixed area has increased about 6 orders in magnitude, but the reliability has also increased by about a factor of hundred. This amazing achievement in semiconductor industry has been achieved not only by a systematic reduction of random failures, but also by a systematic effort to keep, from product generation to product generation, the wear-out phase beyond practical duration.

2. RELIABILITY ASSESSMENT

There are two phases in reliability engineering, i.e., process development phase and manufacturing phase. In the process development phase, each mechanism that can result in device failure is studied. An interesting and very important aspect of accelerated testing is that not all failure mechanisms accelerate in the same way. For instance, gate oxide breakdown is accelerated by increasing temperature, but hot carrier degradation, on the other hands, needs a lower temperature in order to be accelerated.

Therefore, especially in the phase of process development, each mechanism needs to be studied separately. This is typically done on dedicated test structures that are tested on the wafer or are individually packaged. The physics should be understood, and from this the best methods for acceleration and reliability prediction can be determined.

Reliability environmental stress tests are used to simulate the end use (customer) environment and to uncover specific materials and process related marginalities that may be experienced during operational life. Historically, a number of standard tests have been defined to accelerate and capture specific device environments. Military standard environmental stress tests and conditions were developed in the 1940s for ceramic package devices being produced for the U.S. Military. Similarly, BellCor (now Telcordia) developed Network Equipment – Building System (NEBS) test standards for the communications use environment in which the use patterns, fault tolerance and seasonal use ambient environments were accounted for in the tests. Other consortiums such as Joint Electronic Device Engineering Council (JEDEC) and Institute for Printed Circuits (IPC) have adapted, documented and standardized many of the same tests. The following lists many of the types of environmental testing along with brief descriptions or reliability issues intended to be uncovered by the test.

2.1 Burn In

Burn in refers to an electrical overstress used in manufacturing and test and eliminate detect infant mortality failures or process defects that may remain latent until early customer operation. Burn in and high voltage extended life tests (HVELT) bake tests are performed on a functional device at high voltage conditions to discover defect failure modes that may failure in early life or extended operation. Early transistor gate defects and breakdown can be detected in burn-in and HVELT tests. Failure data are converted to a defect probability value that can be used to monitor process health.

2.2 Biased Moisture Test

Accelerated biased moisture tests are used to simulate the reliability of a powered device in an elevated temperature and high humidity environment under at nominal run state static bias. The tests are used to find metal migration, corrosion, dielectric breakdown, de-lamination and adhesion issues. Metal migration occurs by two modes, surface and bulk or interfacial diffusion of metal ions driven by an electric field. Corrosion occurs due to the interaction of heat, moisture and stress in a metal. Polymer dielectric breakdown and de-lamination failures are driven by moisture intrusion into the materials where it acts as a plasticizer by hydrolysis or it interferes the mechanisms of interfacial bonding.

There are two standard forms of the biased moisture test that can be employed. The temperature, humidity bias (THB) test is performed at

2. Reliability Practice

atmospheric pressure and temperatures between 30°C to 85°C and between relative humidity (RH) 50% to 85%. The highly accelerated stress test (HAST) is performed at less than 3 atmospheres of pressure and at static temperatures between 110°C to 156°C.

2.3 Unbiased HAST and Steam Test

The Unbiased HAST is performed to evaluate non-hermetic packaged devices in humid environments. The test employs temperature and humidity to accelerate the penetration of moisture through the external protective material or along the internal build-up or joined interfaces. Bias is not applied in this test to ensure the failure mechanisms potentially overshadowed by bias can be detected (e.g. galvanic corrosion). This test is used to identify failure mechanisms internal to the package and is destructive.

An autoclave test or steam test is similar to the HAST test in that it is performed to evaluate non-hermetic packaged devices in humid environments by accelerating moisture penetration. Steam is different from HAST in four respects; fixed temperature (121°C), fixed RH (100%), controlled ramp up/down and application of bias. Steam test conditions cannot be varied but durations can be altered when not testing to a standard based requirement. While both steam and HAST/THB remain standards based requirements the only HAST/THB can be used for performance characterization and use condition modelling.

2.4 Bake and Extended High Temperature Storage Test

Accelerated temperature bake testing is used to simulate a use environment where a device is continuously powered or stored at high temperature. The test is used to uncover solid-state metal reactions and stress voiding (Al) that result in cracks from volume changes, crystallization or trace shorting due to an affinity for an adjoining metal. Intrinsic polymeric materials degradation and creep can also result from the breaking of bonds and/or deformation at high temperatures because of corrosion or oxidation enhancement. Bake testing is performed under a number of different temperatures although some standards are defined. When determining the bake condition it is important to select temperatures that do not result in solid-state reactions that will not occur in use or that will not result in excessive plastic deformation, which may change the failure mode.

2.5 Electromigration Testing on Devices and Packages

Electromigration in devices results from the "electron wind" effect caused by high current density that is responsible for picking up individual electrons and moving them in the direction of current flow. High current densities (1×10^6 A/cm^2) are commonly found in devices interconnects, package bumps and some package vias and traces. Geometry factors such as bent trace can result in current pinning or high local current density in some traces and vias, which then results in reduce reliability over time. The behaviour was first empirically modelled by Black as following an Arrhenius relation with the addition of geometric factors. The data generated form the EM test allows proper design of metal trace and vias relative to the expected use current and temperature. The test can also be used in the manufacturing environment to monitor process equipment and reproducibility.

2.6 Moisture Sensitivity Test

Moisture preconditioning is required prior to reliability environmental stresses for all parts that are surface mounted to boards. Preconditioning was introduced as a qualification requirement with the release of plastic package because of "popcorning" during surface mount (see Chapter 6 for more details). The popcorn effect results from expanding steam that is evolved during surface mount. The humidity in the local ambient environment contributes to absorbed moisture over time where maximum absorption scales with temperature. The steps for preconditioning include 5-cycles of temperature cycle (–40 to 60°C), Bake (125°C) and temperature-humidity soak to target the needed moisture level and sit time. Preconditioning tests generally insure that the temperature, humidity and/or the shipping requirements are performed before assessing the use reliability.

2.7 Temperature Cycling and Temperature Shock Tests

Temperature cycle testing is used to simulate both ambient and internal temperature changes that result during device power up, operation and ambient storage in controlled and uncontrolled environments. Temperature cycling tests are used to detect thermal-mechanical interactions that result in dielectric or conductor cracking, fatigue and adhesion issues. Cracking, excessive strain, or fatigue results from differences in the coefficients of thermal expansion the various device materials.

Temperature cycle and temperature shock tests may be distinguished by the rate at which temperature is applied to the either the entire device or a

2. Reliability Practice

local area of the device. Temperature shock (TS) and power cycle (PC) occur at very high temperature ramp rates on the entire device or at local areas, respectively. The TS test can be done in the "air to air" or "liquid to liquid" environment while temperature cycle typically occurs only in air. The operating specification for temperature cycling is found in JESD-22-A104B. Both tests can be used to detect cracks that result from very fast coefficient of thermal expansion (CTE) changes in the materials which make up the device. Thermal cycle testing is typically performed at slower ramp rates and longer temperature dwell times to allow mechanical damage such as fatigue and creep to occur.

Table 1. Common thermal cycling test conditions

Environmental Test	Low Temperature (°C)	High Temperature (°C)	Soak Time (minutes)	Cycle Time
TS	−65	165		
	−55	125		
TC	−65	165	15	30
	−55	125	15	30
	−40	85	30	60
	−40	125	30	60
PC	25	80	2	6
	30	80	2	6
	30	110	2	6

2.8 Power Cycle Testing

The power and temperature cycling test is used to assess the capability of a device to withstand alternate power/temperature exposures extremes at operating biases. This test method applies to devices that are subjected to multiple temperature excursions as required by powering the device on and off. Power cycle tests are not accelerated environmental tests but rather they are accelerated in time relative to use operation. The test is best used at interfaces where the time-dependent interactions between materials near the power source need to be assessed. In many ways the test resembles a thermal shock test between constituent materials in the system where one side of an interface heats rapidly and the second interface material is undergoing nearly instantaneous changes in temperature and mechanical state. The test is not intended to simulate slow changes to temperature and the accumulation of damage but rather to test the instantaneous interactions between materials.

2.9 Mechanical Testing

Mechanical shock and vibration testing is commonly used to simulate the transportation and use events that are experienced by components on a system board. Shock is a high rate change in velocity and momentum that occurs when devices are dropped or experience high rates of acceleration. Vibration occurs during all forms of transportation at various vibration and resonant frequencies. The Vibration test is intended to determine the ability of component to withstand moderate to severe vibration as a result of motion produced by transportation or use operation. In addition, design issues with resonant frequency acceleration can be assessed.

The alternative names for the above tests commonly used in semiconductor industry are listed below and summarized in Table 2.

- DHTL: Dynamic High Temperature Life
  168 hours (1 week) to test EFR (Early Failure Rate)
  1008 hours (6 weeks) to test IFR (Intrinsic Failure Rate)
  Few hours: usually called burn-in
- SHTL: Static High Temperature Life
- DLTL: Dynamic Low Temperature Life
- HAST: Highly Accelerated Steam Test
- UHAST: Unbiased Highly Accelerated Steam Test
- TMCL: Temperature Cycle
- PPOT: Pressure Pot Test
- U-POT: Unsaturated Pressure Pot Test
- THBS: Temperature Humidity Bias Stress

Table 2. Typical reliability tests and their conditions

	Temperature	Humidity	Bias on	Typical duration	Comment
D/SHTL	150°C junction	Not applicable	Y	48/168/1008 hours	Sensitive for die faults
DLTL	−40°C ambient	Not applicable	Y	168/100 hours	Seldom used
HAST	132°C	85%	Y	96 hours	Corrosion
UHAST	132°C	85%	N	96 hours	
TMCL	−65°C/150°C	N.A.	N	200 or 500 cycles	Package – die interaction
PPOT	121°C	100%	N	96 hours	Trend is towards unsaturated tests
U-POT	130°C	85%	N	96 hours	See above
THBS	85°C	85%	Y	1008 hours	Too slow: trend is as HAST

2. Reliability Practice

2.10 Design for Manufacturability, Reliability, and Testability (DfMRT)

Design for manufacturability, reliability, and testability (DfMRT) is the systematic process of preventing and eliminating designed-in failures in the early stages before actual physical prototyping. Regardless of product application, DfMRT is an important factor in system design, development, qualification, manufacture, and in-service management.

Design for manufacturability (DfM) is a process to make sure that the design is compatible with the manufacturing process, because process incompatibility not only causes quality problems (yield), but also reliability problems. In order to succeed, the manufacturing process must be fully defined and under good statistical control.

Design for reliability (DfR) is the systematic process of preventing or eliminating designed-in failures. DfR includes:
- Perform a proper Failure Modes and Effects Analysis (FMEA) to discover potential failure modes inherent in the design
- Analyse the failure modes and determine their root cause
- Modify the design as such towards optimal robustness

Design for testability (DfT) concentrates on making the design as such that it can be tested easily and/or cost-effectively in later stages.

Design rules and/or guidelines provide tools to perform DfMRT. For example, to prevent wear-out failure, design rules are generated during process development. As long as designers obey all design rules and users obey the maximum use conditions (like maximum temperature and voltage) wear-out is negligible. Design rules and/or guidelines can take the form of theoretical, empirical, or regulatory models that relate potential failure mechanisms and the average life of the product to the life-cycle requirements of manufacturing, assembly, testing, storage, handling, transportation, operation, and repair.

There are also design rules that make an IC less sensitive for process-induced damage. Plasma etching for instance results in a harsh environment for circuits. Long metal lines are exposed to a plasma and may be charged, and may discharge through the gate oxide of a transistor. This will lead to partially degraded transistors, and therefore a reduced reliability. There are design rules for metal layout that minimize these effects, thereby decreasing the amount of stress on IC's even before they reach a consumer.

In a similar way, there are mechanical stress related rules that prevent too high mechanical stress on the chip in a package. In addition, there are rules to decrease the packaging related yield loss, like die size in relation to die pad to prevent glue short circuit fails.

3. RELIABILITY STATISTICS

To start, here are some useful terms and definitions:
- Reliability Function: R(t), the probability that a device will function for a given time under specific operating conditions. R(t) ranges between 0 and 1.
- Cumulative Distribution Function (CDF): F(t), F(t) = 1 − R(t), the fraction of all failures that occur up until some time, t.
- Probability Density Function (PDF) f(t): the fraction of the initial population which fails within a short interval of time, written as following

$$f(t) = \frac{dF(t)}{dt} = -\frac{dR(t)}{dt} \qquad (1)$$

- Failure rate or hazard function λ(t): the fraction of the remaining population at time t which will fail within the next short period of time, Δt, as following

$$\lambda(t) = \frac{dF(t)}{dt} \bigg/ R(t) = \frac{f(t)}{R(t)} = \frac{-d \ln R(t)}{dt} \qquad (2)$$

There are two commonly used metrics associated with the central tendency of the population. For generalizations of a population performance, a Mean Time to Failure (MTTF) is commonly reported. This is calculated by summing all failure times and dividing by the number of samples. Another method by which to compare performance is by examining the point in time under which half of the population fails. This Median Time to Failure will be denoted as T50. With symmetrical distributions, MTTF and T50 values are similar; however, this is not true for most distributions in reliability due to their skewed tails.

3.1 Life Distributions

In many statistical applications, a Gaussian or "Normal" distribution is commonly used to describe central tendency and variability of the population. However, for reliability assessments three other distributions are commonly used: Exponential, Weibull, and Lognormal.

With the three portions of the bathtub curve defined in Figure 1, the goal of the reliability engineers is to select the life distribution, which most accurately reflects the behaviour of the product in terms of a failure rate.

2. Reliability Practice

None of the typical life distributions used can model the entire bathtub curve at once. Therefore, selection of the life distribution should be strongly coupled with consideration of the region in the product's lifetime (early, stable, or wear-out). The failures observed in the early failure region can be quickly screened out through short operational periods often called "burn-in". Therefore, for this discussion the focus will lie in modelling the stable and wear-out periods. In the following a brief overview of four common life distributions is given.

3.1.1 Normal Distribution

The Normal distribution is completely specified by two parameters: the mean (μ) and the variance (σ^2). The mean of a Normal distribution locates of the center of the density and can be any real number. The variance of a Normal distribution measures the variability of the density and can be any positive real number. The standard deviation (σ) is the square root of the variance and is used more for its interpretability. When $\mu = 0$ and $\sigma = 1$, it is referred to as the standard normal distribution.

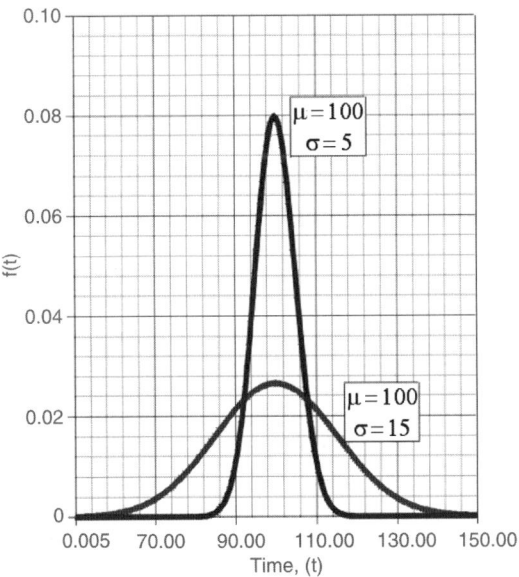

Figure 3. Normal distribution (PDF)

The probability density function (PDF) of a normal distribution is:

$$f(t) = \frac{1}{\sqrt{2\pi\sigma^2}} e^{-(x-\mu)^2/2\sigma^2} \qquad (3)$$

Thus, the CDF of the normal distribution

$$F(t) = \int_{-\infty}^{t} \frac{1}{\sqrt{2\pi\sigma^2}} e^{-(x-\mu)^2/2\sigma^2} dx \qquad (4)$$

Reliability function is

$$R(t) = 1 - \int_{-\infty}^{t} \frac{1}{\sqrt{2\pi\sigma^2}} e^{-(x-\mu)^2/2\sigma^2} dx \qquad (5)$$

The hazard rate function

$$\lambda(t) = f(t)/R(t) \qquad (6)$$

And MTTF

$$MTTF = \mu \qquad (7)$$

3.1.2 Exponential Distribution

The exponential distribution is the simplest of all the life distributions. In requires only one parameter to describe its shape. Its key features include a PDF shape skewed to the right (high rate of failure occurrence initially followed by a step asymptotic decline) and its constant hazard rate. It is this ability to depict a constant rate of failure over time that most commonly lends the exponential distribution to the stable, intrinsic failure portion of the bathtub curve. In reliability, this implies that within this region of time, a "new" part and an "aged" part that both lie in the stable period of time will have equal likelihood of failing in the next small interval Δt. The use of this distribution is typically associated with the wafer fabrication processes and less with the assembly microprocessor packaging processes.

2. Reliability Practice

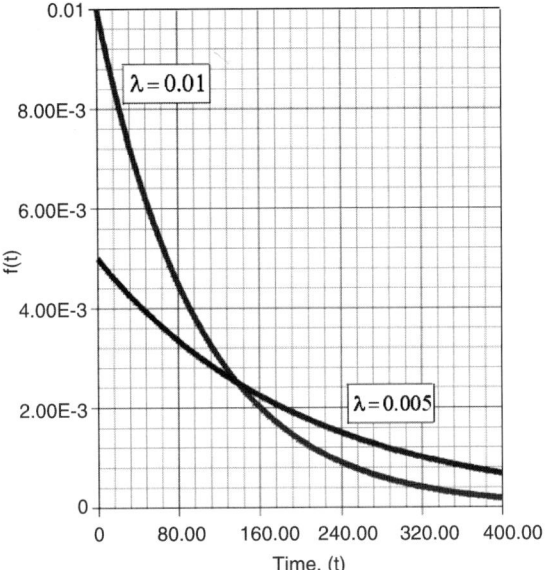

Figure 4. The exponential distribution (PDF)

The PDF of the exponential distribution is expressed as

$$f(t) = \lambda \cdot \exp(-\lambda t) \tag{8}$$

Thus, the CDF of the exponential distribution

$$F(t) = 1 - \exp(-\lambda t) \tag{9}$$

Reliability function is

$$R(t) = \exp(-\lambda t) \tag{10}$$

The hazard rate function

$$\lambda(t) = \lambda \tag{11}$$

And MTTF

$$MTTF = \frac{1}{\lambda} \tag{12}$$

3.1.3 Weibull Distribution

The next distribution of interest is the Weibull distribution. The Weibull distribution describes a "weakest link" mechanism where many potential failure locations exist on a product or system and the product fails when any one of these locations fails. Developed by Dr. E. H. Waloddi Weibull in 1951, this distribution was first used by Pratt & Whitney in airplane engine applications. Its popularity for use in the reliability field stems from its flexibility to describe a wide variety of failure rate curves as illustrated in Fig. 5.

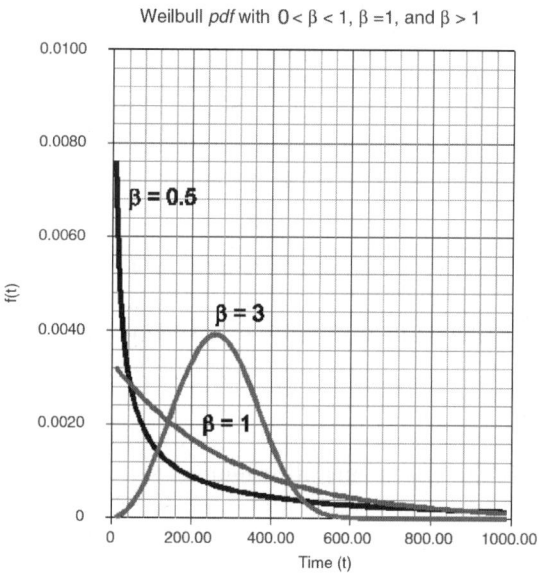

Figure 5. Weibull Distribution (PDF)

The PDF of the Weibull distribution is expressed as

$$f(t) = \frac{\beta}{t}\left(\frac{t}{\alpha}\right)^{\beta} \cdot \exp\left[-\left(\frac{t}{\alpha}\right)^{\beta}\right] \tag{13}$$

Thus, the CDF of the Weibull distribution

$$F(t) = 1 - \exp\left[-\left(\frac{t}{\alpha}\right)^{\beta}\right] \tag{14}$$

2. Reliability Practice

Reliability function is

$$R(t) = \exp\left[-\left(\frac{t}{\alpha}\right)^{\beta}\right] \qquad (15)$$

The hazard rate function

$$\lambda(t) = \frac{\beta}{\alpha} \cdot \left(\frac{t}{\alpha}\right)^{\beta-1} \qquad (16)$$

And MTTF

$$MTTF = \alpha \cdot \Gamma(\frac{1}{\beta}+1) \qquad (17)$$

$$\Gamma(n) = \int_0^{\infty} e^{-x} x^{n-1} dx \qquad (18)$$

The Weibull is a two-parameter model consisting of a scale parameter (alpha) and shape parameter (beta). Alpha is the "characteristic life" of the product and serves as a "location anchor" establishing the x-axis value under which 63.21% of the population occurs. This can be seen more directly by substituting $t = \alpha$ into the Weibull CDF yielding $1 - e^{-1} = 0.6321$. This life distribution can take on other special properties. It is equivalent to the Exponential when $\beta = 1$, Rayleigh when $\beta = 2$, and a symmetrical bell-shape (Gaussian) when $\beta = 3\text{-}4$. A description of the distributions shape, hazard rate characteristics, and applicability to regions of the bathtub curve are summarized in Table 3.

Table 3. Weibull characteristics based on beta values

Shape Parameter	Distribution Description	Use
$0 < \beta < 1$	Exponentially decreasing from infinity.	Early Failure Period (Infant Mortality)
$\beta = 1$	Special Distribution: Exponential Exponentially decreasing from α. Note that this is the exponential distribution with $\alpha = 1/\lambda$.	Stable Failure Period
$1 < \beta < 4$	Rises to peak and then decreases.	Models the early wear-out failure period of a product's life.
$\beta = 2$	Special Distribution: Rayleigh	
$3 < \beta < 4$	Gaussian "Normal" bell-shape shape	
$\beta > 10$	Type 1 Extreme Value shape	End of Life when most wear-out failures occur.

3.1.4 Lognormal Distribution

Another distribution commonly used in reliability is the lognormal distribution. The Lognormal distribution describes a "degradation" mechanism the key product feature is reduced or consumed until failure occurs. Commonly, this is associated with chemical reactions where oxidation or diffusion occurs. This can be derived as a proportional growth model or as the multiplicative efforts of random degradations leading to failure. As the name implies, the lognormal distribution is related to the Normal or Gaussian distribution discovered by Carl Friedrich Gauss in late 1700's. The logarithm of the failure times has a normal distribution, thus analysis and interpretation are eased through a simple transformation. Similar to the Weibull distribution, the lognormal distribution can assume a wide variety of failure rate shapes.

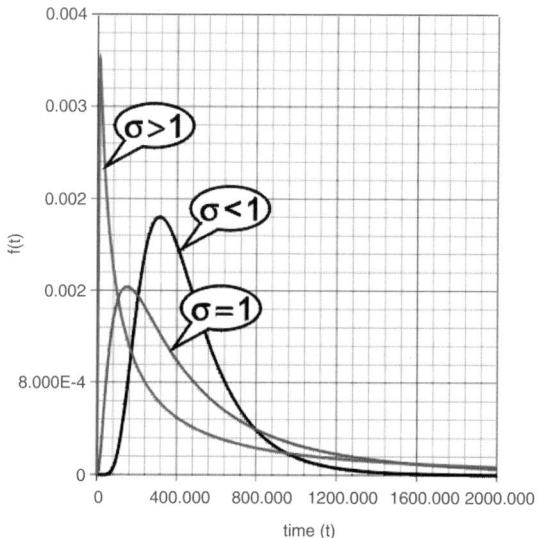

Figure 6. Lognormal distribution (PDF)

The PDF of the lognormal distribution is expressed as

$$f(t) = \frac{1}{\sqrt{2\pi} \cdot \sigma^2 t} \exp\left[-\left(\frac{1}{2\sigma^2}\right) \cdot (\ln(t) - \mu)^2\right] \qquad (19)$$

2. Reliability Practice

Thus, the CDF of the lognormal distribution

$$F(t) = \Phi\left[\frac{\ln(t)-\mu}{\sigma}\right] \tag{20}$$

Reliability function is

$$R(t) = 1 - \Phi\left[\frac{\ln(t)-\mu}{\sigma}\right] \tag{21}$$

where ϕ is a Z-statistic from a standard normal (0,1) distribution. The hazard rate function

$$\lambda(t) = f(t)/R(t) \tag{22}$$

And MTTF

$$MTTF = \exp\left(\mu + \frac{\sigma^2}{2}\right) \tag{23}$$

3.2 Confidence Level

The Confidence Level is defined by

$$\int_0^{F_c} P(F)dF = C \tag{24}$$

from which, one can state that with a confidence level C (0-1), the failed fraction is below Fc. Fc can be found from an expression if one assumes the number of fails as n, and the sample size as N,

$$1 - \sum_{i=0}^{n} \frac{(N+1)!}{i! \cdot (N+1-i)!} (F_c)^i (1-F_c)^{N+1-i} = C \tag{25}$$

4. ACCELERATION FACTOR MODELS

Accelerated life tests are performed to accelerate a physical failure mechanism without inducing new failure mechanisms that do not exist in the use environment. Acceleration factors use time-to-fail at a particular accelerated stress level for a particular failure mechanism to predict the equivalent time to fail at a use or field stress level. Under linear acceleration we define acceleration factor (AF) as:

$$AF = \frac{Time\ to\ fail\ (Stress\ 1)}{Time\ to\ fail\ (Stress\ 2)} \quad where\ Stress\ 2 > Stress\ 1 \qquad (26)$$

Acceleration models are usually based on the science underlying a particular failure mechanism. Successful empirical models are close approximations of a number of complicated physics or kinetics models as to determine when the theory of the failure mechanism is eventually understood. Note that in the case of linear acceleration, while predicting time-to-fail at for the use environment, the type of life distribution (Lognormal, Weibull, Exponential, etc) and the slope parameters do not change but the location parameters change.

Let's discuss some of the acceleration models for failure mechanisms accelerated by temperature, humidity, voltage and temperature cycling.

4.1 Arrhenius Relationship

The Arrhenius model has been used successfully for many chemical and physical failure mechanisms (chemical reactions, diffusion processes or migration processes) accelerated by temperature stress. This empirically based model takes up the following form:

$$R(t) = A' \exp\left(-\frac{Q}{kT}\right) \qquad (27)$$

where
  $R(T)$ = reaction rate
  A' = constant
  Q = activation energy (eV)
  k = Boltzmann's constant (8.617×10^{-5} eV/K)
  T = Temperature in °K (°C + 273.16)

2. Reliability Practice

Typical activation energy (Q) values are in the range of 0.2 to 2.0 eV depending on the failure mechanism. The acceleration factor between accelerated stress and use environment is given by:

$$AF = \exp\left[\frac{Q}{k}\left(\frac{1}{T_{use\,env}} - \frac{1}{T_{stress}}\right)\right] \quad (28)$$

Note that the only parameter unknown in the above equation is the activation energy. By running multiple stress conditions, the activation energy for a specific failure mechanism can be calculated.

4.2 Peck's Model

Corrosion induced by the moisture in the environment is one of the commonly seen failure mechanisms in electronics packaging. Humidity also is responsible for causing some of the interfacial delamination failures induced by hygro-thermal stresses. The acceleration model widely used to model the effect of relative humidity on failures is Peck's model.

$$AF = \left(\frac{RH_{use\,env}}{RH_{stress}}\right)^{-c} \exp\left[\frac{Q}{k}\left(\frac{1}{T_{use\,env}} - \frac{1}{T_{stress}}\right)\right] \quad (29)$$

where
RH = Relative Humidity (%)
c = RH inverse power law coefficient
Q = Activation energy (eV)
k = Boltzmann's constant (8.617×10^{-5} eV/K)
T = Temperature in °K (°C + 273.16)

4.3 Temperature-Voltage-Relative Humidity model (Eyring model)

Sometimes in addition to high temperature and relative humidity, voltage is also influential in driving failures. In such cases, the failure are modeled using:

$$AF = \left(\frac{V_{use\,env}}{V_{stress}}\right)^{-n}\left(\frac{RH_{use\,env}}{RH_{stress}}\right)^{-c} \exp\left[\frac{Q}{k}\left(\frac{1}{T_{use\,env}} - \frac{1}{T_{stress}}\right)\right] \quad (30)$$

where
- A' = constant
- V = Voltage (V)
- n = Voltage inverse power law coefficient
- RH = Relative Humidity (%)
- c = RH inverse power law coefficient
- Q = Activation energy (eV)
- k = Boltzmann's constant (8.617 x 10^{-5} eV/K)
- T = Temperature in °K (°C + 273.16)

4.4 Coffin-Manson Model

A popular model used for thermo-mechanical fatigue effects is the "Coffin-Manson" model. This model was originally used to model metal fatigue subjected to thermal cycling.

$$\frac{\Delta \varepsilon_p}{2} = \varepsilon'_f (2N_f)^c \qquad (31)$$

where
- Δεp = Plastic strain
- ε$_f$' = fatigue ductility coefficient
- Nf = cycles to failure
- c = a material dependent constant

Plastic strain is the thermal strain induced due to the Coefficient of Thermal Expansion (CTE) mismatch between different materials = Δα .T. Typically, within the thermal operating range for most of the electronic devices, the coefficient of thermal expansion doesn't change much with temperature and Coffin-Manson in its simplest form takes into account only the temperature differences.

$$AF = \left(\frac{\Delta T_{use\ env}}{\Delta T_{stress}} \right)^{-n} \qquad (32)$$

where
- ΔT = Entire temperature cycle range within which a device operates
- n = Material dependent inverse power law coefficient (e.g. ~1.9 for eutectic solders)

2. Reliability Practice

Chapter 7 will describe more in details how Coffin-Manson type of model is derived and used for the reliability prediction for solder joints.

4.5 Norris-Landzberg Model

The simple form of the Coffin-Manson Equation doesn't account for the time-dependent material behaviours. To account for the effect of these time-dependent properties (strain rate effects, stress relaxation etc.) on fatigue life, Norris and Landzberg (IBM, 1969) introduced an empirical frequency factor into the original Coffin-Manson equation. In addition, they also observed that specifically for eutectic solders when strain is applied at a continuously changing temperature, the fatigue life would decrease in the upper temperature regions of the cycle due to temperature-related effects such as increased grain boundary sliding. Hence, an empirical peak temperature factor was also introduced. Norris-Landzberg model is given by:

$$AF = \left(\frac{\Delta T_{use\,env}}{\Delta T_{stress}}\right)^{-n} \left(\frac{f_{use\,env}}{f_{stress}}\right)^{m} \exp\left[\frac{Q}{k}\left(\frac{1}{T_{use\,env}} - \frac{1}{T_{stress}}\right)\right] \qquad (33)$$

Where
ΔT = Entire temperature cycle range within which a device operates
n = Material dependent inverse power law coefficient (e.g. ~1.9 for eutectic solders)
f = frequency of cycling
m = frequency inverse power law coefficient (~0.33 for eutectic solders)
Q = Activation Energy (eV)
k = Boltzmann's constant (8.617 x 10^{-5} eV/K)
T = Temperature in the hot zone of temperature cycle in °K (°C +273.16)

5. FAILURE MECHANISMS

5.1 General

This section presents a brief introduction of failure mechanisms in device and package level. Figure 7 shows a cross-section of a transistor with metal connectors, and locations of several associated failure mechanisms.

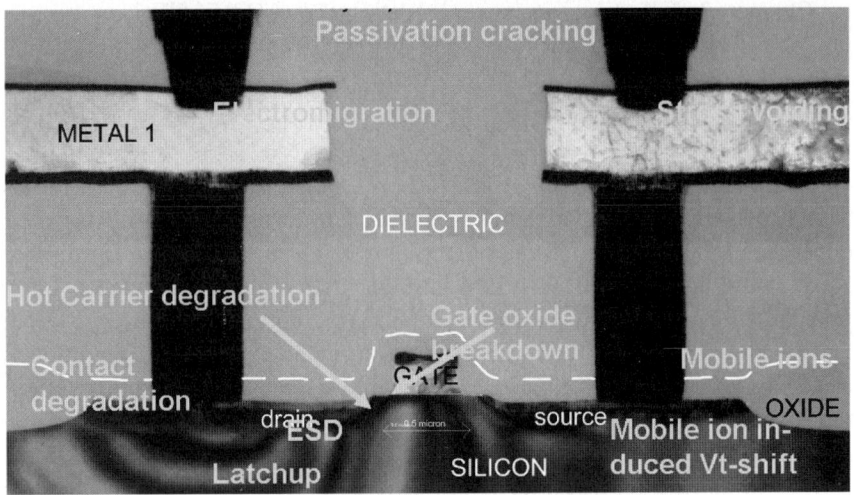

Figure 7. Transmission electro micrograph of a MOSFET with metal connectors, and various locations of possible failure mechanisms

ESD, latchup and passivation cracking are typically induced by thermo-mechanical stress. Electromigration, stress voiding, hot carrier degradation, and oxide breakdown are common failure modes during operating conditions. The underlying physics of these two groups of mechanisms has to be studied, and simulated.

5.2 Examples of Failure Mechanisms

5.2.1 Gate Oxide Breakdown

Gate oxide isolates the gate of a MOSFET from the other terminals. According to a gate oxide breakdown model a discharge will take place though the oxide when the cumulative number of generated traps (generated during leakage of current through the oxide) is enough to form a chain from one side of the capacitor to the other.

The capacitor energy (½CV2) is released through this chain, causing a considerable amount of damage (example shown in Figure 8), which typically leads to a short circuit through the capacitor.

2. Reliability Practice

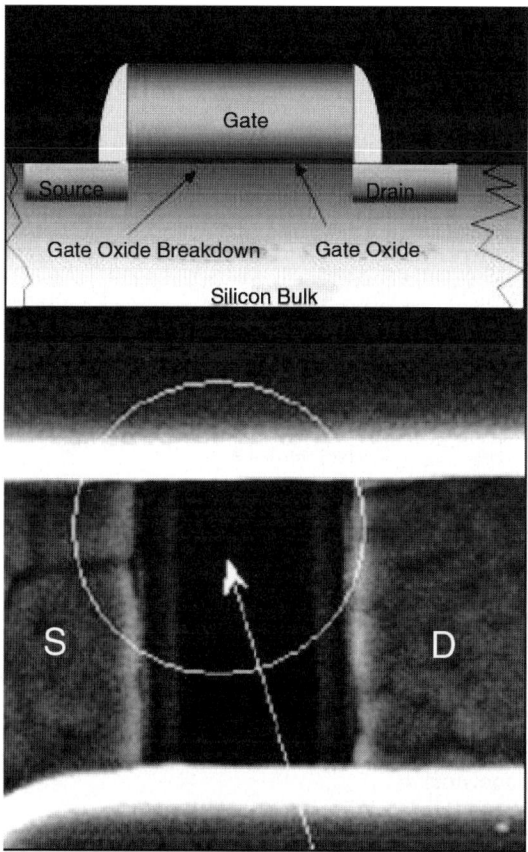

Figure 8. Schematic cross section of a transistor with a gate oxide breakdown, plus a backside planar SEM-view of a broken down gate oxide (black hole in circle). Source and Drain are indicated as well

5.2.2 Hot Carrier Degradation

Charge carriers in a transistor can be accelerated so that collisions with the lattice lead to injection of charge in the insulating gate oxide. This leads to progressive degradation of the transistor performance as seen in a ring oscillator, which runs at a high voltage to accelerate the degradation process. Process optimisation and design rules together are needed to assure hot carrier reliability.

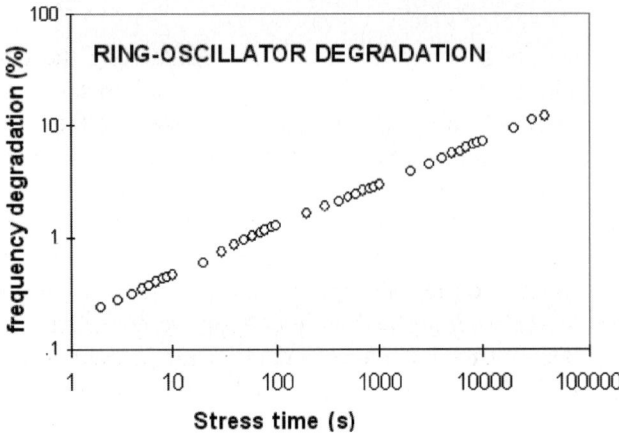

Figure 9. Example of hot carrier induced ring oscillator frequency degradation

5.2.3 ESD (Electro-Static Discharge)

An integrated circuit is very vulnerable when voltage exceeds its supply voltage. In Figure 10 the damage effect of a lethal ESD pulse is shown, where the energy of the ESD pulse causes a melted region between source and drain of a transistor.

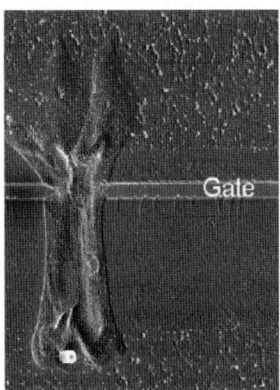

Figure 10. Top view of a molten and re-solidified result of an ESD event in a transistor

2. Reliability Practice

5.2.4 Electromigration

Between transistors, there are metal lines conducting the current. This current can induce electromigration: migration of metal atoms in the direction of "electron wind". The 'holes' might appear in the metal, causing resistance increases, and possible even open connections. Both effects may cause an IC to stop functioning according to its specifications. An example is shown in Figure 11, in which two metal lines are connected with a tungsten plug. Tungsten is a very hard metal and does not migrate, but immediately behind the tungsten, the plug aluminium can migrate away, leaving a hole. Holes in metal can also be formed without applying a current. This is called stress voiding. Due to the difference in coefficients of thermal expansion of metal and glass (the oxide between the metal lines) the metal is usually in a tensile condition, which may lead to atom migration.

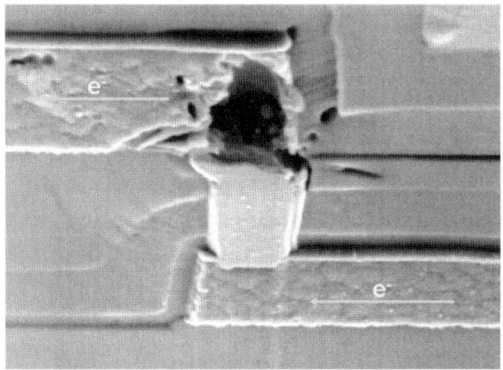

Figure 11. Cross section of two metal layers and a tungsten plug after an electromigration

5.2.5 Die Crack

Surface scratching and cracks in dies may form in packaging processes, such as die scribing, dicing, and/or die-attach stages. If an initial flaw equals to or greater than the critical crack size exist, the die can catastrophically fracture in a brittle manner. For instance, surface cracks due to chemical-mechanical polishing may propagate throughout the complete die during temperature cycling testing. Die fractures may run in vertical direction but also horizontally. Figure 12 illustrates an example of vertical die crack in a HBGA package. In most cases, die crack originates from flaws on the backside due to back grinding operations for wafer thinning.

Figure 12. Example of horizontal die cracking in a HBGA package

5.2.6 Wire Sweep

Wire sweep denotes visible wire deformations, typically an in-plane movement in the direction of mould flow through the cavity. Under this deformation, ball bonds can kink at the connection point as the wire is pulled at during moulding. A consequence of wire sweep is device shortening and/or current leakage. Failures may occur immediately after moulding, but in some cases also after subsequent stress testing. Wire sweep can occur from a number of causes: high resign viscosity, high flow velocity, unbalanced flow in the cavity, void transport, and filler collisions. Figure 13 illustrates an example of wire sweep.

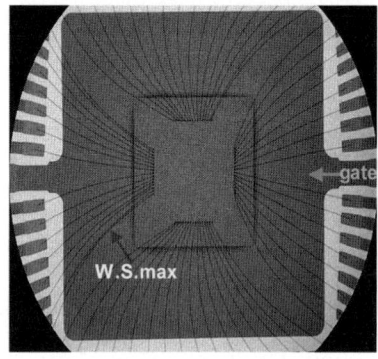

Figure 13. Example of wire sweep problems

5.2.7 Interface Delamination

Interface delamination between two adjacent materials is one of the major problems in packaging industry. One of most common delaminated interface is the one between epoxy and metal materials. Moisture ingress, either through the bulk epoxy or along the interface can accelerate delamination in plastic packages. The adhesion is also degraded by improper assembly processes, for example by oxidation and/or contamination, delamination and micro-cracks introduced at the interfaces. During reflow conditions, moisture at interfaces generates vapour pressures, which will be the driving force for the micro-cracks to propagate. Delamination is the root cause for many other types of failures during subsequent temperature cycle testing. So-called Surface Acoustic Measurements (SAM) can identify delamination. Figure 14 illustrates delamination between leadframe and moulding compound.

Figure 14. Delamination between leadframe and compound

6. CONCLUSIONS

There are many mechanisms that ultimately can and will cause failures of IC's. There are extensive mathematical toolboxes available to describe fails and fail distributions. During process development (both wafer fab process as well as packaging process), reliability of products can be built in by generating knowledge based design rules that address each individual failure mechanism. The rest of book will address some of important failure mechanisms associated with mechanics failures.

7. REFERENCES

[1] Institute of Electrical and Electronics Engineers. IEEE Standard Computer Dictionary: Compilation of IEEE Standard Computer Glossaries. New York, NY: 1990.
[2] J. A. van der Pol, E. R. Ooms, T. van 't Hof, F. G. Kuper, Impact of Screening of Latent Defects at Electrical Test on the Yield-Reliability Relation and Application to Burn-in Elimination, Proc. IRPS, 1998.
[3] Source: ADAC, Germany
[4] F. Kuper, J. van der Pol, E. Ooms, T. Johnson, R. Wijburg, W. Koster, and D. Johnston, Relation between yield and reliability of integrated circuits: Experimental results and application to continuous early failure rate reduction programs, Proc. IRPS, 1996.
[5] M. Ohring, Reliability and failure of electronic materials and devices, Academic Press, 1998.
[6] S. S. Manson, Thermal Stress and Low-Cycle Fatigue, McGraw-Hill Book Company, New York, 1966.
[7] K. C. Norris, and A. H, Landzberg, Reliability of Controlled Collapse Interconnections, IBM Journal of Research and Development, 1969, pp. 266-71.
[8] MIL-HDBK-217, Military Handbook for Reliability Prediction of Electronic Equipment.
[9] http://www.jedec.org/
[10] JESD 47: Stress Driven Reliability Qualification.
[11] JESD 34: Failure Mechanism Driven Reliability Qualification.
[12] JEP148 Reliability Qualification of Semiconductor Devices Based on Physics of Failure Risk and Opportunity Assessment.
[13] http://www.ipc.org/
[14] JESD22-A101-B: Temperature humidity bias life test.
[15] JESD22-A110-B: Test method for performing HAST.
[16] JESD22-A118: unbiased HAST and accelerated moisture test.
[17] JESD22-A102-C: unbiased Autoclave and accelerated moisture test.
[18] JEP139: Guideline for constant temperature aging to characterize Al interconnect metallization for stress induced voids.
[19] J. R. Black, Mass transport of aluminum by momentum exchange with conducting electrons, Proc. 1967 Annual Symp. Reliability Phy., p. 149-51. 1967.
[20] JESD61: iso-thermal electromigration test procedure.
[21] JESD63: Standard method for calculating the electromigration model parameters for current density and temperature.
[22] JESD-22-A113D: Precondition classifications and evaluation procedure.
[23] JESD-47: Manufactures requirements detailing when preconditioning is applied.
[24] J-STD-020: Procedure for characterizing moisture uptake kinetics.
[25] JESD22-A105C: Power and temperature cycling.
[26] JESD22-B104-B: Mechanical shock standard.
[27] JESD22-B103-B: Vibration standard.
[28] ATD-Package Handbook, Intel, in press.
[29] Nicholas P. Mencinger, A Mechanism-Based Methodology for Processor Package Reliability Assessments, Intel Journal of Technology, 2000.

2. Reliability Practice

8. EXERCISES

1. A system comprises 30 IC's. Each IC has a 10 ppm defect level. What is the defect level (due to IC's) of the system?
2. To increase reliability of metal interconnects, a thin layer of a hard alloy metal (like TiN or TiW) is added to an aluminium line. In case there is an electromigration induced void in the metal, the current can still go through this so-called barrier layer. What is the expected impact on the time-to-fail distribution of a set of lines?
3. A product has lifetime distribution (t in hours): $F(t) = 1 - e^{-0.001t}$
   a. What is the name of this life distribution?
   b. What is the probability that a random unit will fail before 1000 hrs?
   c. What is the PDF of this life distribution?
   d. What is the failure rate at 1000 hr?
4. Suppose under stress conditions 1% fails by 10 hours. The acceleration factor is $AF = 50$. Under the assumptions of true and linear acceleration, what is the fraction that fails by 500 hours under use conditions?
5. Suppose you find 10 fails out of a tested group of 1000 IC's after 168 hours use at 150°C. What is the failure rate of this product at a use condition of 85°C? Use 0.7 eV as activation energy.

Chapter 3

THERMAL MANAGEMENT

H.J. Eggink[1], J.H.J. Janssen[1,2]
[1]*Philips Applied Techologies, P.O. Box 218, 5600MD Eindhoven, The Netherlands*
[2]*Philips Semiconductors, P.O. Box 30008, 6503HK Nijmegen, The Netherlands*

Abstract: This chapter gives an introduction in to the basics of heat transfer that a device designer may encounter. A short discussion on the subject of thermal resistance is given. The thermal design of electronic products, linking device and board level, is addressed from the device point of view.

Keywords: Thermal basic theory, heat transfer, thermal resistance, boundary condition independent model.

1. INTRODUCTION

Temperatures and temperature gradients in electronic packages and systems are an important factor in the performance and reliability of electronic products. Because the electrical properties of electronic devices are temperature dependent, a higher operating temperature can lead to thermal drift and loss of performance. A typical issue for power devices is the internal electrical resistance, which increases with temperature. The heat production in the device is proportional to the electrical resistance and the combination of increasing resistance and heat production can lead to thermal runaway of the device.

High temperatures can cause material degradation leading to failure of the device or its direct surroundings, for instance the solder joints or the printed circuit board. Thermal cycling, heating and cooling down during use, is a different source of possible failures, because it introduces fatigue in materials. In particular solder joints are sensitive for this type of thermal loading.

With the increasing demand on performance and reliability, combined with an increased power dissipation and miniaturization of components, thermal management of electronic components is becoming more and more important. An illustration of the power dissipation trend is given in Figure 1 (Azar and Morabito) which gives the power distribution for Intel® microprocessors up to a Pentium® II-300 using approximately 40W.

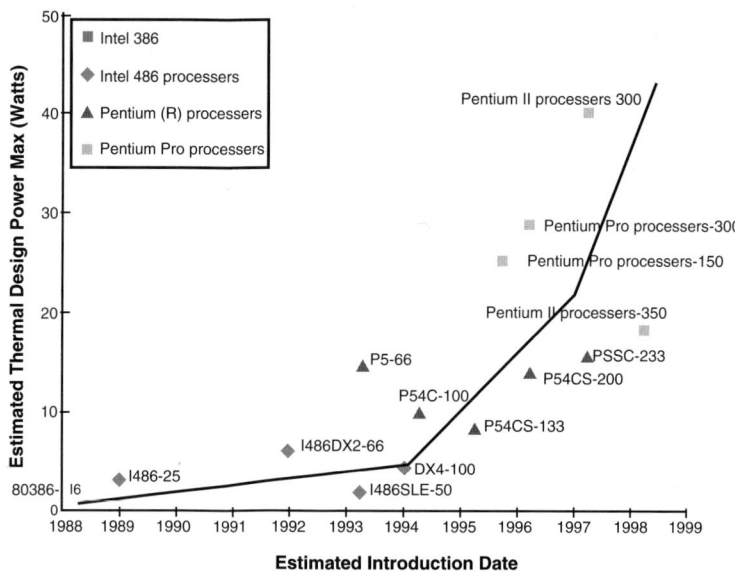

Figure 1. Micro processor power dissipation (Azar and Morabito)

The current Intel® Pentium® 4 processors have a power dissipation of 50 to 90W according to the specifications of Intel®. This power is produced in a package of 35×35 mm², resulting in a power density of 41 to 73kW/m². At the same time the package temperature has to remain below 70°C (specification in data sheet).

The combination of high power (density) and relative low temperatures creates an important task for the thermal management of electronics. Usually three levels of thermal management are distinguished:
- Device level
- Board level
- System level

The "device level" is the area of interest for package designers. The focus is on the internal conduction heat transport from junction to package boundary. Since the heat source, junction, is very small, the local heat flux densities are high which can lead to high temperature gradients. For this

3. Thermal Management

reason the aim is to create sufficient heat spreading inside the package and limit the thermal resistance of the heat path from junction to boundary.

Board and system level are the working areas of the system designers. The division in board and system has its origin in the level of detail in the analysis. Board level is typically dealing with the interaction between packages, board and the direct environment. System level analysis deals with the overall heat transfer in the system, for instance the airflow distribution and air temperatures. Results from a system level analysis give the local thermal conditions for a particular board. It should be noted that it is usually not practical (in terms of CPU efforts) to perform an analysis of an entire system with detailed thermal models of the boards.

Schematically the heat flows in an "exposed pad package on a board" are shown in Figure 2. Heat is produced in the die and is transferred through various paths to the environment. A major fraction of the heat is transferred by conduction to the internal heat spreader through the die-attach layer. The other heat paths are conduction through the body material to the package surface and conduction through bond wires and leads to the board. From the heat spreader the heat is transferred to the board across the gap between package and board. This gap can either be filled with air or gap filler material, which has a higher thermal conductivity than air. Inside the board the heat is conducted in two directions: thickness and in-plane. The board construction (layer structure and materials) has a strong influence on the heat spreading inside the board. At the board surface the heat is transferred to the environment by means of convection and radiation.

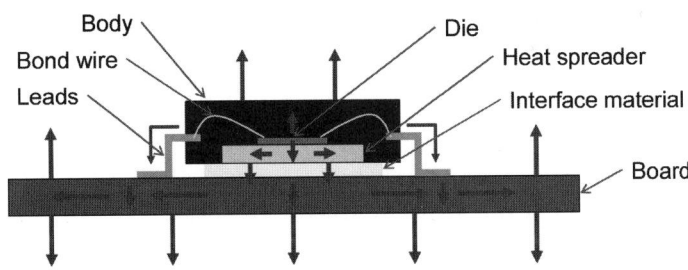

Figure 2. Typical heat flows in exposed pad package on board

The schematic of the heat flow shows the basic principle in the thermal management of packages, i.e., heat spreading. Using heat spreading reduces the heat flux density across interfaces, for instance the gap between package and board, is reduced and thus the temperature rise introduced by the interface is lower. This also implies that the properties of the interface (thermal conductivity and thickness) play an important role.

The intention of this chapter is to introduce the reader in some basic aspects of the thermal management of microelectronics. It should be emphasized that this is only an introduction, because the area of thermal management of microelectronics is very wide. The references at the end of the chapter are a starting point for further reading.

2. HEAT TRANSFER BASICS

Heat transfer is the transport of energy, in a medium or between different media when a temperature difference is present. The transfer is directed in the direction of the lower temperature. Three modes of heat transfer exist: conduction, convection and thermal radiation. Conduction heat transfer occurs in a stationary medium (solid, liquid or gas) under the influence of temperature gradients. Convection heat transfer is the transfer of heat between a surface and a moving fluid (liquid or gas). The third mode, thermal radiation, is the exchange of electromagnetic energy between surfaces. Every surface emits an amount of electromagnetic energy and the result is a net heat flow from that surface.

In the following sections an overview will be given of the basics of heat transfer. More details of general heat transfer can be found in the references.

2.1 Conduction

Conduction heat transfer is quantified by Fourier's law in the form of a rate equation. In Figure 3 a plane wall is shown with a temperature gradient T(x) and length L.

The heat flux q (W/m^2) is given by the equation:

$$q = -k \frac{dT}{dx} \quad (1)$$

In this equation k is the thermal conductivity (W/mK) of the material. For the steady state situation of Figure 3 the temperature gradient is linear and the heat flux q is given by:

$$q = k \frac{T_2 - T_1}{L} \quad (2)$$

The total heat flow through the wall, Q (W), is the product of the heat flux, q, and the area, A, perpendicular to the heat flow:

3. Thermal Management

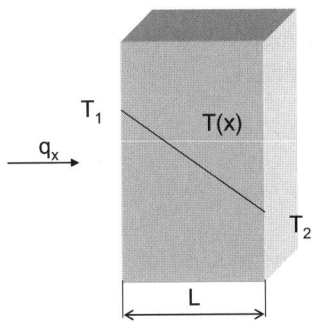

Figure 3. One dimensional conduction

$$Q = Ak\frac{T_2 - T_1}{L} \qquad (3)$$

An alternative formulation for the heat flow Q is using the thermal resistance R_{th} (K/W):

$$Q = \frac{T_2 - T_1}{R_{th}}, \qquad R_{th} = \frac{L}{Ak} \qquad (4)$$

This formulation is often used because of the analogy with electrical networks:
- Heat flow Q corresponds with current, I
- Temperature difference T_2-T_1 corresponds with voltage difference, V
- Thermal resistance R_{th} corresponds with electrical resistance, R.

In this way conduction heat transfer can be described as a "thermal" Ohm's law and thermal networks can be made which can be treated and solved as electrical networks.

The above-derived equations are a special case: one-dimensional Cartesian, steady state conduction. The fundamental equation for heat conduction follows from an energy balance for an infinitesimal volume. In Cartesian co-ordinates this equation is:

$$\frac{\partial}{\partial x}\left(k\frac{\partial T}{\partial x}\right) + \frac{\partial}{\partial y}\left(k\frac{\partial T}{\partial y}\right) + \frac{\partial}{\partial z}\left(k\frac{\partial T}{\partial z}\right) + Q_v = \rho c \frac{\partial T}{\partial t} \qquad (5)$$

The left hand side describes the heat flows and the volume heat production, Q_v (W/m³). The right hand side is the energy storage in the volume. In case of a homogenous isotropic solid with temperature independent conductivity and no heat production, the equation is simplified to the so-called Poisson equation:

$$\nabla^2 T = \frac{1}{\alpha}\frac{\partial T}{\partial t}, \quad \alpha = \frac{k}{\rho c} \tag{6}$$

The constant α is the thermal diffusion coefficient with the dimension m²/s.

For other co-ordinate systems (cylindrical, polar) a co-ordinate transformation has to be carried out. In cylindrical co-ordinates the heat conduction equation for a homogeneous isotropic material with constant thermal conductivity becomes:

$$\frac{1}{r}\frac{\partial}{\partial r}\left(r\frac{\partial T}{\partial r}\right) + \frac{1}{r^2}\frac{\partial^2 T}{\partial \theta^2} + \frac{\partial^2 T}{\partial z^2} + \frac{Q_v}{k} = \frac{1}{\alpha}\frac{\partial T}{\partial t} \tag{7}$$

The heat flux follows from the general formulation of Fourier's law:

$$q = -k\nabla T \tag{8}$$

where q is the heat flux vector and ∇T the gradient of the temperature field.

Solution of the heat conduction equation is the subject of many books and publications. A wide variety of analytical solutions and solution methods can be found in "Carslaw and Jaeger, Conduction of heat in solids". However, heat conduction problems as encountered in electronic packages become very soon complex, because of the geometry and the presence of different materials. In that case analytical methods seize to be useful and one has to use numerical methods like the Finite Difference, Finite Element or Control Volume method. Nevertheless, analytical solutions for simplified or styled problems assist in creating insight in the phenomena and important parameters and should therefore not be abandoned. An illustration of the last statement is given in the next section.

2.1.1 Cooling Fin Performance

As an example of the use of the theory, the thermal performance of a cooling fin will be analysed in this section. Detailed analysis methods of cooling fins, or extended surfaces, can be found in [13]. The geometry is shown in Figure 4 and the assumptions are:

3. Thermal Management

- There is no temperature gradient perpendicular to length of the fin. This is valid for relatively thin fins made of a good thermal conducting material.
- The convection heat transfer along the surface of the fin is constant.
- The environment temperature is set to zero.

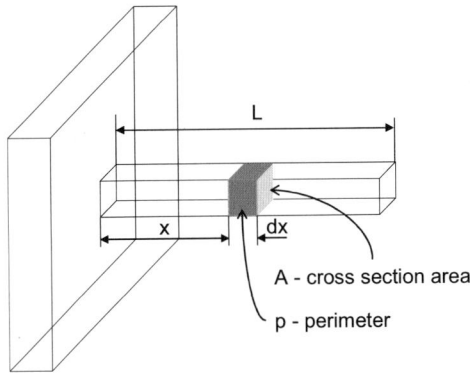

Figure 4. Cooling fin geometry

Setting up the heat balance for the infinitesimal volume leads to the following equation:

$$A k \frac{d^2 T}{dx^2} - p h T = 0 \qquad (9)$$

The first term is the heat gained by the volume and the second term is the heat loss to the environment. The general solution of the equation is:

$$T(x) = c_1 e^{-mx} + c_2 e^{mx}, \quad \text{with} \quad m = \sqrt{\frac{p h}{A k}} \qquad (10)$$

The parameter m is usually called the fin parameter and the constants c_1 and c_2 follow from the boundary conditions. In this case the boundary conditions are:
- At $x = 0$, the fin base, prescribed temperature $T = T_b$.
- At $x = L$, the fin tip, no heat transfer (adiabatic boundary), i.e., $q = 0$.
Applying the boundary conditions gives the following solution:

$$T(x) = T_b \frac{\cosh(m(L-x))}{\cosh(mL)} \tag{11}$$

The heat flow through the base of the fin, Q_b, is given by:

$$Q_b = Ak\, m\, T_b\, \tanh(mL) \tag{12}$$

The tanh(x) function has an asymptotic value of 1 and thus Q_b has a maximum value. Also, the behaviour of the function is that tanh(2) = 0.964 and than(3) = 0.995. This means that increasing the fin length L to increase Q_b has no effect for *mL* values beyond ≈2.5.

2.2 Convection

Convection heat transport takes place between a surface and a moving fluid. Schematically this is shown in Figure 5. The temperature difference between surface, T_s, and fluid, T_f, results in a heat flow from surface to fluid. The heat is stored in the fluid, which causes the fluid to rise in temperature. Obviously the process also works in the other direction. When $T_f > T_s$ the fluid is heating the surface and the fluid temperature drops.

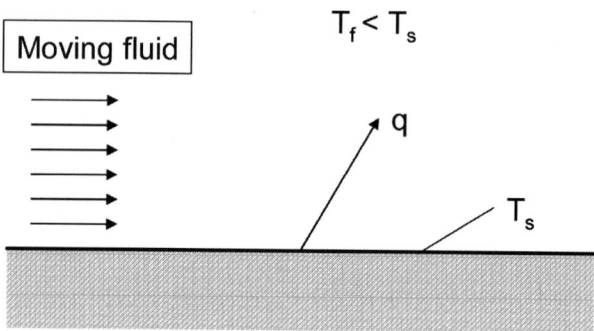

Figure 5. Convection heat transfer

The fluid flow is a major factor in convection heat transfer and determines the heat transfer. For details see the referenced textbooks. When a fluid flows along a surface, a boundary layer is created in which the fluid velocity drops from the free stream velocity to zero at the wall. The heat transfer from surface to fluid takes place in this boundary layer and results in thermal boundary layer. Depending on the flow velocity and geometry a flow can be laminar or turbulent. A laminar flow can be characterized by a

3. Thermal Management

more or less layered structure inside the flow: layers of fluid slide along each other and mixing between layers is low. Turbulent flow is a chaotic movement of fluid particles and there is strong mixing in the flow. Consequently the boundary layer is smaller for turbulent flow and heat transfer from a surface to the fluid will be greater. The thickness of the boundary layers, velocity and temperature, increases in the down stream direction.

The flow field can be a result of an external driving force like a pump or ventilator. In that case one speaks of forced convection. The other driving force is the result of density variations in the liquid caused by the temperature variations. This is called natural convection. Obviously the direction of the flow in natural convection is always against the direction of gravity; hot air rises.

Because of the complexity of convection heat transfer the common approach is to describe the heat flux from surface to fluid by means of a heat transfer coefficient. The equation for the local heat flux is:

$$q_x = h_x (T_s - T_f) \tag{13}$$

with q_x the local heat flux (W/m^2), h_x the local heat transfer coefficient (W/m^{2}K), T_s the surface temperature, and T_f the bulk temperature of the fluid. The bulk temperature is the temperature of the fluid outside the thermal boundary layer in the "undisturbed flow". The local heat flux, and thus the local heat transfer coefficient, varies along the surface. In the downstream direction the thickness of the thermal boundary increases and the heat transfer is reduced. This leads to the definition of the average heat transfer coefficient, h, along the surface A:

$$h = \frac{1}{A} \int_A h_x \, dA \tag{14}$$

The total heat flow from the surface, Q, is then given by:

$$Q = A h (T_s - T_f) \tag{15}$$

In analogy with the thermal resistance for heat conduction one can define the thermal resistance for convection as:

$$R_{th} = \frac{1}{Ah} \tag{16}$$

Determining the value of the heat transfer coefficient has been the subject of many studies, both experimental as numerical by means of computational fluid dynamics. The current practice is that results of these studies are given in the form of so called Nusselt correlations. The Nusselt number, Nu, is defined as:

$$Nu = \frac{hL}{k} \qquad (17)$$

with h the average heat transfer coefficient, L an characteristic length and k the thermal conductivity of the fluid. The Nusselt number is a function of other dimensionless numbers, which are characteristic for the flow and the fluid. For natural convection the characteristic number is the Rayleigh number, while for forced convection the Reynolds and Prandtl numbers are used. In the referenced textbooks correlations are given for various geometries and flow conditions. Some representative values for the heat transfer coefficient are given in Table 1.

Table 1. Typical heat transfer coefficients

Fluid	Configuration	h (W/m²K)
Air	Free flow, vertical plate	5
Air	Forced flow along plate	50
Air	Flow in tube	100
Water	Flow in tube	3000

A warning should be given with respect to the use of correlations. In principle the correlations are a fit of measured or calculated data to describe the heat transfer coefficient for a particular situation. This implies a certain error level, it is a best fit, and errors of 5 to 10% are common. Another limitation lies in the geometry and the heat source. Usually the geometry is relatively simple (flat plate, tube) and the heat source is a uniform temperature or heat flux surface. Obviously this not often encountered in a real electronic system. A board with components is a much more complicated situation. Each component influences the flow, produces a thermal boundary layer and interacts with the other components. Examples of two typical correlations are given in the next section.

2.2.1 Heat Transfer Coefficient Correlations

In the previous section, convection is pointed to Nusselt correlations for estimating the convection heat transfer coefficient. For illustration two correlations will be given:

3. Thermal Management

- Vertical plate with natural convection.
- Flat plate with a forced flow along the surface.

1. *Vertical plate – natural convection*

For a vertical flat plate with uniform surface temperature the heat transfer coefficient can be calculated using the following equations:

$$Ra = \frac{g\beta(T_s - T_f)L^3}{\nu\alpha}$$

$$Nu^T = \overline{C_l}\, Ra^{1/4}$$

$$Nu_l = \frac{2.8}{\ln(1 + 2.8/Nu^T)} \qquad (18)$$

$$Nu_t = C_t^V\, Ra^{1/3}$$

$$Nu = \left[(Nu_l)^m + (Nu_t)^m\right]^{1/m}, \quad m \approx 6, \quad 1 < Ra < 10^{12}$$

$$Nu = \frac{hL}{k}$$

The nomenclature in the equations is:
- Ra, the Rayleigh number (-)
- g, the gravitational acceleration (9.81 m/s^2)
- β, the thermal expansion coefficient of the fluid (1/K)
- T_s, the surface temperature (K)
- T_f, the fluid temperature (K)
- L, the vertical length of the plate (m)
- ν, the kinematic viscosity (m^2/s)
- α, the thermal diffusion coefficient (m^2/s)
- k, the thermal conductivity of the fluid (W/mK)
- h, the average heat transfer coefficient (W/m^{2}K)

The constants $\overline{C_l}$ and C_t^V depend on the Prandtl number of the fluid. For dry air (0 to 100°C) the values are: $\overline{C_l} = 0.515$, $C_t^V = 0.103$.

The fluid properties have to be taken at the average temperature of surface and fluid.

2. *Horizontal plate – forced convection*

The average convection heat transfer coefficient for forced flow along a flat plate with constant surface temperature can be calculated with the flowing formula:

$$\text{Re}_L = \frac{\rho V L}{\mu}, \quad \text{Pr} > 0.6$$

$$Nu_L = 0.664 (\text{Re}_L)^{1/2} \text{Pr}^{1/3} \qquad \text{Re}_L < 5 \times 10^5, \text{ laminar flow}$$
$$Nu_L = 0.0369 (\text{Re}_L)^{0.8} \text{Pr}^{1/3} \qquad \text{Re}_L > 5 \times 10^5, \text{ turbulent flow} \tag{19}$$
$$Nu_L = \frac{h L}{k}$$

The nomenclature in the equations is:
- Re_L, the Reynolds number (-)
- ρ, the fluid density (kg/m^3)
- μ, the fluid dynamic viscosity (Pa·s)
- V, the fluid velocity (m/s)
- L, the plate length (m)
- Pr, the Prandtl number (-); for air $Pr \approx 0.7$
- Nu_L, the Nusselt number for length L (-)
- k, the thermal conductivity of the fluid (W/mK)
  h, the average heat transfer coefficient (W/m^{2}K)

2.3 Radiation

Every surface with a temperature above absolute zero emits electromagnetic energy to the environment. Because electromagnetic waves do not need a transport medium all surfaces in an enclosure receive electromagnetic energy and are automatically connected. When surfaces in an enclosure have different temperatures, a net energy transport between the surfaces takes place. This is in contrast with conduction and convection heat transport in which a medium is necessary to transport the heat. Radiation transport takes place between surfaces, but some materials are more or less transparent for radiation. Glass for instance is transparent for visible light (wavelength 0.3 to 0.7 µm), but absorbs shorter wavelengths (UV) and wavelengths greater than 3.5 µm. Air is transparent for radiation unless the layer is very thick (atmospheric layer). Because most materials encountered in electronics are non-transparent for radiation, only surface radiation will be discussed. For more details on radiation heat transport the reader is referred to the general textbooks.

The energy emitted by a surface is given by Planck's distribution and depends on the temperature of the surface. In Figure 6 the radiation intensity distribution is given for a blackbody with a temperature of 373K and 473K (100 and 200°C). A blackbody is an ideal radiating surface that emits the

3. Thermal Management

maximum amount of energy in all wavelengths and in all directions. The graph shows that with higher temperature more energy is emitted in the shorter wavelengths. For a temperature of 373K approximately 95% of the energy is in the wavelengths > 5 µm and for a temperature of 473K 95% of the energy is in wavelengths > 4 µm.

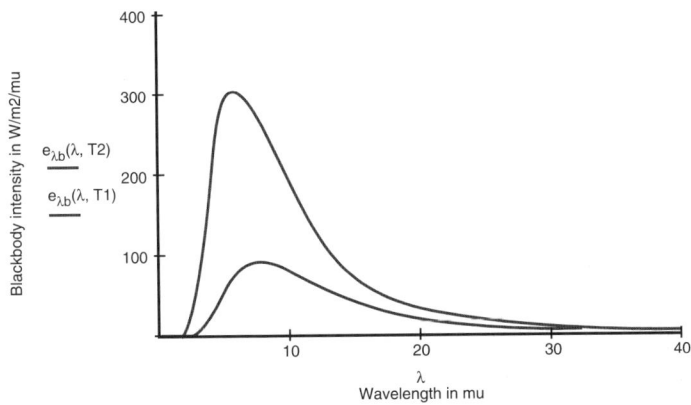

Figure 6. Blackbody intensity distribution for 373 and 433K

The total energy emitted per unit area by a blackbody, q_b, is given by the Stefan-Boltzmann law:

$$q_b = \sigma T^4 \qquad (20)$$

with σ the Stefan-Boltzmann constant (5.67 10^{-8} W/m²K⁴) and T the surface temperature in K. Real surfaces however are usually not ideal blackbody radiators and emit less energy than a blackbody. The emission of real surfaces depends in general on the wavelength and the direction. This non-ideal behaviour is captured in the emission coefficient. The emission coefficient, ε, is defined as the ratio between the real emitted energy and the blackbody energy for a particular surface at a particular temperature. The emitted energy per unit area of a real surface is given by:

$$q = \varepsilon \sigma T^4 \qquad (21)$$

in which ε is the emission coefficient of the surface. From the definition of ε it follows that the value is always between 0 and 1. The emission coefficient used in the last equation is also known as the total hemispherical emission coefficient, thus indicating that it is a value for all wavelengths and all directions.

Besides emission of radiation a surface also absorbs radiation. A blackbody is the ideal absorber, because it absorbs all incoming radiation. Real surface are less perfect and absorb a fraction and reflect the rest. The total hemispherical absorption, α, is defined as the ratio between real absorbed energy and the blackbody absorption. The value lies between 0 and 1.

According to Kirchhoff's law the emission and absorption coefficient are equal for grey surfaces. A grey surface is a surface that has an emission coefficient that is independent of wavelength. This implies that the total hemispherical emission coefficient as defined above is a constant. However, many surfaces are non-grey and this means that the total hemispherical emission coefficient becomes temperature dependent. The same holds for the absorption coefficient. For non-grey surfaces the absorption depends on the wavelength distribution of the incoming radiation. An example is the behaviour of paint layers. In the visible range (0.3 to 0.7 μm) the absorption is different, which results in the surface colour. However, in IR range (> 4 μm) the absorption of white and black paint is almost the same.

Radiation heat transfer on a surface is a combination of direct emission from the surface and incoming radiation from the environment. Schematically this is shown in Figure 7. The incoming radiation is partially absorbed by the surface; the remainder is reflected back to the environment. The result is a net heat flow through the surface. Calculation of the radiation transport between surfaces is based on the enclosure method. This method basically calculates the out-going radiation (direct emission and reflection) for the surfaces in an enclosure.

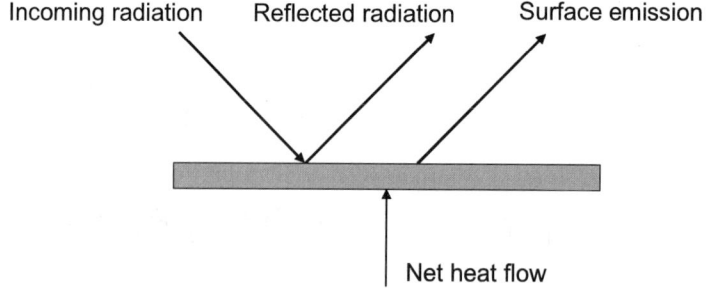

Figure 7. Heat flows on radiating surface

3. Thermal Management

Because all surfaces in an enclosure take part in the transport, this becomes very soon a complex calculation. However, for some simple geometries basic expressions can be derived.

The radiation heat transport between two infinite plates, which have a certain resemblance with parallel circuit boards, is given by the following expression:

$$q = \frac{\sigma(T_1^4 - T_2^4)}{1/\varepsilon_1 + 1/\varepsilon_2 - 1} \quad (22)$$

with q the heat flux density (W/m^2), σ the Stefan-Boltzmann constant, T_1 and T_2 the absolute temperature of the plates and ε_1 and ε_2 the emission coefficients of the plates.

A more general situation is a "small object, surface A_1, in an enclosure, surface A_2". The heat flow, Q in (W), between the surfaces is given by:

$$Q = \frac{A_1 \sigma(T_1^4 - T_2^4)}{1/\varepsilon_1 + A_1/A_2(1/\varepsilon_2 - 1)}, \quad A_1 < A_2 \quad (23)$$

The ratio A_1/A_2 is a measure of the amount of radiation coming from A_2 that strikes A_1. Note that the parallel plate geometry is a special case in which $A_1 = A_2$. A second limiting case occurs when $A_1 \ll A_2$. In that case the expression for the heat flow reduces to:

$$Q = A_1 \varepsilon_1 \sigma(T_1^4 - T_2^4) \quad (24)$$

This is a typical result for a small object in a large environment. In that case the emission coefficient of the environment is no longer important and it seems that the object is placed in a black environment.

In the same way as for convection one can define a radiation heat transfer coefficient, h_{rad}. The following expression is valid for the case of a small object in a large enclosure:

$$Q = A_1 h_{rad} \sigma(T_1 - T_2), \quad h_{rad} = \varepsilon_1 \sigma(T_1^2 + T_2^2)(T_1 + T_2) \quad (25)$$

The radiation heat transfer coefficient can be used in the same way as the convection heat transfer coefficient to define a thermal resistance.

2.4 Remarks on Thermal Resistance

In the previous sections the concept of thermal resistance (R_{th}) was introduced to describe the relation between heat flow (Q) and temperature difference (ΔT) with a simple equation: $\Delta T = R_{th} \times Q$. This simplification finds its origin in the analogy between the electrical and thermal differential equations. Solution of these equations for simple elements or geometries lead to a quantity that is called a resistance: electrical, R, with dimension Ohm and thermal, R_{th}, with dimension K/W. These elementary resistances can be used to build a network representing a complex structure to solve the transport of electricity or heat.

The difference in conductivity between an electrical conductor and an electrical insulator is much bigger than between a thermal conductor and a thermal insulator. The conductivity ratio between an electrical conductor and insulator has order of magnitude 10^{20}, while the ratio for a good thermal conductor and insulator is order of magnitude 10^4. Furthermore, most materials are either a good or poor electrical conductor, but for thermal conductivity such a classification is not possible. The result is that in contract with electrical problems a thermal problem is usually not one-dimensional. The consequence is that defining a "thermal resistance" in the same sense as an "electrical resistance" becomes difficult. To obtain an accurate solution for a particular thermal problem therefore soon leads to a complex network to take into account two or three dimensional heat conduction effects.

The term "thermal resistance" is common practice in the field of thermal management of electronics, but there is a problem. Unfortunately the major limitation (one dimensionality) is often neglected and one uses thermal resistance in the same sense as electrical resistance of a wire. However, a wire is purely one dimensional, but heat flows are rarely one-dimensional. Also the heat flows in devices or assemblies are influenced by the external conditions. The result is that a "thermal resistance" is not a constant.

Part of the origin of this mistake may lie in the measurement practice. To characterize the thermal performance of a system usually the heat production is measured and one or more typical temperatures. With these measured values one can calculate a thermal resistance, for instance a change in heat production, ΔQ, gives a change in temperature, ΔT, which gives $R_{th} = \Delta T/\Delta Q$. However, one must realize that a system (consisting of a device and external conditions) is measured and when the system changes, the measurement may give different results. Therefore it is better to speak of "thermal performance parameter" or "thermal characterization parameter", which is often denoted by the symbol Ψ. Part of the confusion is that the dimension of R_{th} and Ψ is K/W, but the value of Ψ has no physical meaning.

3. Thermal Management

The distinction made above is not yet widely used in the thermal management of electronics. It is up to the designer to carefully look at reported values and note the differences.

An additional complication is that the measurement conditions are not always properly documented, when thermal characterization parameters are reported.

2.5 Typical Thermal Properties

The following tables give an overview of typical values for thermal properties of various materials. Unless otherwise stated the values are at 20°C and 1 bar conditions.

Notes:
- The thermal conductivity may more or less be dependent on the temperature. Silicon is a typical example.
- The composition of alloys has a serious influence on the thermal conductivity.
- The properties of non-metals (ceramics and plastics) depend strongly on composition and processing.

Table 2. Thermal properties of metals and alloys

Metals and alloys	Thermal conductivity W/mK	Density kg/m³	Specific heat J/kgK
Silver, pure	419	10500	234
Copper, pure	390	8960	390
Gold, pure	310	19320	130
Aluminum, pure	237	2700	880
Silicon, pure	140	2330	702
Bronze, 92% Cu, 8% Sn	190	8900	380
Brass (55-70% Cu, 45-30% Zn)	122	8500	380
Steel			
Normal construction steel	50	7800	480
Stainless steel (302)	16	7900	500

Table 3. Thermal properties of non-metals

Non-Metals	Thermal conductivity W/mK	Density kg/m³	Specific heat J/kgK
Aluminum Oxide (CoorsTek AD-85)	16	3420	920
Aluminum Oxide	30	3920	880

Non-Metals	Thermal conductivity W/mK	Density kg/m^3	Specific heat J/kgK
(CoorsTek AD-998)			
Silicon Carbide (CoorsTek SCRB)	125	3100	700
Aluminum Nitride (Morgan AN-985)	175	3300	Not listed
PMMA, general purpose molded	0.19-0.24	1170	1460
Polycarbonate	0.19-0.21	1200	1200
Air at 20°C	0.026	1.188	1007
Air at 60°C	0.029	1.045	1009
Air at 100°C	0.032	0.933	1012

3. THERMAL DESIGN OF ASSEMBLIES

In the design process of electronic products early estimates of component temperatures are of great importance. Based upon these estimates decisions are made which will influence the design during the rest of the process. Practice is that these estimates are being made using the concept of "thermal resistance junction to ambient", R_{thj-a}. With this concept designers can easily calculate the temperature rise of a component for a given heat production and obtain an estimated junction temperature, $T_j = T_{amb} + Q \, R_{thj-a}$. However, investigations have shown that the concept of R_{thj-a} has limited value because it only describes the component behaviour under a specific test set-up. Influences of environments differing from the standardized test set-up cannot be dealt with.

This has led to the development of Boundary Condition Independent (BCI) or compact models of components. These models give a good description of the thermal behaviour by means of a seven or more node thermal resistance network. The network model describes within an accuracy of 5% the thermal behaviour of the component for a wide variety of external conditions. However, there is a penalty for this improved description. The network is more complicated than the R_{thj-a} concept, which makes early estimates more difficult. At the same time the designer has to make estimates for the boundary conditions, which are heat transfer coefficients. In particular the board forms a problem, because the thermal behaviour of a layered structure has to be translated into a heat transfer coefficient.

In principle, the compact models can be used in software tools like Therman™ and Flotherm™. With these tools, board level or system level analyses are possible. However, these tools are mostly used by specialists

3. Thermal Management

and at a later time in the design process. Although the use of these tools is increasing and moving down the design chain towards designers, there is still a problem. On the one side there is an improved component description in terms of compact models, but on the other side there is no easy tool to use these models for early estimates. At present, there is the intention to standardize the description of components using compact models via world wide standardization committees such as JEDEC.

To close this gap, a spreadsheet application, COMIC, is developed within Philips. COMIC stands for Compact Model Integration Code. A spreadsheet is chosen as platform, because the target user group is usually more familiar with spreadsheet applications than with advanced thermal simulation programs. With COMIC it is possible to use the verified compact model description provided by a component supplier to perform preliminary calculations. In this way a designer is not longer using a metric like R_{thj-a}, which is only valid for a specific situation, but uses a verified model to estimate the component temperatures in the actual situation.

A compact model is a network representation of a component that describes the thermal behaviour of the component. The network consists of a number of nodes connected by thermal resistances. For a seven node model these nodes are junction (J), top inner and outer (TI, TO), bottom inner and outer (BI and BO), side (S) and Leads (L). An example of a network is shown in Figure 8 that describes an SQFP package.

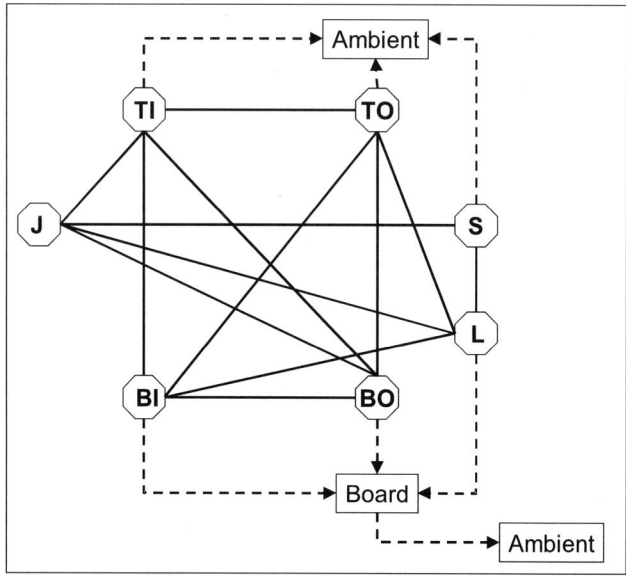

Figure 8. Network of a SQFP package

The junction node represents the heat producing area on the die. The top, bottom and side nodes are assigned to certain areas of the package surface. It should be noted that by this simplification the surface temperatures are average values for that particular region of the package. The leads of the package are lumped into one node, which has the same averaging effect. The calculated lead temperature is an average value for all leads. It gives no information of minimum or maximum values for the lead temperature. Resistance values and surface regions are determined with an optimisation procedure that is carried out by the component or package manufacturer. The result of this optimisation is a network description that describes the thermal behaviour usually within 5%.

In order to use this model the designer must solve the network with boundary conditions imposed. At the junction node the heat production is defined and on the outside convection boundary conditions are applied. Solving this network is not as easy as using the $R_{thj\text{-}a}$ concept, but with straightforward matrix routines this is not a major problem. The real problem is estimating the convection boundary conditions and in particular the behaviour of the board. For this reason a board modeller is made which transforms the thermal behaviour of the board into a heat transfer coefficient that can be applied on the board nodes (BI, BO and L) of the package. Details of calculating the board boundary condition can be found in the references.

Network solver and board modeller are combined into the Compact Model Integration Code, COMIC. The program COMIC is a spreadsheet that contains two worksheets. Figure 9 and Figure 10 show screenshots of these sheets.

The first sheet (Figure 9) is a combined input and output sheet. The user can select a component from a database and import the compact model network into COMIC. A short description of the component and the network resistances are shown in Figure 8. For security reasons the fields are protected to prevent accidental changes of values. The third table is a combined input and output table.

3. Thermal Management

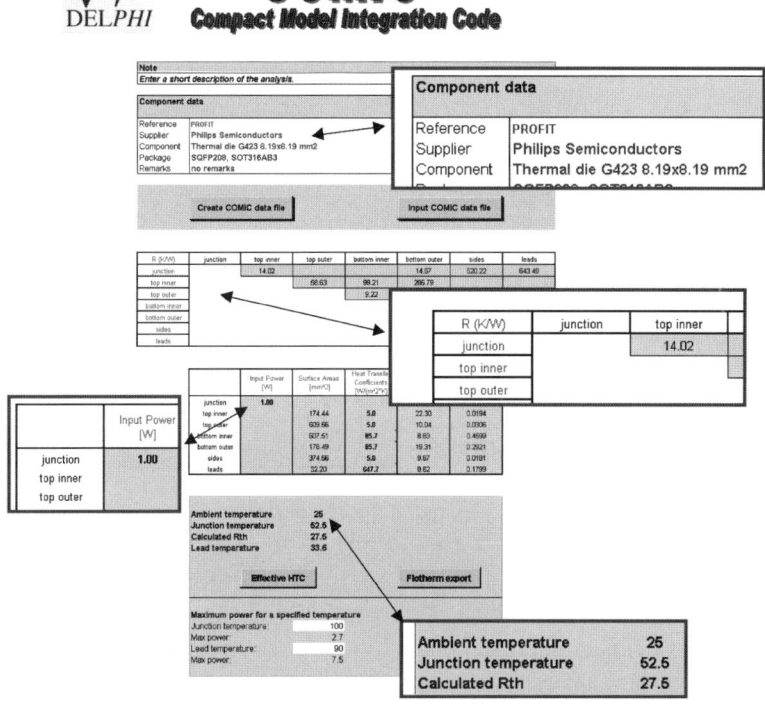

Figure 9. Component sheet of COMIC

The input part is the heat production in the junction and heat transfer coefficients on the component surfaces. The output is the temperature rise of the nodes for the given set of boundary conditions. The bottom region in the component sheet contains some controls, an input field for the ambient temperature and output fields for the calculated junction temperature and $R_{thj\text{-}a}$. It should be noted that this $R_{thj\text{-}a}$ is based on the combined package and board thermal behaviour for the particular boundary conditions and board build-up. It is therefore not the metric as provided by a component manufacturer, but a performance indicator for the package in this particular situation.

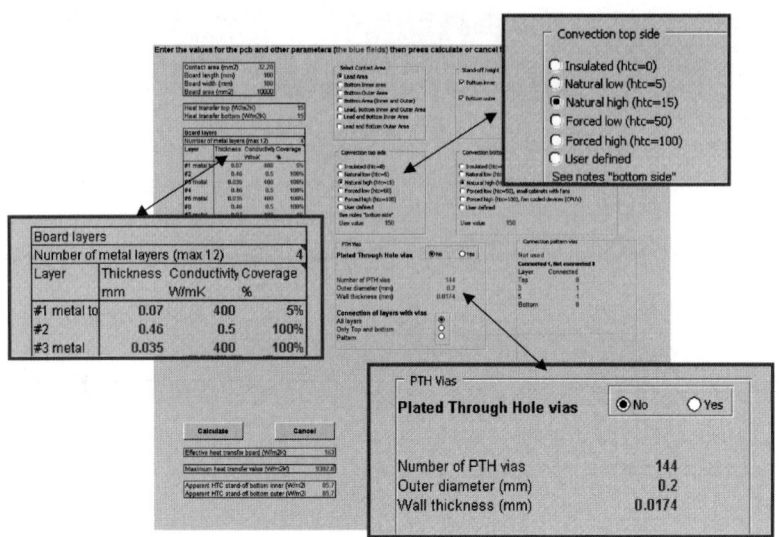

Figure 10. Board sheet of COMIC

The second sheet, shown in Figure 10, is the input for the board modeller. In this sheet the user can define the board build-up: overall size, number of metal layers, thickness, thermal conductivity and coverage. The coverage is an estimate of the amount of solid copper in a metal layer. Actual traces are not modelled. Typical values for convection heat transfer coefficients can be selected or a user-defined value can be given. There is also an option for modelling vias in the board. However, this option is limited to plated trough hole vias with a single connection pattern between the metal layers. This restriction is a result of a necessary simplification in the user interface. The user can also define the standoff height of the package and the thermal conductivity in the gap between component and board. After modelling the board the effective heat transfer of the board is calculated and the results are placed in the heat transfer column of the third region on the first sheet.

With COMIC a designer no longer has to use metrics, which are valid for one particular situation or use educated guess values for board behaviour. The designer can use the validated compact model data from a manufacturer to estimate the thermal behaviour in a realistic situation and obtain specific temperatures of the package. Temperature results are given for the junction, but also on the surface of the package and the leads.

The intended use of COMIC is: "Making early first estimates of temperatures of a package-on-a-board in a particular environment". This specification and the intended user group lead to some limitations in COMIC. The main limitation is that COMIC can only perform calculations for a single component on a board. It is therefore not possible to calculate the influence of a neighbouring component. The question of mutual influence is closely related to available cooling area on the board. With COMIC it is possible to change the dimension of the board and calculate the relation between board dimension and temperature rise. In this way the designer gets an impression of the required cooling area. However, in the opinion of the authors a thorough analysis of multiple components on a board should better be solved with a board level analysis program, which can handle multiple components. A second limitation of COMIC lies in the temperature results, which originates in the compact model description of the package. As stated before, the nodes in the compact model represent a particular region and are lumped temperatures for that region. For instance one can calculate the temperature of the top inner (TI) node, which is the average of the top inner area, but not the maximum value of the top inner area. The same holds for the leads. All leads are lumped into one node, which results in an average temperature for the leads. In the derivation of the compact model the leads are included up to the solder joint. The calculated lead temperature is therefore an averaged value for the temperature at the solder joint.

The previous limitations address also the question: "When to use a compact model and when not?" A compact model should be used when detailed temperature information is not required and average values are sufficient. As soon as one needs detailed temperatures of the in- or outside of the package, a detailed model should be used. A minor limitation of COMIC is that it only uses one ambient temperature for both sides of the board. Applications are possible where a significant difference exists between the ambient temperatures at the two sides of the board. In that case one can use an average value for the ambient temperature to get some indication. A second minor limitation is that COMIC is a linear model. This implies that temperature dependent heat transfers, like radiation, are not taken into account. The workaround is that the user adjusts the entered values of the heat transfer coefficients for the actual temperatures.

4. THERMAL DESIGN FOR A SQFP

In this section COMIC is used to calculate the junction temperature of a SQFP208 package on a 4-layer board with varying board dimension. For comparison the same calculation is performed with Flotherm™. The component is modelled with the Compact Model smart part of Flotherm™; the network resistances from the compact model data are provided by the supplier. The board build-up is listed in Table 4. The 4-layer board represents a standardized JEDEC test board. The total board thickness is 1.59 mm and the epoxy has a thermal conductivity of 0.5W/mK. The distance between the copper layers in the 4-layer board is 460 μm. On all surfaces a heat transfer coefficient of 15W/m²K is assumed. Between component and board a standoff height of 0.35 mm is assumed, which means that only the leads are in direct contact with the board. The gap between component and board is either filled with air (thermal conductivity K = 0.03W/mK) or a material with a thermal conductivity of 30W/mK. The heat production is set on 1W.

Table 4. Build-up 4 layer board

Layer	D (μm)	K (W/mK)	Coverage (%)
1	70	400	5
2	35	400	100
3	35	400	100
4	70	400	1

The calculations are carried out for a square board with dimensions ranging from 50 x 50 mm² to 200 x 200 mm². The calculated junction temperature rise is shown in Figure 11. The graphs show a good agreement between COMIC and Flotherm™ in case of high-conductive gap material between component and board. In case of air between component and board, COMIC tends to over-estimate the temperature rise. The difference for large board sizes is approximately 10%.

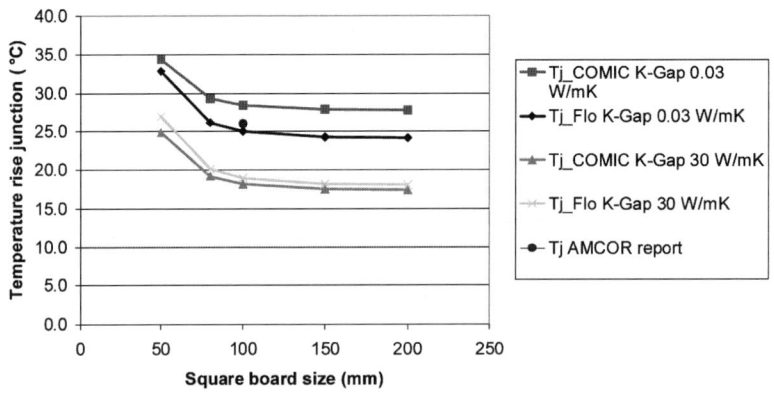

Figure 11. Results SQFP208 on 4-layer board

3. Thermal Management

The reason for this difference lies in the current board modeller. At the time of preparing this paper, work is in progress to improve the performance. In Figure 11 also the value from a subcontractor report is plotted, which predicts a temperature rise of 26°C. With reference to this value, COMIC over-predicts the temperature with 9%, while Flotherm™ under-predicts the temperature with 4%.

Although the absolute values differ, both COMIC and Flotherm™ show the same trend with respect to the board size. Above roughly 100 mm the increase of board area does not result in significantly lower temperatures. Below 75 mm the temperature shows a sharp increase. This indicates that for this combination of package, board and heat transfer the optimum board size is roughly 75 x 75 mm^2.

With COMIC one can also investigate the influence of copper layers in the board. Removing the internal layers (2 and 3) from the 4-layer board, the junction temperature rise increases from 28.4°C to 36.5°C; this is an increase of 29%. When in the two-layer board the bottom layer coverage is increased from 1% to 100%, the temperature rise is 29.3°C instead of 36.5°C or only 3% higher. The additional copper thus increases the heat spreading and compensates the missing two inner layers. This information makes it possible for the designer to start a discussion on board area and copper layers. In this way the board construction is included in the thermal management discussion.

5. HEATSINK DESIGN CHOICES

A question that a designer often has to answer is: "Do I need a heat sink?" This question usually arises early in the design process and has consequences for the following phases, because applying a heat sink implies a volume claim in the design and a significant cost increase. To illustrate the use of COMIC in solving this problem, the following example is given. An HBGA960 package, as shown in Figure 12 with the compact model network, is used in an environment of 75°C with a heat production of 8W. The component is mounted on a 4-layer board with the build-up of Table 4 and dimensions of 100 x 100 mm^2.

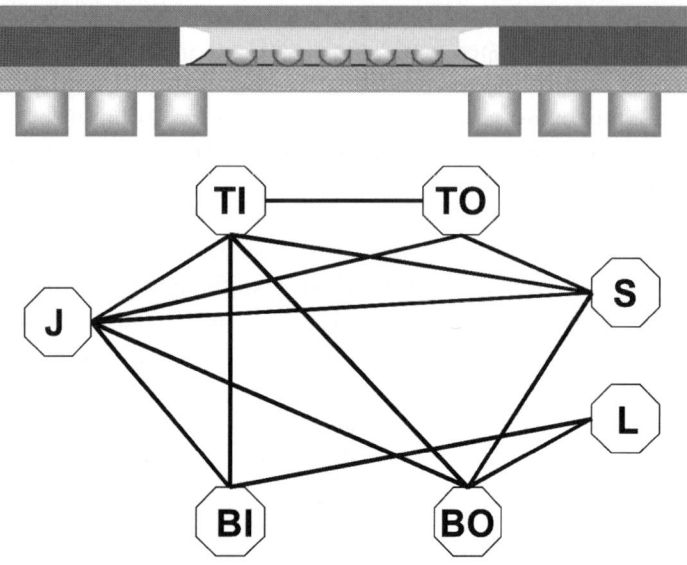

Figure 12. HBGA960 and network

With COMIC the temperatures of junction and leads are calculated for various external heat transfer coefficients to simulate natural and forced convection. Results are listed in Table 5. The specification of the package states that the junction temperature should be lower than 110°C. The values in the table show that for natural convection cooling with heat transfer coefficients in the range 5 to 15W/m^{2}K, the junction temperature is too high. For forced convection with HTC = 50W/m^{2}K, the temperature is just above specification. However, this implies using a ventilator to cool the component and board, which is often not a preferred solution.

Table 5. HBGA960 junction and lead temperature

HTC	Tjunction	Tleads	Remark
5 W/m^{2}K	201°C	185°C	Natural convection low
15 W/m^{2}K	144°C	129°C	Natural convection high
50 W/m^{2}K	116°C	103°C	Force convection

An alternative approach is fitting the component with a heat sink on the top. Assuming that the board is in an environment with a HTC of 15W/m^{2}K, one can change the heat transfer at the top of the package (top inner and top outer area) until the junction temperature is in an acceptable range. For this particular example a top HTC of 150W/m²K results in a junction of 108°C, while a value of 250W/m²K results in a junction of 100°C. Since the required HTC and area (33 x 33 mm²) are known, the required minimum thermal resistance of the heat sink is 3.7K/W. The next step is looking in supplier info to find a matching heat sink.

6. CONCLUSIONS/FINAL REMARKS

Thermal design of electronic systems is an important element of the total design process, because of the impact of temperature on performance and reliability. This implies a strong (thermal) interaction between the device, package or component, the system design and its application. A thermally well-designed device applied in a thermally poor designed system, will still result in a poor total design. Therefore it is important that thermal design on device and system level should be included early in the overall design process. It will be obvious that last minute measures to reduce temperatures and temperature gradients will always lead to a non-optimal design and usually additional costs.

An additional problem is that the thermal design of device and system is usually done in separate worlds, device and system manufacturers, focusing on different issues. This makes it necessary to have some means of communication to link both worlds. For thermal design this can be achieved with compact models of devices. The device manufacturer can supply a "user" with all necessary thermal parameters without revealing specific details. On the other hand, the "user" gets the necessary data without unwanted ballast or the need to perform a device level analysis in a system.

The COMIC program makes verified compact models of components available to designers, sales people and marketers to make early temperature estimates. A designer can define its own component environment, board, and heat transfer to ambient, and calculate the relevant component temperatures. In this way the use of metrics like R_{thj-a}, is no longer necessary. Schematically this is shown in Figure 13.

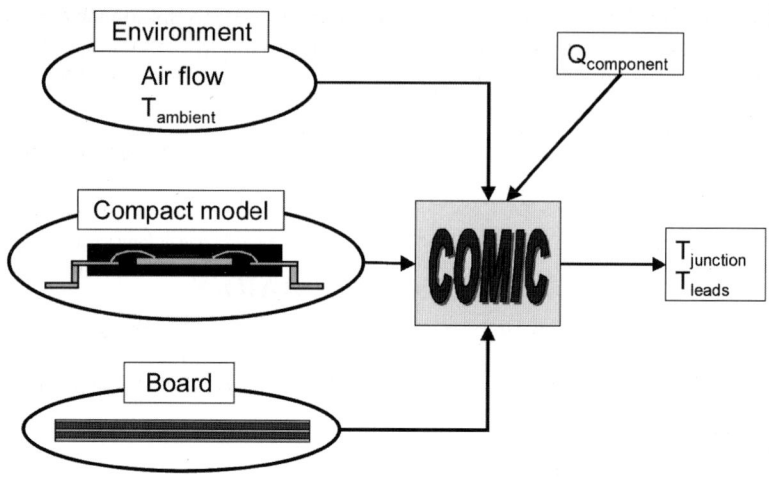

Figure 13. Position/function of COMIC

The system of component, board and environment is split up into the different parts and combined in COMIC. The compact model is supplied by the manufactures and the user can build the system. This is in contrast with the classical metrics, which are a performance indicator for a system defined by the component manufacturer.

Agreement between COMIC, which is a simplified program, and Flotherm™ is reasonable. Although some work still has to be done on the board modeller to improve the accuracy.

Compact models and COMIC are well suited for calculation of "global" component temperatures. These temperatures are lumped or averaged values. For detailed information a fully detailed model of the component is required.

7. REFERENCES

[1] Azar and Morabito, Managing power requirements in the electronics industry, Electronics Cooling Magazine Vol. 6 No. 4 December 2000.
[2] Incropera and DeWitt, Fundamentals of Heat and Mass Transfer, John Wiley & Sons 2002.
[3] Kreith and Bohn, Principles of Heat Transfer, Brooks and Cole 2000.
[4] Bejan, Thermal Design and Optimization, John Wiley & Sons 1996.
[5] Lienhard and Lienhard, A Heat Transfer Textbook 3rd edition, MIT (internet download from web.mit.edu/lienhard/www/ahtt.html).
[6] Carslaw and Jaeger, Conduction of Heat in Solids, Oxford University Press 1980.
[7] Bejan, Convection Heat Transfer, John Wiley & Sons 1984.
[8] Rohsenow, Hartnett and Ganic, Handbook of Heat Transfer Fundamentals, McGraw-Hill 1985.
[9] Siegel and Howell, Thermal Radiation Heat Transfer, McGraw-Hill.

[10] Yeh and Chu, Thermal Management of Microelectronic Equipment, ASME Press, 2002.
[11] Steinberg, Cooling Techniques for Electronic Equipment, John Wiley & Sons 1991.
[12] Lascance, Thermal resistance: an oxymoron?, Electronics Cooling Magazine, Vol. 3 May 1997.
[13] Kraus, Aziz and Welty, Extended Surface Heat Transfer, John Wiley& Sons 2001.
[14] Lee, Song and Moran, Constriction/Spreading Resistance Model for Electronics Packaging, ASME/JSME Thermal Engineering Conference, Vol. 4 1995.
[15] C. A. Harper, High Performance Printed Circuit Boards, McCraw-Hill, 1999.
[16] Eggink, Van Driel, Janssen, Including the board in the junction temperature estimate for electronic packages, Proceedings of the International Conference EuroSimE 2002, Paris, 2002, pp. 171-175.
[17] Eggink, Janssen and Janssen, Using Compact Models in the Early Design of Electronics, Semitherm XIX 2003.
[18] Bruce M. Guenin, Component Thermal Characterization, Electronics Cooling Magazine, Vol. 7 February 2001.
[19] ASME - Journal of Heat Transfer
[20] ASME - Journal of Electronic Packaging
[21] Semi-Therm: www.semi-therm.org
[22] EuroSime: www.eurosime.org
[23] Therminic: tima.imag.fr/conferences/therminic/
[24] ECTC: www.ectc.net
[25] ITherm: www.itherm.org
[26] Electronics Cooling: www.electronics-cooling.com
[27] Cooling Zone: www.coolingzone.com
[28] Microelectronics Heat Transfer Laboratory: www.mhtl.uwaterloo.ca
[29] Material properties: www.matweb.com
[30] Thermal resource center K&K Associates: www.tak2000.com
[31] Standardization committee JEDEC: www.jedec.org

8. EXERCISES

1. What are the three modes of heat transfer?
2. Explain the concept of "thermal resistance". What is the major limitation of this concept?
3. Compare the heat transfer coefficient for natural convection and radiation for a black surface at 100°C in an environment of 20°C. Assume that the vertical length of the surface is 100 mm.
   a. How important is the radiation heat transfer in this case?
   b. What is the outcome when the air flows with a velocity of 2m/s along the surface?

   Note: Relevant air properties can be found in for instance [2] or [5].
4. Cooling fins are usually made of aluminium.
   a. What is the useful length for a fin with 3mm diameter in a natural convection flow (typical $h = 10W/m^2K$)?
   b. What is the influence of applying forced convection cooling with typical $h = 100W/m^2K$?

5. The printed circuit board plays a role in the thermal management of devices, because it acts as an extended surface with convection on the surfaces. An indication of the performance can be obtained by applying the cooling fin formulation to a printed circuit board.
   a. Estimate the useful length for a board of 1.6 mm thick epoxy with two copper layers of 70 mm, when the effective heat transfer on both sides is 15W/m^{2}K?
   b. Considering the above estimate, how close can devices be placed to each other?
6. Thermally enhanced packages like a HQFP128 or SOT760 have an exposed pad to create a thermal contact between package and board by means of soldering. Typical dimension of the exposed pad is 21 x 21 mm and the distance (stand-off) between package and board is 0.25 to 0.50 mm. What is the influence of poor soldering, resulting in 50% contact, on the thermal resistance between package and board?

Chapter 4

INTRODUCTION TO ADVANCED MECHANICS

J. Zhou[1] and G.Q. Zhang[2,3]
[1]*Department of Mechanical Engineering, Lamar University, Beaumont, Texas 77710, USA*
[2]*Philips Semiconductors, HTC 60, 5656AG Eindhoven, The Netherlands*
[3]*Delft University of Technology, Mekelweg 2, 2628CD Delft, The Netherlands*

Abstract: This chapter serves as a brief guide and introduction of the basics of mechanics needed for the mechanics analysis in microelectronics. Theory of stress and strain, thermal stress and strain, and the fundamental principles and equations of thermal mechanics are presented first. Several common nonlinear constitutive laws of materials such as hyperelasticity, plasticity and creep are introduced. Failure criteria for static loading and cyclic loading are described. Then, the fundamentals of fracture mechanics including linear fracture mechanics, mixed-mode fracture, and elasto-plastic fracture mechanics are presented. Interface fracture mechanics is also discussed. In the part of computational mechanics, the basics of finite element theory and numerical implementation are briefed and the focus is on the advanced topics such as the treatment of geometric and material nonlinearity in finite element implementation, extraction of fracture parameters, sub-structural and sub-modelling methods, and adaptive meshing and element birth and death.

Key words: Mechanics, stress, strain, principal stress, thermal, expansion, coefficient of thermal expansion, CTE, plasticity, failure, brittle, ductile, creep, fracture, stress intensity factor, J-integral, energy release rate, failure criterion, cyclic loading, fatigue, finite element, FEA, nonlinearity, sub-modelling, sub-structuring, superelement.

1. INTRODUCTION

Mechanical issues always occur in microelectronics, in both front-end and back-end processes, and in reliability qualification test. For example, an electronic package assembly is a typical example of composite structure that undergoes thermal loading. It is comprised of various conducting and insulating materials, and is subjected to non-uniform temperature distributions. Due to geometry, material construction, and thermal expansion mismatch of different parts of package, thermal stress can occur inside the packaging system while it is being manufactured and while it is being used.

Figure 1 illustrates a cross-sectional view of a plastic flip-chip ball grid array (FC BGA) package assembly, where a silicon chip is mounted on an organic substrate and the package is attached to a printed circuit board (PCB) through solder ball interconnections to form a final second level assembly. In addition, a metal heat sink is attached to the package to dissipate the excessive heat. The numbers shown in Figure 1 indicate a typical *coefficient of thermal expansion* (CTE) value of each material in $10^{-6}/°C$. When the chip is powered up so that the package is subjected to a temperature change, each material deforms at a different rate. This non-uniform CTE distribution produces thermally induced mechanical stresses within the package assembly.

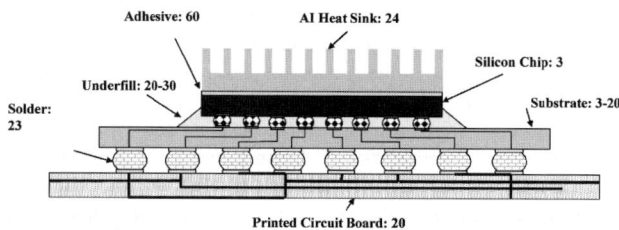

Figure 1. CTE distribution in a typical electronic package

It is not an easy task to determine the thermal stresses in electronic packaging. Closed form and semi-closed form solutions can be derived for very simple geometries and temperature loadings, but very limited in applications and difficult to obtain. The finite element analysis is one of the best candidates for obtaining numerical results for thermal strains and stresses in electronics packages. However, many of the finite element analyses performed are not properly executed due to a limited understanding of the principles of mechanics for electronics packaging.

In the present chapter, the fundamentals of mechanics needed for mechanics analysis in microelectronics are introduced. The chapter starts from the analysis of stress and strain, and thermal strain and thermal stresses. The basic and governing equations of thermal-mechanics are presented. Failure criteria for static loading and cycling loading are presented in Section 3. Introduction of fracture mechanics and interface fracture mechanics are described in Section 4. Section 5 outlines the finite element methods, with

4. Introduction to Advanced Mechanics

emphasis on the treatment of nonlinearity including geometric and material nonlinearity, the advanced finite element techniques such as sub-modelling and sub-structuring methods.

2. STRESS AND STRAIN

2.1 Analysis of Stress

2.1.1 Stress

Stress is defined as force per unit area. In a part subjected to some forces, stress is generally distributed as a continuously varying function within the continuum of material. Every infinitesimal element of material can conceivably experience different stresses at the same time. Thus, we must look at stresses as acting on vanishingly small elements within the part. These infinitesimal elements are typically modelled as cubes, shown in Figure 2. The stress components are considered to be acting on the faces of these cubes in two different manners. *Normal* stresses act perpendicular (i.e., normal) to the face of the cube and tend to either pull it out (tensile normal stress) or push it in (compressive normal stress). *Shear* stresses act parallel to the faces of the cubes, in pairs (couples) on opposite faces, which tends to distort the cube into a rhomboidal shape. These normal and shear components of stress acting on an infinitesimal element make up the terms of a *tensor*.

Stress is a tensor of order two and thus requires nine values or components to describe it in three dimensions. The 3-D stress tensor can be expressed as the matrix:

$$\begin{bmatrix} \sigma_{xx} & \tau_{xy} & \tau_{xz} \\ \tau_{yx} & \sigma_{yy} & \tau_{yz} \\ \tau_{zx} & \tau_{zy} & \sigma_{zz} \end{bmatrix} \quad (1)$$

where the components σ_{xx}, σ_{yy}, and σ_{zz} are the normal stresses, so-called because they act, respectively, in directions normal to the x, y, and z surfaces of the cube. The components τ_{xy}, and τ_{xz}, for example, are shear stresses that act on the x face and whose directions of action are parallel to the y and z axes, respectively. The sign of anyone of these components is defined as positive if the signs of its surface normal and its stress direction are the same, and as negative if they are different. Thus the components shown in

Figure 2 are all positive because they are acting on the positive faces of the cube and their directions are also positive.

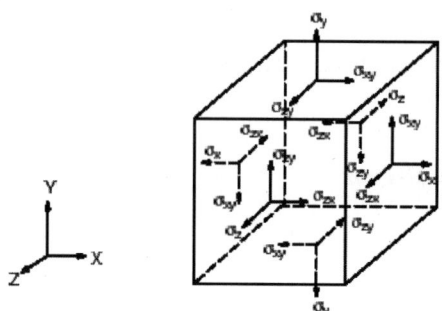

Figure 2. The stress components

In Equation (1) each stress component contains three elements, a magnitude (either σ or τ), the direction of a normal to the reference surface (first subscript) and a direction of action (second subscript). It can be shown that the stress tensor is symmetric, which means that

$$\tau_{yx} = \tau_{yx}$$
$$\tau_{yz} = \tau_{zy} \qquad (2)$$
$$\tau_{zx} = \tau_{xz}$$

This reduces the number of stress components to be calculated as six.

2.1.2 Equilibrium Equations

If the body force components per unit volume are represented as X, Y, and Z, the equilibrium equations of the stress components are:

$$\frac{\partial \sigma_x}{\partial x} + \frac{\partial \tau_{yx}}{\partial y} + \frac{\partial \tau_{zx}}{\partial z} + X = 0$$
$$\frac{\partial \tau_{xy}}{\partial x} + \frac{\partial \sigma_y}{\partial y} + \frac{\partial \tau_{zy}}{\partial z} + Y = 0 \qquad (3)$$
$$\frac{\partial \tau_{xz}}{\partial x} + \frac{\partial \tau_{yz}}{\partial y} + \frac{\partial \sigma_z}{\partial z} + Z = 0$$

2.1.3 Principal Stresses

The axis systems taken in Figure 2 are arbitrary and are usually chosen for convenience in computing the applied stresses. For a general state of stress at any point in a structure there exist three mutually perpendicular planes at that point on which the shear stresses vanish. The remaining normal stress components on these three planes are called *principal normal stresses*. These three planes are called principal planes and the three mutually perpendicular axes that are normal to the three planes are called principal axes, i.e., the principal axes coincide with the three principal stress directions. Consequently, the principal stresses are, by definition, the stresses that act perpendicularly to the principal planes. There will also be another set of mutually perpendicular axes along which the shear stresses will be maximal. The *principal shear stresses* act on a set of planes that are at 45° angles to the planes of the principal normal stresses.

These principal normal stresses can be determined by solving

$$\sigma^3 - I_1\sigma^2 + I_2\sigma - I_3 = 0 \tag{4}$$

where

$$I_1 = \sigma_x + \sigma_y + \sigma_z$$
$$I_2 = \sigma_x\sigma_y + \sigma_y\sigma_z + \sigma_z\sigma_x - \tau_{xy}^2 - \tau_{yz}^2 - \tau_{zx}^2 \tag{5}$$
$$I_3 = \sigma_x\sigma_y\sigma_z + 2\tau_{xy}\tau_{yz}\tau_{zx} - \sigma_x\tau_{yz}^2 - \sigma_y\tau_{zx}^2 - \sigma_z\tau_{xy}^2$$

The three roots of Equation (1-4) will give the three principal stresses (σ_1, σ_2, σ_3) of the given stress field (σ_x, σ_y, σ_z, $\tau_{xy} = \tau_{yx}$, $\tau_{yz} = \tau_{zy}$, and $\tau_{zx} = \tau_{xz}$). It can be shown that σ_1, σ_2, and σ_3 are not only orthogonal but also real. The three principal stresses are usually ordered such that $\sigma_1 > \sigma_2 > \sigma_3$.

The principal direction of the principal stress σ_i (i = 1, 2, 3) can be determined by the following simultaneous equations:

$$(\sigma_i - \sigma_x)l_i - \tau_{xy}m_i - \tau_{zx}n_i = 0$$
$$-\tau_{xy}l_i + (\sigma_i - \sigma_y)m_i - \tau_{yz}n_i = 0$$
$$-\tau_{zx}l_i - \tau_{yz}m_i + (\sigma_i - \sigma_z)n_i = 0 \tag{6}$$
$$l_i^2 + m_i^2 + n_i^2 = 1$$

The solutions (l_i, m_i, n_i) of these equations will give the direction of the principal stress σ_i. In the above Equations, l_i, m_i, and n_i are the direction cosines of a unit normal vector coinciding with the principal axis of σ_i.

The principal shear stresses can be found from the values of principal normal stresses using

$$\tau_{13} = \frac{|\sigma_1 - \sigma_3|}{2}$$

$$\tau_{21} = \frac{|\sigma_2 - \sigma_1|}{2} \tag{7}$$

$$\tau_{32} = \frac{|\sigma_3 - \sigma_2|}{2}$$

If the principal normal stresses have been ordered as above, then $\tau_{max} = \tau_{13}$. The directions of the planes of the principal shear stresses are 45° from those of principal normal stresses and also mutually orthogonal.

For a two-dimensional plane stress state (σ_x, σ_y, τ_{xy}) in the *xy*-axes, the stress components $\sigma_{x'}$, $\sigma_{y'}$, $\tau_{x'y'}$ with respect to new x'y'-axes rotated counter clockwise through an angle θ with respect to the *xy*-axes are given by

$$\sigma_{x'} = \frac{\sigma_x + \sigma_y}{2} + \frac{\sigma_x - \sigma_y}{2}\cos 2\theta + \tau_{xy}\sin 2\theta$$

$$\sigma_{x'} = \frac{\sigma_x + \sigma_y}{2} - \frac{\sigma_x - \sigma_y}{2}\cos 2\theta - \tau_{xy}\sin 2\theta \tag{8}$$

$$\tau_{x'y'} = -\frac{\sigma_x - \sigma_y}{2}\sin 2\theta + \tau_{xy}\cos 2\theta$$

The principal normal stresses are given by

$$\sigma_1 = \frac{\sigma_x + \sigma_y}{2} + \sqrt{\left(\frac{\sigma_x - \sigma_y}{2}\right)^2 + \tau_{xy}^2}$$

$$\sigma_2 = \frac{\sigma_x + \sigma_y}{2} - \sqrt{\left(\frac{\sigma_x - \sigma_y}{2}\right)^2 + \tau_{xy}^2} \tag{9}$$

and the principal direction of principal normal stresses are given by

$$\theta_p = \frac{1}{2}\tan^{-1}\frac{2\tau_{xy}}{\sigma_x - \sigma_y} \tag{10}$$

4. Introduction to Advanced Mechanics

where the angle θ_p is measured counter clockwise from the positive x-axis. There are two values of θ_p and they are 90° apart, one between 0 and 90° and the other between 90° and 180°.

2.2 Analysis of Strain

The elongation or contraction of a line segment per unit of length is referred to as *normal strain*. The change in angle that occurs between line segments that were originally perpendicular to one another is referred to as *shear strain*. Similar to stress tensor, strain is also a second-order tensor and can be expresses as:

$$\begin{bmatrix} \varepsilon_{xx} & \gamma_{xy} & \gamma_{xz} \\ \gamma_{yx} & \varepsilon_{yy} & \gamma_{yz} \\ \gamma_{zx} & \gamma_{zy} & \varepsilon_{zz} \end{bmatrix} \tag{11}$$

where ε represents either a normal or a shear strain, the two being differentiated by their subscripts. The strain tensor is also symmetric. We will also simplify the repeated subscripts for normal strains to ε_x, ε_y, and ε_z, for convenience while retaining the dual subscripts to identify shear strains.

The strain-displacement relations have six equations for three unknown displacement component functions u, v, and w:

$$\begin{aligned} \varepsilon_x &= \frac{\partial u}{\partial x} \\ \varepsilon_y &= \frac{\partial v}{\partial y} \\ \varepsilon_z &= \frac{\partial w}{\partial z} \\ \gamma_{xy} &= \frac{\partial u}{\partial y} + \frac{\partial v}{\partial x} \\ \gamma_{yz} &= \frac{\partial v}{\partial z} + \frac{\partial w}{\partial y} \\ \gamma_{zx} &= \frac{\partial w}{\partial x} + \frac{\partial u}{\partial z} \end{aligned} \tag{12}$$

They will not have a single valued solution in general if the strain component functions are arbitrarily assigned. In order to have a single-valued solution for u, v, w, these strain components have to satisfy the following equations of compatibility:

$$2\frac{\partial^2 \varepsilon_x}{\partial y \partial z} = \frac{\partial}{\partial x}(-\frac{\partial \gamma_{yz}}{\partial x}+\frac{\partial \gamma_{xz}}{\partial y}+\frac{\partial \gamma_{xy}}{\partial z})$$

$$2\frac{\partial^2 \varepsilon_y}{\partial x \partial z} = \frac{\partial}{\partial y}(\frac{\partial \gamma_{yz}}{\partial x}-\frac{\partial \gamma_{xz}}{\partial y}+\frac{\partial \gamma_{xy}}{\partial z})$$

$$2\frac{\partial^2 \varepsilon_z}{\partial x \partial y} = \frac{\partial}{\partial z}(\frac{\partial \gamma_{yz}}{\partial x}+\frac{\partial \gamma_{xz}}{\partial y}-\frac{\partial \gamma_{xy}}{\partial z})$$

$$\frac{\partial^2 \varepsilon_x}{\partial y^2}+\frac{\partial^2 \varepsilon_y}{\partial x^2}=\frac{\partial^2 \gamma_{xy}}{\partial x \partial y} \quad (13)$$

$$\frac{\partial^2 \varepsilon_y}{\partial z^2}+\frac{\partial^2 \varepsilon_z}{\partial y^2}=\frac{\partial^2 \gamma_{yz}}{\partial y \partial z}$$

$$\frac{\partial^2 \varepsilon_z}{\partial x^2}+\frac{\partial^2 \varepsilon_x}{\partial z^2}=\frac{\partial^2 \gamma_{xz}}{\partial x \partial z}$$

Stress and strain relations for an isotropic linearly elastic material are related by Hooke's law:

$$\varepsilon_x = \frac{1}{E}(+\sigma_x - v\sigma_y - v\sigma_z)$$

$$\varepsilon_y = \frac{1}{E}(-v\sigma_x + \sigma_y - v\sigma_z)$$

$$\varepsilon_z = \frac{1}{E}(-v\sigma_x - v\sigma_y + \sigma_z)$$

$$\gamma_{xy} = \frac{\tau_{xy}}{G} \quad (14)$$

$$\gamma_{yz} = \frac{\tau_{yz}}{G}$$

$$\gamma_{zx} = \frac{\tau_{zx}}{G}$$

where Young's modulus E, shear modulus G, and Poisson's ratio v are not independent. They satisfy the relation

$$G = \frac{E}{2(1+v)} \quad (15)$$

The generalized Hooke's law for anisotropic materials can be found in Section 2.4.

2.3 Thermal Strain and Thermal Stress

Substances expand when their temperature is raised and contract when cooled. The deformation of expansion or contraction due to the temperature changes in the absence of mechanical loads is called *thermal strain*. The thermal strain is not exactly linear with temperature change, but for first-order approximation and small temperature changes, this strain can be described as proportional to the temperature change. This proportionality is expressed by the *coefficient of thermal expansion*, which is defined as the change in length that a bar of unit length undergoes when its temperature is changed by one degree. The thermal strain for isotropic material can be expressed as

$$\varepsilon_x^{th} = \alpha(T - T_0)$$
$$\varepsilon_y^{th} = \alpha(T - T_0)$$
$$\varepsilon_z^{th} = \alpha(T - T_0)$$
$$\gamma_{xy}^{th} = \gamma_{yz}^{th} = \gamma_{zx}^{th} = 0$$
(16)

where α is the coefficient of thermal expansion (CTE). T is the current temperature, and T_0 is the reference temperature. Thermal expansion is a volumetric expansion so that the thermal shear components are zeros.

Thermal stresses occur when any portion of the thermal expansion or contraction in structure is constrained. Basically, there are two different set of constraints under which thermal stresses occur: external constraints and internal constraints. Thermal stresses due to external constraints are easily apparent. The most familiar example is a bar fixed at both ends and subject to a temperature rise. In this case, the bar is at a state of compression except at the fix ends, which are at a very complex state of stress. However, thermal stresses due to internal constraints are not so obvious. A structure made of one material may be free to expand but still have thermal stresses due to a non-uniform temperature distribution. On the other hand, a structure made of more than one material, such as a composite structure like an assembly of electronic package, may be uniformly heated, have no external constraints, and still have thermal stresses due to different coefficients of expansion and material prosperities. When thermal strain exists, stress and strain relations for an isotropic linearly elastic material are related by generalized Hooke's law:

$$\varepsilon_x = \frac{1}{E}(+\sigma_x - \nu\sigma_y - \nu\sigma_z) + \alpha(T - T_0)$$

$$\varepsilon_y = \frac{1}{E}(-\nu\sigma_x + \sigma_y - \nu\sigma_z) + \alpha(T - T_0)$$

$$\varepsilon_z = \frac{1}{E}(-\nu\sigma_x - \nu\sigma_y + \sigma_z) + \alpha(T - T_0) \qquad (17)$$

$$\gamma_{xy} = \frac{\tau_{xy}}{G}$$

$$\gamma_{yz} = \frac{\tau_{yz}}{G}$$

$$\gamma_{zx} = \frac{\tau_{zx}}{G}$$

2.4   Thermoelasticity

2.4.1   Fundamental Equations of Thermoelasticity

Let us consider a linear elastic package subjected to heating and external forces. We assume that the package is stress-free at a uniform reference temperature T_0 when all external forces are removed. Furthermore, the velocity and displacement of every particle of the package is assumed to be small so that it can be treated as infinitesimal strains and the material derivative is the same as the partial derivative with respect to time.

Based on the foregoing assumptions, the fundamental equations of thermoelasticity for the package are [1-7]:

1. Infinitesimal strain-displacement relations

$$\varepsilon_{ij} = \frac{1}{2}(\frac{\partial u_i}{\partial x_j} + \frac{\partial u_j}{\partial x_i}) ; \qquad (18)$$

2. Constitutive equation (Duhamel-Neumann law)

$$\sigma_{ij} = C_{ijkl}\varepsilon_{kl} - \beta_{ij}(T - T_0) ; \qquad (19)$$

3. Conservation of mass (continuity equation)

$$\frac{\partial \rho}{\partial t} + \frac{\partial \rho v_i}{\partial x_i} = 0 ; \qquad (20)$$

4. Introduction to Advanced Mechanics

4. Conservation of momentum (Newton's law)

$$\rho \frac{\partial v_i}{\partial t} = \frac{\partial \sigma_{ij}}{\partial x_j} + X_i; \tag{21}$$

5. Conservation of energy

$$\frac{\partial \varepsilon}{\partial t} = T \frac{\partial \varphi}{\partial t} + \frac{1}{2\rho} \sigma_{ij} (\frac{\partial v_i}{\partial x_j} + \frac{\partial v_j}{\partial x_i}); \tag{22}$$

6. Rate of change of entropy

$$\rho \frac{\partial \varphi}{\partial t} = -\frac{1}{T} \frac{\partial h_i}{\partial x_i} + \frac{W}{T}; \tag{23}$$

7. Heat conduction (Fourier's law)

$$h_i = -k_{ij} \frac{\partial T}{\partial x_j}; \tag{24}$$

8. Definition of specific heat (if $\partial \varepsilon_{ij} / \partial t = 0$)

$$-\frac{\partial h_i}{\partial x_i} = \rho C_v \frac{\partial T}{\partial t}, \tag{25}$$

where T is instantaneous absolute temperature, x_i the space coordinates, u_i the components of the displacement vector, C_v the heat capacity per unit mass, T_0 the reference temperature, X_i the body force components per unit volume, β_{ij} the thermal moduli, σ_{ij} the component of stress tensor, ε_{ij} the component of strain tensor, k_{ij} the heat conduction coefficients, C_{ijkl} the elastic moduli, ρ the mass density, W the heat generation per unit time per unit volume, ε the internal energy per unit mass, φ the entropy per unit mass, h_i the components of heat flux vector, and v_i the components of the velocity vector. All the indices range over 1, 2, 3 and the summation convention for repeated indices is used.

2.4.2   Governing Equations of Thermoelasticity

We further assume that 1) the interaction between strain and temperature is ignored and the effects of elasticity (change in dimensions on the package) on the temperature distribution are negligible; 2) small variations of temperature with time, that is, the inertia terms may be neglected in the equation of motion. Then by combining equations in the Section 2.4.1, we have the following equations of uncoupled, quasi-static thermoelasticity or the theory of thermal stresses:

$$\frac{\partial}{\partial x_i}\left(k_{ij}\frac{\partial T}{\partial x_j}\right) = \rho C_v \frac{\partial T}{\partial t} - W \qquad (26)$$

$$\frac{1}{2}\frac{\partial}{\partial x_j}\left[C_{ijkl}\left(\frac{\partial u_k}{\partial x_l}+\frac{\partial u_l}{\partial x_k}\right)\right] = -X_i + \frac{\partial}{\partial x_j}(\beta_{ij}T) \qquad (27)$$

2.4.3 Theory of Isotropic Thermal Stresses

Fundamentally, most materials in microelectronics are anisotropic. For practical applications, however, it is possible to gain some insights by assuming the materials to be isotropic. In that case, the governing equations of theory of isotropic thermal stresses are

$$k_{ij}\delta_{ij}\frac{\partial}{\partial x_i}\left(\frac{\partial T}{\partial x_j}\right) = \rho C_v \frac{\partial T}{\partial t} - W \qquad (28)$$

$$Gu_{i,\mu\mu} + (\lambda + G)u_{\mu,\mu i} = -X_i + \beta\frac{\partial T}{\partial x_i} \qquad (29)$$

where

$$\sigma_{ij} = \lambda\varepsilon_{\mu\mu}\delta_{ij} + 2G\varepsilon_{ij} - \beta\delta_{ij}(T-T_0) \qquad (30)$$

$$\beta = \frac{\alpha E}{1-2\nu} \qquad (31)$$

4. Introduction to Advanced Mechanics

$$\lambda = \frac{E\nu}{(1+\nu)(1-2\nu)} \tag{32}$$

$$G = \frac{E}{2(1+\nu)} \tag{33}$$

and where α is the coefficient of thermal expansion (CTE), k is heat conductivity, E is Young's modulus, ν is Poisson's ratio, λ is Lame's constant, G is shear modulus, and δ_{ij} is Kronecker delta.

For the special steady heat flow case, i.e., $\partial T/\partial t = 0$, we have

$$\delta_{ij}\frac{\partial}{\partial x_i}\left(\frac{\partial T}{\partial x_j}\right) = -\frac{W}{k} \tag{34}$$

$$Gu_{i,\mu\mu} + (\lambda+G)u_{\mu,\mu i} = -X_i + \beta\frac{\partial T}{\partial x_i} \tag{35}$$

These are the equations for isotopic thermal stresses with steady heat flow. If there is no heat source, then $W = 0$ in the foregoing equations.

For engineering applications, Equations (28), (29), and (30) are rewritten as equations (36), (37), and (38), respectively.

$$\nabla^2 T = \frac{\rho C_v}{k}\frac{\partial T}{\partial t} - \frac{W}{k} \tag{36}$$

$$\frac{\partial^2 u}{\partial x^2} + \frac{\partial^2 v}{\partial x \partial y} + \frac{\partial^2 w}{\partial x \partial z} + (1-2\nu)\nabla^2 u + \frac{X_x}{(\lambda+G)} = 2(1+\nu)\alpha\frac{\partial T}{\partial x}$$

$$\frac{\partial^2 v}{\partial y^2} + \frac{\partial^2 u}{\partial x \partial y} + \frac{\partial^2 w}{\partial y \partial z} + (1-2\nu)\nabla^2 u + \frac{X_y}{(\lambda+G)} = 2(1+\nu)\alpha\frac{\partial T}{\partial y} \tag{37}$$

$$\frac{\partial^2 w}{\partial z^2} + \frac{\partial^2 u}{\partial x \partial z} + \frac{\partial^2 v}{\partial y \partial z} + (1-2\nu)\nabla^2 u + \frac{X_z}{(\lambda+G)} = 2(1+\nu)\alpha\frac{\partial T}{\partial z}$$

$$\sigma_x = \frac{\lambda}{\nu}\left[(1-\nu)\frac{\partial u}{\partial x} + \nu\left(\frac{\partial v}{\partial y} + \frac{\partial w}{\partial z}\right)\right] - \beta(T - T_0)$$

$$\sigma_y = \frac{\lambda}{\nu}\left[(1-\nu)\frac{\partial v}{\partial y} + \nu\left(\frac{\partial w}{\partial z} + \frac{\partial u}{\partial x}\right)\right] - \beta(T - T_0)$$

$$\sigma_z = \frac{\lambda}{\nu}\left[(1-\nu)\frac{\partial w}{\partial z} + \nu\left(\frac{\partial u}{\partial x} + \frac{\partial v}{\partial y}\right)\right] - \beta(T - T_0)$$

$$\tau_{xy} = G\left(\frac{\partial u}{\partial y} + \frac{\partial v}{\partial x}\right)$$

$$\tau_{yz} = G\left(\frac{\partial v}{\partial z} + \frac{\partial w}{\partial y}\right)$$

$$\tau_{xz} = G\left(\frac{\partial w}{\partial x} + \frac{\partial u}{\partial z}\right)$$

(38)

where

$$\nabla^2 = \frac{\partial^2}{\partial x^2} + \frac{\partial^2}{\partial y^2} + \frac{\partial^2}{\partial z^2} \tag{39}$$

and in equation (38), the linear strain-displacement relations of Equation (12) has been substituted.

2.4.4 Boundary Value Problems

Heat transfer and thermal stress are usually applied in two stages, transient (power on/off) and steady state (during operation). In both cases, for theory of isotropic thermal stresses and strains, the temperature distribution $T(x_i,t)$ is calculated by solving the heat conduction equation, Equation (36), with the prescribed initial and boundary conditions. The displacement components (u, v, and w) everywhere are then determined by solving Equation (37) with the prescribed stress-displacement boundary conditions and with the calculated temperature distribution as an imposed boundary condition. This temperature distribution is shown mathematically on the right hand side of equation (37), and is a known function. The thermal stresses ($\sigma_x,\sigma_y,\sigma_z,\tau_{xy},\tau_{yz},\tau_{zx}$) and strains ($\varepsilon_x,\varepsilon_y,\varepsilon_z,\gamma_{xy},\gamma_{yz},\gamma_{zx}$) everywhere are then calculated by equations (38) and (12), respectively.

2.5 Geometric Nonlinearity

When a microelectronics system is under very large deformations, e.g., if the maximum displacement of a printed circuit board (PCB) is larger than the thickness of the PCB, the above equations of strain-displacement relations have to be modified in order to account for the excessive deformations.

The most popular strain tensor for large deformations was introduced by Green and St. Venant (Green's strain tensor) and is often referred to as a strain tensor in Lagrangian coordinates as below.

$$\varepsilon_x = \frac{\partial u}{\partial x} + \frac{1}{2}\left[\left(\frac{\partial u}{\partial x}\right)^2 + \left(\frac{\partial v}{\partial x}\right)^2 + \left(\frac{\partial w}{\partial x}\right)^2\right]$$

$$\varepsilon_y = \frac{\partial v}{\partial y} + \frac{1}{2}\left[\left(\frac{\partial u}{\partial y}\right)^2 + \left(\frac{\partial v}{\partial y}\right)^2 + \left(\frac{\partial w}{\partial y}\right)^2\right]$$

$$\varepsilon_z = \frac{\partial w}{\partial z} + \frac{1}{2}\left[\left(\frac{\partial u}{\partial z}\right)^2 + \left(\frac{\partial v}{\partial z}\right)^2 + \left(\frac{\partial w}{\partial z}\right)^2\right] \quad (40)$$

$$\gamma_{xy} = \frac{\partial u}{\partial y} + \frac{\partial v}{\partial x} + \frac{\partial u}{\partial x}\frac{\partial u}{\partial y} + \frac{\partial v}{\partial x}\frac{\partial v}{\partial y} + \frac{\partial w}{\partial x}\frac{\partial w}{\partial y}$$

$$\gamma_{yz} = \frac{\partial v}{\partial z} + \frac{\partial w}{\partial y} + \frac{\partial u}{\partial y}\frac{\partial u}{\partial z} + \frac{\partial v}{\partial y}\frac{\partial v}{\partial z} + \frac{\partial w}{\partial y}\frac{\partial w}{\partial z}$$

$$\gamma_{zx} = \frac{\partial w}{\partial x} + \frac{\partial u}{\partial z} + \frac{\partial u}{\partial z}\frac{\partial u}{\partial x} + \frac{\partial v}{\partial z}\frac{\partial v}{\partial x} + \frac{\partial w}{\partial z}\frac{\partial w}{\partial x}$$

It should be pointed out that the displacement components (u, v, w) in the above equations are considered as functions of (x, y, z), the positions of points in the structure in the *unstrained configuration*.

The other popular way to define the strain tensor for larger deformations was introduced by Cauchy for infinitesimal strains and by Almansi and Hamel for finite strains (Almansi's strain tensor) and is often referred to as a strain tensor in Eulerian coordinates:

$$\varepsilon_x = \frac{\partial u}{\partial \bar{x}} - \frac{1}{2}\left[\left(\frac{\partial u}{\partial \bar{x}}\right)^2 + \left(\frac{\partial v}{\partial \bar{x}}\right)^2 + \left(\frac{\partial w}{\partial \bar{x}}\right)^2\right]$$

$$\varepsilon_y = \frac{\partial v}{\partial \bar{y}} - \frac{1}{2}\left[\left(\frac{\partial u}{\partial \bar{y}}\right)^2 + \left(\frac{\partial v}{\partial \bar{y}}\right)^2 + \left(\frac{\partial w}{\partial \bar{y}}\right)^2\right]$$

$$\varepsilon_z = \frac{\partial w}{\partial \bar{z}} - \frac{1}{2}\left[\left(\frac{\partial u}{\partial \bar{z}}\right)^2 + \left(\frac{\partial v}{\partial \bar{z}}\right)^2 + \left(\frac{\partial w}{\partial \bar{z}}\right)^2\right] \quad (41)$$

$$\gamma_{xy} = \frac{\partial u}{\partial \bar{y}} + \frac{\partial v}{\partial \bar{x}} - \frac{\partial u}{\partial \bar{x}}\frac{\partial u}{\partial \bar{y}} - \frac{\partial v}{\partial \bar{x}}\frac{\partial v}{\partial \bar{y}} - \frac{\partial w}{\partial \bar{x}}\frac{\partial w}{\partial \bar{y}}$$

$$\gamma_{yz} = \frac{\partial v}{\partial \bar{z}} + \frac{\partial w}{\partial \bar{y}} - \frac{\partial u}{\partial \bar{y}}\frac{\partial u}{\partial \bar{z}} - \frac{\partial v}{\partial \bar{y}}\frac{\partial v}{\partial \bar{z}} - \frac{\partial w}{\partial \bar{y}}\frac{\partial w}{\partial \bar{z}}$$

$$\gamma_{zx} = \frac{\partial w}{\partial \bar{x}} + \frac{\partial u}{\partial \bar{z}} - \frac{\partial u}{\partial \bar{z}}\frac{\partial u}{\partial \bar{x}} - \frac{\partial v}{\partial \bar{z}}\frac{\partial v}{\partial \bar{x}} - \frac{\partial w}{\partial \bar{z}}\frac{\partial w}{\partial \bar{x}}$$

It should be pointed out that the displacement components (u, v, w) in equation (39) are considered as functions of ($\bar{x}, \bar{y}, \bar{z}$), the positions of points in the structure in the *strained configuration*. For infinitesimal deformations, the quadratic terms in Equations (40) and (41) are neglected. Then the distinction between the Lagrangian and Eulerian strain tensor disappears and equations (40) and (41) reduce to Equation (12).

2.6 Material Nonlinearity

Fundamentally, most microelectronics materials are nonlinear, although for practical applications it is sometimes possible to gain some insight into microelectronics problems by assuming the materials to be linear. In order to have an optimal design and perform failure analysis of microelectronics components and systems, however, nonlinear behaviour of materials has to be considered. In this section, nonlinear theories of materials such as hyperelasticity, plasticity, viscoelasticity, viscoplasticity, and creep are briefly discussed. Detailed descriptions can be found in ref [1, 8-11].

2.6.1 Hyperelasticity

The objective of the theory of hyperelasticity is to offer a mathematical description of the nonlinear mechanical behaviour of materials in the *large-strain range*. It is usually applied to solid propellant, elastomer, rubber, and other rubber like materials. In microelectronics applications, for example, the rubber like materials can be used as interface structures between the cap and the substrate of a thermal conduction module.

4. Introduction to Advanced Mechanics

A hyperelastic material is characterized by the existence of a strain energy W, measured per unit volume of the undeformed state, which is a function of the deformation gradient. The form of the strain energy as a function of the strain components is restricted by any symmetry properties that the material may possess in the undeformed state. For an isotropic material (no preferred directions) the form of W does not depend on the orientation of the coordinate system used to describe the initial locations of the particles and depends only on the strain components through the three principal extension (stretch) ratios $\lambda_1, \lambda_2, \lambda_3$ (i.e., ratio of the deformed dimension to the undeformed dimension).

$$W = \sum_{i+j=1}^{N} C_{ij}(J_1 - 3)^i (J_2 - 3)^j + \sum_{i=1}^{N} \frac{1}{D_i}(J - 1 - R)^{2i} \tag{42}$$

where

$$\begin{aligned} J &= \lambda_1 \lambda_2 \lambda_3 \\ J_1 &= J^{-2/3}(\lambda_1^2 + \lambda_2^2 + \lambda_3^2) \\ J_2 &= J^{-4/3}\left[(\lambda_1 \lambda_2)^2 + (\lambda_2 \lambda_3)^2 + (\lambda_3 \lambda_1)^2\right] \\ R &= (1 + \alpha \Delta T)^3 - 1 \end{aligned} \tag{43}$$

and α is the coefficient of thermal expansion of the material, ΔT is the temperature change, R is the nominal volumetric strain associated with ΔT, J is the ratio of current to original volume, C_{ij} and D_i are constants, and N is an integer usually less than 3.

For materials such as elastomer and rubber, which undergo little change in volume at stress levels that cause severe deformations, the material can be treated as incompressible ($J = 1$) and equation (42) becomes (with $N = 2$)

$$W = C_{01}(J_1 - 3) + C_{10}(J_1 - 3) + C_{11}(J_1 - 3)(J_2 - 3) \\ + C_{02}(J_2 - 3)^2 + C_{20}(J_1 - 3)^2 \tag{44}$$

The material constants $C_{01}, C_{10}, C_{11}, C_{02}, C_{20}$ can be determined by experiment, e.g., the simple tension test, pure shear test, and equibiaxial tension test.

For the rubber like materials over a moderate range of strain, for example the strains smaller than 450-500 percent, we can use the so-called Mooney-Rivlin law to adequately represent:

$$W = C_1(J_1 - 3) + C_2(J_2 - 3) \tag{45}$$

However, experiments on natural rubbers indicate a departure from the Mooney-Rivlin law. In those cases, equation (44) is necessary.

2.6.2 Plasticity

The objective of the theory of plasticity is to offer a mathematical description of the nonlinear mechanical behaviour of materials in the plastic range, which is beyond the elastic range. Unlike elastic deformation, plastic deformation is not a reversible process, and depends not only upon the initial and final states of loading but also upon the loading path by which the final state is achieved. Several aspects of real material behaviour, such as the Bauschinger effect (i.e., the yield stresses in tension and compression are not the same, Figure 3), cyclic hardening, plastic anisotropy, elastic hysteresis, etc., can be modelled at different levels of sophistication by the theories of plasticity. In this section, some well-established and most often used theories are briefly discussed.

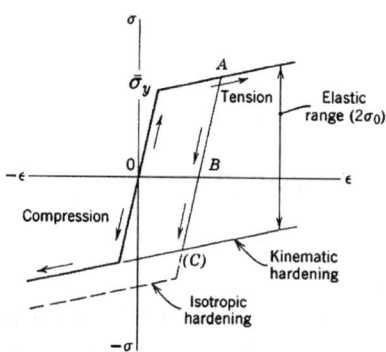

Figure 3. A stress-strain curve with Bauschinger effect

2.6.2.1 Yield Surface

The yield surface is defined as the surface in stress space with stress components as coordinates. Within the yield surface the stress vector may change without any plastic strain increment; stress increments beginning from points in the surface, if directed toward the exterior, imply plastic strain increments.

2.6.2.2 Initial Yield Surface

For isotropic plasticity, the initial yield surface must be independent of the orientation of the reference axes. By choosing the axes of the principal

4. Introduction to Advanced Mechanics

stresses (σ_1, σ_2, σ_3) as the reference axes, the initial yield surface may be expressed in terms of the principal stresses and represented by a surface in a stress space with σ_1, σ_2, and σ_3 as coordinate axes. Thus the initial yield function may appear as

$$f(\sigma_1,\sigma_2,\sigma_3) = 0 \qquad (46)$$

Furthermore, experiment indicates that the hydrostatic pressure has very small effect on the plastic deformation. Hence the initial yield condition may be expressed in terms of the deviatoric stress invariants in the form

$$f(\bar{I}_2,\bar{I}_3) = 0 \qquad (47)$$

where

$$\bar{I}_2 = \frac{1}{2} S_{ij} S_{ij}$$

$$\bar{I}_3 = \frac{1}{3} S_{ij} S_{jk} S_{ki} \qquad (48)$$

$$S_{ij} = \sigma_{ij} - \frac{1}{3}\sigma_{nn}\delta_{ij}$$

$\bar{I}_2$ and $\bar{I}_3$ are the deviatoric stress invariants and S_{ij} is the deviatoric stress tensor.

2.6.2.3 Von Mises Yield Condition (Distortion Energy Theory)

This is a simple yield surfaces conditions for the initial yield of isotropic materials with isotropic hardening (i.e., the tensile and compressive yield stresses are equal at all times) that has provided highly useful descriptions of many real materials. It is based on the assumption that yielding occurs when the second deviatoric stress invariant attains a prescribed value k.

$$f = \bar{I}_2 - k^2 = \frac{1}{2} S_{ij} S_{ij} - k^2 = 0 \qquad (49)$$

or

$$f = \frac{1}{6}\left[(\sigma_1-\sigma_2)^2 + (\sigma_2-\sigma_3)^2 + (\sigma_3-\sigma_1)^2\right] - k^2 = 0 \qquad (50)$$

or

$$f = \frac{1}{6}\left[(\sigma_x - \sigma_y)^2 + (\sigma_y - \sigma_z)^2 + (\sigma_z - \sigma_x)^2 + 6(\tau_{xy}^2 + \tau_{yz}^2 + \tau_{zx}^2)\right] - k^2 = 0 \qquad (51)$$

where k may be a function of plastic strain for strain-hardening materials and the relation of k to test data follow $k = \tau_y$ (τ_y is the yield stress in pure shear), or $k = \sigma_y / \sqrt{3}$ (σ_y is the yield stress in uniaxial tension).

2.6.2.4 Subsequent Yield Surface

Continued loading beyond the initial yield surface leads to plastic deformation, which may be accompanied by changes in both size and shape of the yield surface. For perfect plasticity the yield surface does not change during plastic deformation and the initial yield surface remains valid. For isotropic hardening, however, the size of the yield surface increases but the shape remains the same during loading, shown in Figure 4. To take such changes into account it is necessary to modify the initial yield surface and to define the subsequent yield surface, also known as the loading surface. A general form for the loading surface is given by

$$f(\sigma_{ij}, \varepsilon_{ij}^p, k) = 0 \qquad (52)$$

which depends not only upon the stresses σ_{ij} but also upon the plastic strain ε_{ij}^p and the work-hardening characteristics represented by the parameter k. Differentiating f by the chain rule of calculus, we have

$$df = \frac{\partial f}{\partial \sigma_{ij}} d\sigma_{ij} + \frac{\partial f}{\partial \varepsilon_{ij}^p} d\varepsilon_{ij}^p + \frac{\partial f}{\partial k} dk \qquad (53)$$

where ∂f, $d\sigma_{ij}$, $d\varepsilon_{ij}^p$, and dk represent time differentials. If $f = 0$ and $df < 0$, a condition leading to an elastic state is implied, and it must follow that $d\varepsilon_{ij}^p = dk = 0$. Thus

$$f = 0, \qquad \frac{\partial f}{\partial \sigma_{ij}} d\sigma_{ij} < 0 \qquad (54)$$

is defined as *unloading* and

$$f = 0, \qquad \frac{\partial f}{\partial \sigma_{ij}} d\sigma_{ij} = 0 \qquad (55)$$

4. Introduction to Advanced Mechanics

is defined as *neutral loading,* since it implies that the stress-point remains on the initial yield surface; and

$$f = 0 \qquad \frac{\partial f}{\partial \sigma_{ij}} d\sigma_{ij} > 0 \tag{56}$$

is defined as *loading,* since it implies that the stress-point is moving outward from the current yield surface. For perfectly plastic materials, plastic flow occurs for $f = 0$, $(\partial f/\partial \sigma_{ij})d\sigma_{ij} = 0$, and the case $f = 0$, $(\partial f/\partial \sigma_{ij})d\sigma_{ij} > 0$ does not exist.

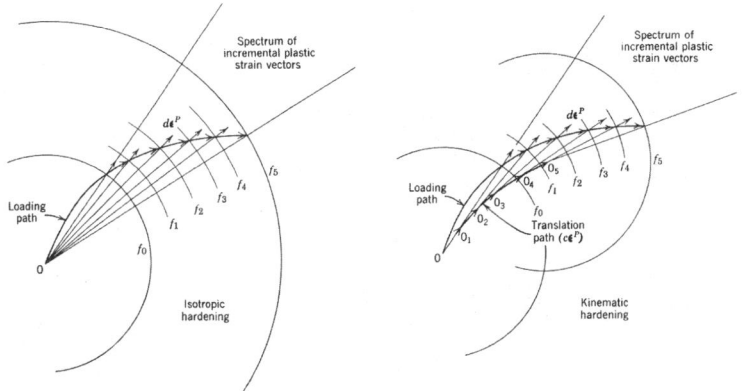

Figure 4. Isotropic hardening vs. kinematic hardening

2.6.2.5 Constitutive Equation: Incremental Theory

A general equation for determining the plastic stress-strain relation for any yield condition was proposed by Drucker. Based on his definition of work-hardening materials ($d\sigma_{ij}d\varepsilon_{ij} > 0$ upon loading; $d\sigma_{ij}d\varepsilon_{ij}^p > 0$ on completing a cycle), he stated that the plastic strain increment vector must be normal to the yield or loading surface at a smooth point on that surface, in Figure 4, and must lie between adjacent normals at a corner point, i.e.,

$$d\varepsilon_{ij}^p = d\lambda \frac{\partial f}{\partial \sigma_{ij}} \tag{57}$$

Equation (57) is called the normality principle of plasticity and can be applied to plastic anisotropic hardening; $d\lambda$ is a function that may depend on stress, strain, and strain history.

For isotropic hardening, the von Mises yield condition is given by

$$f = \frac{1}{2}S_{ij}S_{ij} - \frac{1}{3}\sigma_y^2 \tag{58}$$

Substituting Equation (58) into Drucker's equation (57), we have

$$d\varepsilon_{ij}^p = d\lambda S_{ij} \tag{59}$$

Defining the effective or equivalent stress

$$\bar{\sigma} = \sqrt{(3/2)S_{ij}S_{ij}}$$

$$\bar{\sigma} = \frac{\sqrt{2}}{2}\sqrt{(\sigma_1 - \sigma_2)^2 + (\sigma_2 - \sigma_3)^2 + (\sigma_3 - \sigma_1)^2} \tag{60}$$

$$\bar{\sigma} = \frac{\sqrt{2}}{2}\sqrt{(\sigma_x - \sigma_y)^2 + (\sigma_y - \sigma_z)^2 + (\sigma_z - \sigma_x)^2 + 6(\tau_{xy}^2 + \tau_{yz}^2 + \tau_{zx}^2)}$$

and the effective or equivalent plastic incremental strain is

$$d\bar{\varepsilon}_p = \sqrt{(3/2)d\varepsilon_{ij}^p d\varepsilon_{ij}^p}$$

$$d\bar{\varepsilon}_p = \frac{\sqrt{2}}{3}\sqrt{(d\varepsilon_1^p - d\varepsilon_2^p)^2 + (d\varepsilon_2^p - d\varepsilon_3^p)^2 + (d\varepsilon_3^p - d\varepsilon_1^p)^2} \tag{61}$$

$$\bar{\sigma} = \frac{\sqrt{2}}{3}\sqrt{(d\varepsilon_x^p - d\varepsilon_y^p)^2 + (d\varepsilon_y^p - d\varepsilon_z^p)^2 + (d\varepsilon_z^p - d\varepsilon_x^p)^2 + 1.5(d\gamma_{xy}^{p\,2} + d\gamma_{yz}^{p\,2} + d\gamma_{zx}^{p\,2})}$$

then Equation (59) becomes

$$d\lambda = \frac{3}{2}\frac{d\bar{\varepsilon}_p}{\bar{\sigma}} \tag{62}$$

Suppose there exists a universal stress-strain curve (which coincides with a uniaxial true stress vs. true plastic strain curve for the material)

$$\bar{\sigma} = H\int d\bar{\varepsilon}_p \tag{63}$$

expressing an equivalent stress $\bar{\sigma}$ as a function H of an equivalent plastic strain increment $d\bar{\varepsilon}_p$ integrated over the strain history. Then the slope (H') of the equivalent stress $\bar{\sigma}$ vs. equivalent plastic strain $\int d\bar{\varepsilon}_p$ curve is given by

4. Introduction to Advanced Mechanics

$$H' = \frac{d\bar{\sigma}}{d\bar{\varepsilon}_p} \tag{64}$$

Substituting Equation (64) into Equation (62), then Equation (59), we have

$$d\lambda = \frac{3}{2} \frac{d\bar{\sigma}}{\bar{\sigma} H'}$$

and

$$d\varepsilon_{ij}^p = \frac{3}{2} \frac{d\bar{\sigma}}{\bar{\sigma} H'} S_{ij} \tag{65}$$

Equation (65) is referred to as the Levy-Mises equation and applied to problems of mostly plastic deformation (the elastic deformation is very small and is neglected). However, in most of microelectronics problems the elastic strains cannot be neglected. In that case the incremental total strain is given by

$$d\varepsilon_{ij} = d\varepsilon_{ij}^e + d\varepsilon_{ij}^p \tag{66}$$

where $d\varepsilon_{ij}^e$ is the incremental elastic strain tensor and can be obtained from Equation (17), and $d\varepsilon_{ij}^p$ is given by Equation (65). Thus Equation (66) becomes

$$d\varepsilon_{ij} = \frac{dS_{ij}}{2G} + (1-2\nu)\delta_{ij} \frac{d\sigma_{ij}}{3E} + \frac{3}{2} \frac{d\bar{\sigma}}{\bar{\sigma} H'} S_{ij} \tag{67}$$

Equation (67) is called the Prandtl-Reuss equation.

Equation (67) can only be applied for isotropic hardening materials. For kinematic hardening materials that obey the von Mises yield condition for the initial yield surface, we can use the Reuss-Prager equation

$$d\varepsilon_{ij} = \frac{dS_{ij}}{2G} + (1-2\nu)\delta_{ij} \frac{d\sigma_{ij}}{3E} + \frac{9}{4} \frac{[S_{mm} - 1.5H'\varepsilon_{mm}^p]dS_{mm}}{\bar{\sigma}_y^2 H'} \tag{68}$$

$$(S_{ij} - 1.5H'\varepsilon_{ij}^p)$$

The qualitative difference between isotropic and kinematic hardening can be illustrated for the same loading path in the π-plane ($\sigma_1 + \sigma_2 + \sigma_3 = 0$) in the stress space, Figure 4. It can be seen that, for kinematic hardening, the initial yield surface translates in the π-plane without rotation and without change in size (the centre of the yield surfaces f_o moves from 0 to 0_1 for the subsequent yield surface f_1, etc.) On the other hand, isotropic hardening assumes a uniform expansion of the initial yield surface in the π-plane. Also, the resultant of the incremental plastic strain vectors predicted by isotropic hardening lags farther behind the resultant stress vector than the strain resultant predicted by Prager's kinematic hardening.

2.6.2.6   Viscoelasticity

Viscoelastic materials (e.g., polymers, silicon) display a pronounced influence of the *rate of straining or stressing* and possess *time-dependent* material properties. Under a constant loading in time, viscoelastic materials may exhibit an initial elastic strain followed by *creep* where the materials continue to strain at a rate that depends on the magnitude of the applied load. Upon removal of the load there may be a partial instantaneous elastic recovery followed by a decrease of strain until the strain recovery is complete (delayed elasticity). On the other hand, if viscoelastic materials are subjected to a constant deformation in time, the materials will relax in the sense that the stress decreases with time *(stress relaxation)* from its initial value. Consequently, for viscoelastic materials, the current stress is a function of the entire history of deformation, and conversely, the current strain is a function of the entire history of loading. For linear viscoelastic materials, the stress-strain law of *relaxation type* viscoelastic materials is given by

$$\sigma_{ij}(x,t) = \int_{-\infty}^{t} G_{ijkl}(x,t-\tau) \frac{\partial \varepsilon_{kl}}{\partial \tau}(x,\tau) d\tau \qquad (69a)$$

and the stress-strain law of *creep type* viscoelastic materials is given by

$$\varepsilon_{ij}(x,t) = \int_{-\infty}^{t} J_{ijkl}(x,t-\tau) \frac{\partial \sigma_{kl}}{\partial \tau}(x,\tau) d\tau \qquad (69b)$$

where G_{ijkl} and J_{ijkl} are fourth-order tensors and are called the *tensorial relaxation function* and *tensorial creep function*, respectively.

By means of simple models built of springs (to produce instantaneous deformations in response to the applied load) and dashpots (to produce velocity response to the applied load at any instant), the stress-strain relaxations, Equations (69a, 69b), may be put into the form of differential equations.

4. Introduction to Advanced Mechanics

2.6.2.7 Creep

Creep is a mathematical model for *rate-sensitive elastoplastic* materials operating at *elevated temperature*. Creep strain may be broadly defined as elastoplastic time-dependent deformation under constant load at high temperature. Since different materials have different melting temperatures, it is convenient to define a homologous temperature (the ratio of the test or use temperature to the melting temperature on an absolute temperature scale). In general, creep becomes of engineering significance at a homologous temperature greater than 0.5. For many materials used in microelectronics such as solder, creep deformation becomes important even at room temperature.

Figure 5 shows a typical creep curve that can be obtained from long-time uniaxial test at constant load and constant temperature. The slope of this curve ($\dot{\varepsilon} = d\varepsilon/dt$, ε is strain and t is time) is referred to as the creep rate ($\dot{\varepsilon}$). At the first stage, the creep rate is undefined and the initial creep strain consists of either entirely elastic strain or partially elastic strain and partially plastic strain. During the second stage, the creep rate decreases with time because the effect of strain hardening is greater than that of annealing (recovery). These two effects are in equilibrium (balance) during the third stage and the creep rate reaches essentially a steady state and changes very little with time. However, during the fourth stage, the creep rate increases rapidly with time until fracture occurs at point D. This is because the reduced cross-sectional area (either due to necking or to internal void formation) causes an increase in stress.

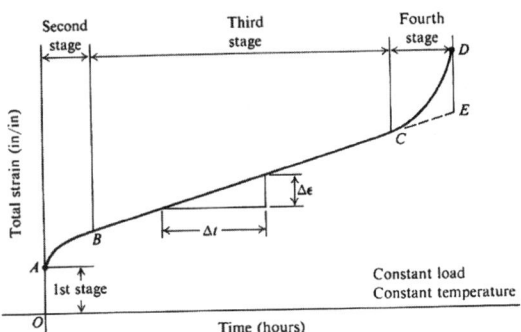

Figure 5. Typical creep curve

The second stage of creep is called primary creep. During this period, the primary creep is dominated by transient creep. For low temperatures and stresses, as in the creep of lead at room temperature, primary creep is the predominant creep process. The third stage of creep is called secondary creep or steady-state creep. The average value of the creep rate during secondary creep is called the minimum creep rate. Log-log plots of stress vs. minimum creep rate for various temperatures are essentially straight lines for many metals. The fourth stage of creep is called tertiary creep, which is often associated with microstructural changes such as coarsening of precipitate particles, recrystallization, or diffusional changes in the phases that are present.

Some of the commonest empirical uniaxial constitutive equations for describing the different stages of creep can be found from the related references [1,10,11].

2.7 Contact Nonlinearity

Boundary non-linearity occurs in most contact problem, in which two surfaces come into or out of contact. The displacements and stresses of the contacting bodies are usually not linearly dependent on the applied loads. This type of non-linearity may occur even if the material behaviour is assumed linear and the displacements are infinitesimal, due to the fact the size of the contact area is usually not linearly dependent on the applied loads, i.e. doubling the applied loads does not necessarily produce double the displacements. If the effect of friction is included in the analysis, then a stick-slip behaviour may occur in the contact area, which adds a further nonlinear complexity that is normally dependent on the loading history.

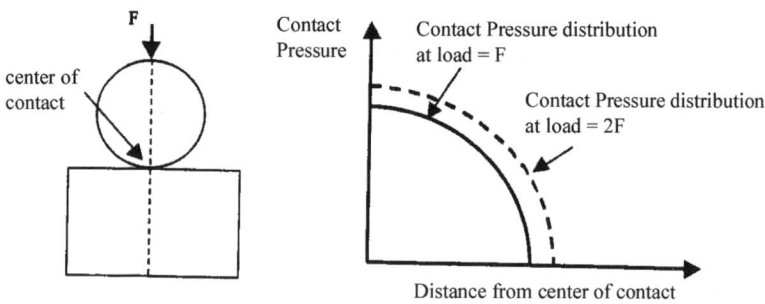

Figure 6. Examples of boundary non-linearity problems - contact of a roller on a plane

4. Introduction to Advanced Mechanics

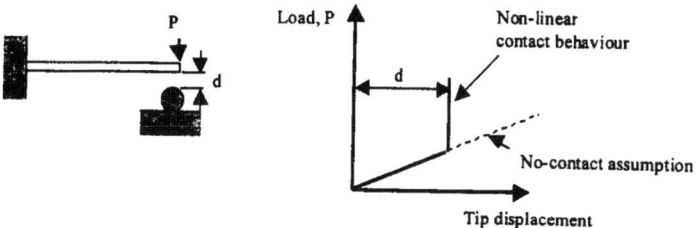

Figure 7. Examples of boundary nonlinearity problems - a cantilever in contact with a rigid support

Figure 6 shows a typical contact problem of a cylindrical roller on a flat plane. Initially the contact is at a single point, and then spreads as the load is increased. The increase in the contact area and the change in the contact pressure are not linearly proportional to the applied load. Another example is shown in Figure 7 where the tip of the cantilever comes into contact with a rigid surface.

There are various possible conditions between the two contacting surfaces, both normal and tangential to the surface lines. The possible conditions are now defined.

- Normal Contact

The normal contact conditions are those that relate to the behaviour of the two surfaces normal to their common line. The conditions are defined at any point on the surface by two quantities, the normal pressure P and the relative normal displacements, u. The conditions are then

- Closed Contact $P < 0$; $u = 0$
- Open Contact $P = 0$; $u > 0$, and these should be satisfied at every point on the surface.
- Sliding Contact

The tangential contact conditions are those that relate to the behaviour of the two surfaces parallel to their common line. These conditions depend upon the friction between the two surfaces and must be defined in terms of the normal pressure, P, the coefficient of friction, μ, the maximum shear stress on the surface, S, and the relative sliding displacement, v. The sliding conditions are then

- Open Contact with Free Sliding $P = 0$; $S = 0$; $|v| > 0$
- Closed Contact with No Sliding $P < 0$; $|S| < |\mu P|$; $|v| = 0$
- Closed Contact with Sliding $P < 0$; $|S| = -\text{sign}(v)|\mu P|$; $|v| > 0$ and these conditions must be satisfied at any point on the surface.

3. FAILURE CRITERIA

3.1 Failure of Ductile Materials

While ductile materials will fracture if statically stressed beyond their ultimate tensile strength, their failure is generally considered to occur when they yield under static loading. The yield strength of a ductile material is appreciably less than its ultimate strength.

Historically, several theories have been formulated to explain this failure: the maximum normal-stress theory, the maximum normal-strain theory, the total strain-energy theory, the distortion-energy (von Mises-Hencky) theory, and the maximum shear-stress theory. Of these only the last two agree closely with experimental data for this case, and of those, the von Mises-Hencky theory is the most accurate. We will discuss only the last two in detail, starting with the most accurate and preferred approach.

3.1.1 The Von Mises-Hencky or Distortion Energy Theory

The microscopic yielding mechanism is now understood to be due to relative sliding of the materials atoms within their lattice structure. This sliding is caused by shear stress and is accompanied by distortion of the shape of the part. The energy stored in the part from this distortion is an indicator of the magnitude of the shear stress present.

The distortion-energy theory predicts that failure by yielding occurs when the total strain energy in a unit volume reaches or exceeds the strain energy in the same volume corresponding to the yield strength in tension or in compression.

The distortion energy per unit volume u_d can be obtained as

$$u_d = \frac{1+v}{3E}\left[\frac{(\sigma_1-\sigma_2)^2+(\sigma_2-\sigma_3)^2+(\sigma_3-\sigma_1)^2}{2}\right] \quad (70)$$

For the simple tensile test, at yield, $\sigma_1 = S_y$, and $\sigma_2 = 0$, $\sigma_3 = 0$, where S_y is the yield strength of the material. The distortion energy per unit volume is

$$u_d = \frac{1+v}{3E}S_y^2 \quad (71)$$

So for the general state of stress given by equation (38), yield is predicted if equation (38) equals or exceeds equation (39). This gives

4. Introduction to Advanced Mechanics

$$\frac{(\sigma_1-\sigma_2)^2+(\sigma_2-\sigma_3)^2+(\sigma_3-\sigma_1)^2}{2} \geq S_y \qquad (72)$$

The left of the above equation can be thought of as a equivalent or effective stress for the entire state of stress given by σ_1, σ_2, and σ_3. This effective stress is usually called the *von Mises stress*, σ^{vm} or σ'. Therefore yielding is predicted to occur when

$$\sigma' \geq S_y \qquad (73)$$

where the von Mises stress is

$$\sigma' = \frac{(\sigma_1-\sigma_2)^2+(\sigma_2-\sigma_3)^2+(\sigma_3-\sigma_1)^2}{2}. \qquad (74)$$

3.1.2 Maximum Shear Stress (Tresca) Theory

The maximum shear stress theory states that yielding begins whenever the maximum shear stress in any element equals or exceeds the maximum shear stress in a tension test specimen of the same material when that specimen begins to yield.

Many theories are postulated on the basis of the consequences seen from tensile tests. As a strip of a ductile material is subjected to tensions, slip lines form at approximately 45° with the axis of the strip. These slip lines are the beginning of yield, and when loaded to fracture, fracture lines are also at angles approximately 45° with the axis of tension. Since the shear stress is maximum at 45° from the axis of tension, it makes sense to think that this is the mechanism of failure. Actually, it was shown that there is a little more going on than this. However, it turns out that the maximum shear stress theory is an acceptable but conservative predicator of failure.

Recall that for simple tensile test, the maximum shear stress occurs on a surface 45° from the tensile surface with a magnitude of $\tau_{max} = \sigma/2$. So the maximum shear stress at yield is $\tau_{max} = S_y/2$. For a general state of stress, the maximum shear stress is $\tau_{max} = (\sigma_1 - \sigma_3)/2$. Thus, for a general state of stress, the maximum shear stress theory predicts yielding when

$$\tau_{max} = \frac{\sigma_1 - \sigma_3}{2} \geq \frac{S_y}{2} \qquad (75)$$

Note that this implies that the yield strength in shear is given by

$$S_{sy} = 0.5S_y \tag{76}$$

which actually is about 15% conservative than the yield strength predicted by the distortion energy theory.

3.2 Failure of Brittle Materials – Maximum Normal Stress Theory

The word *brittle* is relating to the materials whose true strain at fracture is 0.05 or less. Brittle materials fracture rather than yield. Maximum normal stress theory is applicable to the failure of brittle materials. The maximum normal stress theory states that failure occurs whenever one of the three principal stresses equals or exceeds the strength, which is the failure occurs whenever

$$\sigma_1 \geq S_{ut} \quad \text{or} \quad \sigma_3 \leq -S_{uc} \tag{77}$$

where S_{ut} and S_{uc} are the ultimate tensile and compressive strengths, respectively, given as positive quantities.

There are some other theories applicable to the brittle materials, such as the Coulomb-Mohr theory and the Modified-Mohr theory.

3.3 Fatigue Failure

In Sections 3.1 and 3.2, we considered failure subjected to static loading. The behaviour of materials is entirely different when they are subjected to time-varying loading. Often, elements are found to have failed under the action of repeated or fluctuating stresses; yet most careful analysis reveals that the actual maximum stresses were well below the ultimate strength of the material, and quite frequently even below the yield strength. The most distinguishing characteristic of these failure is that the stresses have been repeated a very large number of times. Hence the failure is called a *fatigue failure*.

There are three fatigue failure models in current use and each has a place and a purpose. They are the *stress-life (S-N)* approach, the *strain-life (ε-N)* approach and the *linear elastic fracture-mechanics* (LEFM) approach.

3.3.1 Fatigue Regimes

Based on the number of stress or strain cycles that the part is expected to undergo in its lifetime, it is relegated to either a low-cycle fatigue (LCF) regime or a high-cycle fatigue (HCF) regime. There is no sharp dividing line

4. Introduction to Advanced Mechanics

between the two regimes and various investigators suggest slightly different divisions. It usually defines HCF as starting at around 10^3 cycles of stress/strain variation with the number varying with the material.

3.3.2 The Stress-life Approach

This is the oldest of the three models and is the most often used for high-cycle fatigue (HCF) applications where the assembly is expected to last for more than about 10^3 cycles of stress. It works best when the load amplitudes are predictable and consistent over the life of the part. It is a stress-based model, which seeks to determine a fatigue strength and/or an endurance limit for the material so that the cyclic stresses can be kept below that level and avoid failure for the required number of cycles. The part is then designed based on the materials fatigue strength (or endurance limit) and a safety factor. In effect, this approach attempts to keep local stresses in notches so low that the crack-initiation stage *never begins*. The assumption (and design goal) is that stresses and strains everywhere remain in the elastic region and no local yielding occurs to initiate a crack.

This approach is fairly easy to implement and a large amount of relevant strength data are available due to its long-time use. However, it is the most empirical and least accurate of the three models in terms of defining the true local stress/strain states in the part, especially for low-cycle fatigue (LCF) finite-life situations where the total number of cycles is expected to be less than about 10^3 and the stresses will be high enough to cause local yielding. On the other hand, with certain materials, the stress-life approach allows the design of parts for infinite life under cyclic loading.

3.3.3 The Strain-life Approach

Because the initiation of a crack involves yielding, a stress-based approach cannot adequately model this stage of the process. A strain-based model gives a reasonably accurate picture of the crack-initiation stage. It can also account for cumulative damage due to variations in the cyclic load over the life of the part, such as overloads that may introduce favourable or unfavourable residual stresses to the failure zone. Combinations of fatigue loading and high temperature are better handled by this method because the creep effects can be included. This method is most often applied to LCF, finite-life problems where the cyclic stresses are high enough to cause local yielding. It is the most complicated of the three models to use and requires a computer solution. Test data are still being developed on the cyclic-strain behaviour of various engineering materials. In Chapter 7, the strain-life approach will be applied to predict fatigue life of various solder materials.

3.3.4 The Linear-Elastic Fracture Mechanics Approach

Fracture-mechanics theory provides the best model of the crack-propagatin stage of the process. This method is applied to LCF, finite-life problems where the cyclic stresses are known to be high enough to cause the formation of cracks and is most useful in predicting the remaining life of cracked parts in service. It is often used in conjunction with non-destructive testing (NDT) in a periodic service-inspection program, especially in the aircraft/aerospace industry. Its application is fairly straightforward but relies on the accuracy of the expression for the stress-intensity geometry factor and on the estimation of initial crack size required for the computation. One approach is to assume that a crack smaller than the smallest detectable crack already exists in order to begin the calculation. It gives more accurate results when a detectable and measurable crack already exists. In the subsequent section, fracture mechanics will be discussed in detail.

4. FRACTURE MECHANICS

There are a multitude of interfaces in any microelectronic systems. Studies have shown that delamination and cracking are common in all types of microelectronics systems. It is known that the mismatch of the coefficient of thermal expansion between the different layers could results in high interfacial stresses. Cracks may form in the interface from manufacturing defects, material flaws, damage caused by applied loads, or fatigue caused by temperature cycling. This crack may then propagate through the interface and destroy the functionality of device. With smaller sizes, higher performance requirements, and new material systems that are dissimilar, thermally induced interfacial delamination or cohesive failure is a cause for concern in these microelectronic systems.

4.1 Linear Elastic Fracture Mechanics

Fracture mechanics deals with the structure with cracks. Most studies on fracture mechanics have dealt with situations and materials for which the use of linear elasticity is valid or constitutes a good approximation. This is the case when the region of inelastic deformation that inevitably arises around the crack tip is limited to a neighbourhood of the crack tip much smaller than

4. Introduction to Advanced Mechanics

the size of the crack. Under this condition the material is said to be under a small-scale yielding regime and linear elastic fracture mechanics (LEFM) concepts are applicable. Furthermore, if small-scale yielding is assumed, then it can be expected that similar fracture behaviour will occur for different crack configurations if these have similar local deformation/stress fields.

4.1.1 Stress Intensity Factors

Fracture can be seen as relative movement, a separation, between the surfaces of a geometrical discontinuity following the application of load. This separation of the cracks is always associated with three different types of deformation, as presented in Figure 8. These modes of loading are usually referred to by Roman numbers I, II and III. Other descriptions used are opening mode or tension mode for mode I in-plane, shear mode for mode II, and out-of-plane shear mode or tearing mode for mode III.

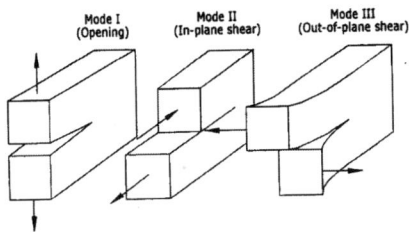

Figure 8. The crack loading mode

We consider linear isotropic solids with Young's modulus E and Poisson's ratio v. Results from crack mechanics for 2D, plane strain elasticity will be cited.

Consider the generic body in Figure 9. Take the origin of the (x_1, x_2) axes at the crack tip and align the x_1-axis such that it is parallel to the crack. (If the crack is curved, x_1 is aligned with the tangent to the crack at the tip). Let (r, θ) be associated planar polar coordinates. The body is loaded in the plane. For any loading, the crack tip fields as the crack tip approaches (for either plane strain or plane stress) have the universal form (as $r \to 0$)

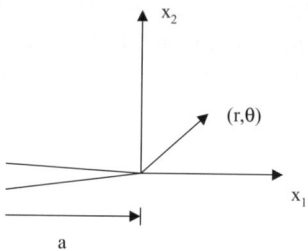

Figure 9. Generic plain strain crack problem

$$\sigma_{ij} = \frac{K_I}{\sqrt{2\pi r}} \tilde{\sigma}_{ij}^{\ I}(\theta) + \frac{K_{II}}{\sqrt{2\pi r}} \tilde{\sigma}_{ij}^{\ II}(\theta) \tag{78}$$

where K_I and K_{II} are the mode I and II stress intensity factors. The stress intensity factors have dimension stress × length$^{1/2}$. It is worth noting that stress intensity factors are a measure of the strength of the singularity $1/\sqrt{r}$ of the stress field, and according to linear elastic fracture mechanics theory they completely determine the crack behaviour. It can then be said that under small-scale yielding conditions, similar K factors result in similar facture behaviour.

4.1.2 Griffith's Criterion for Crack Growth

The phenomenon of crack growth cannot be explained only by the singular behaviour of the stress fields in the crack tip region. On the contrary, it would seem that infinite stress would lead to unstable and rapid crack growth followed by failure for the smallest of the applied loads according the failure criteria described in Section 3. The Griffith energy criterion for fracture states that unstable crack growth occurs if the total energy of the body remains constant or decreases as the crack length increases. If W is the strain energy contained in an elastic solid and U is the energy required for crack growth, then according to Griffith the necessary condition for crack growth can be expressed as

$$\frac{dW}{da} \geq \frac{dU}{da} \tag{79}$$

It is usual to replace dW/da by the so called strain energy release rate or crack extension force G, and dU/da by G_C, the crack resistance, i.e.,

$$G \geq G_c \tag{80}$$

4. Introduction to Advanced Mechanics

The mode I, II, III energy release rates G_I, G_{II} and G_{III} can be defined, and are related to the stress intensity factors by following

$$G_I = \begin{cases} K_I^2/E & \text{plane stress} \\ K_I^2(1-v^2)/E & \text{plane strain} \end{cases} \qquad (81)$$

$$G_{II} = \begin{cases} K_{II}^2/E & \text{plane stress} \\ K_{II}^2(1-v^2)/E & \text{plane strain} \end{cases} \qquad (82)$$

$$G_{III} = \frac{K_{III}^2(1+v)}{E} \qquad (83)$$

$$G = G_I + G_{II} + G_{III} \qquad (84)$$

The criterion can also be expressed in terms of K in the following way

$$K \geq K_c \qquad (85)$$

where K_c is the fracture toughness of the material.

In summary, a fracture parameter, called the stress intensity factor, can be defined along a crack. In a centre-cracked plate under an applied loading stress of σ, this is given by

$$K_I = \sigma\sqrt{a\pi} \qquad (86)$$

provided the crack length a is small within the plate. In a general cracked body of complicated shape, the right hand side of Equation (86) is multiplied by a geometric factor, so that K_I has to be calculated from a stress analysis such as offered by finite element method. In section 5, several methods are introduced to extract the stress intensity factors from finite element solutions.

For a given material at a given temperature, a critical value of stress intensity factor exists as K_{Ic}, such that when the applied load is large enough and the calculated K_I becomes equal to K_{Ic}, then fast, usually catastrophic, fracture will occur. K_{Ic} is known as the fracture toughness. Thus, for the given material and temperature, having evaluated K_{Ic} from a laboratory test specimen, it may be used for direct comparison with the stress intensity factor calculated at crack tips in any structural components, since geometry independence is assumed.

4.1.3 Crack Tip Plasticity

Linear elastic stress analysis of sharp cracks predicts infinite stresses at the crack tip. In real materials, however, stresses at the crack tip are finite because the crack tip radius must be finite. Inelastic material deformation, such as plasticity in metals, leads to further relaxation of crack tip stresses.

The elastic stress analysis becomes increasingly inaccurate as the inelastic region at the crack tip grows. Simple corrections to LEFM are available when moderate crack tip yielding occurs. For more extensive yielding, one must apply alternative crack tip parameters that take nonlinear material behaviour into account.

The size of the crack tip yielding zone can be estimated by two methods: the Irwin approach, where the elastic stress analysis is used to estimate the elastic-plastic boundary, and the strip yield model.

4.1.3.1 The Irwin Approach

The simplest model to take into account the development of a small plastic region around the crack tip is due to Irwin. In this model the size of the crack is corrected, enlarged, to cater for "weakening" of the material.

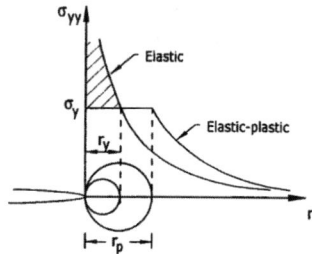

Figure 10. First-order and second-order estimates of plastic zone size

A first approximation to the plastic zone size, r_y, can be obtained by determining the distance r_y from the crack tip at which the elastic stress is equal to the yield stress. For a crack of a given length a, loaded under Mode I and plane stress conditions, r_y can easily be determined and solving for r:

4. Introduction to Advanced Mechanics

$$r_y = \frac{1}{2\pi}\left(\frac{K_I}{\sigma_y}\right)^2 \tag{87}$$

A second better estimation for the size of the plastic region, which follows the load redistribution, is given by $r_p = 2r_y$. If this parameter, r_p, is small compared to the crack length, then the stress fields as determined by elasticity remain valid, but only in the neighbourhood of the crack outside the yielding region. The stress intensity factor is therefore still valid and a new effective value for it can be expressed as:

$$K_{eff} = Y\sigma\sqrt{\pi(a + r_p)} \tag{88}$$

4.1.3.2 The Strip Yield Model

The strip yield model, which is illustrated in Figure 11(a) was first proposed by Dugdale. He assumed a long, slender plastic zone at the crack tip in a non-hardening material in plane stress. The strip yield zone is modelled by assuming a crack of length 2a + 2ρ, Where ρ is the length of the plastic zone with a closure stress equal to σ_y applied at each crack tip as depicted in Figure 11(b).

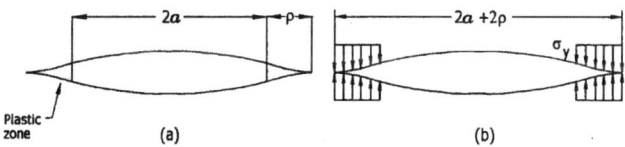

Figure 11. The strip yield model

Since stresses are finite in the strip yield zone, there cannot be a stress singularity at the crack tip. Therefore, the loading terms in the crack tip field that varies with $1/\sqrt{r}$ must be zero. The plastic zone length, ρ, must be chosen such that the stress intensity factors from the remote tension and closure stress cancel each other. Following this approach, the expression for the effective stress intensity factor given by Dugdale is

$$K_{eff} = \sigma\sqrt{\pi a \sec\left(\frac{\pi a}{2\sigma_y}\right)} \tag{89}$$

4.2 Elasto-Plastic Fracture Mechanics

Linear fracture mechanics (LEFM) is valid only as long as nonlinear material deformation is confined to a small region surrounding the crack tip. In many cases, it is virtually impossible to characterize the fracture behaviour with LEFM, and alternative fracture mechanics regime is required.

Elastic-plastic fracture mechanics (EPFM) applies to materials that exhibit nonlinear behaviour (i.e., plastic deformation). Two elastic-plastic parameters are introduced: the crack tip opening displacement (CTOD) and J-integral. Both parameters describe the crack tip conditions in elastic-plastic materials and each can be used as a fracture criterion.

4.2.1 Crack Tip Opening Displacement (CTOD)

One of the consequences of yielding in the crack tip region is the separation of the crack faces at the original crack tip. This was first recognized by Wells who proposed the CTOD (crack tip opening displacement δ) as the measure for the plastic strains at the crack tip, shown in Figure 12(a). The CTOD was initially obtained from consideration of Irwin's plastic zone correction as shown Figure 12(b). The strip yield model provides an alternate means for analysing the CTOD. The CTOD can be defined in this case as the crack opening displacement at the end of that strip yield zone, as Figure 12(c) illustrates.

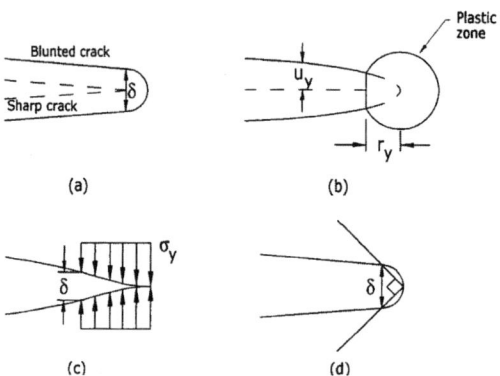

Figure 12. Crack tip opening displacement; alternative definitions

An alternative definition of CTOD is given by 90° intercept as depicted in Figure 12(d). This definition is commonly used to infer CTOD in finite

4. Introduction to Advanced Mechanics

element measurements. Note that definitions (a) and (d) are equivalent if the crack blunts in a semicircle.

By using the strip yield model criterion to define CTOD, it is possible to demonstrate the relationship between the CTOD, G and K. The more general form of this relationship can be expressed as follows:

$$\delta = \frac{K_I^2}{m\sigma_y E} = \frac{G}{m\sigma_y} \tag{90}$$

where m is a dimensionless constant that is approximately 1.0 for plane stress and 2.0 for plane strain.

4.2.2 The J-Integral

The J-integral proved popular as facture characterizing parameter for nonlinear materials. By idealizing elastic-plastic deformation as nonlinear elastic, Rice provided the basis for extending fracture mechanics methodology well beyond the validity of LEFM.

Rice applied deformation plasticity (i.e. nonlinear elasticity) to the analysis of a crack in a nonlinear material. He showed that a nonlinear energy release rate, J, could be written as a path-independent line integral. For general contour around the crack tip such as one depicted in Figure 13, the form of the J integral is as follows

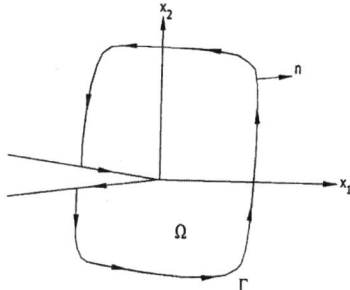

Figure 13. Definition of the J-integral contour around the crack tip

$$J = \int_\Gamma \left(W n_1 - t_i u_{i,1} \right) d\Gamma \tag{91}$$

where Γ is a closed contour around the crack tip, n_1 is the component of the outward normal to the contour along to the crack axis, t_i are the traction components, $u_{i,1}$ are the derivatives of the displacement components u_i with respect to the crack axis x_1, and W is the strain energy density given by

$$W = \int_0^\varepsilon \sigma_{ij} d\varepsilon_{ij} \tag{92}$$

Rice showed that the J-integral is, for elastic materials (both linear and nonlinear), equal to the variation of the potential energy for virtual crack extension da, that is

$$J = -\frac{d\Pi}{da} \tag{93}$$

where Π is the potential energy of the body in the absence of body forces. This is expressed as the balance of the strain energy and the work done by the applied load,

$$\Pi = \int_\Omega W d\Omega - \int_{\Gamma_t} t_i u_i d\Gamma \tag{94}$$

$$-\frac{d\Pi}{da} = \int_\Gamma (W n_1 - t_i u_{i,1}) d\Gamma \tag{95}$$

This integral can be proved to be path-independent by choosing two separate integration paths and using the divergence theorem.

Note that in the case of a linear material $-d\Pi/da$ is equal to G, the energy release rate; then it follows that

$$J = G = \frac{K^2}{E'} \tag{96}$$

where E' =E for plane stress and E' =E/(1-v^2) for plain strain state.

4.3 Fatigue Crack Propagation

The material in the preceding sections has dealt with static or monotonic loading of cracked bodies. This section considers crack growth in the presence of cyclic stresses, which can lead to an increase in crack length at each step, even though the maximum stress intensity factor may be much less than the critical one. By characterizing fatigue crack growth using linear elastic fracture mechanics parameters, it is possible to predict crack growth rates under cyclic loading.

4. Introduction to Advanced Mechanics

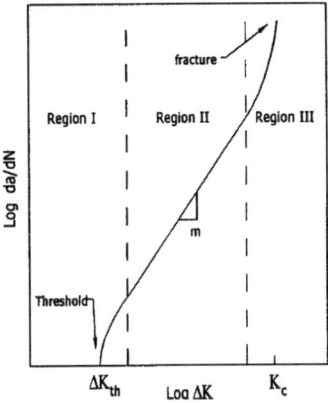

Figure 14. Typical fatigue crack growth behavior in metals

Figure 14 is a schematic log-log plot of the rate of crack growth per load cycle, *da/dN*, as a function of the applied stress intensity factor range, $\Delta K = K_{max} - K_{min}$. The sigmoidal curve contains three distinct regions. In the first region, crack growth goes asymptotically to zero as ΔK approaches a threshold value (ΔK_{th}). This means that for stress intensity factors below ΔK_{th} there is no crack growth, i.e. there is a fatigue limit. The threshold effect is believed to be caused by a number of different processes, which lead to crack blocking. In region II, the log *da/dN* tends to vary linearly with respect to the log of ΔK, finally to accelerate dramatically in region III as K_{max} approaches K_c, the fracture toughness of the material.

Paris et al., developed an empirical formula, which relates the rate of growth per load cycle, *da/dN*, to the stress intensity factor range ΔK in the linear region

$$\frac{da}{dN} = f(\Delta K)^n \tag{97}$$

Variations to this general form were presented by different researchers.

4.4 Mixed-Mode Fracture

There are already several criteria for the specification of the direction of crack growth under in-plane mixed-mode loading. The most important and the most commonly applied theories are: (1) maximum principal stress criterion, and (2) strain energy density criterion. In the case of small or medium mixed-mode ratios, the criteria predict kink angles of almost the

4.4.1 Maximum Principal Stress Criterion

Maximum principal stress criterion is due to Erdogan and Sih, and was initially proposed for two dimensions. This theory states that crack growth takes place in the direction perpendicular to the maximum principal stress. If a polar local coordinate system (r, θ) is taken on the crack front (see Figure 15), the stresses $\sigma_{\theta\theta}$, and $\sigma_{r\theta}$ can be written as:

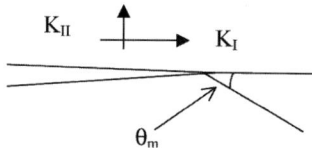

Figure 15. Mixed-mode cracking

$$\sigma_{\vartheta\vartheta} = \frac{1}{\sqrt{2\pi r}} \cos\left(\frac{\theta}{2}\right)\left[K_I \cos^2\left(\frac{\theta}{2}\right) - \frac{3}{2} K_{II} \sin\theta\right] \qquad (98)$$

$$\sigma_{r\vartheta} = \frac{1}{\sqrt{2\pi r}} \cos\left(\frac{\theta}{2}\right)\left[K_I \sin\left(\frac{\theta}{2}\right) - K_{II}(3\cos\theta - 1)\right] \qquad (99)$$

Hence, the crack growth direction according to this theory is obtained from the condition $\sigma_{r\theta} = 0$, i.e., $K_I \sin\theta_0 - K_{II}(3\cos\theta_0 - 1) = 0$

$$\tan\frac{\theta_0}{2} = \frac{1}{4}\left[\frac{K_I}{K_{II}} \pm \sqrt{\left(\frac{K_I}{K_{II}}\right)^2 + 8}\right] \qquad (100)$$

When the criterion is adapted to three dimensions, the Mode I stress intensity factor in the above equation is replaced by an equivalent Mode I stress intensity factor K_{eq} which combines Mode I and Mode II stress intensity factors, as following

4. Introduction to Advanced Mechanics

$$K_{eqv} = \cos\frac{\theta_m}{2}[K_I \cos^2\frac{\theta_m}{2} - \frac{3}{2}K_{II}\sin\theta_m] = K_{Ic} \qquad (101)$$

Experimental investigations have shown that the maximum principal stress criterion gives an adequate description of the kink angle. The maximum principal stress criterion has the advantages that there is no need to assume a virtual crack increment. Equation (100) gives the solution to calculate the kink angle and the failure criterion under the mixed mode loading

4.4.2 Minimum Strain Energy Density Criterion

The minimum strain energy density criterion, or S-Criterion, was formulated by Sih. Strain energy W is defined as the energy stored in the system due to changes in the deformation state. It can be written, in a linear elastic body, as

$$W = \int_V \frac{1}{2}\sigma_{ij}\varepsilon_{ij} \, dV \qquad (102)$$

where V is the volume of the body. The strain energy density defined as dW/dV can be derived as

$$\frac{dW}{dV} = \frac{1}{2}\sigma_{ij}\varepsilon_{ij} \qquad (103)$$

Using the constitutive relations gives

$$\sigma_{ij} = 2\mu\varepsilon_{ij} + \frac{2\mu\upsilon}{1-2\upsilon}\varepsilon_{kk}\delta_{ij} \qquad (104)$$

together with the near crack front expressions, the explicit expression for the strain energy density (103) around the crack front can be written as

$$\frac{dW}{dV} = \frac{S(\theta)}{r\cos\phi} + O(1) \qquad (105)$$

$$S(\theta) = a_{11}K_I^2 + 2a_{12}K_I K_{II} + a_{22}K_{II}^2 + a_{33}K_{III}^2 \qquad (106)$$

where

$$a_{11} = \frac{1}{16\pi \mu}(3-4v-\cos\theta)(1+\cos\theta) \qquad (107)$$

$$a_{12} = \frac{1}{8\pi \mu}\sin\theta(\cos\theta-1+2v) \qquad (108)$$

$$a_{22} = \frac{1}{16\pi \mu}\left[4(1-v)(1-\cos\theta)+(3\cos\theta-1)(1+\cos\theta)\right] \qquad (109)$$

$$a_{33} = \frac{1}{4\pi \mu} \qquad (110)$$

in which μ is the shear modulus of elasticity and v is the Poisson ratio. It can be seen from Equation (105), S/cosϕ represents the amplitude or intensity of the strain energy density field and it varies with angles ϕ and θ. It is apparent that the minimum S/cosϕ always occurs in the plane normal to the crack front, namely $\phi = 0$. Then the term S is known as the strain energy density factor and can be used in a similar way to the stress intensity factor on crack growth analysis.

4.5 Crack Closure

After the Paris law gained wide acceptance as a fatigue crack growth criterion, many researchers came to the realization that this simple expression was not universally applicable, as experimental evidence showed that fatigue crack growth rates exhibit a dependence on the load ratio $R = K_{min}/K_{max}$.

A discovery by Elber provided at least a partial explanation for the R effect. Elber postulated that crack closure decreased the fatigue crack growth rate by reducing the effective stress intensity range. Figure 16 illustrates the closure concept. When a specimen is cyclically loaded between K_{max} and K_{min} the crack faces are in contact below K_{op}, the stress intensity at which the

4. Introduction to Advanced Mechanics

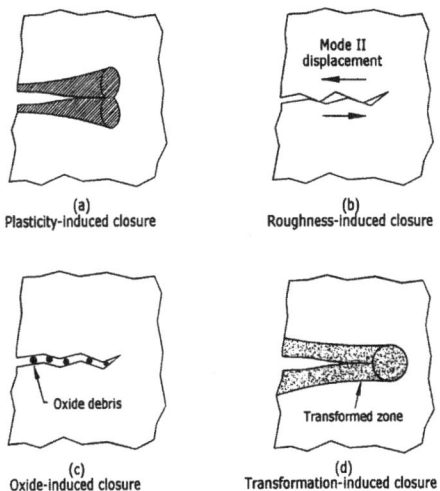

Figure 16. Fatigue crack closure mechanism in metals

crack opens. Elber assumed that the portion of the cycle, which is below K_{op} does not contribute to fatigue crack growth. He defined an effective stress intensity range as follows:

$$\Delta K_{eff} = K_{max} - K_{op} \tag{111}$$

and also introduced an effective stress intensity ratio:

$$U = \frac{\Delta K_{eff}}{\Delta K} = \frac{K_{max} - K_{op}}{K_{max} - K_{min}} \tag{112}$$

Elber also proposed a modified Paris law equation:

$$\frac{da}{dN} = C\left(\Delta K_{eff}\right)^m \tag{113}$$

Equation (113) has been reasonably successful in correlating fatigue crack growth data at various R ratios.

Since Elber's original study, numerous researchers have confirmed that crack closure does in fact occur during fatigue crack propagation. Many mechanisms have identified for fatigue crack closure; some of them are illustrated in Figure 16. Plasticity-induced closure, Figure 16(a), results from residual stresses in the plastic wake, while roughness-induced closure, which is illustrated in Figure 16(b), is influenced by microstructure. Oxide-induced closure, Figure 16(c), is usually associated with an aggressive environment. In this mechanism, oxide debris or rather corrosion products act as a wedge between crack faces. Another mechanism is given by a stress-induced transformation at the tip of the growing crack (see Figure 16(d)), which results in a process zone where residual compressive stresses can lead to crack closure.

4.6 Singularity of Angular Corner of a Homogeneous Material

As shown in Figure 17, the stresses may be singular at a wedge corner of a homogeneous material. A crack in a homogeneous material can be treated as a special case of this wedge with $\theta_0 = 0$. For a general case, the order of the singularity is dependent on the angle θ_0, and can be determined by [16]

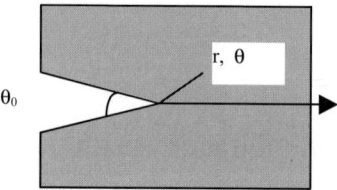

Figure 17. Angular corner of a homogenous material

$$\frac{\sin p(2\pi - \theta_0)}{p(2\pi - \theta_0)} = \frac{\sin(2\pi - \theta_0)}{2\pi - \theta_0} \tag{114}$$

with the stress field at wedge tip

$$\sigma_{ij} = r^{p-1} f_{ij}(\theta, p) \tag{115}$$

It can be verified from this equation that with $\theta_0 = 0$, $p = 1/2$ for a true crack, which is the solution for classical fracture mechanics. p increases when θ_0 increases. There is no stress singularity when $\theta_0 \geq \pi$.

4. Introduction to Advanced Mechanics

4.7 Interface Fracture Mechanics

4.7.1 Mechanics of Interfacial Delamination

A fracture may start in an interface but will only propagate along it if that is the path of least resistance. The most favourable path depends upon both the energy release rate and the interfacial toughness, which is strongly influenced by mode-mixity. However, in this section we will only consider fractures that do not kink out of the interface.

For any bimaterial system, in which the materials are considered as isotropic and linear elastic, four elastic constants, i.e., two Young's moduli and two Poisson's ratios are involved. However, it has been proven that under traction-specified boundary conditions the solution to plane problems of elasticity depends on only two-dimensional combinations of the elastic moduli, namely, Dundurs parameters defined by [17]

$$\alpha = \frac{(1-v_2)/\mu_2 - (1-v_1)/\mu_1}{(1-v_2)/\mu_2 + (1-v_1)/\mu_1} \tag{116}$$

$$\beta = \frac{1}{2}\frac{(1-2v_2)/\mu_2 - (1-2v_1)/\mu_1}{(1-2v_2)/\mu_2 + (1-2v_1)/\mu} \tag{117}$$

or by the connection,

$$\varepsilon = \frac{1}{2\pi}\ln\frac{1-\beta}{1+\beta} \tag{118}$$

where v is Poisson's ratio, μ is shear modulus, and subscripts 1 and 2 refer to materials across the interface. The α measures the relative stiffness of the two materials. Material 1 is stiffer than material 2 if $\alpha > 0$. The so called oscillating index, β or ε, which is responsible for oscillating stress behaviours at crack tip in some cases as to be discussed shortly, can hardly be interpreted intuitively. In general, though $\beta \neq 0$ it is usually small (<0.25), and often approximated to zero. When $\beta = 0$, the singularity at the tip of an interface crack has the same as that in an isotropic homogeneous solid, and the normal definition of stress intensity factor can be used. The only difference being that stress distribution on an interface is usually mixed mode. If $\beta \neq 0$ the stress ahead of a crack tip oscillates and there can be contact between the interface surfaces. As Rice [18] pointed out, the most difficult problem that the oscillatory stress distribution presents is in understanding the mode-mixity which varies with distance ahead of the crack tip. The stress intensity when $\beta \neq 0$ is most usefully defined following Rice, who defines the stresses, a distance r ahead of the crack tip as

$$\sigma_{22} + i\sigma_{12} = \frac{Kr^{i\varepsilon}}{\sqrt{2\pi r}} \tag{119}$$

where K is the *complex* stress intensity factor with dimension $FL^{-3/2}L^{-i\varepsilon}$. The complex stress intensity factor K has real and imaginary components (K_1, K_2) which have a similar role to the mode I and II stress intensity factors in homogenous materials to which they degenerate when $\beta = 0$. The energy release rate in terms of the stress intensity factor components is

$$G = \frac{1-\beta^2}{2}\left(\frac{1-v_1^2}{E_1} + \frac{1-v_2^2}{E_2}\right)(K_1^2 + K_2^2) \tag{120}$$

Many experiments have shown that the interfacial toughness depends strongly on the mode mixity. Figure 18 shows the effect of mode-mixity on a microelectronic interface. Near the tip of an interfacial crack a local, position dependent mode-mixity is defined by σ_{12}/σ_{22} and what has been termed the local phase angle, ψ_{local}, is given by

$$\psi_{local} = \tan^{-1}\left[\frac{\sigma_{12}}{\sigma_{22}}\right] = \tan^{-1}\left[\frac{\text{Im}(Kr^{i\varepsilon})}{\text{Re}(Kr^{i\varepsilon})}\right] = \psi + \varepsilon \ln(L/r) \tag{121}$$

where L is material characteristic length or a specimen length such as film thickness. The former definition is more appropriate to the effect of mode-mixity on interface toughness. Since ε is small, for over a hundred material pairs studied by Suga et al. [19], the variation in mode-mixity angle with the chosen characteristic length is not large. The characteristic length must be greater than the contact zone for equation (119) to be valid. Rice has shown that if the radius of contact is to be less than $0.01L$ then the phase angle must be greater than -69 degree if $\varepsilon = 0.08$. This restriction will usually be met in practice. There is little accuracy lost by assuming $\beta = 0$.

4. Introduction to Advanced Mechanics

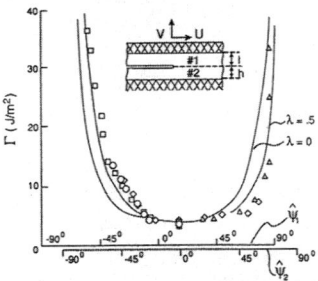

Figure 18. Toughness dependency on the phase angle

4.7.2 Stress Singularity of a General Bimaterial System

In spite of complexity of geometry and material combinations, all stress singularities arising in bimaterial wedge configurations in electronic packages can be grouped into two categories, i.e., an angular corner of bimaterial wedge, and a bimaterial wedge with adhesion, as shown in Figure 19 (a) and (b), respectively [16].

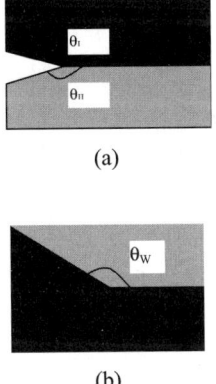

Figure 19. General cases of bi-material wedges

The singular stress field is generally expressed as following

$$\sigma_{ij} = r^{p-1} f_{ij}(\theta, p) \tag{122}$$

where r, and θ are polar coordinates at wedge tip, and p the order of singularity. If $0 < p < 1$, the stress field is singular. The order of singularity p

is not only dependent on the material properties (α and β or ε), but also dependent on the geometry, such as θ_I and θ_{II}, or θ_W for two different cases in Figure 19. It should be noted that equation (122) is valid only when p is a real value. In some cases, p appears to be complex value, which introduces oscillating stresses in the vicinity of wedge tip. The stress field in this case should be written in the form of $\sigma_{ij} = r^{\text{Re}p-1} f_{ij}(\theta, p, \text{Im } p \ln r)$. The term *Implnr* gives rise to the oscillating behaviour of stresses, but the stress singularity still remains as Re(p-1). In the following sections, the determination of the singularity order p is outlined for two cases respectively.

Figure 19(a) shows a general configuration for an angular corner of a bimaterial. The configuration is reduced to a standard interface crack, which refers to a crack lying along the interface in dissimilar material ($\theta_I = \pi$, $\theta_{II} = \pi$). A free edge of bimaterial corresponds to $\theta_I = \pi/2$, $\theta_{II} = \pi/2$. With $\theta_I = \pi/2$, $\theta_{II} = 3\pi/2$, the configuration corresponds to a corner (e.g., chip/underfill in Figure 1) with delamination along one interface.

The general characteristic equation for determining the order of the singularity is

$$16[\sin^2(\theta' p) - p^2 \sin^2 \theta'][\sin^2(\theta'' p)$$
$$- p^2 \sin^2 \theta'']\beta^2$$
$$+ 16 p^2 \{\sin^2 \theta'[\sin^2(\theta'' p) - p^2 \sin^2 \theta'']$$
$$+ \sin^2 \theta''[\sin^2(\theta' p) - p^2 \sin^2 \theta']\}\alpha\beta$$
$$+ \{16 p^2 (p^2 - 1) \sin^2 \theta' \sin^2 \theta''$$
$$- 4[p^2 \sin^2(\theta'' + \theta') - \sin^2 p(\theta'' + \theta')]\}\alpha^2 \quad (123)$$
$$+ 16 p^2 [\sin^2 \theta' \sin^2(\theta'' p) - \sin^2 \theta'' \sin^2(\theta' p)]\beta$$
$$+ 8\{-2 p^2 [\sin^2 \theta' \sin^2(\theta'' p) - \sin^2 \theta'' \sin^2(\theta' p)]$$
$$+ [\sin^2(\theta'' p)$$
$$- p^2 \sin^2 \theta''] - [\sin^2(\theta' p) - p^2 \sin^2 \theta']\}\alpha$$
$$- 4 p^2 \sin^2(\theta' - \theta'') + 4 \sin^2 p(\theta' - \theta'') = 0$$

The solution can be obtained by solving this equation numerically by Newton-Raphson method. In the case of a standard interface crack, i.e., $\theta_I = \pi$, $\theta_{II} = \pi$, the well-known complex stress singularity is found, see Rice or equation (119) with

$$p = 1/2 + i\varepsilon \quad (124)$$

4. Introduction to Advanced Mechanics

The imaginary part of the stress singularity at above equation introduces oscillating stresses in the vicinity of the crack tip and a possible overlapping of both materials. However, in the case of $\theta_I + \theta_{II} < 2\pi$, stress field around the tip does not exhibit an oscillating singularity. For instance, at the free edge of bimaterial, the stress field is singular, but showing no oscillating behaviour.

Figure 19(b) shows a bimaterial wedge in which two bonded interfaces are involved. The characteristic equation of the order of singularity for this situation is derived by ref. [16]

$$p^4(1-\Gamma)^4\sin^4\theta - \hat{x}(1-\Gamma)^2\sin^2\theta\{(1+\Gamma\kappa_1)^2\sin^2(p\theta) \\ +(\Gamma+\kappa_2)^2\sin^2(p(\theta-2\pi))-2\Gamma(1+\kappa_2)\sin^2(p(\theta-\pi))) \\ +\{(1+\Gamma\kappa_1)(\Gamma+\kappa_2)\sin(p\theta)\sin(p(\theta-2\pi)) \\ -\Gamma(1+\kappa_1)(1+\kappa_2)\sin^2(p(\theta-\pi))\}^2 = 0 \tag{125}$$

where

$$\kappa_1 = \begin{cases} 3-4v_1 & \text{plane strain} \\ (3-v_1)/(1+v_1) & \text{plane stress} \end{cases}$$

$$\kappa_2 = \begin{cases} 3-4v_2 & \text{plane strain} \\ (3-v_2)/(1+v_2) & \text{plane stress} \end{cases} \tag{126}$$

$$\Gamma = \mu_2/\mu_1$$

It was found that the order of stress singularity is always considerably smaller than the same wedge system with one delaminated interface (obtained from equation (124), with $\theta_I + \theta_{II} = 2\pi$). This implies that the stress is released when delamination is present at one interface for a bimaterial wedge. For most cases in Figure 19(b), the stress field does not have an oscillating singularity.

4.7.3 Delamination Criterion for Interfaces at Bimaterial Wedge

For a crack in homogeneous material, the well-known mode I, II, and III stress intensity factors (K_I, K_{II}, and K_{III}) were found to be the only parameters that characterize the magnitude of the singular stress field at the crack tip for each basic failure mode, respectively. For a standard interface crack in a dissimilar material (i.e., $\theta_I = \pi$, $\theta_{II} = \pi$ in Figure 19(a)), the total energy release rate or J-integral can be used as fracture parameters, while the separated mode fracture parameters like K_I and K_{II} do not exist except for special material combinations with $\beta = 0$. For a bimaterial wedge with two bonded interfaces, however, it is impossible to find a unique parameter (or a few) to represent the whole singular field of the wedge tip. In the following analysis, without lack of the generality, a bimaterial wedge with two bonded interfaces, which are placed horizontally and vertically, respectively, is considered (see Figure 20). The singular stress field for this situation can be represented in the form of equation (122). The singularity shows no oscillating behaviour and the order is much less than ½. With the introduction of the polar coordinate system, the hoop stress σ_θ and $\tau_{r\theta}$ at the tip along the different angular direction can be expressed as

$$\sigma_\theta = f(\theta)/r^{1-p}, \quad \tau_{r\theta} = g(\theta)/r^{1-p} \qquad (127)$$

Unlike the stress field in a crack tip of a homogeneous material, where the stress intensity factors (K_I, K_{II}, and K_{III}) are found to be only parameters governing the crack tip behaviours, equation (127) virtually gives no such kind of solution. However, the opening and shearing mode 'stress intensity factors' can be defined as function of the angular position at the corner tip by

$$K_o = f(\theta), \quad K_s = g(\theta) \qquad (128)$$

The introduction of opening and shearing stress intensity factors in Equation (128) is important, since in the mixed-mode crack propagation of homogeneous material, the cracking is present in the direction of $K_s = 0$ or maximum K_o. For a bimaterial wedge considered here in Figure 20(a), the opening and shearing stress intensity factors along the two bonded interfaces can thus be introduced as the parameters for delamination prediction as follows

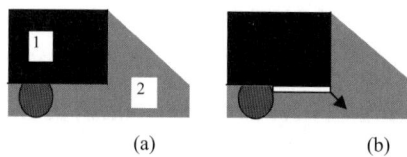

(a) (b)

Figure 20. Bi-material wedge: with and without delamination

4. Introduction to Advanced Mechanics

$$K^h = [(K_s^h)^2 + (K_o^h)^2]^{1/2}, \quad \varphi^h = \tan^{-1}(K_s^h / K_o^h) \tag{129}$$

$$K^v = [(K_s^v)^2 + (K_o^v)^2]^{1/2}, \quad \varphi^v = \tan^{-1}(K_s^v / K_o^v) \tag{130}$$

where K^h, φ^h, and K^v, φ^v are the combined stress intensity factors and phase angle, and the superscript h and v represent the horizontal and vertical interfaces, respectively. The delamination behaviour along the interface can be characterized by these two parameters completely. The delamination initiates when the stress intensity factor reaches its critical value, i.e., $K(\varphi) = K_{ad}(\varphi)$, where $K_{ad}(\varphi)$ is considered to be a material property as function of phase angle and is termed as the interface adhesion strength for this wedge configuration. Since two bonded interfaces are involved, the delamination will be along the horizontal interface when

$$\frac{K^h(\varphi^h)}{K_{ad}^h(\varphi^h)} > \frac{K^v(\varphi^v)}{K_{ad}^v(\varphi^v)} \tag{131}$$

is satisfied. Delamination will occur along the vertical interface if the inequality is reversed.

5.   FINITE ELEMENT METHOD

5.1   Introduction

It is not an easy task to determine the thermal stresses in microelectronics systems. Closed form and semi-close form solutions can be derived for very simple geometries and temperature loadings, but very limited in applications and difficult to obtain. The finite element analysis is one of the best candidates for obtaining numerical results from thermal strains and stresses in complex structures.

The finite element method is a numerical analysis technique for obtaining approximate solutions to a wide variety of engineering problems [14, 15]. The solution of a continuum problem by the finite element method always follows an orderly step-by-step process.
1. *Discretize the continuum.* The first step is to divide the continuum or solution region into elements.

2. *Select interpolation function.* The next step is to assign nodes to each element and then choose the interpolation function to represent the variation of the field variable over the element.
3. *Find the element properties.* Once the finite element model has been established (that is, once the elements and their interpolation functions have been selected), we are ready to determine the matrix equations expressing the properties of the individual elements.
4. *Assemble the element properties to obtain the system equations.* To find the properties of the overall system modelled by the network of elements we must assemble all the element properties.
5. *Impose the boundary conditions.* Before the system equations are ready for solution they must be modified to account for the boundary conditions of the problem.
6. *Solve the system equations.* The assembly process gives a set of simultaneous equations that we solve to obtain the unknown nodal values of the problem.
7. *Make additional computational if desired.* Many times we use the solution of the system equations of calculate other important parameters.

With the advances of various kinds of commercial finite element software, most of procedures described above have been automated. The challenges arise that the general-purpose finite element software is used to specific problems such as in microelectronics with advanced analysis techniques and methods.

In this section, the general treatment on various nonlinear problems in finite element analysis is introduced. This includes the treatment of material nonlinearity, geometric nonlinearity, creep algorithm, and contact non-linearity. Various methods in extracting the fracture parameters are described. The mostly widely used sub-modelling and sub-structural modelling techniques in microelectronics are introduced. The adaptive meshing technique and element death and birth are described.

5.2 Treatment of Nonlinearity in Finite Element Analysis

5.2.1 Newton-Raphson Method

Newton-Raphson method is an iterative method for finding the roots of equations. Reader may be familiar with this for the solution of the non-linear equation with one unknown, $f(x) = 0$, as:

$$x_{i+1} - x_i = \Delta x = -\frac{f(x)}{(\partial f/\partial x)_i} \tag{132}$$

4. Introduction to Advanced Mechanics

This process is illustrated schematically in the figure below (Figure 21).

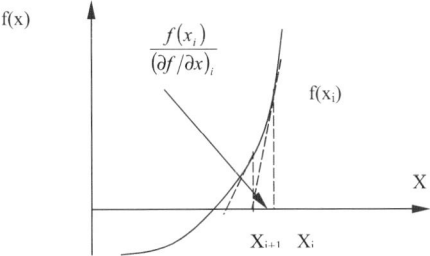

Figure 21. Newton-Raphson Method

Commonly outlined for one unknown parameter, it can be extended by analogy to the solution of nonlinear FEA equations with many unknowns as:

$$[a]_{i+1} - [a]_i = \Delta[a]_i = -\left[\frac{\partial[\psi]}{\partial[a]}\right]_i^{-1} [\psi(a_i)] \qquad (133)$$

The inverted matrix above is known as the Jacobian matrix of $[\psi]$.

5.2.2   Outline of a Non-Linear Solution

Because the stiffness and the load vectors may both be functions of the displacements in nonlinear problems, equilibrium equation can be written as:

$$[K][x] - [F(x)] = [\psi(x)] = 0 \qquad (134)$$

The product of $[K][x]$ can be thought of as internal forces $[P]$, which, in an equilibrium state, will balance the external force $[F]$. Non-linear FEA generally involves starting from a position where the internal and external loads do not balance. This imbalance in the loads is referred to as the residual $[\Psi]$, which is reduced to an acceptably small value through a series of iterations. This process is illustrated in Figure 22 on a nonlinear load displacement response.

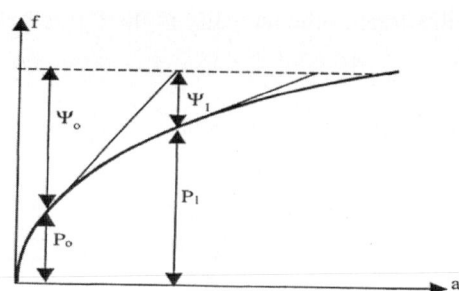

Figure 22. Illustrating an iterative reduction of residual ψ

Although there are a number of possible schemes to solve such equations, the one usually adopted in FEA is based on the Newton-Raphson method described above. Most structural, non-linear problems are solved in an incremental manner. An increment of the load is applied and then iteration is undertaken until a converged solution has been achieved. Following this, the next load increment can be applied. There usually two reasons for such an approach:
1. To apply Newton-Raphson technique, it is necessary to start from a good guess. The end of a converged increment forms a good starting guess for the next increment. This is illustrated in Figure 23 where it can be seen that if a solution lies outside the "radius of convergence" it can diverge (a) whereas it can converge if it starts within this radius (b).
2. Usually the material laws are expressed so that increments of stress are related to increments of strain ($[d\sigma] \propto [\varepsilon]$), and not total stress to total strain.

When solving in an incremental manner, [F] will often be consistent. After some manipulation, it is possible to write the Newton-Raphson equations for the nonlinear problem defined in Equation (134) as:

$$\Delta x = [K_t]^{-1}[\psi(x)] \qquad (135)$$

where the matrix $[K_t]$ is known as the tangent stiffness matrix.

4. Introduction to Advanced Mechanics

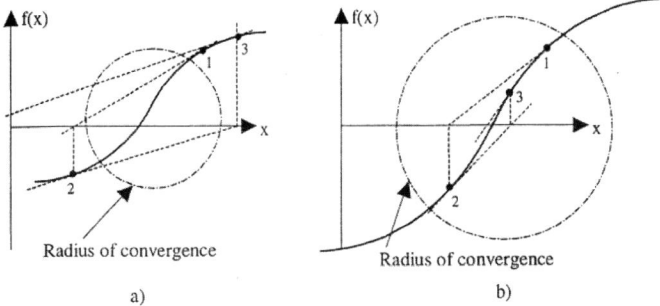

Figure 23. Illustrating the effect of the solutions starting a) outside the radius of convergence (diverging) and b) inside the radius of convergence (converging)

5.2.3 Nonlinear Matrix Equations

Nonlinear algebraic equations are usually solved by an iteration procedure. Consider a typical equation from a nonlinear set of equations of the form of equation (134). We write

$$F_i(x_1,x_2,\ldots,x_n) = \sum_{j=1}^{n} k_{ij}(x_1,x_2,\ldots,x_n)x_j - R_i(x_1,x_2,\ldots,x_n),$$
$$j = 1,2,\ldots,n$$
(136)

where $F_i(x_1,x_2,\ldots,x_n)$ is the unbalanced load in the ith equation, and n is the number of equations. If the vector of nodal variables $x_1,x_2,\ldots,x_n$ is the exact solution, then $F_i(x_1,x_2,\ldots,x_n) \equiv 0$ for $i=1,2,\ldots,n$. Usually, however, we cannot compute the exact solution, but we obtain an approximate solution such that the unbalance in a typical equation is smaller than a specified tolerance. We do this with Newton-Raphson iteration. We begin with an approximate solution vector and use a Taylor series expansion of F_i in n variables to obtain an improved solution vector. Thus

$$F_i(x_1+\Delta x, x_2+\Delta x_2,\ldots,x_n+\Delta x_n)$$
$$F_i(x_1,x_2,\cdots,x_n) + \sum_{j=1}^{n} \frac{\partial F_i}{\partial x_j}(x_1,x_2,\cdots,x_n)\Delta x_j + \cdots, \quad i=1,2,\cdots n \quad (137)$$

Now as an approximation we neglect all higher-order terms and require for a solution that the left-hand side vanish. Then we obtain a set of linear algebraic equations with unknowns Δx_j;

$$\sum_{j=1}^{n} \frac{\partial F_i}{\partial x_j}\left(x_1, x_2, \cdots; x_n\right) \Delta x_j = -F_i\left(x_1, x_2, \cdots; x_n\right) \tag{138}$$

This set of equations is the basis for the Newton-Raphson iteration algorithm, which we write in matrix notation as

$$[J]^m \{\Delta x\}^{m+1} = -\{F\}^m \tag{139}$$

$$\{x\}^{m+1} = \{x\}^m + \{\Delta x\}^{m+1} \tag{140}$$

where the superscript m denotes the mth iteration, and $[J]^m$ is the system Jacobian at the mth iteration. The elements of $[J]^m$ are

$$J_{ij}^m = \frac{\partial F_i}{\partial x_j}\left(x_1^m, x_2^m, \cdots, x_n^m\right) \tag{141}$$

We proceed by computing the Jacobian and unbalanced node load vector and solving a set of linear algebraic equations at each iteration. When used with a reasonable initial guess, the Newton-Raphson method shows an excellent convergence rate, but in some problems if the initial guess is a poor approximation, the method may be computationally expensive for large sets of equations. Large computational costs have led some analysts to use a modified Newton-Raphson method wherein the Jacobian is formed and factored only once and is held constant throughout the balance of the iteration process. More iterations are required with the modified Newton-Raphson method, but usually net computational costs are reduced.

5.2.4   FE Treatment of Geometric Nonlinearity

There are many algorithms specifically devised to deal with difficult geometrically nonlinear behaviour, such as limit points in load-displacement curve. The following is a summary of a number of these special algorithms.

5.2.4.1   Arc-Length Method

This is an iterative technique used to solve non-linear problems at or near limit points, where there is a change in sign of the slope of the load-displacement curve. The solution is advanced by a specified arc-length along the load path, rather than a load or displacement increment. Expressed

4. Introduction to Advanced Mechanics

simply, the residual to be reduced to zero over the solution increment (which contains a fraction of the total external load vector $[f_0]$ can be expressed as

$$[\psi]_{n+1} = [P(a_{n+1})] - \lambda_{n+1}[f_\circ] \tag{142}$$

with

$$[a]_{n+1} = [a]_n + [\Delta a]_n \tag{143}$$

$$\lambda_{n+1} = \lambda_n + \Delta \lambda_n \tag{144}$$

5.2.4.2 Load and Displacement Control

In the Newton-Raphson iterative procedure, the solution is usually driven by load control. This does not work when a snap-through behaviour is encountered, because the load-displacement tangent becomes horizontal and the load must subsequently remain constant or decrease in order to follow the load-displacement curve. Therefore, to follow the load path correctly in such problems, displacement control should be used where the displacement of a specified node or a set of nodes is increased by a small increment. This is similar to the procedure adopted in some laboratory tests of components in which the displacement of the specimen is controlled. Unfortunately, this approach cannot cope with snap-back behaviour, i.e., when the load-displacement tangent becomes vertical. One way of avoiding these shortcomings is for the FE code to use a mixture of load control and displacement control steps, with the program automatically switching from one to another when a limit point is encountered, or about to be encountered, in the loading path.

The user of an FE code may be able to switch from prescribed load increments to prescribed displacement increments depending on the type of problem encountered. Alternatively, the user can prescribe artificial springs at a certain node to restrain the structure at or near limit points. The FE code may avoid the solution of the equilibrium equations around the limit point.

5.2.4.3 Line Search Technique

This is a technique for accelerating incremental-iterative solution procedures. This is achieved by applying a scaling to the corrective displacements. Consider a possible scaling of the corrective displacement as

$$[\delta a^*]_i = (1 + \eta_i)[\delta a]_i \tag{145}$$

The corresponding residual can be evaluated as

$$[\psi^*]_i = [\psi(a_i + \delta a_i)] \qquad (146)$$

A scalar measure of this residual can be taken as its projection on the search direction (δa_i) as

$$G_i = [\delta a_i]^T [\psi_i^*] \qquad (147)$$

By calculation of G_i in this manner, it is possible to estimate the value of η_i that will result in a zero value. This is the value the scaling parameter that is finally used to determine the updated corrective displacement that will be used in the iteration.

5.2.5　FE Treatment of Material Nonlinearity

5.2.5.1　Outline of Finite Element Plasticity Algorithm

In elasto-plastic applications, both load increment and iterations are necessary to arrive at the final solution. Within each load increment, the plasticity relationships, which depend on the current state of loading, are satisfied and variables updated ready for the next load increment. Iterations are required within each load increment in order to ensure that the dual requirements of satisfying the constitutive plasticity equations and the overall structural equilibrium are satisfied. The iterations are terminated when the solution has converged according to a suitable convergence criterion. The number of iterations depends on the size of the load increment, and whether the stiffness matrix has been updated to reflect the current stress-strain state.

Typical FE algorithms for modelling elastic-plastic behaviour may adopt the following procedure:

1. Apply the load assuming elastic behaviour everywhere. The overall stiffness matrix is obtained, and the following equations solved to obtain the displacement vector:

$$[K_e][u] = [F] \qquad (148)$$

where $[K_e]$ is the elastic stiffness matrix, and $[F]$ is the full load vector.

2. Calculate the effective stress at all Gauss (integration) points, and check the maximum value against the yield stress, σ_{ys}. If it does not exceed σ_{ys}, then there is no need for a plasticity analysis. If it does, scale down the magnitudes of all nodal displacements such that the node or Gauss point

4. Introduction to Advanced Mechanics

with highest effective stress is just yielding. The scaling factor is used to determine the fraction of the applied loads that causes initial yielding.

3. Divide the remainder of the applied load into small increments, either as specified by the user, or according to a suitable automatic scheme.
4. Apply one load increment, and re-solve the equations to obtain the new displacements corresponding to this load vector. Use either the initial (elastic) stiffness matrix [K_e], or the tangent stiffness matrix, [K_{ep}], which is updated to contain the current state of plasticity, as follows:

$$[K_{ep}][\Delta u] = [\Delta F] \qquad (149)$$

Many FE codes use the initial stiffness matrix and then update it after a few load increments or iterations to reflect the current state of plasticity.

5. Perform iterations to ensure that the solution is acceptable, i.e. it satisfies both equilibrium conditions and the plasticity material laws. From the computed [Δu], the total and plastic strain increments can be calculated, and then the corresponding stress. An out-of-balance or residual force vector, [R], is calculated by integrating the stresses over all the elements and subtracting the internal forces from the external forces. If convergence is not achieved, i.e. [R] is not smaller that a specified tolerance, an iteration is commenced by solving a new set of equations, as follows:

$$[K_{ep}][\Delta u_{correction}] = [R] \qquad (150)$$

where [$\Delta u_{correction}$] is a correction to the displacement vector in order to balance the residual force vector. Using this correction, the displacement vector is improved and the iteration repeated to obtain a new residual force vector. The displacement corrections and the residual forces should be getting smaller with each iteration until convergence is achieved.

6. Store the computed increments of displacements, strains and stresses at each node, and update the existing values.
7. Apply the next load increment and perform iterations as necessary. Terminate the computation when the final load is reached.

5.2.5.2   Outline of Finite Element Creep Algorithms

FE formulations for creep problems based on the initial strain approach usually adopt the following procedure:

1. Apply the mechanical loads and displacements, and solve the equations to determine the nodal displacements and the elastic (or elastic-plastic) stress distribution at t = 0.

2. For a small time increment, Δt, assume that the stresses remain constant, and determine the creep strain rate and the creep strain increments (initial strain) using the multi-axial creep law for the material.
3. From the creep strain increment, determine a creep load vector and solve the equations to determine a displacement increment Δu, from which an increment in stress $\Delta \sigma$ is determined. Physically, the "creep loads" represent the internal forces necessary to resist the creep strains and have to be in equilibrium with the loads generated by the elastic (or elastic-plastic) behaviour of the structure.
4. The resulting $\Delta \sigma$ must be suitably small when compared to the existing stresses at the beginning of the time increment. If not, the time increment is reduced and the above procedure is repeated until $\Delta \sigma$ is sufficiently small.
5. Add the increments $\Delta \sigma$ and Δu to the existing stresses and displacements to obtain the starting values for the next time increment. Similarly, the creep strains at the end of the time increment are updated using the creep strain increments.
6. Apply another time increment and repeat the above steps until the final time is reached.

Solution accuracy can be assessed by limiting the changes in effective stress in each time interval to an acceptable range. If the changes in effective stress are deemed to be too large, the time increment is reduced and the calculation repeated. Automatic checking of the solution at each time step and, if necessary, reduction of the size of the time interval are usually incorporated in FE codes.

5.2.6 Finite Element Treatment of Contact Nonlinearity

The first necessary step of modelling contact is to detect penetration. An inexperienced FE user may not realize that unconnected finite elements can simply pass through one another. This, of course, is physically inadmissible and it is necessary to take steps to prevent such penetration. In commercial FE codes, contact can be considered as a) node to node b) node to surface or c) surface to surface. Where one contact surface is very stiff it can often be taken to be rigid. The profile of this rigid surface can be defined analytically, without the use of nodes and elements. Each set of nodes, or surface, can be considered as slave or master (or even both). In a contact problem the loading is applied incrementally. At each increment, iterations are performed until convergence is achieved. At each iteration, the slave is checked for contact with the master. Such contact is determined by geometric proximity. If contact occurs, equal and opposite pressures are applied to the contacting surfaces to prevent penetration. If friction is being modelled, then shearing

4. Introduction to Advanced Mechanics 157

stresses will also be generated to resist slipping. These are generally related to the contact pressure through the coefficient of friction. In addition to requiring a suitably small value of the residual, convergence can only occur when there is no charge of contact state between successive iterations.

5.2.7 Convergence

Convergence of the solution is achieved if equilibrium is satisfied, i.e., when no out-of-balance forces exist. In practice, the FE solution is acceptable if the nodal values of residual out-of-balance forces are negligible, i.e., below a certain tolerance, which is specified either by the user or automatically by the FE code. Typically, convergence is checked by checking the magnitude of the norm of the out-of-balance vector, which is defined as the square root of the sum of the square of all the nodal values.

If the residual force is not smaller than the specified tolerance, then another iteration is necessary to correct the displacement vector. The residual force can be used to obtain a correction to the displacement, which can be used to obtain a new (improved) value of the displacement vector. This correction procedure is continued until the displacement corrections and the residual force vector become negligible.

Several procedures exist for accelerating the rate of convergence. Standard incremental-iterative procedures require that the initial trial solution is sufficiently close to the exact solution, and the load-deformation slope remains of the same sign during the loading path. In situations where the slope may change sign, e.g., in a material softening or a snap-through behaviour, more sophisticated numerical algorithms should be employed.

So far it has been stated that the iterative process continues until a convergence is achieved without being very specific as to what constitutes the criteria for convergence. There are various possibilities for this, typically the maximum displacement could be tracked or some mean square displacement of the complete structure could be used. The displacements are not generally used for convergence purposes largely because these can be relatively insensitive to the effects of non-linearity.

The more usual convergence criterion is taken from the out of balance force at each stage of the iteration. This is given by

$$\text{Out of balance force} = (R - R_j) \tag{151}$$

and is a measure of how well equilibrium is satisfied. This is a good measure for the finite element method since compatibility is almost always enforced exactly within the finite element method so that errors in the equilibrium terms give a good indication of the total error in the solution. The convergence criteria require that a single number is defined to test for convergence. The usual form for this is taken the norm of the out of balance force vector in the form

$$\|(R-R_j)\| = \sum (R_j - R_{ij})^2 \qquad (152)$$

where R_j is the j'th term in the R vector. This norm is always positive but tends to zero as the solution convergences and satisfies equilibrium better. One problem with this test is the norm will never become exactly zero since the solution will never be that exact. This means that it must be compared with some other expression so that its relative size can be tested. The usual convergence criteria is then of the form

$$\|(R-R_i)\| \leq \varepsilon (\|R\| + \|R_i\|) \qquad (153)$$

where ε is some small number typical 0.01. The smaller the value for ε then the more iterations will be required and the cost of the analysis will be correspondingly greater. If ε is made too large then equilibrium will not be satisfied very well and the solution will not be accurate. There is always the possibility that since equilibrium is only being satisfied to the level defined by the tolerance ε for each iteration then the solution will drift away from correct solution as the number of load steps increase. This would not seem to be a real problem in practice.

5.3 Finite Element Implementation in Fracture Mechanics

There are several numerical methods available for extracting fracture parameters. Four widely used methods will be described in the following.

5.3.1 Modified J-integral Method

The strain energy release rate due to thermal loading G_t is equal to the modified J-integral J_t due to resulting thermal stresses. An alternative expression had been obtained by Wilson and Yu [20]. The corresponding modified J-integral J_h was obtained by Lin and Tay [21-23] as

$$J_h = J_{\Gamma,h} - \int_{\Gamma^+ + \Gamma^-} T_i \frac{\partial u_i}{\partial x_1} ds - \frac{E\beta}{1-2\nu} \int_{A_0} [\frac{1}{2} \frac{\partial}{\partial x_1}(C\varepsilon_{ii}) - \varepsilon_{ii} \frac{\partial C}{\partial x_1}] dA \qquad (154)$$

where $J_{\Gamma,h}$ is the J-line integral computed along a contour around the crack-tip, T_i the surface tractions along the crack surfaces Γ^+ and Γ^-, u_i the displacement, x_1 the Cartesian coordinate along the interface, E the elastic modulus, ν the Poisson ratio, ε the strain and A_c an arbitrary domain around the crack tip.

The magnitude K of the complex interfacial stress intensity factor $\boldsymbol{K}$ is related to the value of the strain energy release rate G and the modified J by

$$G = J = \frac{(1-\beta_D^2)}{E^*} K^2 \tag{155}$$

A difficulty arises here as it is necessary to obtain the separate components K_I and K_{II} of both $\boldsymbol{K}_t$ and $\boldsymbol{K}_h$ in order to obtain the composite $\boldsymbol{K}_{tot}$. This difficulty can be resolved by obtaining the ratio K_{II}/K_I using an extrapolation method suggested by Yuuki and Cho [24], where

$$\frac{K_{II}}{K_I} = \lim_{r \to 0} \frac{1-(\delta_2/\delta_1)H}{(\delta_2/\delta_1)+H} \tag{156}$$

where

$$H = \frac{\lambda - 2\varepsilon}{1 + 2\varepsilon\lambda} \tag{157}$$

$$\lambda = \tan(\varepsilon \ln r) \tag{158}$$

Once the ratio K_{II}/K_I and K are determined, the K_I and K_{II} components of both $\boldsymbol{K}_t$ and $\boldsymbol{K}_h$ can be calculated in order to obtain the composite hygrothermal $\boldsymbol{K}_{tot}$.

5.3.2 The Crack Flank Displacement Extrapolation Method

The Crack Flank Displacement Extrapolation Method (CFDEM) is a variation of the extrapolation method proposed by Yuuki and Cho [24]. Instead of extrapolating for the ratio K_{II}/K_I the values of K_I and K_{II} themselves can be extrapolated from

$$\begin{aligned} K_I &= P[\delta_1(\lambda - 2\varepsilon) + \delta_2(1 + 2\varepsilon\lambda)] \\ K_{II} &= P[\delta_1(1 - 2\varepsilon\lambda) + \delta_2(\lambda + 2\varepsilon)] \end{aligned} \tag{159}$$

$$P = E^* \sqrt{\pi/32r(1-\beta_D^2)(1+\lambda^2)} \tag{160}$$

The purpose of extrapolation is to avoid the oscillatory behaviour close to the crack tip. Values of K_I and K_{II} should be calculated for various distances r from the crack tip and straight-line extrapolations performed to obtain the values of K_I and K_{II} at the crack tip.

5.3.3 The Virtual Crack Closure Method

The strain energy release rate G is a measure of the energy available for an increment of crack extension. Irwin's virtual crack-closure method (VCCM) is based on the principle that the work necessary to extend the crack from a to $a + \Delta$ is the same as that necessary to close the crack tip from $a + \Delta$ to a. This procedure will provide a set of formulae for G that depend on the crack opening displacement and the nodal forces at and ahead of the crack tip. Rybicki and Kanninen [25] gave the expression of G that depends upon the crack opening displacements and the nodal forces ahead of the crack tip. Raju [26] continued to develop the procedure for calculating G for higher order elements and singular elements. With reference to Figure 24, he showed that the mode I and mode II components of G are given by

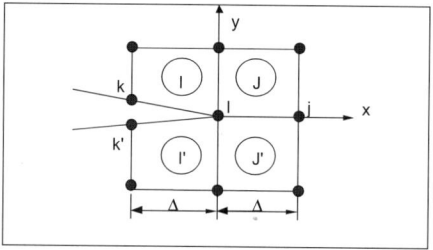

Figure 24. Mesh of 4–noded elements around crack tip

$$G_I = \lim_{\Delta \to 0}\left\{-\frac{1}{2\Delta}\left[F_{y_i}(v_k - v_{k'})\right]\right\} \qquad (161)$$

$$G_{II} = \lim_{\Delta \to 0}\left\{-\frac{1}{2\Delta}\left[F_{x_i}(u_k - u_{k'})\right]\right\} \qquad (162)$$

where F_{xi} and F_{yi} are the x- and y-components of the nodal force at node i, u_k and v_k are the displacements at node k in the x- and y- directions, respectively. K_I and K_{II} can then be obtained from Equations (161) and (162).

4. Introduction to Advanced Mechanics

For the quarter-point (singular) element, the mode I strain energy release rate can be expressed as (refer to Figure 25)

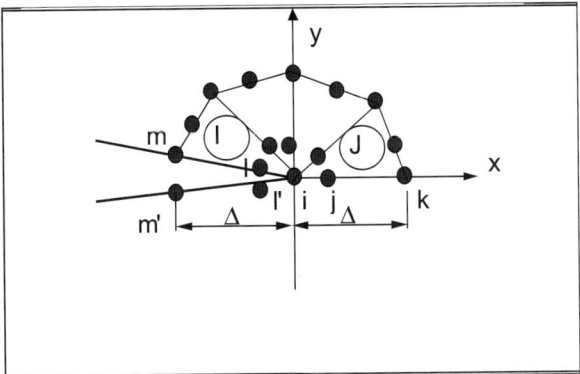

Figure 25. Mesh of quarter-point elements around crack tip

$$G_I = \frac{-1}{2\Delta} \begin{bmatrix} F_{y_i}\{t_{11}(v_m - v_{m'}) + t_{12}(v_l - v_{l'})\} + \\ F_{y_j}\{t_{21}(v_m - v_{m'}) + t_{22}(v_l - v_{l'})\} + \\ F_{y_k}\{t_{31}(v_m - v_{m'}) + t_{32}(v_l - v_{l'})\} \end{bmatrix} \quad (163)$$

where

$$t_{11} = 14 - \frac{33\pi}{8}; \quad t_{12} = -52 + \frac{33\pi}{2},$$

$$t_{21} = -\frac{7}{2} + \frac{21\pi}{16}; \quad t_{22} = 17 - \frac{21\pi}{4}, \quad (164)$$

$$t_{31} = 8 - \frac{21\pi}{8}; \quad t_{32} = -32 + \frac{21\pi}{2}$$

A similar equation was obtained for G_{II}, where F_y is replaced with F_x and v is replaced by u. As is evident from the relative simplicity of the above equations, the virtual crack closure technique developed by Raju [26] is an easy way to obtain the magnitude of the strain energy release rate and the mode mixity. However, the mesh around the crack tip has to be sufficiently fine in order to get good accuracy. The use of singular elements at the crack tip can improve the accuracy by about 10%. One basic assumption used in the above analysis is that the crack is an open crack. If the crack is closed, then contact surface elements must be defined for the crack surfaces and G_I will be zero.

5.4 Advanced Techniques in Finite Element Analysis

5.4.1 Sub-Modelling

Sub-modelling or global-local-modelling is a finite element technique used to get more accurate results in a region of interest. Often in finite element analysis, the finite element mesh may be too coarse to produce satisfactory results in a region of interest, such as a stress concentration region in a stress analysis. The results away from this region, however, may be adequate.

To obtain more accurate results in such a region, there are two options: (a) reanalyse the entire model with greater mesh refinement, or (b) generate an independent, more finely meshed model of only the region of interest and analyse it. Obviously, option (a) can be time-consuming and costly (depending on the size of the overall model). Option (b) is the sub-modelling technique.

Sub-modelling is also known as the cut-boundary displacement method or the specified boundary displacement method. The cut boundary is the boundary of the sub-model, which represents a cut through the coarse model. Displacements calculated on the cut boundary of the coarse model are specified as boundary conditions for the sub-model.

Sub-modelling is based on St. Venant's principle, which states that if an actual distribution of forces is replaced by a statically equivalent system, the distribution of stress and strain is altered only near the regions of load application. This implies that stress concentration effects are localized around the concentration; therefore, if the boundaries of the sub-model are far enough away from the stress concentration, reasonably accurate results can be calculated in the sub-model.

Aside from the obvious benefit of giving more accurate results in a region of the model, the sub-modelling technique has other advantages:
- It reduces, or even eliminates, the need for complicated transition regions in solid finite element models.
- It makes easier to experiment with different designs for the region of interest (different fillet radii, for example).
- It helps in determining the adequacy of mesh refinements.

Some restrictions for the use of sub-modelling are:
- It is valid only for solid elements and shell elements.

4. Introduction to Advanced Mechanics

- The principle behind sub-modelling assumes that the cut boundaries are far enough away from the stress concentration region. It is necessary to verify that this assumption is adequately satisfied.

5.4.2   Sub-Structure Modelling

Sub-structuring is a procedure that condenses a group of finite elements into one element represented as a matrix. This single matrix element is called a *superelement*.

One of the reasons for substructuring is to reduce computer time. Examples of this reason are nonlinear analyses and analyses of structures that contain repeated geometrical patterns. In a nonlinear analysis, one can substructure the linear portion of the model so that the element matrices for that portion need not be recalculated every iteration. In a structure with repeated patterns (such as the four legs of a table), one can generate one superelement to represent the pattern and simply make copies of it at different locations, thereby saving a significant amount of computer time. Another reason is to allow solution of very large problems with limited computer resources. An example of this reason is an analysis that is too large for the computer in terms of wave front size or disk space requirements. In such a situation, one can analyse the model in pieces, where each piece is a superelement small enough to fit on the computer.

The advantages of substructuring are:
- Separates linear and nonlinear parts of the model.
- Allows repetition of symmetrical or identical parts of the model for linear elastic analysis.
- Separates large models into multiple, moderate-size models.
- Separates fixed model parts from parts of the model that may undergo design changes.

The disadvantages of substructuring are the large amount of data that must be stored on the database, and the sub-structure must be linear.

In finite element implementation, the superelement generation step is done for every superelement at a certain level. The use of superelements in subsequent finite element code runs is done at the highest level. The recovery of solutions within a certain superelement can or cannot be done for every superelement.

5.4.3   Adaptive Mesh Generation

The adaptive mesh generation capability increases the number of elements and nodes to improve the accuracy of the solution. The capability is applicable for both linear elastic analysis and for nonlinear analysis. The capability can be used for lower-order elements, 3-node triangles, 4-node quadrilaterals, 4-node tetrahedrals, and 8-node hexahedral elements. When used in conjunction with the elastic parameter for linear analysis, the mesh is adapted and the analysis repeated until the error criteria is satisfied. When used in a nonlinear analysis, an increment is performed. If necessary, this increment is followed by a mesh adjustment, which is followed by the analysis of the next increment in time. While this can result in some error, as long as the mesh is not overly coarse, it should be adequate.

The adaptive meshing procedure works by dividing an element and internally tying nodes to insure compatibility. Figure 26 shows the process for a single quadrilateral element.

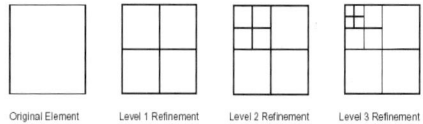

Figure 26. Single Quadrilateral Element Process

A similar process occurs for the triangles, tetrahedrons, and hexahedrons elements. One can observe that for quadrilaterals the number of elements expands by four with each subdivision; similarly, the number of elements increases by eight for hexahedrals. If full refinement occurs, you observe that the number of elements is 2(level x 2) for quadrilaterals and 2(level x 3) for hexahedrons elements.

5.4.4 Element Birth and Death

If material is added to (or removed from) a system, certain elements in the model may become existent (or nonexistent). The element birth and death options can be used to deactivate or reactivate selected elements in such cases. This feature can be useful in analysing excavation (as in mining and tunnelling), staged construction (as in shored bridge erection), sequential assembly (as in fabrication of layered computer chips), and many other applications in which one can easily identify activated or deactivated elements by their known locations.

To achieve the element death effect, the finite element program does not actually remove killed elements. Instead, it deactivates them by multiplying their stiffness (or conductivity, or other analogous quantity) by a severe reduction factor. This factor is set to 1.0E-6 by default, but can be given other values. Element loads associated with deactivated elements are zeroed out of the load vector, however, they still appear in element-load lists.

4. Introduction to Advanced Mechanics

Similarly, mass, damping, specific heat, and other such effects are set to zero for deactivated elements. The mass and energy of deactivated elements are not included in the summations over the model. An element's strain is also set to zero as soon as that element is killed.

In like manner, when elements are born, they are not actually added to the model; they are simply reactivated. To add an element, the element group is deactivated first, and then reactivated at the proper load step. When an element is reactivated, its stiffness, mass, element loads, etc. return to their full original values. Elements are reactivated having no record of strain history (or heat storage, etc.). However, initial strain defined as a real constant will not be affected by birth and death operations. Also, unless large-deformation effects are turned on, some element types will be reactivated in their originally specified geometric configuration (large-deformation effects should be included to obtain meaningful results). Thermal strains are computed for newly activated elements based on the current load step temperature and the reference temperature. Thus, newborn elements with thermal loads may not be stress-free as intended.

6. REFERENCES

[1] J. H. Lau, Thermal Stress and Strain in Microelectronics Packaging, Van Nostrand Reinhold, New York, 1993.
[2] A. P. Boresi and R. J. Schmidt, Advanced Mechanics of materials, 6th ed. Wiley, New York, 2002.
[3] Y. C. Fung, Foundation of Solid Mechanics, Prentice-Hall, New Jersey, 1965.
[4] S. P. Timoshenko and J. N. Goodier, Theory of Elasticity, McGraw-Hill, New York, 1970.
[5] L. E. Malven, Introduction to the Mechanics of a Continuous Medium, Prentice Hall, Englewood.
[6] W. Nowacke, Thermoelasticity, Noordhoff International Publishing, Leyden, 1975.
[7] A. E. Green, and W. Zerna, Theoretical Elasticity, 2nd ed, Oxford University Press, London, 1968.
[8] R. Hill, Mathematical Theory of Plasticity, Pergamon Press, Oxford, 1982.
[9] A. Khan and S. Huang, Continuum Theory of Plasticity, Wiley, New York, 1995.
[10] F. Carofalo, Fundamentals of Creep and Creep-Rupture in Metal, The Macmillan Company, New York, 1965.
[11] J. Gittus, Viscoelasticity and Creep Fracture in Solids, John Wiley-Halsted Press, New York, 1975.
[12] J. E. Shigley, C. R. Mischke, and R. G. Budynas, Mechanical Engineering Design, 7th ed, McGraw-Hill, New York, 2004.
[13] R. L. Norton, Machine Design, 2nd ed, Prentice-Hall, New Jersey, 1998.
[14] O. C. Zienkiewicz, The Finite Element Method in Engineering Science, McGraw-Hill, London, 1971.
[15] T. Hughes, The Finite Element Method, Prentice-Hall, New Jersey, 1987.

[16] X. J. Fan, H. B. Wang and T. B. Lim, Investigation of the underfill delamination and cracking in flip-chip modules under temperature cyclic loading, IEEE Transactions and Packaging Technologies, 24(1), pp. 84-91, 2001.
[17] J. W. Hutchinson and Z. Suo, Mixed Mode Cracking in Layered Materials, Advances in Applied Mechanics, pp. 63-191, 1992.
[18] J. R. Rice, Elastic fracture concepts for interfacial cracks, J. Appl. Mech. 55, pp. 98-103, 1998.
[19] T. Suga, E. Elssner, and S. Schmauder, Journal Composite Materials, 22, pp. 917, 1988.
[20] W. K. Wilson and I. W. Yu, The use of the J-Integral in thermal stress crack problems, International Journal of Fracture, Vol. 15. No. 4, 1979.
[21] A. A. O. Tay, Modelling of interfacial delamination in plastic IC packages under hygrothermal loading, Proceeding of 3rd International Conference on Benefiting from Thermal and Mechanical Simulation in Microelectronics, ESIME 2002, pp. 195-206, 2002.
[22] T. Y. Lin and A. A. O. Tay, A J-integral criterion for delamination of bi-material Interfaces incorporating hygrothermal stresses, ASME EEP-Vol. 19-1, Advances in Electronic Packaging, pp. 1421-1428, 1997.
[23] T. Y. Lin and A. A. O. Tay, Dynamics of moisture diffusion, hygrothermal stresses and delamination in plastic IC packages, ASME EEP-Vol. 19-1, Advances in Electronic Packaging, pp. 1429-1436, 1997.
[24] R. Yuuki and S. B Cho, Efficient boundary element analysis of stress intensity factors for interface cracks in dissimilar materials, Engineering Fracture Mechanics, Vol. 34, pp. 179-188, 1989.
[25] E. F Rybicki and M. F, Kanninen, A finite element calculation of stress intensity factors by a modified crack closure integral, Engineering Fracture Mechanics, Vol. 9, pp. 931-938, 1977.
[26] I. S. Raju, Calculation of strain-energy release rates with higher order and singular finite elements, Engineering Fracture Mechanics Vol. A8, No. 3, 1987, pp. 251-274.

7. EXERCISES

1. How many components in a stress tensor? How many independent components in the stress tensor?
2. The stress field of a body is given as

$$\sigma_x = ax^4, \quad \sigma_y = 6ax^2y^2, \quad \sigma_z = cxyz^2,$$
$$\tau_{xy} = bx^3y, \quad \tau_{yz} = cx^2yz, \quad \tau_{zx} = cx^3y,$$

where a, b, c are constants (whose units are Pa/m^6). In addition, the body force is known to be zero throughout the body. Under what conditions do these stresses represent a state of equilibrium at all points of the body?
3. A body is deformed such that the displacements at any points are given by

4. Introduction to Advanced Mechanics

$$u = \frac{1}{B}\left[x^3 y \mathbf{i} + 3y^4 \mathbf{j} + (z-4)^2 xy \mathbf{k}\right], \quad 0 \le x, \; y \le 1$$

where $B \gg 1$, and is a constant. Determine the strain components at all points of the body.

4. A thin sheet made of an aluminium alloy having $E = 67$ GPa, $G = 256$ GPa and $v = 1/3$ was used for two dimensional surface strain measurements. The measurements provided the strains as $\varepsilon_{xx} = 10.5e-5$, $\varepsilon_{yy} = -20e-5$, and $\gamma_{xy} = 240e-5$. Determine the corresponding stress in Cartesian coordinates.
5. What are the principal stresses? Let the state of stress at a point be given by $\sigma_x = 20$ MPa, $\sigma_y = 40$ MPa, $\sigma_z = -20$ MPa, $\tau_{xy} = -50$ MPa, $\tau_{yz} = 20$ MPa, $\tau_{zx} = -60$ MPa. Determine the three principal stresses and the direction associated with the three principal stresses.
6. For the state of stress in Problem 5, determine the three maximum shear stresses and the direction associated with the three maximum shear stresses.
7. Does thermal expansion contributes to shear strain components?
8. Explain how thermal stresses occur when applied to internal constraints and external constraints.
9. The stress state at a point in a packaging component is $\sigma_x = 200$ MPa, $\sigma_y = 100$ MPa, $\sigma_z = -50$ MPa, $\tau_{xy} = -50$ MPa, $\tau_{yz} = \tau_{zx} = 0$. Use the von Mises yield condition to calculate if the component at that point exhibits yield.
10. What is the difference between the failure of ductile materials and brittle materials under static loading? What kinds of failure criteria applied for the ductile material under static loading?
11. A hot-rolled steel has a yield strength of $S_y = 100$ kpsi and a true strain at fracture is $\varepsilon_f = 0.55$. Estimate the factor of safety for the following principal stress states: (a) 0, –30, –70 kpsi; (b) 30, 70, 0 kpsi. Using both distortion theory and maximum shear stress theory to calculate and compare the results.
12. How to define the high-cycle fatigue regime? Which fatigue failure model is most often used for this fatigue regime?
13. What is the stress intensity factor and how to calculate it?
14. A steel bar is subjected to a uniaxial tension of $\sigma_y = 140$ MPa. It is 8 mm thick and 50 mm wide with an initial single-edge crack of 10 mm long. (a) Determine the stress intensity factor K_I. Is the crack stable? (b) Determine the critical crack size, and (c) Determine the critical load. Assume the fracture toughness of the material is $K_{Ic} = 80 MPa\sqrt{m}$.
15. In what circumstances do we need to use J-integral in the fracture mechanics?

Chapter 5

THERMO-MECHANICS OF INTEGRATED CIRCUITS AND PACKAGES

W.D.van Driel[1,2], G.Q. Zhang[2,3], and X.J. Fan[4]
[1]*Philips Semiconductors, P.O.Box 30008, 6503HK Nijmegen, The Netherlands*
[2]*Delft University of Technology, Mekelweg 2, 2628CD Delft, The Netherlands*
[3]*Philips Semiconductors, HTC 60, 5656AG Eindhoven, The Netherlands*
[4]*Intel Corporation, CH5-263, 5000 W Chandler Blvd, Chandler, AZ 85226 USA*

Abstract: This chapter describes the state-of-the-art thermo-mechanical based prediction methodologies and techniques for integrated circuits (ICs) and packages. The problems of thermo-mechanics and their solutions are presented in three levels, i.e., IC backend process, packaging process, and the interaction between packaging and backend processes. For each part, both simulations and experimental techniques are described. The chapter starts with a general introduction of IC backend and packaging processes and testing conditions. At the end of the chapter, two case studies are included, which represent the strength of current thermo-mechanical prediction techniques.

Key words: Design for reliability, material characterization, finite element modelling, stress, strain, warpage, intrinsic stress, backend, thermo-mechanics, fracture mechanics, visco-elastic, multi-scale modelling, global-local modelling, cohesive zones, fracture toughness, wire failure, pattern shift, Cu/low-k, delamination, virtual prototyping.

1. INTRODUCTION

At present, thermo-mechanical reliability of Integrated Circuits (IC's) and packages is one of major concerns in microelectronic industry. Based on root cause analyses from observed failures of microelectronic components, it is found that these reliability problems are often triggered by various thermal and mechanical loadings associated with manufacturing processes. For example, during the backend process (metal depositions, etching, and chemical vapour deposition) stresses are developed in IC layer stacks. The packaging manufacturing processes (such as die-attach and moulding) further induce additional stresses in IC stacks. These deformations and stresses can become critical for product assembly, and may eventually endanger the targeted lifetime of the product during operations. It has shown that the approximately 65% of all failures can be attributed to thermo-mechanical related effects during manufacturing processes. In most cases, these thermo-mechanical reliability problems originate from the design phase of product and process. Examples of failures are cracks, voids, delamination, buckling, hillocks, wire fatigue, and many more. Thermo-mechanical reliability is becoming one of the major bottlenecks for both current and future microelectronics technologies.

The major trends in microelectronics industry are miniaturization and function integration, which not only increase the density of interconnects, but also increase power dissipation density, thus temperature. The smallest feature sizes on today's IC's are already falling to 90 nm and beyond. The International Technology Roadmap for Semiconductors (ITRS) projects that in 15 years the smallest feature size will be even smaller than 10 nm. Comparing it with the well-known quote from the Scientific American in 1977, stating that *'present technology can routinely reproduce elements a few micrometers across, and it appears possible to reduce the smallest features to about one micrometer'*, one instantly grasps the enormous development over years.

Material characterization at elevated temperatures at micro-level or nano-level becomes one of challenges to understand the thin film material behaviours during IC backend and packaging process. Many new techniques have been developed such as nano-indentation method, digital image correlation (DIC) technique, and laser interferometer method. Using multi-level and multi-scale approaches and multi-physics principles with the advancement of finite element (FE) techniques are widely applied to predict the deformations and stresses and their evolution during IC processes, packaging manufacturing processes and reliability testing [1]. This chapter presents the state-of-the-art methodologies and techniques to predict thermo-mechanically induced failures in microelectronics. The combined simulation methods and experimental techniques are introduced in three levels:
- IC backend processes,
- Packaging processes and testing,
- Coupled analysis between IC and package processes.

2. MANUFACTURING PROCESSES AND TESTING METHODS

2.1 IC Backend Processes

The semiconductor industry has seen the continuous development of new and improved processes leading to highly integrated and reliable circuits. These improvements led to the complementary metal oxide semiconductor (CMOS) process, but CMOS is not the only technology available. In this section a short introduction will be given on how these CMOS devices are manufactured, for more detailed information readers are referred to reference [2].

An IC is a layered stack of substrate and thin films with thicknesses ranging from approximately 100 nm to 1μm. For a typical CMOS process these films include:
- Semiconductors (as active part),
- Metal interconnects,
- Via plugs (as carrier for current),
- Dielectrics (for electrical isolation),
- Passivation layers (for mechanical protection).

The relatively thick single crystal silicon substrate serves as ground material and as a mechanical carrier during processing. Total IC processing can be divided into 2 serial processes:
1. *Frontend process*: semiconductors are constructed on the silicon substrate. This is done through so-called doping processes leading to positive (p-type) and negative (n-type) regions in the silicon substrate.
2. *Backend process*: metal interconnect lines, dielectrics, via plugs and passivation layers are deposited on the frontend wafer. This multi-layered stack of ductile and brittle thin films are deposited by cycles of:
   - Deposition of the thin film material (and further treatment if necessary). The growth of the number of thin film materials has resulted in a number of deposition techniques. Still, the majority of the films are deposited by a Chemical Vapour Deposition (CVD) technique. Chemicals containing the atoms or molecules required in the final film are mixed and reacted in a deposition chamber to form a vapour. The atoms or molecules deposit on the wafer surface and build up or form a film. During the process the deposited film grows until the required thickness is obtained. Typical process temperature is 400°C.
   - Lithographically patterning. Photolithography is one of the most critical operations in semiconductor processing. It is the process that

sets the surface dimensions on the various parts of the devices and circuits. The required pattern is formed by using reticles or photomasks and takes place in two steps. First, the pattern on the reticle or mask is transferred into a layer of photoresist. Exposure to light will change the photoresist from a soluble condition to an insoluble one, which enables the formation of the pattern. Typical process temperature is 450°C.

- Removal of the photoresist. The second step in the patterning is the removal of the soluble photoresist material. The chemistry of photoresists is such that they do not dissolve in the chemical etching solutions.
- Annealing if necessary. Disruption of crystal structures and/or metal alloying is obtained by a heat treatment between 450-1000°C called annealing.

Figure 1 shows a schematic representation of the step-wise photomasking process. Remind that in this case seven steps are indicated to create one metal layer, where current advanced CMOS process may include up to six metal layers.

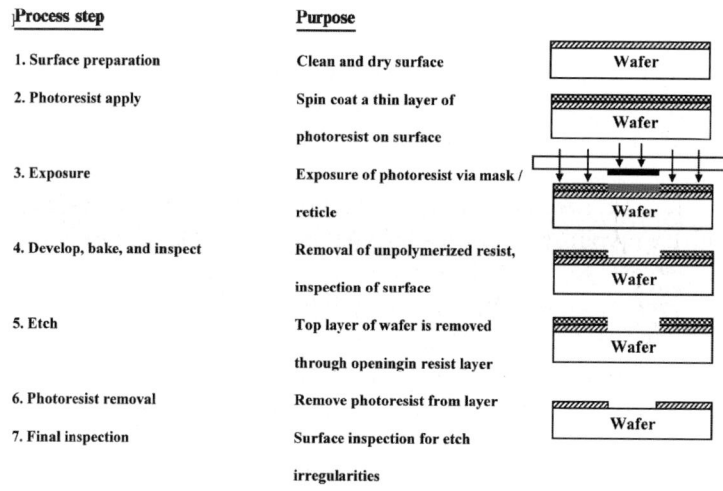

Figure 1. Schematic representation of the step-wise photomasking process

Backend processes are carried out in a waferfab, where continuous monitoring is carried out. Monitoring regularly takes place after each process step, regarding film thickness, warpage of the wafer, and interference of moisture and dust. The measured bending of the wafer (warpage) derives a biaxial stress state in a thin film bonded to a thick

5. Thermo-Mechanics of Integrated Circuits and Packages

substrate. During and after the backend process, significant stress levels are observed in IC's. The total stress level in a particular film is considered as the result of two stresses; intrinsic stresses and thermo-mechanical stresses. Intrinsic stresses are induced at the deposition process due to the non-equilibrium microstructure of the film. Subsequently, during cool down from the deposition temperature, thermo-mechanical stresses occur due to thermal mismatch between the different materials. Failures observed during and after backend processes can be attributed to these stress levels, they will be dealt with in chapter 3. Typical backend related materials are listed in Table 1. Figure 2 shows a cross-section of a typical IC.

Table 1. Typical backend materials

Application	Material
Substrate	Silicon (Si)
Metal interconnects	Stack of (Ti), TiN, Al(Cu), TiN, Cupper (Cu)
Via plugs	Tungsten (W)
Dielectric	Silicon oxide (SiO_2)
	TetraEthylOxiSilane (TEOS)
	Polymers (FPI, FPAE)
	BoroPhosphosilicate glass (BPSG)
	Black Diamond (BDI, BDII)
Planarisation	Spin on glass (SOG)
	Silicon Nitride (Si_3N_4)
	PhosphoSilicate Glass (PSG)
	Benzocyclobutene (BCB)

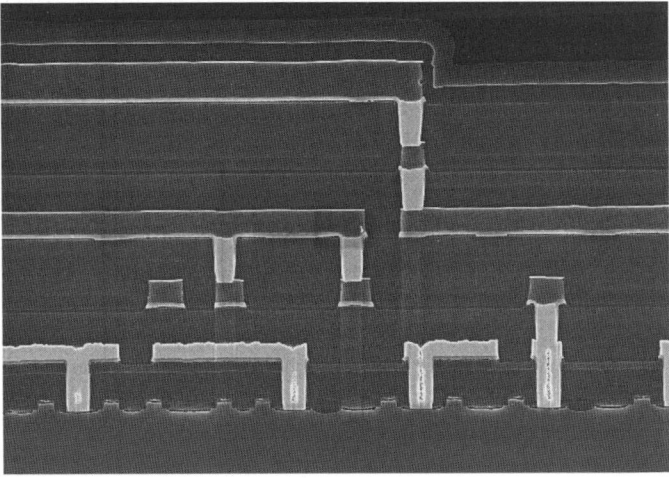

Figure 2. Cross-section of a typical IC

2.2 Packaging Processes

After IC manufacturing in the waferfab, packaging is the next step in the assembly process. The major functions of a package are:
- Allow an IC to be handled for PC Board assembly,
- Mechanical and chemical protection against the environment,
- Enhance thermal and electrical properties,
- Allow standardization (footprints).

Like IC's, many different packages exist with different classifications. International organisations like JEDEC (Joint Electron Device Engineering Council), IPC (Institute for Interconnection and Packaging Electronic Circuits), IEC (International Electro technical Commission) allow for standardization of the different packages developed by the assembly companies. Most used classification is the distinction between (see Figure 3):
- Through Hole Mount IC Packages;
- Surface Mount IC Packages;
- Contactless Mount IC Packages.

Examples of packages are ceramic packages, Quad Flat Packages (QFP), Ball Grid Arrays (BGA), Flip Chip packages (FC), Dual In Line packages (DIL), etc. In this section a short introduction will be given on how packages are manufactured, for more detailed information readers are referred to references [3-5].

Packages are manufactured through a series of processes, widely using polymers in various forms such as in encapsulants, adhesives, underfills, moulding compounds, insulators, dielectrics, and coatings. For a surface mounted package the manufacturing process involves:

1. *Grinding, Etching, and Sawing of the wafer.* IC's are cut out of the wafer and can be used for further 'single' processing. This is done at room temperature using a circular sawing or a laser cutting process.
2. *Die-attach.* The single IC is attached on a carrier material, here a metal, by using some kind of glue. Typical process temperature is 150-175°C and strongly depends on the die-attach type. Many different die-attach types exist; all are related either to their function or either to specific processing demands. For example, die-attach may be thermally and/or electrically conductive (by adding silver flakes), non-bleeding, snap-cured or oven-cured, etc.
3. *Wire bond.* Connections are made between the IC and the carrier by using thin wires. Wire bonding is still the major form of first level interconnection in the world today. At present, over 95% of the manufactured packages (in volume) are wire bonded. Wire bonding technology is changing due to the increasing demands placed on wire

5. Thermo-Mechanics of Integrated Circuits and Packages

bond pad pitch. The techniques that are being used today are wedge bonding and ball bonding. Ultrasonic and thermosonic technologies are used to create contact between the ball and the bondpad and between the stitch and the lead. At present, the majority of wire bonding is done with thermosonic gold ball bonding. The mostly used wire material is gold (Au) but is being replaced (because of material costs) by copper (Cu). Typical process temperature is 200-220°C.

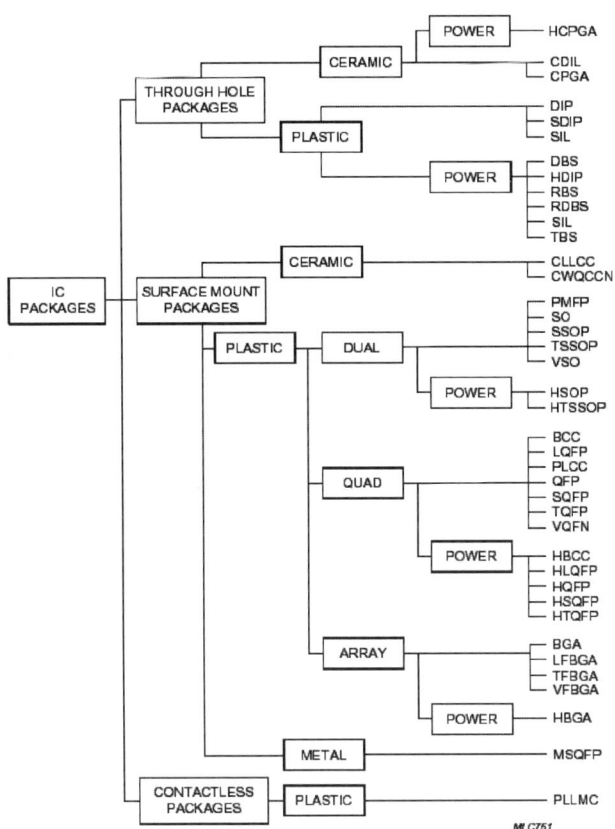

Figure 3. Package classification

4. *Chipcoat.* To protect the IC top surface a chipcoat material can be used. Chipcoat is a highly viscous liquid or paste that encapsulates the IC - mostly epoxies or silicones, with some inorganic fillers. Typical process temperature is 150°C.

5. *Mould.* IC, wires and carrier are encapsulated by an epoxy. Moulding compounds are very complex mixtures of epoxy resin(s), hardener(s), mixture of accelerator(s), (very) high filler loadings, adhesion promoters, release agents, flow additives, carbon black, ion trapping agents, stress absorbers, flame retarders, etc. The chemistry of widely used moulding compounds can be described by a combination of different building blocks with epoxy and hydroxyl reactive groups. Phenol novolac and cresol novolac based resins and hardeners are common, but also newcomers such as biphenyl-, multi-aromatic, and DCPD based precursors and mixtures thereof are being used regarding environmentally green and/or very good MSL-performing materials when it comes to 260°C reflow soldering conditions. The moulding compound is injected under high pressure and a process temperature of about 175°C, followed by a 3 to 4 hours curing step at 175°C.
6. *Solder plate.* The leads are treated as to obtain a better contact with the Printed Circuit Board (PCB) in later manufacturing stages. Plating is done at room temperature in a plating bath.
7. *Mark, Trim & Form.* Package is marked and by a trim & form process redundant material is removed.
8. *Final test.* Package is electronically tested whether it fulfils the aimed function.
9. *Pack.* A number of packages are packed in either tubes, or reals. This is done for reasons of easy transport to the (end-) customer.

Figure 4 shows the flowchart of the assembly process for a typical leadframe based package. Typical packaging related materials are listed in Table 2. Figure 5 shows pictures of a leadframe, a QFP cross-section, and a 3D view of a package.

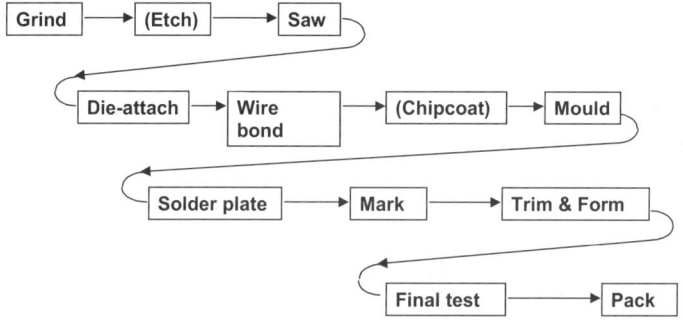

Figure 4. Assembly process flowchart for a leadframe based package

5. Thermo-Mechanics of Integrated Circuits and Packages

Table 2. Typical packaging materials

Application	Material
Leadframe and Heatsink	Copper (Cu)
	Copper-alloys ($CuNi_3$)
	Iron ($FeNi_3$)
Substrate (BGA-like)	BT (bismaleimide triazine) based
	Flame Retardant Type 4 (woven glass reinforced epoxy resin)
	Tape, teflon or polyimide based (PTFE, PI)
	Ceramic
Die-attach	Conductive and non-conductive adhesive
	Metal-filled epoxies (thermoset) or polyimide siloxanes
	Underfill (silica-filled epoxy resins)
Wire bond	Gold (Au), Copper (Cu), Aluminium (Al)
Compound	Granulated and powdered resin, with hardener, accelerator, fillers, flame retardents and other modifiers
Solder	Lead containing (Pb-5Sn, Pb-10Sn, etc)
	Lead free (Sn69.5/Ag3.5, Sn96.3/Ag3.2/Cu0.5, etc)

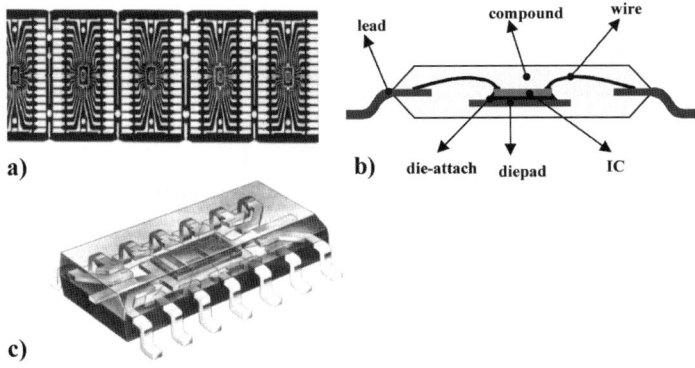

Figure 5. Picture of a) leadframe, b) QFP cross-section, and c) 3D view of a package

2.3 Reliability Testing for IC Packages

Reliability is defined as the probability that a product in operation will survive under certain conditions during a certain period of time, see Chapter 2. Semiconductor products are qualified using specially designed tests to ensure that they have sufficient life so that failures do not occur during the normal usage period. These test are called reliability tests and the specific purpose includes:
- Determination of the failure distributions,
- Evaluation of new designs, components, processes and materials,

- Discovering problems with safety,
- Collecting reliability data,
- Reliability control.

Reliability tests are classified in the so-called Qualification System under various names according to the test format, purpose, method of applying stress, and other factors. In reliability tests, environmental conditions (temperature, moisture) are extrapolated such as to accelerate the circumstances under which the product could fail. Currently classified reliability tests that can be related to thermo-mechanical conditions as are:

1. *High Temperature Operation Life.* The IC package is subjected to a temperature of 150°C and will switch on and off continuously. In such a HTOL test, the operational life is accelerated and the product is qualified if it can sustain 1000 to 2000 hours under these conditions.
2. *Temperature cycling.* During temperature cycling, IC packages are subjected to a temperature change from –65°C to 150°C for a number of cycles. Typical numbers of cycles are 200 to 500, and depend on the application. For instance, the demand for IC packages aimed for an automotive application (under the hood of a car the temperature change is large) is higher than those aimed for an end-customer application (for instance a mobile phone).
3. *Moisture sensitivity.* Moisture is believed to be hazardous for interface delamination in an IC package. Given the different humidity conditions throughout the world, IC packages will be subjected to moisture during assembly and transport. As such, moisture sensitivity levels (MSL) are introduced and for each IC package this level should be deducted. MSL levels are 1 to 6, with 1 being the most severe and 6 being the less severe. MSL levels are obtained via a MSL assessment, which can be coupled with the actual floor life before further assembly on the printed circuit board (PCB). Examples are:
    - MSL1: unlimited floor life under a 85% humidity and a temperature lower then 30°C. In a MSL assessment, the IC package should withstand experimental conditions of 85%RH/85°C for a period of 168 hours.
    - MSL3: limited floor life of 168 hours under a 60% humidity and a temperature lower then 30°C. In a MSL assessment, the IC package should withstand experimental conditions of 60%RH/30°C for a period of 168 hours.
    - MSL6: limited floor life of 6 hours under a 60% humidity and a temperature lower then 30°C. In a MSL assessment, the IC package should withstand experimental conditions of 60%RH/30°C for a period of 6 hours.

5. Thermo-Mechanics of Integrated Circuits and Packages

These examples clearly show the cost impact of moisture conditions on the handling of the IC package, in some severe cases all packages need to be dry-baked before further assembly can take place. The impact of moisture on the reliability of packages will be discussed in Chapter 6.

4. *Combined moisture and temperature cycling.* Before temperature cycling, the IC package is subjected to its subsequent moisture sensitivity level to explore the reliability hazards under this combined loading.

All above testing conditions are standardized throughout the complete micro-electronic industry via international organisations like JEDEC, IPC, and IEC. For each developed package the tests will be performed to investigate whether the long-term reliability demands can be met. From IC and packaging modelling point of view, these tests form part of the thermal and moisture loading regime that is subjected to it.

3. THERMO-MECHANICS OF IC BACKEND PROCESSES

Failure mechanisms in IC's are described in Chapter 2. These failures may originate at several locations. Examples are electromigration, hot carrier degradation, breakdown of dielectrics, pattern shift, IC cracking, and/or delamination. The last three failure types are strongly thermo-mechanical related:
- In the case of pattern shift (or metal shift) metal lines are shifted as an effect of thermal loads and interface delamination.
- In the case of IC cracking due to thermal loads cracks may occur in materials such as dielectrics, passivation and/or metal interconnect lines.
- In the case of delamination adherence between two interfaces may be lost and/or buckling may occur at interfaces and via plugs.

Compared with package modelling, less effort is carried out on IC backend modelling. This is mainly due to fact that most failure types are believed to originate from the packaging process. However, it could well be possible that some failures originate from the backend process. This leads to the challenging task of predictive IC modelling, taking into account the full backend process and integrate it with the packaging processes. In this case the stress evolution during the full manufacturing process of an electronic package, starting from a silicon substrate and ending with a packaged IC, can be optimised to acquire reliable IC's. This section deals with the prediction of warpage, stress and strain levels in the deposited layers during the backend processes.

3.1   IC Material Characterization

The total mechanical stress σ_{total} in a brittle film without plasticity deposited on a Si-substrate consists of two components, the thermal stress $\sigma_{thermal}$ and the intrinsic stress $\sigma_{intrinsic}$:

$$\sigma_{total} = \sigma_{thermal} + \sigma_{intrinsic} \qquad (1)$$

The thermal stress is due to the difference in coefficient of thermal expansion (CTE) between the film and the substrate. During cool down from deposition temperature to room temperature the thermal stress develops according to the material properties of the film and the substrate. A simple expression for the (maximum) thermal stress is given by:

$$\sigma_{thermal} = \frac{E_f}{1-v_f}(\alpha_s - \alpha_f)\Delta T \qquad (2)$$

where α_f and α_s are the CTE of the film and substrate, respectively. E_f and v_f are the Young's modulus and Poisson ratio of the film, and ΔT is the temperature difference between deposition temperature and the temperature of interest.

The total stresses in the film σ_{total} are calculated using the warpage R of the processed wafers measured at room temperature. Assuming that the film is thin compared to the thickness of the wafer the total film stress can be calculated using the modified Stoney's equation [6]:

$$\sigma_{total} = \frac{E_s}{6(1-v_s)} \frac{t_s^2}{t_f} \left(\frac{1}{R_2} - \frac{1}{R_1} \right) \qquad (3)$$

with E_s and v_s the Young's modulus and Poisson ratio of the substrate, respectively, t_s the substrate thickness, and t_f the film thickness. R_2 is the radius of curvature after processing and R_1 the initial radius of curvature of the wafer before film deposition. For thickness ratios $t_f / t_s < 0.1$, the error in Stoney's equation remains below 5%. For thicker films, a more rigorous analysis shows that a correction factor equal to $(1 + \gamma\delta^3)/(1 + \delta)$ is required [7], where $\delta = t_f / t_s$ and $\gamma = M_f / M_s$, with the biaxial modulus: $M = E/(1-v)$. For thick films ($0.1 < \delta < 0.4$) with unknown film properties, the film modulus contribution can be neglected and $1/(1 + \delta)$ can be used as a correction factor [8]. Others presented a correction factor, which includes the effects of a finite film thickness and film modulus as well as the effect of a large deflection [9].

Intrinsic stresses are the result of the depositioning process of the layer. These intrinsic stresses are already present at depositioning temperature and are influenced by the parameters of the depositioning process. The intrinsic stresses remain constant with changing temperatures if the properties of the film are not influenced by the temperature (for example viscous flow at high temperatures) or the environment (influence of humidity). The intrinsic stresses are believed to depend on the thickness of the film. For thin films (<300 nm) an increase of intrinsic stresses is observed where above this level the intrinsic stresses are constant with thickness. This influence of thickness is the result of nucleation phenomena of the film on the substrate. The intrinsic stresses can play a significant role and for some type of films the intrinsic stresses are comparable or even larger than the thermal stresses. Modern equipment allows control over the intrinsic stress, which gives the process technicians a method of controlling the overall warpage of the wafer at room temperature. It is important that the intrinsic stress is accounted for in FE simulations. Measuring the intrinsic stresses is difficult due to the fact that the measured total stress is usually a combination of thermal and intrinsic stresses. Intrinsic stresses can be both tensile and compressive. There are several mechanisms for forming intrinsic stresses, such as [10]:

- Density of film-building atoms in absorbed state,
- Absorption of by products,
- Increased surface mobility of reactive species,
- Ion bombardment (PECVD),
- Cross-linking.

For SiO_2 based layers the intrinsic stress is not a constant value. The as-deposited intrinsic stress is influenced by several factors, which can play a large factor in the total stresses in the SiO_2 based layer. For oxynitride, metal or ceramic layers this effect is usually not present. The different factors, which influence this change in intrinsic stresses, are:

- *Absorption of water in the layer.* CVD deposited PSG films generally exhibit as-deposited tensile intrinsic stresses. This is due to absorbed foreign atoms at deposition temperature, which prohibit the formations of a dense film. The film has a reduced density with elongated bonds and/or microcavities and this leads to tensile stresses in the layer. However, if this film is subjected to humidity it will absorb the water atoms and swell in the process. This will decrease the tensile stress and eventually lead to compressive intrinsic stresses. PSG films with lower tensile stresses or compressive stresses will show less reactivity with water [11]. This effect of water absorption plays a large role when measuring the film stresses. In order to get the true as-deposited intrinsic stress the measurement should be made directly after depositioning. The effect of water absorption doesn't play a role when in the process cycle

the film is covered by another film, which prevents the water from reaching the PSG film.
- *Densification of the layer due to temperature increase.* For SiO_2 based films increasing the temperature causes the formation of additional Si-O-Si bonds. This leads to densification of the film and the development of bond strains. These bond strains result in intrinsic tensile stresses in the film. After cooling to room temperature these tensile stresses do not disappear but remain in the film. This effect is present for PSG films but also for TEOS-PECVD films. The result of this phenomenon is the occurrence of hysteresis during thermal cycling of the film. This thermal cycling plays of course a role in practice due to the deposition of additional films [12]. The stress development due the densifications of the layer can be such that the maximum occurring tensile stress is in the order of 200-500MPa.
- *Visco-elastic relaxation of the stresses at temperatures above the softening point.* SiO_2 based films are usually amorphous, glass type structures which exhibit a time-dependent, visco-elastic behaviour. For temperatures above the softening point of the material the viscosity of the material will be low enough for the occurrence of flow phenomena and rearrangement of molecules. The temperature of the softening point will depend on the chemical composition of the layer. For pure SiO_2 films this temperature will be higher than for SiO_2 films with impurities (such as PSG). At temperature above the softening point the film stresses will relax due to the viscous behaviour of the film. At sufficient high temperatures or for longer holding time at certain temperature the film stress will decrease to zero. During cool down from these high temperatures to room temperature only thermal stresses will develop. The complete stress behaviour during a temperature cycle to high temperatures shows initially the development of tensile stresses due to the densification. At higher temperatures these stresses will relax and decrease to zero [12].

Through FE simulations the stress and strain levels in IC layers can be predicted, however, it is important to know the properties of the different materials and the mechanisms, which cause stresses in the film to develop. Currently, a number of techniques are available able to measure the (intrinsic /thermal) stress levels and/or the properties of IC materials. These techniques include:
1. Wafer warpage as a function of temperature,
2. Nano-indentation,
3. IC interface toughness characterization,
4. Other techniques like the bulge test, impulsive stimulated thermal scattering, and X-ray diffraction.

3.2 Wafer Warpage as a Function of Temperature

Measuring the warpage of processed wafers at room temperature is a regular method and provides a value for the total stresses in the layer at room temperature. This value is the sum of the thermal and intrinsic stresses and is used by process technicians as a controlling parameter for the depositioning process. By measuring the warpage as function of temperature the intrinsic stress levels can be obtained, since at deposition temperature no thermal stresses exist and the warpage at this temperature is a direct result of the intrinsic stresses. From the slope of the stress versus temperature it is theoretically possible to calculate the CTE of the film. This is only possible if the intrinsic stresses remain constant with temperature. This is not usually the case, and only valid during cooling to room temperature. The expression for the slope follows from differentiating Equation 2 with respect to temperature:

$$\frac{d\sigma}{dT} = \frac{E_f}{1-v_f}(\alpha_s - \alpha_f) \qquad (4)$$

For the CTE of the film this can be rewritten to:

$$\alpha_f = \alpha_s - \frac{d\sigma}{dT}\frac{1-v_f}{E_f} \qquad (5)$$

with the material parameters and the slope, the CTE of the film can be calculated. Experiments with two different substrates, on the other hand, allow the determination of both α_f and M_f [13]. The substrate curvature method can be further extended to extract the Poisson ratio by using films with periodic line patterns [13].

The out-of-plane thermal expansion coefficient can, in principle, be obtained by monitoring the thickness change as a function of temperature. Care should, however, be taken with the interpretation of this data since the substrate constraining effect may result in an increase of the CTE with a factor of 2 to 3. The apparent constraint CTE can be expressed in terms of the Poisson ratio and the free film expansion coefficients [14]:

$$\alpha_f^{constr} = \left(\frac{1+v_f}{1-v_f}\right)\alpha_f - \left(\frac{2v_f}{1-v_f}\right)\alpha_s \qquad (6)$$

The substrate curvature (or curved beam) method is a useful tool to determine stresses in multi-layer structures. The curvature of a multi-layer structure, κ_{multi} results from a linear superposition of bending effects originating from the stresses in the individual film layers [15, 16]:

$$\kappa_{multi} \cong \frac{6}{M_s h_s^2} \sum_i M_{fi} h_{fi} (\alpha_{fi} - \alpha_s) \Delta T \tag{7}$$

with $M = E/(1-v)$, the biaxial modulus. The measurement of substrate curvature and/or wafer thickness can be achieved with either a stress analyser that is equipped with a laser, a hot stage in dry nitrogen gas atmosphere or a laser interferometer [17, 18, 19]. Below gives an example of a calculation of wafer stress levels and how to determine layer properties from wafer curvature measurements.

A layer of 300 nm TEOS is deposited on a 725 μm Silicon wafer at a temperature of 400°C. For the Silicon, $E = 165.7$GPa, $v = 0.23$ and CTE equals 3.0ppm/°C. The warpage of this wafer at room temperature is measured at –330m, where the warpage of the Silicon wafer only is measured at –3500m. Using Equation 3 the total mechanical stress in deposited layer can be calculated:

$$\sigma_{total} = \frac{165.7E3}{6(1-0.23)} \frac{0.725^2}{300E-6} \left(\frac{1}{-330} - \frac{1}{-3500} \right) = -171 MPa \tag{8}$$

Figure 6 shows the measure warpage as a function of temperature for the wafer. The wafer is subjected to a temperature cycle from room temperature to 400°C and back. The heating rate is 5°C/min. The intrinsic stress in the film can be found from the measured stress at deposition temperature. The intrinsic stress for these TEOS films is approximately –60 to –70MPa (compressive intrinsic stress). The slope of the stress-temperature curve is positive which indicates that the CTE of TEOS is lower than the CTE of the silicon. With the fitted line a CTE for the TEOS can be predicted according to Equation 5:

$$\alpha_{TEOS} = 3.0E-6 - (0.20)\frac{1-0.25}{70e3} = 0.86 ppm/°C \tag{9}$$

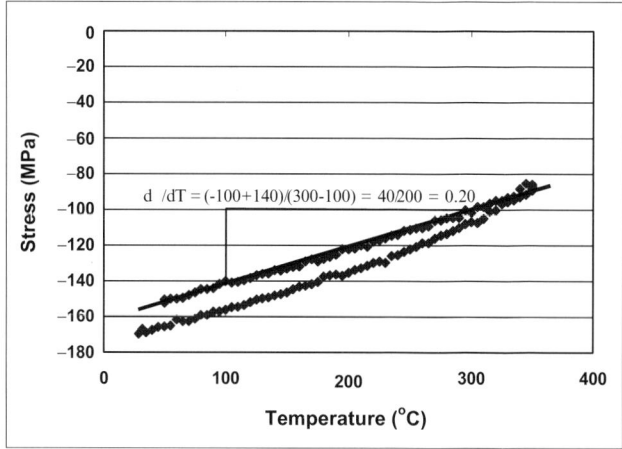

Figure 6. Measured warpage as a function of temperature for a 300 nm TEOS layer deposited on a 725 μm Silicon wafer

Above example is related to stress levels and material properties for the dielectric materials. Wafer warpage measurements can also be used to determine the material properties of the metals used in IC's. Drawback in this case is that the material properties of the metal lines in IC's are different than the bulk properties. This is caused by the small dimensions, the texture, the deposition process and the presence of constraining volumes (substrate, dielectrics). Measurements on free standing films may give an indication of the intrinsic properties of the metal film, revealing lower Young's modulus than the bulk value and a higher yield strength [20-24]. Keep in mind that the actual configuration of aluminium lines consists of a stack of Ti/TiN/Al. The mechanical response of such a stack is different than that of the pure aluminium line. Also, aluminium is being replaced by copper in the latest and to be developed CMOS technologies (90 nm up to 45 nm line width).

3.3 Nano-Indentation

Nano-indentation is derived from the hardness test for bulk materials in which a diamond tip is forced into the test surface at a selected rate and to a selected maximum force. During both the loading and the unloading operation the force and the tip displacement are recorded. A detailed analysis of the force-displacement curve then yields material parameters such as the elastic modulus and the hardness. A good overview of the theory on nano-indentation can be found in [25]. An example of a nano-indentation experiment and a typical loading-unloading experiment is shown in Figure 7. The principle of the experiment is straightforward: the material surface is indented with a tip loaded with a force P, resulting in a penetration depth h. During indentation, the force P and the penetration depth h are recorded as function of time, and in this way the load-displacement relationship is obtained. The hysteresis between the loading and unloading curves is caused by plastic deformation effects.

For the elastic part of the loading (and unloading) curves a generalized analytical solution is derived [26]:

$$P = \alpha(h - h_f)^m \tag{10}$$

where α contains geometric constants, the sample elastic modulus, the sample Poisson's ratio, the indenter elastic modulus, and the indenter Poisson's ratio, h_f is the final unloading depth and m is a power law exponent that is related to the geometry of the indenter (m = 1 for flat-ended cylindrical punch, m = 1.5 for a parabolic of revolution, m = 2 for a cone). The experimentally observed power law exponents are lower than the theoretical values as a consequence of the complex elastic-plastic material behaviour [27]. The interpretation of the loading curve data is complicated by the presence of plastic deformation. Therefore the unloading data is analyzed according to a model which relates the force-displacement slope at maximum loading, dP/dh, and the contact area at peak load, A, to the elastic modulus:

$$S = \frac{dP}{dh} = \frac{2}{\sqrt{\pi}} E_r \sqrt{A} \tag{11}$$

Corrections for non-rigid indenters are effectively accounted for by using the reduced modulus E_r:

$$\frac{1}{E_r} = \frac{1 - v_f^2}{E_f} + \frac{1 - v_s^2}{E_s} \tag{12}$$

where E_s and E_f are the Young's moduli of the substrate and the film and v_s, v_f are the respective Poisson's ratios. Other corrections exist for the shape of the deformed surface (factor of 1.05 to 1.10) [28], for the effect of tip rounding [29], and for the substrate effect [30].

A second material property, which can be obtained from nano-indentation measurements, is the surface hardness, H, which is given by:

$$H = \frac{P_{max}}{A} \tag{13}$$

where P_{max} is the peak indentation load. One of the problems associated with nano-indentation is to obtain a reliable estimate of the contact area A. Direct imaging the small indentations by electron microscopy is elaborate, time-consuming, and sometimes even impossible due to location difficulties. In order to obtain the contact depth from the experimental data the following equation is suggested [26]:

$$h_c = h_{max} - \frac{\varepsilon P_{max}}{S} \tag{14}$$

where ε is a geometrical constant which ranges from 0.72 to 1.0. With modern indentation equipment load resolutions of 0.01 mN and displacement resolutions of 0.1 nm are achieved. Due to the many assumptions and uncertainties in the analysis procedure, however, the absolute values of the calculated elasticity may vary as much as 10-20%.

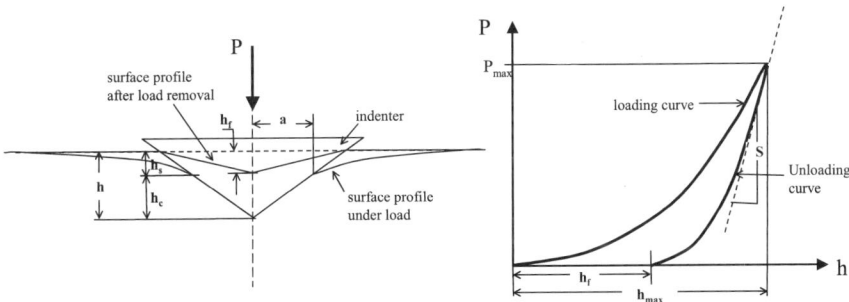

Figure 7. Schematic presentation of nano-indentation (left) and a typical resulting loading and unloading curve (right)

Recently, nano-indentation is used to determine the visco-elastic properties of the currently introduced low-k dielectrics [31, 32, 33, 34]. In such an indentation-creep experiment, the indentation load P is fixed in time while the displacement h(t) is measured. From the measured indentation-creep curves, see Figure 8, the creep compliance and the relaxation modulus for the indented material can be obtained. Assuming thermo rheologically simple material behaviour and time-temperature shift according to the Williams-Landel-Ferry (WLF) form, a 10th order Maxwell model is fitted to the measured data:

$$E(t) = E_0 + \sum_k E_k \exp\left(\frac{-t}{\tau_k}\right) \tag{15}$$

where the characteristic time τ_k and the relaxation moduli E_k are listed in Table 3.

Table 3. Maxwell constants for a low-k dielectric linear visco-elastic material model, obtained using nano-indentation experiments

k	1	2	3	4	5
τ_k	0.1	1	10	100	1000
E_k	200.2	178.6	159.3	142.1	126.8
k	6	7	8	9	10
τ_k	10000	100000	1E+6	1E+7	1E+8
E_k	113.0	100.9	80.3	71.6	

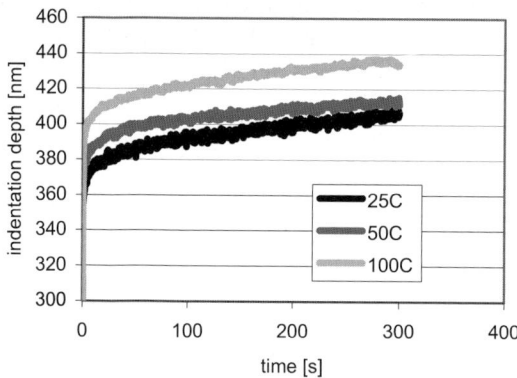

Figure 8. Measured indentation-creep curves on a 8 μm thin low-k dielectric film at different temperatures

3.4 IC Interface Toughness Characterization

Interface toughness characterization for IC materials predominantly focuses on the four-point bending with pre-notch crack tests. This test can be used to measure the adhesion strength between different IC layers. Initially, this adhesion test was first developed as a method for characterizing the fracture resistance of bi-material interfaces [35]. Recently, the adhesion strength of low-k materials with its adjacent materials is measured by using this technique [36]. A schematic representation of the test is visualized in Figure 9. The sample consists of a bi-material sample with an initial notch loaded in a four-point bending test. The characteristics of this test are:
- Mixed-mode loading conditions at the crack tip comparable to practical loading condition,

- Stable crack propagation,
- Simple sample geometry,
- Well-established testing routing with a minimum of alignment influences.

The stable crack propagation results in a constant load during delamination, which simplifies the determination of the fracture resistance because it is independent on the delamination length. A typical ideal load-displacement response of this test is visualized in Figure 9. Analytical Equations for the resulting energy release rate can be determined. Several characteristics and dependencies of the test have been analysed such as the dependency of the mode mixity (phase angle) on the thickness ratio and the Young's modulus ratio of the two materials [35]. A problem with this set-up that might occur is vertical cracking of the debonding layer, which prohibits the determination of the interface strength. To solve this problem an additional stiffening layer can be used [37]. This additional layer suppresses the segmentation of the (brittle) layer and increases the stored energy in the sample and therefore the driving forces for delamination. In general, the evaluation of the fracture toughness from this test requires numerical calculations, however with special geometric assumptions, an analytical solution for the energy release rate G_i is possible:

$$G_i = \frac{21(1-v^2)P^2L^2}{16Eb^2h^3} \tag{16}$$

with E and v the Young's modulus and Poisson's ration of the substrate, respectively, h is the thickness of the substrate, b is the width of the specimen, and L is the distance between the inner and outer support points. A more general analytical evaluation of the four-point bending test can be found in [35, 37]. Some points should be noted:

- Steady-state crack growth can be obtained under the condition that the debonding crack length is significantly larger than the substrate thickness and the debonding crack length is smaller than a quarter of the inner span.
- During debond cracking, a combination of mode I (normal stress) and mode II (shear stress) will act on the crack. For the four-point bend test geometry, the ratio of shear to normal stress, called phase angle or mode mixity, is between 40° and 70° depending on geometry and materials properties.
- Plastic or visco-elastic dissipation that occurs in the layer stack itself will also lead to an increased measured energy release rate. The measured value is then not the intrinsic adhesion energy, but an effective value including other dissipating mechanisms.

- To draw any meaningful conclusions, the fracture surfaces of the specimens should be chemically analysed after adhesion testing to assess the locus of failure.
- Alternative tests exist to measure adhesion strength, such as the wedge test, (button) shea/pull test, and the dual cantilever beam test. They are described in the next section.

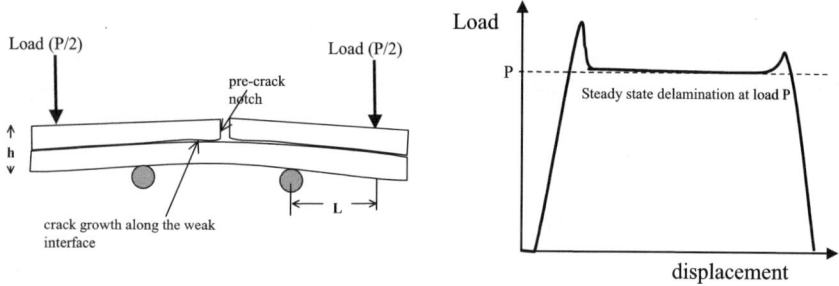

Figure 9. Schematic presentation of 4-point bending with pre-notch crack (left) and a typical resulting load-displacement curve showing the steady state delamination growth (right)

3.5 Other IC Material Characterization Techniques

1. *Bulge test.* The bulge test is used for the determination of the residual stresses and the Young's modulus of a film. In this technique, the deflection of a suspended film is measured as a function of applied pressure [38]. It has been applied for metals [39], semiconductors, dielectric materials, and polymer thin films [40]. In the bulge test, an uniform pressure is applied to one side of a free standing thin film window, causing it to deflect outwards. From this test, stress and strain in the film can de determined from measurements of pressure, the window deflection, the film's thickness, and the window dimension. The bulge test is a flexible technique, allowing characterization of elastic, plastic and time-dependent deformation of a film.
2. *The impulsive stimulated thermal scattering.* The impulsive stimulated thermal scattering (ISTS) is a time resolved optical spectroscopy method that has been used extensively for characterization of bulk samples, determination of elastic constants and thermal diffusion rates [41, 42]. A modification of this technique allows real-time characterization of acoustic waves propagating on a thin film [43], determination of elastic moduli, thickness, and thermal diffusion rate [44-46]. This laser based ultrasonic method is a fast, non-invasive, non-contact in-situ evaluation

of thin films. In this method, two short excitation laser pulses are crossed spatially and temporally in the film sample to create an interference or grating pattern. This gives rise to sudden (impulsive) heating and thermal expansion with the sample spatially periodic geometry. The resulting acoustic and thermal responses of the sample are observed by monitoring the time-dependent diffraction of a probe laser beam off the generated material. Elastic properties, namely the shear and longitudinal velocities, Young's modulus, and Poisson's ratio can be determined. This method is also applicable for the determination of the film thickness [47].

3. *X-ray diffraction.* X-ray diffraction is used to measure the evolution of residual strain in a metal substrate layer covered by a polymer coating. The method is simple, non destructive, and does not require special sample preparation [48].

3.6 Typical IC Material Properties

Typical material properties for IC materials is listed in Table 4. This table includes ranges for Young's modulus, Poisson ration, CTE, intrinsic stress levels, fracture strength, yield stress (for the metals), and interface strength J-values (for low-k) for the given materials as reported in the literature [36, 43, 44, 49, 50, 51, 52].

Table 4. IC material data

Material	E [GPa]	ν [–]	CTE [ppm/°C]	Intrinsic stress [MPa]	Fracture strength [GPa]
Si	150 to 169	0.23 to 0.25	2.5 to 3.0	-	8 to 10
PSG	60 to 80	0.23 to 0.25	2.8 to 3.2	+50 to +90	1 to 1.2
SION	60 to 70	0.23 to 0.25	2.7 to 2.9	–100 to –140	1.5 to 2.0
Si_3N_4	150 to 190	0.23 to 0.25	1.8 to 2.2	–100 to –140	1.5 to 2.0
TEOS	100 to 110	0.23 to 0.25	0.5 to 0.6	–70 to +70	0.5 to 2.0
Low-k	5.0 to 10.0	0.38 to 0.40	10 to 30	-	J-value 5 to 10 N/m with adjacent materials
Al	60 to 80	0.30 to 0.35	20 to 25	+100 to +200	Yield stress: 0.1 to 0.2
Cu	100 to 130	0.30 to 0.35	16 to 18	-	Yield stress: 0.2 to 0.6

3.7 Finite Element Modelling for Backend Processes

Finite Element (FE) analysis is a powerful method to take into account complicated geometries and non-linear material behaviour. Some efforts are done on FE analysis to predict stress/strain levels in patterned structures, including aluminium and copper metal lines [53, 54]. Still, reliable thin film

FE modelling is a challenging task due to enormous size discrepancy between the thin films and the thick substrate, e.g., 0.5-1.0 µm thick dielectric versus the ~0.5 mm thick substrate [55]. This often leads to inefficient models with regard to element mesh and convergence. When modelling backend processes the following techniques should be addressed:
- Large displacement theory should be applied (see Chapter 4).
- Both 2D and 3D models can be used, for 2D general plain strain element type is preferred, but axi-symmetric can also be used.
- To overcome the enormous size discrepancy, techniques to use are:
  o Substructuring (see Chapter 4), however the majority of the comercially available FE software do not allow the use of non-linear material behaviour (such as temperature dependent Young's Modulus).
  o Submodelling (see Chapter 4), also denoted with global-local or multi-level modelling. This technique is quite powerful and used to model the interaction between IC layers and packaging materials.
  o Contact bodies. Contact bodies can be used to overcome different mesh sizes and to allow interaction with small features such as metal lines with large features such as the package.
  o Insert or links. Links can be used to overcome different mesh sizes.
- Sufficient mesh coarseness should be applied, obtained via a mesh sensitivity analysis. For instance, when using 3D models, in vertical direction, at least 3 nodes are needed to correctly describe the bending behaviour. More accurate bending behaviour of the 3D solid elements can be obtained by using an extra set of interpolation functions.
- Homogenization should be used to correctly describe the behaviour of the layers including the metal lines. Volume based homogenization, stiffness smearing (or smeared properties), and local models are options to generate equivalent properties.
- Intrinsic stresses in TEOS, PSG, and SiON should be included, preferably as an initial biaxial stress at the deposition temperatures.
- All (relevant) process steps should be incorporated, be it thermal and/or mechanical originated.
- Element birth and death (see Chapter 4), also denoted with element activation and deactivation, can be used to add and/or remove material layers as an effect of deposition processes.

3.7.1 Prediction and Verification of Process Induced Wafer Warpage

To predict wafer warpage levels, both analytical and numerical modelling techniques, combined with experimental verifications will gain

5. Thermo-Mechanics of Integrated Circuits and Packages

insight into the physics of the backend processes [56]. 3D FE modelling results are verified by a series of experiments using specially designed backend wafers to monitor the warpage during manufacturing. Figure 10 shows a 3D FE model representing a silicon wafer with diffused metal and dielectric layers. Because of symmetry, only one quarter is necessary. The model is meshed with 3600 solid 8-noded brick elements. Appropriate boundary conditions are used along both symmetry axes. Six silicon wafers from a 0.5 μm CMOS process with a diameter of 200 mm are used from experimental side. Layer thicknesses are 800 μm bare Si, 0.5 μm TEOS, 1 μm Al, 1.0 μm TEOS, 0.5 μm PSG, and 0.5 μm SiON. The last three are passivation layers. An analytical model is developed to predict the in-plane layer stresses and the warpage of the wafer during the building up of the thin films. For each method, analytical, numerical, or experimental all process steps are taken into account, see Figure 11.

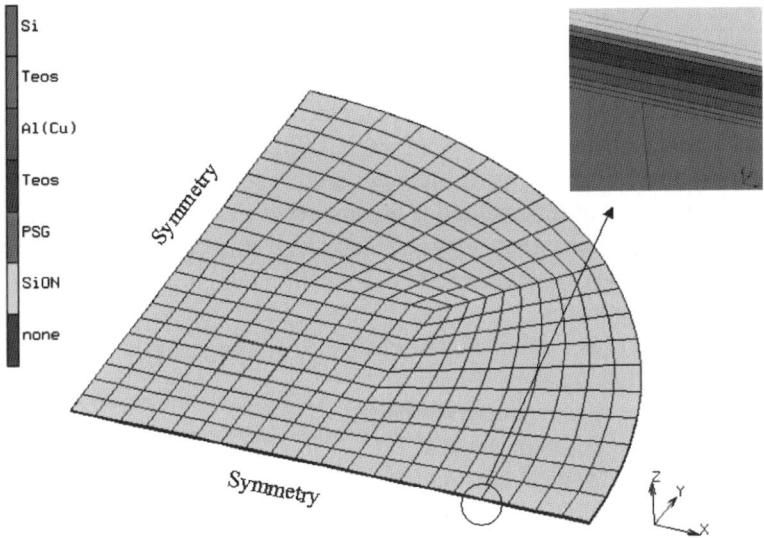

Figure 10. 3D FE wafer model including diffusion layers

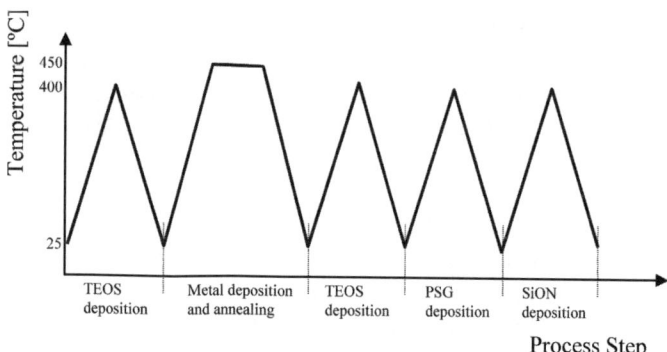

Figure 11. Process steps of successive deposited thin films on the bare Silicon

As mentioned before, element birth and death methods can be used to activate deposited layers correctly. However, the deposited layer should have a changed geometry but exert no stress and strain, since deposition is done at high temperatures. To simulate this successive build-up correctly a special procedure is developed using stiff beam elements. These beam elements transport the nodes for the solid elements that are not yet activated. For not yet deposited layers, the solid elements are deactivated and the beam elements (located at the edges of each solid elements) are activated. By doing so, the deformations of the layer are properly taken into account. An important feature of this method is that the solid elements are in a stress- and strain free condition when they are activated, this cannot be achieved by using a trick based on temperature dependent material properties. Figure 12 illustrates the approach.

For the six wafer samples, the change in warpage $\Delta\kappa$ is measured after each of the following steps:
- Step 0: Bare silicon (initial warpage),
- Step 1: After TEOS deposition,
- Step 2: After Metal deposition,
- Step 3: After TEOS, PSG, and SiON deposition (passivation layers).

The effect of metal deposition is significant: metal deposition changes the curvature of the wafer from convex (n-shape or grumpy, denoted with + sign) to concave (u-shape or smiley, denoted with – sign). The metal patterning reduces the warpage change with approximately 50%. This reduction is driven by the coverage of the metal, which is approximately 50% in the wafer samples.

5. Thermo-Mechanics of Integrated Circuits and Packages

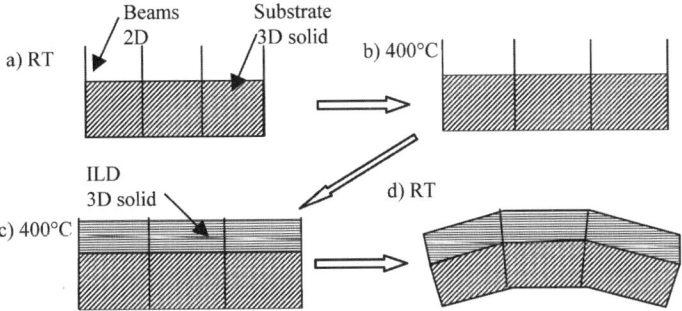

Figure 12. Element birth and death approach added with stiff beam elements to correctly describe deformed structures of deposited layers. During heating-up, beam elements running from the top of the substrate to the top of the last layer are activated, and will therefore move in x and y direction (b). At the deposition temperature, the solid elements of the layer are activated in the present deformed state (c). During cooling down to room temperature, the beams in the layer are deactivated and have therefore no influence on the behaviour of the solid elements (d). The approach is continued during the complete build-up of the wafer

The experimental warpage change of the wafer after thin film deposition is used to calculate the stress level in the film. Both the warpage change and the analytically calculated stress levels are compared with the numerical predictions. Table 5 lists the results of the experimental, analytical and numerical predictions. The following is concluded:

1. For the warpage, a good agreement between experiment and numerical FE prediction is obtained. Large deviation, approximately 30%, is found for the warpage change due to thermo-mechanical effects of the metal. Several explanations can be found for this difference:
   - The metal behaviour is quite complex including plasticity, phase transformation and position dependent material properties.
   - Annealing of the wafer is observed to influence the warpage significantly.
   - In the FE model no time dependency is taken into account.
2. For the stress levels, again, a good agreement is found between the analytical and numerical models. Note that in the analytical prediction both no intrinsic stresses and metal yielding are included.
3. The FE simulations showed that the stress in a thin film is hardly influenced by the warpage changes caused by subsequent processes. Also, the total film stresses are a summation of thermo-mechanical and intrinsic stress levels. The added intrinsic stresses hardly relax, because the films of TEOS, PSG, and SiON are very thin compared to the substrate. However, adding initial stresses in a thin film does influence the warpage of the wafer. Tensile intrinsic stress leads to a more positive

warpage change (upward bending) and compressive intrinsic stress leads to a more negative warpage change (downward bending).
4. Resulting stress values from (non-linear) FE analysis showed that the Von Mises stress in the metal exceeds the (temperature dependent) yield stress. Stresses in all other layers except for the substrate, are not influenced by this non-linear metal behaviour.

This 3D example shows that FE modelling techniques are capable of simulating the wafer build-up processes and to achieve quantitatively reliable predictions, both the intrinsic stresses and the process dependent material properties should be taken into account.

Table 5. Resulting warpage and stress levels after thin film deposition; experiments vs. analytical and numerical predictions

Process Step	Experimental	Analytical	Numerical
TEOS			
ΔK [m^{-1}]	−0.00168		−0.00169
σ_{total} [MPa]	-	−87	−62
Metal			
ΔK [m^{-1}]	0.00872		0.01139
σ_{total} [MPa]	-	679	200
PV 1, 2, and 3			
ΔK [m^{-1}]	−0.00188		−0.00225
σ_{total} [MPa]	-	−87 (PV1)	−62 (PV1)
		+4 (PV2)	+54 (PV2)
		−7 (PV3)	−56 (PV3)

3.7.2 Prediction of Local IC Stress Levels

Cracks in the dielectric and/or passivation layers on an IC are often observed in the micro-electronic industry [57]. In order to investigate the crack failure behaviour of the final product, the defected ones are opened and observed using microscope techniques such as SEM. Perpendicular cracks are observed in the passivation layers close to metal lines. Usually those cracks start at the bottom of the dielectric layers corresponding to the bottom of multi-metal layers or at the top corners of the passivation layers, as shown in Figure 13 [58]. A major cause for these failures is the use of materials with different properties.

5. Thermo-Mechanics of Integrated Circuits and Packages

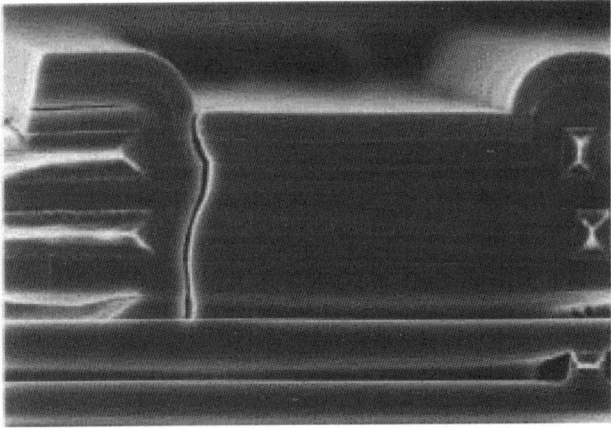

Figure 13. Example passivation crack from field observations

Both micro-scale structures with dimensions of micrometers and the macro-scale of the surrounding package with dimensions of millimeters contribute to the total stress distribution near the small metal lines. However, the largest contribution is caused by the local geometry on a micro-scale. The small metal lines themselves tend to shrink much more than the dielectric surroundings, resulting in high local stress levels. As such, 2D FE models can be used to estimate the crack growth on this micro-scale. Crack propagation in the brittle dielectric materials on a micro-scale, may lead to electrical shortage and/or pattern shift on a macro-level in subsequent manufacturing steps. For predicting crack growth in materials, fracture mechanics theory is often employed. Stress intensity factors (or strain energy release rates) are generally used to predict the formation of cracks in layered structures. In this example, the J-integral technique (see Chapter 4) is used to calculate the effect of initial cracks of different lengths, positions and directions. Two-dimensional plane strain FE models are used to perform the simulations. The evaluated geometry is visualized in Figure 14 (top) which shows a typical part of an interconnect structure on the IC. Six layers are present on top of the silicon substrate consisting of metal and dielectric materials in the interconnect structure and a compound layer on top. The crack location, length and direction are varied within the region bounded by the metal lines, the compound and the IC.

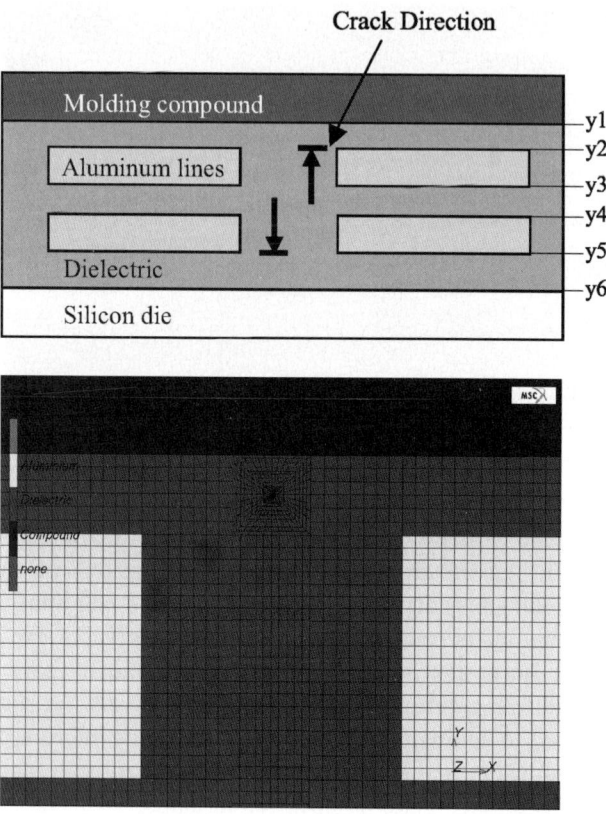

Figure 14. 2D geometry of interconnect structures (top) and crack tip mesh (bottom)

Cracks may initiate at a position right of the centre with a downwards propagation direction; cracks may also initiate at a position left of the centre with an upward propagation direction. The crack starting position is also changed, it may start from the upper passivation layer and propagates through the passivation layer to the bottom dielectric layer and vice versa, denoted in Figure 14 (top, y1 to y6). The cracking part has 1020 elements with a size of 30 nm x 50 nm. The mesh used in the crack region is shown in Figure 14. Thermal loading is used, simulating the manufacturing process of the IC and the package.

Some typical results are shown in Figure 15, including the effect of crack starting positions and direction, a) for downward propagation and b) for

5. Thermo-Mechanics of Integrated Circuits and Packages

upward propagation, and the effect of the crack location in horizontal direction (i.e., changing the distance between the crack and the left metal line) for a down crack length of 1.014 μm and up a crack length 2.5 μm in c). In this figure, also, the effect of including the intrinsic stress levels is visualized. From Figure 15, the following can be concluded:

- For the same crack length, the cracks with direction downward have larger J-values. As such, a crack is eager to propagate in downward direction.
- Cracks starting at the top of the passivation layer, location y1, have larger J-values than with other starting points. As such, a crack is eager to occur at the top IC layers.
- Cracks located at positions close to metal lines have larger J-values. As such, a crack is eager to occur near metal lines.
- For cracks growing upwards, longer crack lengths have larger J-values, which can result in unstable crack propagation. For downward cracks, larger cracks tend to have smaller J-values, resulting in stable crack propagation, which requires additional loading for further propagation.
- A tensile intrinsic stress will increase the J-integral values. For intrinsic stress levels of 100 MPa an increase of approximately 50% is observed.

Simulations increasing the number of metal layers showed that more metal layers would increase the J-values, mainly for short crack lengths. When crack length becomes larger than a certain value, some trends change because of co-operational effects of metal line layers.

The finite element simulations using the J-integral approach show that cracks can occur after the de-positioning processes; cracks are easier to propagate downwards than upwards; cracks are easier to occur close to the metal lines; tensile intrinsic stresses increase the sensitivity to crack propagation; and increasing numbers of metal layers will increase the sensitivity to crack propagation.

The above mentioned conclusions match with field observations in actual IC interconnect structures. This methodology can be used to derive design rules for IC interconnect structures [59].

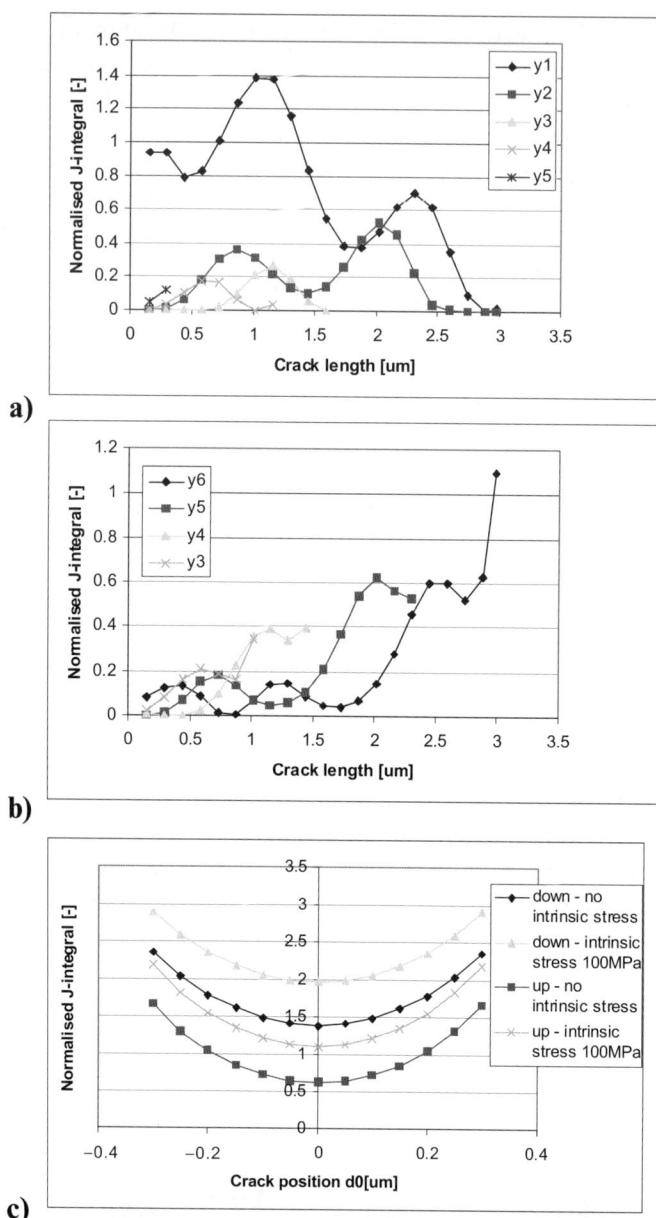

Figure 15. J-values for a) effect of crack starting positions with downward propagation and b) effect of crack starting positions with upward propagation, and c) effect of the crack location in horizontal direction

3.7.3 Prediction of Cu/Low-k Bond Pad Delamination

Current issues in the development of new Cu/low-k CMOS technologies is the thermo-mechanical reliability of bond pad structures due to the bad thermal and mechanical integrity of the low-k materials and associated interfaces [60]. The resulting forces due to wire bonding process and subsequent qualification tests can easily result in reliability problems like bond pad delamination and low-k cracking. In the packaging industry, typical wire bond reliability tests concern wire pull and wire shear. The configuration of the wire pull qualification test consists of a wire that is pull at by a hook until failure occurs. The wire should withstand a certain force level, which depends on the wire diameter, before failure of the wires is allowed to occur. In traditional CMOS technologies always the ball bond failed, but when using the so-called low-k dielectrics the failure modes shifted to what is called metal peal off (MPO). An example of MPO, or bond pad delamination, of a CMOS090 Cu/low-k process due to this wire pull qualification is illustrated in Figure 16. The thermo-mechanical reliability of the (low-k) bond pads depends strongly on the metal and via layout of the backend layers beneath this bond pad. The copper metal lines and vias act as mechanical support in the relative weak low-k matrix. But due to the complex interaction of multiple processes, materials and interfaces an optimum metal layout design is difficult to design up front.

In order to investigate the structural integrity of the bond pad, 3D finite element analysis is used, based on a multilevel sub-modelling approach, see Chapter 4. For analysing and comparing different backend structures, we have developed a novel method, the so-called Area Release Energy [61]. This value predicts the change of delamination of critical interfaces without knowing a priori the exact location of the delamination. The amount of energy is calculated that is released upon delamination for any position along a critical interface.

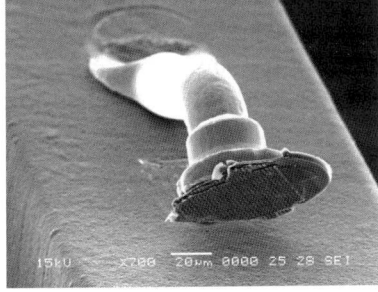 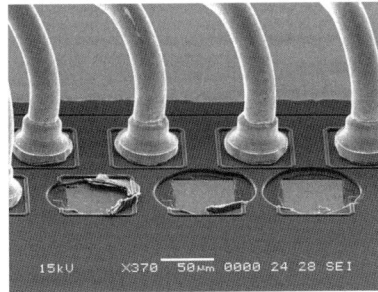

Figure 16. Example of the metal peal off in Cu/low-k structures

Area Release Energy

Due to the brittle nature of the dielectric films, it is assumed that the use of (linear elastic) fracture mechanics is valid for describing the interface delamination in Cu/low-k backend structures [62, 63]. Within FEA the most commonly used methods to predict the crack driving force are J-integral calculations and Virtual Crack Closure Techniques (VCCT). The method proposed here does not require any presupposed position of any initial crack. Instead, at any desired positions within the specimen, an area energy release value is calculated which basically results from releasing an area (having a defined dimension) around each point in the specimen. Use of this 'Area Release Energy' has some important advantages compared with traditional approaches:

- It is an energy based failure criterion, which is more accurate for describing delamination damage than stress-based approaches.
- It is not required to assume a pre-existing delamination at a specific location with a specific size, as is the case in traditional fracture mechanics based approaches, such as J-integral and Virtual Crack Closure Techniques (VCCT).
- It allows an instant overview of the critical areas through a contour map and also makes comparison between different structures possible.

To explain the method, a two-dimensional example is used, see Figure 17. Here, for node *i*, a set of nodes are depicted that fall within the area that will be released, defined by *l*. Clearly, the reason behind the usage of releasing a predefined area instead of simply releasing surrounding nodes is to prevent mesh dependency of the energy release values, i.e., no convergence of the energy values would be obtained upon mesh refinement. For every node *i*, the resulting area release value is calculated as:

$$G_i = \frac{1}{2A_i} F_j u_j \text{ with } j = 1, \cdots, n \qquad (17)$$

in which *n* equals the number of nodes that belong to the released area. For 3D analyses, the area equals $A = \pi \ell^2$, meaning that a circular area surrounding each node is released. The Double Cantilever Beam (DCB) NAFEMS delamination benchmark [64] was used for verification of the Area Energy Release methodology.

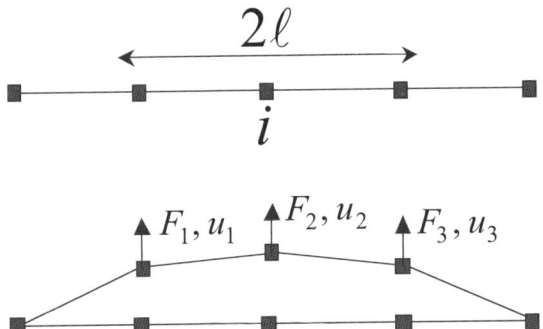

Figure 17. 2D Schematic representation of the area release method

Finite Element Modelling

To bridge the length scale difference between the wire bonding and interconnect levels, a 3D multi-scale submodelling method has been employed [61]. At the global (packaging) level, the simulations are performed using homogenized (effective) stiffness matrices obtained from the local (bondpad) level, to give effective anisotropic stiffness matrices. The coefficients of the stiffness matrix are determined by imposing six independent deformation modes on the local model (unit cell model). These deformation modes are prescribed on the unit cell by periodic boundary conditions. To formulate these periodic boundary conditions, shape vectors for the front and the rear surface of the unit cell are formulated. The assumption of periodicity dictates that the shapes of opposite surfaces are equal and remain equal during deformation of the unit cell. To prevent rigid body modes, all displacements of one vertex have been suppressed. The effective coefficients of thermal expansion are determined by calculating the resulting thermal strains due to a temperature step on the unit cell. The homogenisation and localization steps have been verified with 'direct' simulations, i.e., simulations in which several structures, including local details, are completely modelled. The obtained results showed a good agreement, which indicates that the method can be used to predict the global and local behaviour more efficiently. The localization step consists of prescribing the calculated displacements at global level on the unit cell. The displacements for the nodes in the local model that do not coincide with the global nodes, are simply interpolated from the displacements of the nearest neighbouring nodes. It is thus possible to calculate the local state variables, such as stresses and damage values, in the unit cell. Figure 18 further explains the principle of our developed multi-scale methodology.

The global model consists of a bond pad with an opening of 70 x 70 μm^2 that is covered with an aluminium layer, passivation layers, homogenized properties of the backend structure, a gold wire of 25 μm diameter, the ball bond and the silicon die. Due to symmetry half the bond pad is simulated and the wire pull qualification test is simplified by assuming a pure tensile force on the wire. The force applied on the global model equals 3gram, which is a typical qualification loading of such a ball bond. Further boundary conditions consist of a full constraint of the bottom on the silicon in z-direction, symmetry in y-direction and full constraints in x-direction on the left side of the model.

Three different local geometries are considered: (i) metal plates with holes; (ii) a full metal plate geometry, which equals (i) except that the 'hole' in the metal plate is removed; (iii) a full metal plate geometry including four symmetrically positioned vias, to illustrate the effect of using vias. The layout of the unit cells is conform the CMOS090 design: layers of PSG, hard mask, metal and dielectric materials. The dimensions of the unit cell are assumed as: 4.00 x 4.00 μm^2 and are equal for all unit cells.

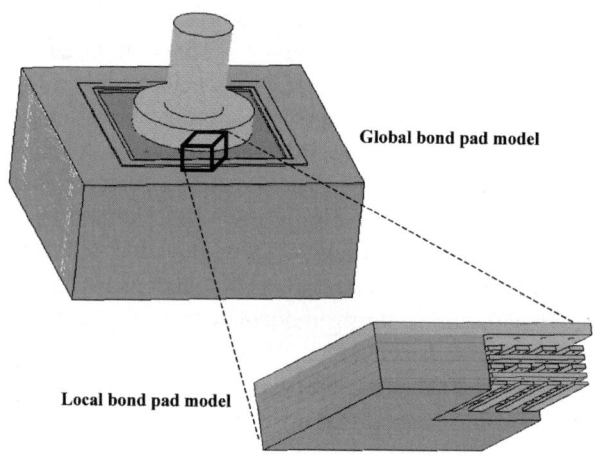

Figure 18. Multi-scale modelling methodology for local bond pad structures

Results

The displacements are calculated at the global level, using the effective stiffness matrices determined from the three different local models. Figure 19 shows the resulting deformed structure as a result of the wire pull forces. Imposed to this deformed structure, the Von Mises stress response is depicted. Clearly, due to the pull forces both the wire and the metal cap will exhibit yielding behaviour. Notice that due to the pull forces, the wire imposes a rotational force on the bond pad structure. This is a result of not

5. Thermo-Mechanics of Integrated Circuits and Packages

perfectly in-line pulling but rather by using a hook at a certain distance from the bond pad.

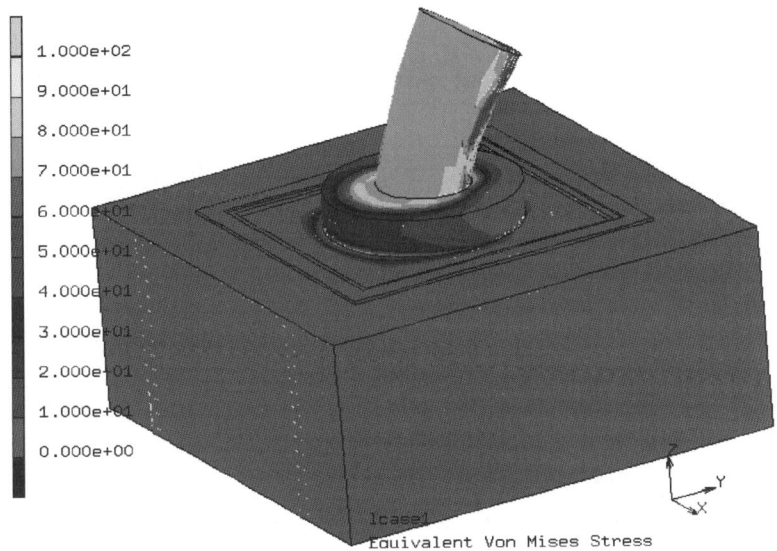

Figure 19. Deformed global structure with Von Mises stress response

The area energy release values are calculated at the (x,y)-surface, situated at the interface between the 4th level of hardmask and low-k dielectrics. From experimental data, it is known that this surface is a critical location at which delamination often initiates.

Figure 20 shows the values of the area release energy as contour maps in the surface, for the three different bond pad structures. Notice that the values are normalized with respect to the highest values. Of course, higher values are more critical. For the full plate geometry, the asymmetry is a result from the global deformation and clearly shows that the unit cell is loaded more in the left bottom corner. Second, some form of edge effect is present. This can be explained by the fact that the unit cell is indeed bounded by its edges, which basically means that it is more sensitive to failure at the edges. This is demonstrated by the higher values of the area release values. In reality, the unit cell is not really bounded by its edges (the bond pad structure is of course continuous). One way to prevent this (numerical) phenomenon is to enlarge the dimensions of the unit cell artificially and zoom in the area that is situated at a distance from approximately of the edges of the unit cell, with being the characteristic size of the edge effects. As can be inferred from Figure 20, the plate with holes geometry is less favourable with regard to values near the edge when compared with the full metal plate layout. However, notice also that in the plate with holes pad, the influence of the 'hole' in the metal plate is clearly visible, yielding lower values at the position of the hole. The result for the full plate with vias, indicates that this geometry results in the lowest values for the energy release. The vias effectively lower the area release values of the full plate. The presence of vias also reduces the edge effects.

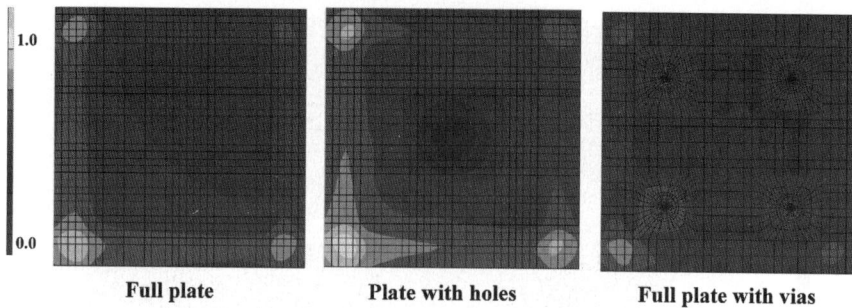

Figure 20. Normalized Area release contour map for three different bond pad structures

The Area Release Energy method is able to pinpoint at failure sensitivity of different 3D structures. No initial cracks or imperfections need to be inserted in the models. The comparisons of the three different model geometries indicated the effect of metal and via layout on the chance of delamination. At this moment, experiments of certain bond pad designs are in progress to align the presented modelling results.

4. THERMO-MECHANICS OF PACKAGING PROCESSES

Failure mechanisms in IC packages originate at several locations due to thermo-mechanical loadings during the packaging processes [65]. Typical examples are (see Figure 21):
- Delamination at all possible interfaces. Delamination is caused by combined mismatch in material properties, such as CTE, E-modulus, Poisson ratio, and insufficient or decreased interface strength.
- Die cracks. Due to excessive warpage, stress levels in the package may exceed the tensile strength of the silicon substrate causing it to break.
- Wire related problems, such as wire fatigue, ball bond lift, wire break, etc.

5. Thermo-Mechanics of Integrated Circuits and Packages

- Popcorn cracking and body cracks. Due to moisture uptake and vapour pressure during reflow conditions may lead to cracks in the package. This will be described in Chapter 6.
- Voiding. Due to thermal shrinkage of the materials, voids may occur.

Many studies have been published on package modelling [1, 65-122]. This section deals with the prediction of warpage, stress and strain levels in the IC package.

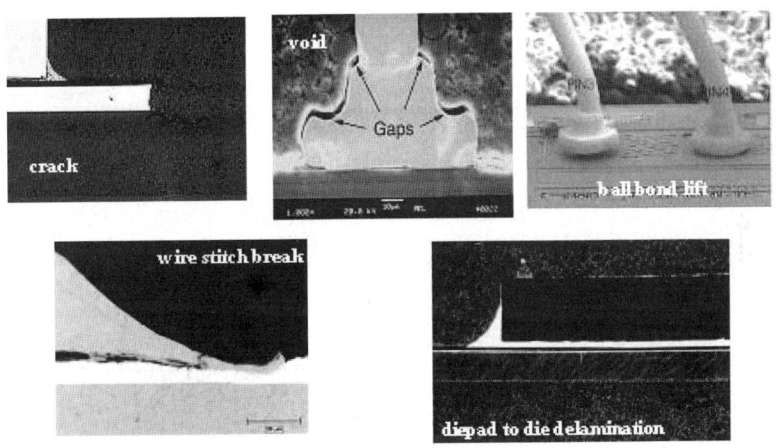

Figure 21. Examples of package failures

4.1 Packaging Material Characterization

As mentioned in section 2.2, typical packaging materials involve a wide range including silicon, metals (copper, aluminium), moulding compounds, ceramics, solders, die-attach, etc. Silicon is a crystal having a diamond like crystallographic structure. Due to this structure, anisotropic material behaviour is to be expected [66]. For metals, typical material characterization tests include tensile/bending tests and the results are reported in many handbooks and\or websites [67]. Off course, this data refers to bulk properties, where thin films may behave significantly different. For such thin films, tests as described in the previous sections can be applied. For metals, typically, elasto-plastic material behaviour is used. Characterization and modelling of solder materials, like Pb63Sn37 or Pb-free (SnAgCu), will be described in Chapter 7.

Characterization of thermosetting resins like moulding compounds, die-attach material, and underfill needs advanced characterization methods and require specially designed measuring apparatus, such as Dynamic Mechanic Analyses (DMA), Thermal Gravimetric Analyzer (TGA), and Thermal Mechanical Analyzer (TMA) [68]. These apparatus are used to measure the following properties:

1. *Under dry conditions.*
   - Glass temperature T_g and the Coefficient of Thermal Expansion (CTE) using TMA;
   - Stiffness as a function of frequency and temperature using DMA.
2. *Under moisture conditions.*
   - Moisture Diffusivity D and Saturated Moisture Concentration C_{sat}, using a weight balance;
   - Coefficient of Moisture Expansion (CME) by combining TGA with TMA.

Chapter 6 describes the experimental procedures for the measurement of the material properties under moisture conditions. Section 4.3 describes the characterization of epoxy systems under dry conditions.

Given the fact that different materials exist in a package, also, different interfaces may occur, such as moulding compound - leadframe, die-attach and silicon, etc. Different techniques are available to measure the interface strength between these materials, including:

1. Button shear/tensile test,
2. 4-Point bending with pre-notch crack, as described in the previous section,
3. Dual or double cantilever beam test,
4. Wedge test,
5. Modified ball-on-ring test (or blister test),
6. Any other, either combinations or deviations of above.

Section 4.4 describes the different features and results of these interface characterization tests.

4.2 Silicon Anisotropy

In most of the publications addressing the thermo-mechanical behaviour of IC packages, the silicon is modelled as temperature independent and isotropic. However, given the nature of the silicon material, which is a crystal, the assumption of isotropy is not true. In fact, due to the orientation of the silicon crystal, a diamond like crystallographic structure, anisotropic material behaviour is to be expected. The temperature dependence of the stiffness values are negligible, for instance the in plane value at 473 Kelvin is only 0.5% lower. Also, the temperature independence of the thermal expansion coefficient (CTE) for the silicon crystal is valid within a certain temperature range [65].

5. Thermo-Mechanics of Integrated Circuits and Packages

For a linearly elastic material, a general equation relating the stress tensor σ_i to the strain tensor ε_j can be expressed as follows:

$$\sigma_i = C_{ij}\varepsilon_j \quad ;(i,j=1,...,6) \tag{18}$$

where C_{ij} is a stiffness tensor. For the silicon crystal, due to the cubic symmetry of its structure in the orthogonal ([100] [010] [001]) coordinate system, the anisotropic properties of the stiffness can be presented by a 6 × 6 matrix with three independent components:

$$C_{ij} = \begin{bmatrix} C_{11} & C_{12} & C_{13} & 0 & 0 & 0 \\ C_{21} & C_{22} & C_{23} & 0 & 0 & 0 \\ C_{31} & C_{32} & C_{33} & 0 & 0 & 0 \\ 0 & 0 & 0 & C_{44} & 0 & 0 \\ 0 & 0 & 0 & 0 & C_{55} & 0 \\ 0 & 0 & 0 & 0 & 0 & C_{66} \end{bmatrix} \tag{19}$$

The components of C_{ij} were obtained from the ultrasound measurements [60] as:

$$\begin{aligned} C_{11} &= C_{22} = C_{33} = 165700 \, MPa \\ C_{12} &= C_{21} = C_{13} = C_{31} = C_{23} = C_{32} = 63900 \, MPa \\ C_{44} &= C_{55} = C_{66} = 79600 \, MPa \end{aligned} \tag{20}$$

However, these measurements were performed on a silicon crystal lying in the [001] plane. On the other hand, the circuit on the wafer is normally aligned along the [110] plane. These values should be rotated to the orthogonal coordinate system used in the FE simulations. Therefore, the stiffness matrix undergoes the following transformation:

$$[C'_{ij}] = [T][C_{ij}][T]^T \tag{21}$$

where $[T]$ and $[T]^T$ are the transformation matrix and its transverse, respectively. If [001] is chosen as the z direction, and [110] as x and y directions, after undergoing rotation from the crystal coordinate system to the FE coordinate system, the stiffness matrix C'_{ij} in the FE simulation coordinate system becomes:

$$C_{ij}' = \begin{bmatrix} 194400 & 35200 & 63900 & 0 & 0 & 0 \\ 35200 & 194400 & 63900 & 0 & 0 & 0 \\ 63900 & 63900 & 165700 & 0 & 0 & 0 \\ 0 & 0 & 0 & 79600 & 0 & 0 \\ 0 & 0 & 0 & 0 & 79600 & 0 \\ 0 & 0 & 0 & 0 & 0 & 50900 \end{bmatrix} \qquad (22)$$

The impact of the assumption of anisotropic behaviour can be quite significant, see Table 6. This table shows calculated die stress levels during manufacturing and testing of a typical QFP package under the assumption of Silicon isotropy and anisotropy. The results show that when the die is considered to be anisotropic, the stress levels at the top of the die are about 15% lower than those obtained by using isotropic properties. If this stress level is used for the failure index for vertical die cracking, it can be seen that the die failure prediction using isotropic properties is more likely to lead to crack than the one using anisotropic properties.

Table 6. Die stress during QFP manufacturing & testing; isotropic vs. anisotropic assumption

Process	Top Die Stress σ_{xx} [MPa] Isotropic	Top Die Stress σ_{xx} [MPa] Anisotropic [110]
Die-attachment (T = 175°C)	0.0	0.0
Cool down (T = 80°C)	–6.2	–5.5
Wire bonding (T = 240°C)	–9.3	–1.3
Cool down (T = 25°C)	11.7	–7.2
Moulding (T = 175°C)	–2.7	–2.9
Cool down (T = 25°C)	–399.2	–349.0
Thermal cycle (T = –65°C)	–627.9	–548.1
Thermal cycle (T = 150°C)	–82.8	–73.3

4.3 Characterization of Thermosetting Resins

Thermosetting resins, like other polymeric materials, have strong time- and temperature-dependent mechanical properties even if they are filled with a high percentage of filler. The creep and relaxation of the packaging material during packaging processes and/or testing will cause a redistribution of stress and strain levels in the IC. However, for the reason of simplicity, in most of the thermo-mechanical packaging simulations the visco-elasticity of the moulding compound is totally or partially neglected. As a consequence, the predicted stress levels and its evolution during packaging processes and/or testing may not be representative for the reality. For a good overview on constitutive modelling for moulding compounds refer to [68].

5. Thermo-Mechanics of Integrated Circuits and Packages

Characterization of thermosetting resins under dry conditions involves:
1. Measurement of the glass temperature T_g and the Coefficient of Thermal Expansion (CTE) using TMA.
2. Measurement of the stiffness as a function of frequency and temperature using DMA.

For both tests, specially created specimens of the moulding compound are needed. These can be prepared by melting and subsequently pressing the solid tablets of the compound into a (special) rectangular mould and cured at 170-180°C for 60-90 seconds, followed by a post curing at 175°C for 3-4 hours. The specimens are rectangular shaped with typical sizes of 90.0 x 4.5 x 2.0 mm^3 or 50 x 9.0 x 2.0 mm^3.

1. *T_g and CTE measurements using TMA*. The thermal expansion coefficient, $\alpha(T)$, and also the glass transition temperature, T_g, of the moulding compound are determined through a Thermal Mechanical Analysis (TMA) measurement. A piece of the specimen is probed for the changes of its thickness while the temperature increases from room temperature to 250°C. To anneal the sample, an initial temperature swing to 200-250°C is required. A typical result is shown in Figure 22. From the resulting dimension change, both the CTE and T_g can be determined, in this case 17.0ppm/°C below a T_g of 189°C and 62ppm/°C above that T_g.

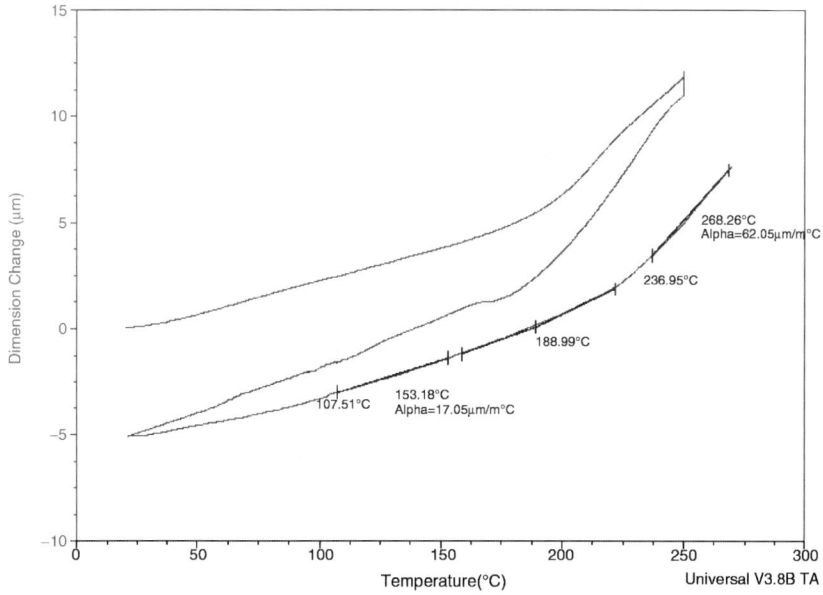

Figure 22. Example TMA result to obtain values of T_g and CTE

2. *Time and temperature dependent properties using DMA.* The time and temperature dependent Young's, shear and bulk moduli of the moulding compound are obtained through bending and torsion Dynamic Mechanical Analysis (DMA) tests. For the bending tests, a piece of the compound specimen is clamped as a single cantilever beam. A sinusoidal deflection is applied to one end of the beam at various frequencies ranging from 0.1 Hz to 100 Hz. For the torsion tests, a piece of the compound specimen is clamped at both ends. A cyclic twist is applied to one end of the bar specimen, and the torque responded by the specimen is measured. Frequency-temperature sweep DMA tests are performed in order to determine the frequency dependence of the moduli at any constant temperature. The result of such a frequency sweep experiment is presented in Figure 23.

In order to obtain master curves, the moduli at a certain temperature are taken as a reference. The master curves are fitted by a fractional series, which is the conversion of the exponential Prony series into the frequency domain:

$$E'(\omega) = \sum_{n=1}^{N} E_n \frac{(\omega \tau_n)^2}{1+(\omega \tau_n)^2} \qquad (23)$$

where ω is the frequency in rad/s. For the fitting of the fractional series the relaxation times τ_n are given as the selected values, and the coefficients E_n are obtained through the fitting procedure. The moduli curves at the temperatures below and above the reference are horizontally shifted until they meet the master curve, by applying the time-temperature conversion principles, as:

$$\tau_n(\alpha, T) = \tau_n^{ref} \cdot a_T \qquad (24)$$

where τ^{ref} are the reference relaxation times, a_T is the shift factor related to temperature. Different equations exist to described the shift factor a_T:
- Arrhenius equation (or Generalized Vogel equation):

$$\log(a_T) = \frac{\Delta H}{R}(\frac{1}{T} - \frac{1}{T_{ref}}) \qquad (25)$$

- The Williams-Landel-Ferry (WLF) equation:

$$\log(a_T) = \frac{-C_1(T-T_g)}{C_2+(T-T_g)} \tag{26}$$

- Others, for instance a 3rd order polynomial:

$$\log(a_T) = A_1(T-T_{ref}) + A_2(T-T_{ref})^2 + A_3(T-T_{ref})^3 \tag{27}$$

ΔH (enthalpy), C_1, C_2, A_1, A_2, and A_3 are fitting parameters and R the Bolzmann constant (8.314 Jmol^{-1}K^{-1}). Following the procedure described above, the master curves for the relaxation Young's, shear, and bulk modulus can be obtained, an example is shown in Figure 22. The (linear) visco-elastic model to describe thermosetting resin behaviour is currently implemented in many of the commercially available nonlinear finite element codes, such as Marc, Ansys, and Abaqus.

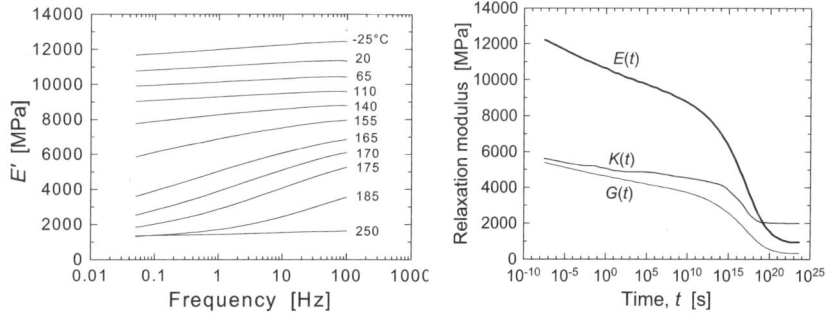

Figure 23. Typical DMA results, frequency sweep (left) and master curves (right)

In [65, 69, 70] it is reported that by using a visco-elastic model for the moulding compound, the predicted stresses and deformations are closer to the real situations. Figure 24 shows die shear stress levels during thermal cycling of a QFP package when linear and visco-elastic properties are used for the moulding compound. Significantly lower die stress levels are found when the visco-elastic behaviour of the moulding compound is included in the packaging process modelling. The visco-elastic effect of the moulding compound causes stress re-distributions and relaxation in the chip, resulting in lower stress levels. Using only linear elastic models for the moulding compound could largely over-predict the stress state in the IC package. Therefore, the time-dependent effect of IC constituents can not simply be neglected if reliable thermo-mechanical stress predictions are desired.

At present, work is in progress to measure and model the cure-dependent behaviour of the thermosetting resins. A cure-dependent visco-elastic constitutive model is established to describe the evolution of material properties during the curing process of a thermosetting polymer [71]. The need for such a constitutive model is the upcoming package concepts such as QFN and BGA, which will be assembled in map systems instead of the traditional strip systems. One critical issue for manufacturing for such map systems is the warpage induced during the moulding process. First results show that the curing process has significant contribution on the total warpage [71].

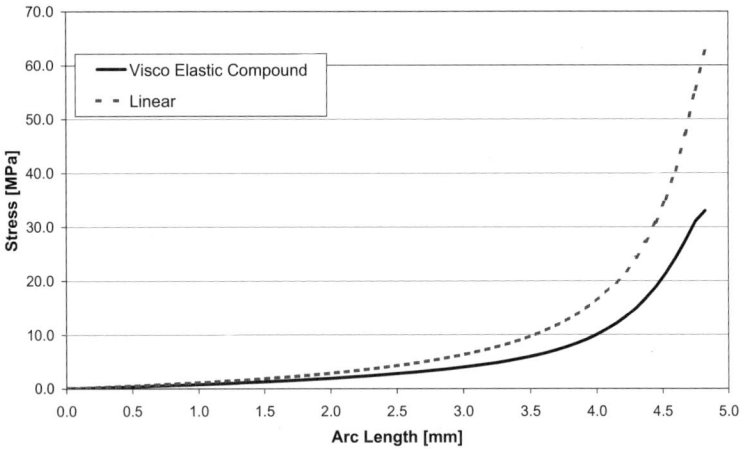

Figure 24. Shear stress between IC and compound during thermal cycling for a QFP

4.4 Advanced Experimental Techniques for Packaging Stresses and Deformations

Different techniques are currently available to measure package stresses and deformations. The results of such measurements can be used to verify simulation techniques and/or results. Below a list of such advanced techniques.
- *Micro-Raman Spectroscopy.* Micro-Raman spectroscopy is used to measure Silicon stress levels. Important feature of micro-Raman spectroscopy is the fact that the frequency of the Silicon Raman peak depends on the level of mechanical stress. In order to determine this frequency, the peak is fitted with a Lorentzian function. The plasma

lines of the laser are used as reference. Their frequency is not influenced by stress in the Silicon. They are fitted using a Gauss function to determine their frequency. Variations in the position of the plasma line are due to instrumental variations (laser, room temperature, detector temperature, etc.). The Raman spectra are corrected for these variations using the plasma lines. If one assumes uni-axial stress, σ, in the Silicon, the relation between the shift of the Silicon Raman peak frequency and this stress is given by:

$$\sigma(MPa) = -434 \cdot \Delta\omega \quad (cm^{-1}) \tag{28}$$

For biaxial stress, the relation between the Raman shift and the stress is given by:

$$\sigma_{xx} + \sigma_{xyy}(MPa) = -434 \cdot \Delta\omega \quad (cm^{-1}) \tag{29}$$

So, if tensile stress is present ($\sigma > 0$), the Silicon Raman frequency will shift downwards ($\Delta\omega < 0$). For compressive stress ($\sigma < 0$), it will shift upwards ($\Delta\omega > 0$). For more complicated stress distributions, the relation between the stress tensor components and the measured Raman shift is not so straightforward, and special modelling is necessary if quantitative stress values are required. For a good overview of this technique refer to [72, 73].

- *Interferometry*. Interferometry is mainly used to measure package warpage. Many different techniques exist, such as:
  o Laser profile Interferometry, using laser beam that scans the surface of the sample;
  o Projection / Shadow Moiré Interferometry, a well know technique for out-of-plane deformation measurement;
  o Twyman/Green Interferometry, a well known technique for in-plane deformation measurement;
  o Holograpic Interferometry;
  o Speckle Interferometry/Electronic Speckle Pattern Interferometry. Basically all above techniques can be used to measure package deformation with different levels of resolutions and/or sensitivity in the order of 0.3-0.5 μm/fringe. Besides warpage, the technique can also be used to measure the CTE of materials. An example of Interferometry measurements combined with FE modelling is described in section 4.7.
- *Digital image speckle correlation (μ-DISC)*. Digital image speckle correlation is a technique, which correlates a pair of digital speckle patterns obtained at two different loading conditions. Next, it searches

for the location of any one point within one speckle pattern spreading all over another one speckle pattern, by maximizing the correlation coefficient of the pair of digital speckle patterns, and determines the deformation of the specimen subjected to any loading. A number of correlation algorithms, such as coarse-fine, cross-search and Fourier transform, etc., have been developed to correlate the digital images or speckle patterns for different applications. The system can be used as an effective experimental tool to characterize properties of electronic materials and verify findings of theoretical and/or numerical models. For a good overview of this technique refer to [74].
- *Other.* Other techniques that can be used are, for instance, classical strain gauges that are able to measure strain levels at certain locations in the IC Package.

4.5 Package Interface Toughness Characterization

The goal of the adhesion tests is to obtain a quantitative value for the interface strength which can be used as input/comparison with Finite Element results. Techniques available to quantify the interface strength between package materials are shortly described in this section. For a theoretical background on J-integral, mode mixity, mode I, II, III, please refer to Chapter 4.

1. *Button shear/tensile test.* For measuring the adhesion properties of moulding compounds on silicon, leadframe, FR4-substrates etc., simple pull and shear tests are often performed on small studs made of compound. These traditional adhesion tests have many weaknesses, including poor repeatability; sensitivity to variables that are unrelated to adhesion, or unduly complicated analysis. While such techniques can be useful for making qualitative comparisons of the adhesion in similar material systems, it is difficult to obtain quantitative information about dissimilar systems. Furthermore, these interface strength values are not applicable as input in quantitative simulations. Therefore, these tests are often used as qualitative comparison. Although the set-up of the tests is relatively simple, the obtained results are not very reliable due to the following characteristics:
    - Misalignment has a very large influence, mainly for the pull test.
    - Results obtained are a maximum force, translated to average stress levels. Fracture mechanics based data can't be extracted from these tests.
    - Initial defects and de-adhesion has a large influence on obtained results.

- The shear test doesn't result in pure mode-II but an undefined mixed-mode failure [75, 76].
- No information on crack length, but this can be resolved when the C-Mode Scanning Acoustic Microscope (C-SAM), is used for the characterisation of the amount of interface delamination [77].

Figure 25 shows a sketch of the button test and a typical fore-time recording. The force-time recording shown relates to measurements on samples of passivation material and epoxy moulding compound [77]. Tensile strength at 25°C is about 50% weaker as shear strength. Also, a strong relation between temperature and shear strength is found, revealing a 25% decreased shear strength at 75°C.

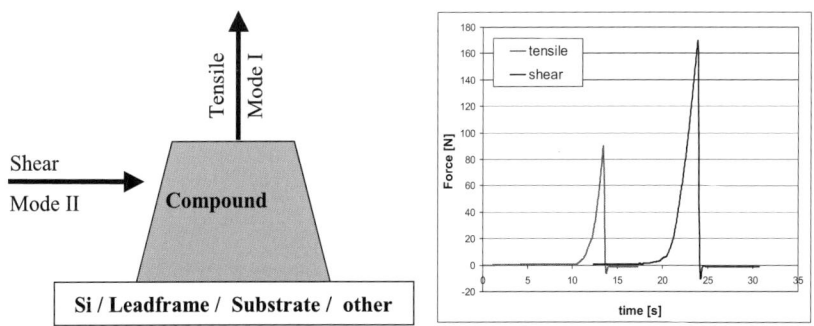

Figure 25. Button shear/tensile test sample (left) and typical force-time recording (right)

2. *4-Point bending with pre-notch crack.* The theory of this adhesion test and application to the thin IC layers is described in section 3. This test method also applies to interface strength measurement in packages. An example is given below [78].

The four-point bending test is used to determine the interface fracture toughness between moulding compound-substrate, moulding compound-solder resist, and moulding compound-copper traces as a function of temperature and moisture content. To trigger the interfacial delamination, a notch of 85% of the moulding compound layer thickness was created. J-integral values are calculated at the interface using FE simulations that mimic the experiments. Figure 26 shows an example of observed crack growth along the substrate-solder resist interface (right) and a typical J-integral mesh used to calculate J-values. The interface strength values extracted from the combined experimental/numerical method are listed in Table 7. For some interfaces, such as MC - FR4, no significant decreasing strength due to increased temperature is found whereas for others, such as

MC - SR the strength may decrease with 50%. For all possible interfaces, moisture content may decrease the strength by 50% or more.

 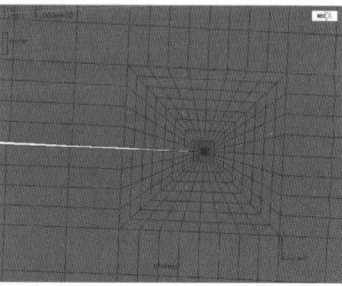

Figure 26. Observed crack along the substrate-solder resist interface (left, MC = moulding compound, SR = solder resist, FR4 = substrate) and typical J-integral FE mesh (right)

Table 7. J-values for the interface strength in BGA packages (MC = moulding compound, SR = solder resist, FR4 is core substrate material, Cu = copper)

	Interfacial Strength [N/m]	
Interface	Under dry conditions	Under wet conditions
MC - FR4	600-800 (20°C to 150°C range)	400-600 (20°C to 150°C range)
MC - SR	>500 (20°C)	200-300 (20°C to 150°C range)
	200-300 (>100°C range)	
SR - FR4	400-500 (20°C to 150°C range)	200-300 (20°C to 150°C range)
MC - Cu	5-7 (20°C to 150°C range)	2-3 (20°C to 150°C range)

3. *Dual or double cantilever beam test.* The Dual Cantilever Beam (DCB) test method is a well-known method for determining mode-I fracture toughness of materials and interfaces. Samples used are sandwich like specimens where on both ends cantilevers are connected to apply a vertical load. The interface fracture energy can be measured at a phase angle $\psi = 0°$ (nearly mode I) with the double cantilever beam specimen. An alternative version is the Tapered Double Cantilever Beam (TDCB), which is designed so that, over a large range of values of crack length, the rate of change of compliance with crack length is constant and independent of the value of crack length. Good examples of this method applied to packaging interfaces are found in [79, 80]. In this study, a series of DCB and TDCB experiments are conducted to investigate the adhesion of the underfill-silicon, underfill-substrate, and die-attach-copper leadframe interfaces. Different passivation materials are used, one being silicon nitride (Si_3N_4) and one being Benzocyclobutene (BCB) and different substrates are used, one being Polyimide (PI) and one being

5. Thermo-Mechanics of Integrated Circuits and Packages

Flame Retardant Type 4 (FR4) based. Different die-attach curing temperatures are used to investigate the adhesion between die-attach and leadframe. The results are listed in Table 8. Depending on the type of underfill, different adhesion values are found where FR4 based substrates revealed the highest J-values. Clearly, the weakest interface is the one between underfill and IC passivation. This in contrary to the results presented in the previous section, where the weakest interface in BGA packages is the one between moulding compound and the copper metal traces. The die-attach however, adheres very well to the leadframe.

Table 8. Interfacial adhesion results using the double cantilever beam method [79, 80]

Interface	Interfacial Strength [N/m] (depending on UF-type)
Underfill - FR4 substrate	127-208
Underfill - PI substrate	40-73
Underfill - PI	40-90
Underfill - Silicon SI_3N_4 passivation	8-13
Underfill - Silicon BCB passivation	6-17
Die-attach - Cu leadframe	450-600

4. *Wedge test.* In recent years, the Boeing Wedge test has been widely used to evaluate surface treatments under adverse environmental conditions as a means of determining the durability of bonded joints. The test introduces a known tension in an adhesive joint. This fracture test is an ASTM standard (ASTM D 3762) and utilizes a mode I specimen configuration [81]. The stress is produced by elastic deformation of two adherent plates through the introduction of a wedge. The test consists of creating an initial crack by inserting a wedge, and then following the propagation of the crack with time. The driving force for he propagation of crack comes primarily from the stiffness of the beams separated by the wedge and this driving force decreases as the crack propagates. It is important to note that in this test the cracked specimen also experiences simultaneous environmental attack at the crack site (when the specimens are placed in that environment). The length of the crack at equilibrium gives both the effective fracture energy and the peel strength of the adhesive assuming no plastic yielding in the specimen. Some points on this test should be noted:
- It is difficult to quantify the exact crack length.
- The standard test uses the rule of thumb height/width aspect ratio of 1:10 for hard, brittle adhesion layers, for elastic, soft layers this aspect ratio has to be modified.

- The test is still primarily used in a qualitative manner as the exact crack length is difficult to monitor and the exact state of environmental penetration is difficult to assess.

An example of this method applied to packaging interfaces is found in [82]. In this study, the adhesion of different silver filled die-attach materials to FR4 are tested using the (falling) wedge test. Values reported are in the order of 150 to 180 N/m for this interface.

5. *Modified ball-on-ring test (or blister test).* The modified ball-on-ring or shaft-loaded-blister test is a typical mode I interfacial strength measurement technique [83]. Figure 27 shows the experimental set-up of this test and a typical force recording. A stainless steel cylindrical shaft with a concave end is attached to the load-cell of a universal-testing machine. The specimen with the hole facing up is put on a ring support so that the path of the shaft will not be obstructed. A steel ball is placed inside the blind hole and the shaft is adjusted to just touch the steel ball. A crosshead speed is set on the universal testing machine. The applied load versus shaft displacement is recorded simultaneously throughout the entire loading process. An example of this method applied to packaging interfaces is found in [83]. In this study, the adhesion of moulding compound to copper leadframe is tested using the blister test as a function of temperature and moisture. Values reported are in the order of 2-3N/m for this interface at room temperature, which can drop down to < 1N/m at reflow temperature of 240ºC. Our own measurements performed on this interface revealed values of 3-4N/m at room temperature. Clearly, the adherence of moulding compound to leadframes is not very good.

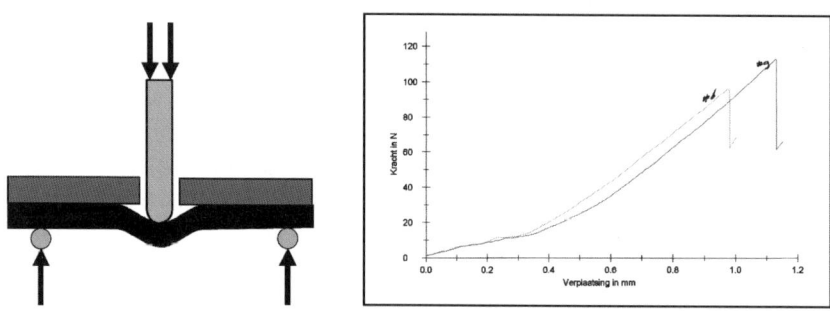

Figure 27. Set-up of the blister test (left) and typical recording (right)

6. *Any other, either combinations or deviations of above.* Other adhesion test methods reported in the literature are, for instance, the 90º peel test [84], three point bending with pre-crack [83, 85], and the mixed mode bending test [86, 87].

5. *Thermo-Mechanics of Integrated Circuits and Packages*   221

The 90° peel test can be used to measure the adhesion in an adhering system. In the peel test, thin films are attached to a Silicon substrate and peeled off in an 90° angle.

The three point bending with pre-crack test is identical to the four point bending variant (item 2), with the difference that a pre-crack is put between both materials by using a specific releasing agent. Using this method, the interface strength of different moulding compounds with leadframe materials is measured as a function of moisture content [85]. Reported values are 1-2N/m at room temperature for the moulding compound - copper interface, which may drop to 0.5-1.0N/m due to moisture and even to 0.2-0.5N/m due to combined moisture and increased temperature.

The mixed mode bending (MMB) test can be regarded as a superposition of the DCB test, which is pure mode I, and the end notched flexure (ENF) test, which is pure mode II. The MMB test provides stable crack growth over the full range of mode angles. Via a hinge system, the position of the force acting on the sample can be changed and, thus, the mode mixity. An advantage of the MMB test is that only one sample geometry is needed. Recently, the MMB test is developed for adhesion between IC layers [86] and packaging materials [87].

4.6   Typical Packaging Material Properties

An overview of material properties for typical packaging materials is listed in Table 9. This table includes ranges for Young's modulus, Poisson ratio, CTE, yield stress (for the metals), and interface strength J-values for the given materials as reported in the literature [65-87].

Table 9. Some typical packaging material data, interface strength at 25°C (PV = passivation, MC = moulding compound, D/A = die-attach, Cu = copper, SR = solder resist, UF = underfill, PI = polyimide)

Material	Constitutive Relation	E [GPa]	ν [-]	CTE [ppm/°C] T_g [°C]	Interface strength J-value[N/m]
Si	Anisotropic	150-169 (main direction)	0.23-0.25	2.5-3.5	PV-MC > 600 PV-UF 8-13
Die-attach	Linear visco-elastic	1-6 (25°C) 0.8-2 (>200°C)	0.2-0.3	50-100 < T_g 100-200 > T_g T_g: 30-80	D/A-Cu 450-600
Compound	Linear visco-elastic	10-25 (25°C) 0.1-0.8 (>200°C)	0.2-0.3	6-50 < T_g 50-100 > T_g T_g: 80-200	MC-Cu 4-5 MC-SR >500
Substrate FR4	Orthotropic In-plane:	15-20 (25°C) 8-10 (>200°C)	0.15-0.2	14-17 50-60 < T_g	FR4-MC 600-800 FR4-Cu 5-10 FR4-SR 400-500

Material	Constitutive Relation	E [GPa]	ν [-]	CTE [ppm/°C] T_g [°C]	Interface strength J-value[N/m]
Underfill	Out-plane: Linear visco-elastic	4-5 (25°C) 1-2 (>200°C) 5-10 (25°C) 0.1-0.5 (>200°C)	0.2-0.3	200-250 > T_g T_g: 180-250 6-50 < T_g 50-100 > T_g T_g: 80-200	FR4-UF 127-208 UF-PI 40-73
Leadframe CuNi$_3$ or CuFe2P	Elasto-plastic	120-130	0.30-0.35	16-18	Yield stress: 200-600 MPa
FeNi$_{42}$	Elasto-plastic	140-150	0.30-0.35	5-6	Yield stress: 400 to 600 MPa

4.7 Finite Element Modelling for Packaging

Package FE modelling is able to quantitatively predict the package warpage and the stress and strain levels in the different materials. Important techniques that should be addressed are:
- Element birth and death to activate, for instance, the moulding compound at the correct temperature.
- The total manufacturing process should be included in the model to correctly predict the stress/strain history.
- Time and temperature dependent properties, E(T) and/or E(f, T), should be taken into account.
- Linear fracture mechanics, using J-integral, area release energy, virtual crack closure, or any other method to calculate fracture energy to predict the occurrence of delamination within the package.
- Contact elements and/or contact bodies to simulate effect of delamination within the package.
- Verify the modelling results with respect to warpage, stress, and/or strain levels by using verification techniques.

Four cases on the package level will be discussed in the following sections.

4.7.1 Prediction and Verification of Process Induced Package Warpage

In this example, 3D FEM models are developed to predict the thermal deformations of certain electronic packages and naked die samples under packaging and testing loading [88]. For all the package constituents, appropriate material properties and models are used, including temperature-dependent visco-elasticity, anisotropy, and temperature-dependent elasticity and plasticity. To verify the developed FE models, a series of optical

5. Thermo-Mechanics of Integrated Circuits and Packages

metrology tests are performed. A compact 3D Interferometry testing system that can measure simultaneously out-of plane and in-plane deformations has been developed [89]. Thermal deformation measurements are performed on samples of both real electronic packages and naked dies attached on a leadframe. Parametric 3D FE models representing both a naked die and a full package geometry are developed. Because of symmetry, only one quarter of the sample is modeled. The effect of element sensitivity is explored by using several distributions and/or discretisations. In total, the naked die model consists of approximately 4,000 8-noded elements, the package model of approximately 12,000 elements. Sample geometry and the FE models are presented in Figure 28 and consist of:
- Naked dies attached to a leadframe. A 7.5 x 7.5 mm^2 IC is attached to a 9.0 x 9.0 mm^2 diepad.
- Real electronic packages. A 9.8 x 9.8 mm^2 IC is attached to a 11.0 x 11.0 mm^2 diepad, moulded in a 28.0 x 28.0 mm^2 body.

In both FE models, along the xz-symmetry axis all nodes are constrained in y-directions, along the yz-symmetry axis all nodes are constrained in x-directions. To avoid rigid body movements, one node is constrained in z-direction.

Figure 28. 3D FE models (top) and sample geometry (bottom) for the naked die (left) and the package (right) samples

The single crystal silicon die is modelled as temperature independent anisotropic [65]. For the compound and die-attach materials, linear visco-elastic models are used [65]. For both sample types, the complete time and temperature profile during manufacturing, i.e., die-attach, wire bonding, moulding, post mould cure, and marking, is used in the simulations, followed by the experimental thermal cycle. To develop reliable and efficient

non-linear thermo-mechanical prediction models for the product/process designs of electronic packaging, various justified simplifications and assumptions are needed:
- The curing of the die-attach material is investigated in a heat production test. In such a test, the heat generated in the material is continuously monitored during temperature increase. The heat production measurements on the die-attach material indicated at a warpage/stress free temperature of 150°C. Therefore, the warpage/stress free state for the silicon die - die-attach - leadframe assembly is set at 150°C. For the compound material a warpage/stress free temperature of 175°C is assumed.
- Isothermal loading conditions are used for both the manufacturing processes and the experimental testing conditions.
- The initial IC warpage/stresses are neglected.
- Perfect adhesion is assumed between die and leadframe, and between die and compound.
- The curing process induced stresses are neglected.

An effectively integrated, compact 3D testing system is set up, based on the Moiré and Twyman/Green Interferometry, to verify the predicted warpage of both samples. An optical fibre induces the light source. The laser is coupled into a single-mode fibre. The single-mode fibre is split into two branches; one provides the light source for the sub-system of in-plane deformation measurement (2D Moiré system) and another for the sub-system of out-of-plane deformation measurement (Twyman/Green system). The loading system is an important accessory part of the 3D testing system. This system is composed of a mini-mechanical loading frame and a mini-thermal loading chamber, which is specifically designed for micro-electronic products. The mini-loading frame and thermal chamber are sited on a six-dimensional adjustable base, which can move or rotate the sample along the x-, y-, and z-axis. With this system, the 3D deformation of samples can be measured simultaneously during thermal cycling and/or mechanical loading conditions to give an accurate and 3D characterization of the thermo-mechanical behaviour of electronic packages. At the start of the experiment, the sample is placed, in a fixed way, in the mini thermal chamber. Starting from a temperature of 25°C the sample is heated at a rate of 8°C/minute. When it reaches 175°C, the temperature is held for about 5 minutes, and then it is cooled down naturally with the chamber door closed. The temperature in the thermal chamber is monitored by a thermocouple, placed close to the sample. For one experiment, the temperature of 175°C was held for about 1 hour, to explore the relaxation effect of the samples. However, this experiment did not give other results. Deformations of the sample are recorded during heating and cooling, respectively. Two temperature cycles are recorded for each sample.

5. Thermo-Mechanics of Integrated Circuits and Packages

Figure 29 shows an example of measured versus calculated deformations for the naked die sample (w-field/z-direction) and the package sample (v-field/y-direction). This figure clearly shows identical patterns for both the measured and simulated results.

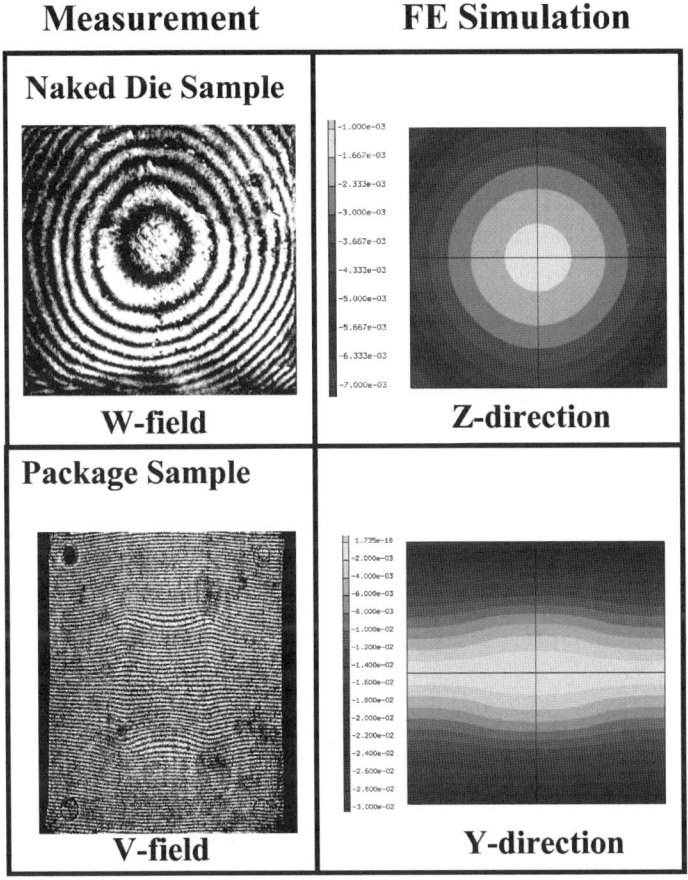

Figure 29. Measured versus calculated deformation patterns; w-field/z-direction for the naked die sample (top) and v-field/y-direction for the package sample (bottom)

Figure 30 shows the deformations of both samples during cool down as a function of temperature. For the naked die sample the warpage (z-deformation) is depicted clearly showing that the mismatch between measured and simulated results is within 15%. For the package sample the horizontal and vertical deformation from centre to edge after cool-down to 25°C is also depicted in Figure 30. Measured results are compared with those

from the FE model using visco-elastic properties versus linear elastic properties for the compound and die-attach materials. Linear elastic properties for both materials are based on supplier data. This figure shows that the mismatch is very large when linear elastic properties are used in the FEM models. The figure clearly demonstrates that the time-dependent effects in IC materials cannot simply be neglected for quantitatively reliable prediction of deformations in electronic packages. The largest deviation between measured and simulated results is found nearby the interface between the diepad and the compound. Quantitatively, Figure 30 shows that the mismatch between the measured and simulated results is within 15%. This difference can be attributed to:
- Time effects such as degradation and relaxation,
- Asymmetric placement of the die on the diepad,
- Non-uniform die-attach thickness below the die,
- Possible interface delamination.

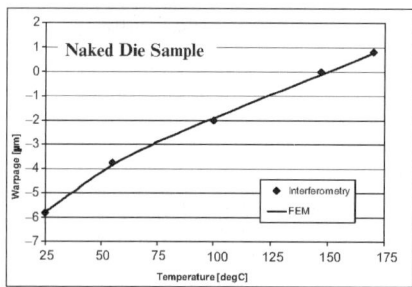 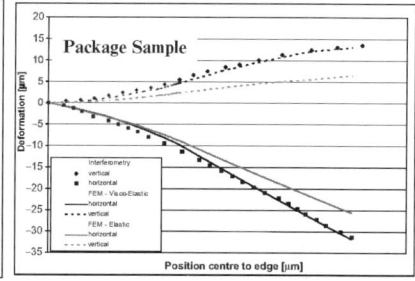

Figure 30. Comparison between measured and calculated deformations as a function of temperature; warpage of the naked die sample during cool down (left), horizontal and vertical deformation of the package sample from centre to edge after cool-down to 25°C

The results presented by this example show that the Interferometry technique can be used for both in-plane and out-of-plane deformation measurements of IC packages, with quantitatively reliable accuracy. Thermal deformations obtained from the non-linear FEM models match well with measured deformations for both the naked die samples and the real electronic packages.

4.7.2 Prediction of Wire Failures in Micro-Electronic Packages

Wire bonding is still the dominant form of first level interconnection in the world today. At present, over 95% of the manufactured packages (in volume) are wire bonded. Wire bonding technology is changing due to the increasing demands placed on wire bond pad pitch. Due to the trend of

miniaturization, function integration and cost reduction, the ability to predict and eventually to prevent thermo-mechanical failures of microelectronics is becoming increasingly important. Wire bond related failures, such as wire sweep and wire break, account for more than 25% of the total reliability problems of electronic packages under manufacturing and testing conditions. The currently used wire is gold with a typical diameter of about 30 µm. According to the technology roadmap, the thickness will be reduced to beyond 15 µm, and copper may replace gold for many applications. These changes generate extra challenges for the prediction and prevention of wire bond failures. Some studies already addressed the influence of different low-k materials and interconnection materials [90, 91, 92], but still unknown influences for thermo-mechanically related wire bond failures exist. Different modelling methods can be used to investigate the wire failure mechanisms to explore:
- The interaction between wire and moulding compound;
- The influence of manufacturing processes and testing;
- The influence of interfacial delamination.

Wire bonding is a complex process; the types that are being used today are wedge bonding and ball bonding. Ultrasonic and thermosonic technologies are used to create contact between the ball and the bondpad and between the stitch and the lead. The choice of wire bond process, either thermosonic ball or ultrasonic wedge bonding is made based on the pad pitch, device characteristics and throughput requirements. For pad pitches of 60-100 µm, the options are thermosonic gold ball or ultrasonic aluminium wedge bonding. At present, the majority of wire bonding is done with thermosonic gold ball bonding. Once the bond has been formed, the quality of it is critical for long-term reliability. To evaluate the strength/quality of the bond, two standard tests are being used by the micro-electronic industry, i.e., the bond pull and shear tests. In packages, the second bond (the stitch) may have a small/thin body near the lead, thus, making strength a critical issue at this region. The stitch strength is mainly determined by the outer radius of the capillary; the larger the outer radius, the higher the strength (due to a thicker body near the lead). The standard bond pull technique is not very suitable to measure the stitch strength. To evaluate the ball bond strength, the shear strength measurement is applied. The shear force depends on the contact area of the ball, which differs considerably from its wire diameter. The larger the contact area, the higher the force. Typical forces required for pull and shear tests, used to ensure the quality, range from 3.0 gr (minimum wire pull strength requirement for 25 µm wire diameter) to 6.0 gr (minimum ball shear strength requirement for 30 µm ball diameter). Although the quality of the bond should be guaranteed by the required pull and shear strengths, still wire bond related failures account for more than

25% of the total reliability problems of electronic packages under manufacturing and testing conditions. Some examples are wire sweep, stitch break, ball lift, wire break, and/or wire fatigue. Examples are shown in Figure 16. Due to the ongoing trend of miniaturization, material changes, and cost reduction, wire bond related failures are becoming increasingly important.

A 2D plain strain FE model is developed to investigate the above-mentioned interactions and influences. Contact algorithms are used to model the detailed wire in the plastic package. These algorithms include the use of contact bodies that are attached to each other or are able to delaminate. In order to predict strains/forces on a 25 μm wire, the wire is modelled as a frictionless contact body within the compound, allowing the wire to move within the compound (tunnelling effect). This simulates the lack of adhesion due to the inert properties of the gold with respect to the moulding compound. At the position of the wire duplicate elements are used on top of each other where one set simulates the wire and the other set of element the moulding compound. Thicknesses of the 2D elements are set to represent the appropriate cross section area. The bond-pad pitch is chosen 100 μm and the thickness of the 25 μm round gold wire in the 2D square representation equals 22 μm. Inelastic strains within the wire are calculated since these will eventually lead to a wire fatigue break. Figure 31 shows the 2D FE model.

The electronic package consists of a silicon die, attached to the leadframe by using die-attach material and in a later packaging stage encapsulated by a moulding compound. The wire loop profiles are taken from cross-sections. The single crystal silicon die is modelled as an anisotropic material, the leadframe material as an ideally elasto-plastic material, for the die-attach and moulding compound linear visco-elastic properties are used with temperature dependent coefficient of thermal expansion. The gold wire is modelled as an ideally elasto-plastic material with a yield stress of 150MPa. For the studied package, both assembly processes (die-attachment, moulding, board assembly) and temperature cycling testing conditions are considered. To develop reliable and efficient non-linear thermo-mechanical prediction models for the product/process designs of electronic packaging, various justified simplifications and assumptions are commonly used, such as:

- The stress free state for each material is set at its process temperature.
- During the simulations, all elements of a specific material are deactivated, and they are activated during the relevant process.
- Isothermal loading conditions are used for the modelling.
- The initial die warpage/stress is neglected.
- Contact is assumed between the wire and the compound, and between the leadframe and moulding compound. The wire remains attached to the die and leadframe regardless whether delamination between compound and leadframe exists at that place.

5. Thermo-Mechanics of Integrated Circuits and Packages

The simulation results are described below.

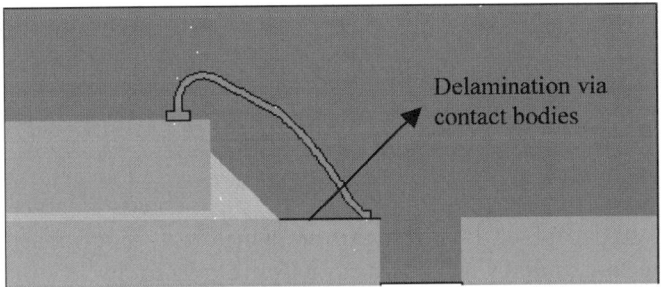

Figure 31. 2D plane strain FE model with wire included

- *The interaction between wire and moulding compound.* The interaction of the wire with the moulding compound is investigated by using contact algorithms. Figure 32 shows the interaction between the wire and the compound at two different process temperatures where a red colour indicates full contact. Clearly, at 25°C the compound fully encloses the wire, simply an effect of shrinking due to cool down. However, at higher temperatures, 150°C, the compound releases the wire at some places allowing the wire to move free. As a result of this interaction, the compound grasps the wire and moves it around during processing and testing. Only the friction between compound and wire will restrict its movement.

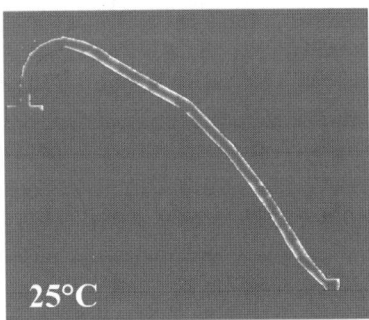 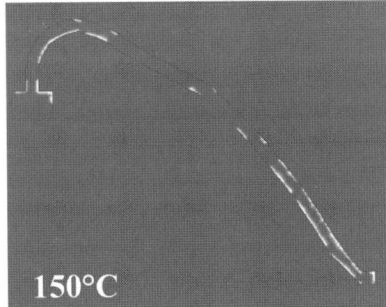

Figure 32. Interaction between wire and compound at 25°C and 150°C, White indicates full contact

- *The influence of manufacturing processes and testing.* Figure 33 shows the (normalized) forces on the wire stitch after moulding, after board assembly, and during temperature cycling testing both on a package-only and a package-on-PCB level. To simulate the effect of the PCB, a 4-layer FR4-based board is included in the model. The figure clearly shows that the wire forces increase during processing and testing:
    o A factor of 1.55 when assembled to the board;
    o A factor of 2.11 due to package testing;
    o A factor 1.92 due to the board level testing.

The effect of the package testing on the stitch forces is larger than the effect of the package-on-PCB testing from which is it can be concluded that 'package testing covers wire failure during further assembly'.

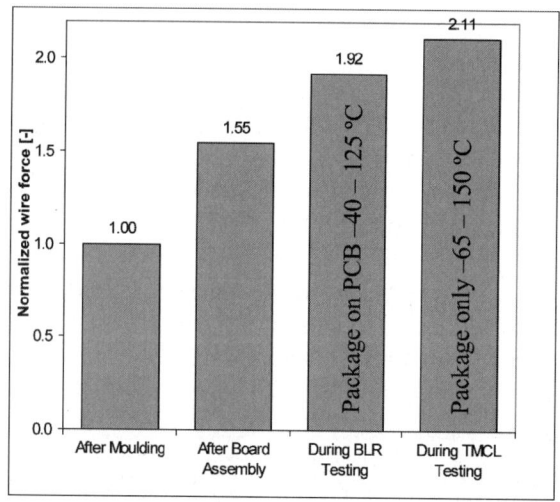

Figure 33. Normalized forces subjected to the wire as an effect of processing

- *The influence of interfacial delamination.* Figure 34 shows the (normalized) forces on the wire stitch as an effect of delamination between the leadframe and the compound during processing/testing. The figure shows that the effect of delamination is more than a factor 3.5. Processing and testing influences are much less than those of the delamination. This indicates that delamination is the key trigger for wire failures. The (increased) forces on the wire will result in inelastic deformations within the wire and due to the cyclic nature of the force during temperature changes this eventually leads to wire fatigue. Figure 35 shows the inelastic strains in the wire with and without delamination

5. Thermo-Mechanics of Integrated Circuits and Packages

and compares a small package with a large one. The following can be concluded from this figure:

- Without delamination, there are no significant differences between a small and a large package;
- Delamination significantly increases both wire forces and inelastic strains;
- With delamination, wire inelastic strain are consumed in the delaminated area, in this case the wire stitch;
- With delamination increasing package size results in increased inelastic strains.

Or in other words: *delamination is the key trigger for wire failures*.

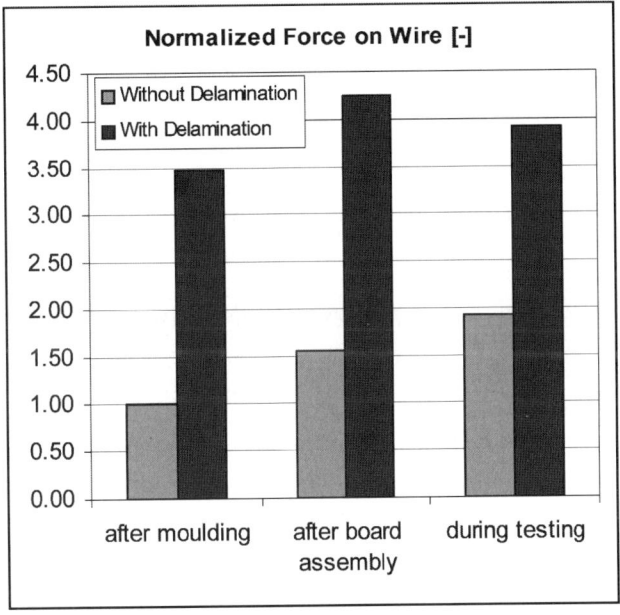

Figure 34. Normalized forces subjected to the wire as an effect of delamination

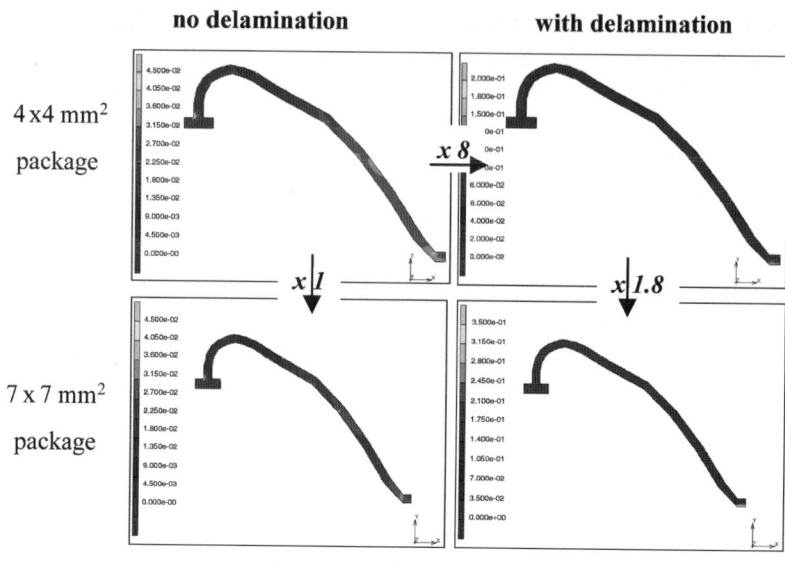

Figure 35. Wire inelastic strains for a small and large package; with and without delamination

4.7.3 Prediction of Delamination Related Problems in Exposed Pad Packages

Many reliability problems are triggered by delamination at given frontend and/or backend material interfaces. Examples are die-lift (delamination between die-attach and leadframe) and downbond stitch breaks associated with compound to diepad delamination [94]. This example highlights our results to find the driving mechanisms for delamination-related reliability problems in micro-electronic packages using state-of-the-art virtual prototyping and/or qualification techniques. Novel numerical prediction techniques are explored able to predict the occurrence of interfacial delamination as function of manufacturing and testing conditions in micro-electronic packages. At present, numerous numerical techniques are available to predict delamination:
1. *Virtual Crack Closure.* The Virtual Crack Closure Technique (VCCT) is a known technique able to predict initiation of cracks and/or interface delamination. The technique can be used to calculate the energy release rate both in 2D and 3D [95]. Since the critical energy release rate is a function of mode mixity, the components of the energy release rate corresponding to the three basic fracture modes I, II, and III can be separately determined.

Software codes apply dedicated crack tip elements to calculate these values. Others have used the technique to predict energy release rates for interface cracks at specific positions and with a specified length [47, 96]. The major disadvantages of the use of VCCT are:
- Crack locations need to be fixed in the model, in other words the exact location of the delamination needs to be a priori known.
- Due to the above is no information on crack initiation, only on crack propagation.

2. *J-integral Method.* J-integral methods are widely used in rate-independent quasi-static fracture analysis to characterize the energy release rate associated with crack growth. The technique was introduced by Cherepanov [97] and Rice [98] in 1967 and 1968, respectively. It found worldwide interest and applications in the 70s, and with increasing capabilities of computers and finite element methods, J-based elastic-plastic fracture mechanics became also an issue of numerical computations [99]. The J-integral method needs a so-called crack-tip mesh. Through a path-independent contour integral, the available energy to delaminate the given interface can be calculated. It is shown that the J-integral is identical with the energy release rate. Big advantage of this method is the relative short calculation time; major disadvantages are identical to the ones for the VCCT method.

3. *Cohesive Zones/Interface Elements.* The cohesive zone method is based on a dedicated element-description for the given interface. As such, the interface is mimicked itself by so-called interface elements which properties describe the adhesion between the two materials. Due to this nature, cohesive zone methods require more (material) input need, for instance, the (initial) interface stiffness. Dedicated experiments are required to obtain this input. Besides this, when using cohesive zone techniques long calculation times are to be anticipated, especially when many and brittle interfaces are present. This is due to the fact that when describing the fracture properly, high mesh densities are needed. Major advantage of this method, besides the fact that it is easy to use (relative to the other techniques), is the ability to predict initiation and propagation of interface delamination.

4. *Area Release Method.* For analysing and comparing different back end structures, a novel failure index, the so-called Area Release Energy is recently developed [61], see also section 3.7. This value predicts the change of delamination of critical interfaces without knowing a priori the exact location of the delamination. The amount of energy is calculated that is released upon delamination for any position along a critical interface. The validity and usability of the developed Area Energy Release methodology was illustrated based on comparisons with simulation results obtained from fracture mechanics. The Area Release method does not require any

presupposed position of any initial crack. Instead, at any desired positions within the specimen, an area energy release value is calculated which basically results from releasing an area (having a defined dimension) around each point in the specimen.

In reality, a combination of the above mentioned techniques should be used to predict initiation and progression of interface delamination. For fast calculation of interface(s) at risk, the J-integral method can be used. For the given interface(s) at risk, a cohesive zone simulation can be used to predict the delamination pathway.

Prediction of Delamination in Exposed Pad Packages

A 2D non-linear FE model, based on plane strain assumptions, including isotropy for silicon [65], visco-elasticity for moulding compound and die-attach [65], elasto-plasticity for the copper leadframe constructed. A multi-physics FE methodology is used which can take into account the moisture and thermo-mechanical related mechanisms. The effects of hygro-swelling, vapour pressure, and thermal expansion on the failures in the exposed pad family are taken into account [100-103].

For predicting delamination growth of an existing delamination, both the J-integral and the cohesive zone method are used. J-integral values are calculated at different locations within the package, see Figure 36. As nominal model, the HLQFP package is selected with body size 20 x 20 x 1.4 mm^3; pad size 5.6 x 5.6 x 0.125 mm^3; die size 4.5 x 4.5 x 0.38mm^3 and given material combination and Moisture level MSL3.

Through a solid mesh sensitivity analysis the eventually used mesh size is fixed. Different integration-paths are analysed to fix the path to calculate the eventual J-value.

For the cohesive zone model, dedicated elements are put between the epoxy to leadframe parts to mimic this interface. Adhesion strength values and initial stiffness values are taken from measured results [93].

The loading scheme includes the thermal, moisture, and vapour pressure loading.

5. Thermo-Mechanics of Integrated Circuits and Packages

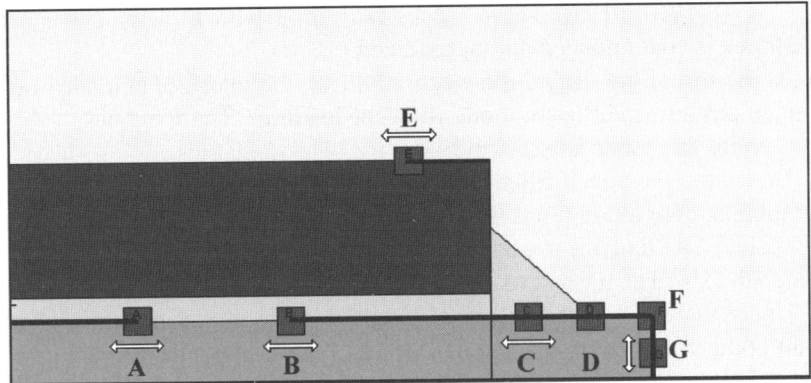

Figure 36. At locations A - G the J-integral values are predicted

Nominal Model Results

Figure 37 shows the locally deformed structure after moulding and MSL loading conditions. It is clear that the die-attach pulls at the exposed pad and high J-values are expected in this location. Due to the moisture loading, the swelling of the compound and die-attach decreased these local deformations, and thereby, closes any interface present.

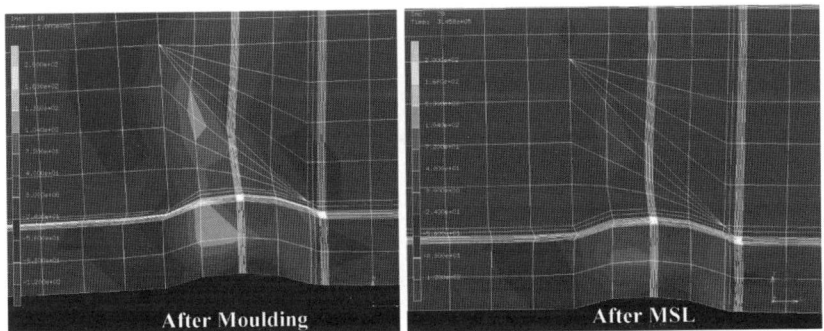

Figure 37. Deformed structure at the die edge in the exposed pad and die-attach area after moulding (left) and MSL (right) loading conditions

Mode mixity is an indicator for the opining mode of the interface. Figure 38 shows the mode mixity along the diepad for the nominal model, 0% is the starting point at the side of the pad, 20% is the corner point, 40% is the point of the die-attach fillet, 60% is exactly below the die, 100% is at the symmetry line.

At the diepad side (between 0-20%), the interface is loaded under a mode mixity of 20°. The interface toughness value for compound-leadframe at

20°C is ~8-10N/m [93]. These values may drop with a factor 3 when the leadframe is contaminated and/or oxidized.

At the top of the diepad (between 20-40%), the interface is loaded under a mode mixity of 90° (pure mode II, shear loading). The toughness value at 20°C under this mode is >25N/m [93].

Under the die-attach fillet (between 40-60%), the mode mixity drops to 10° (almost pure mode I, tensile loading) and rises to 60° when it reaches the die corner. The toughness values for this interface under this mode would be about 10-25N/m at 10° and rising to 100-125N/m at 60° [93].

Under the die (between 60 to 100%), the mode mixity remains constant at 60°, where a toughness value of 100-125 N/m is expected (probably even higher but no data is available/measured at this mode).

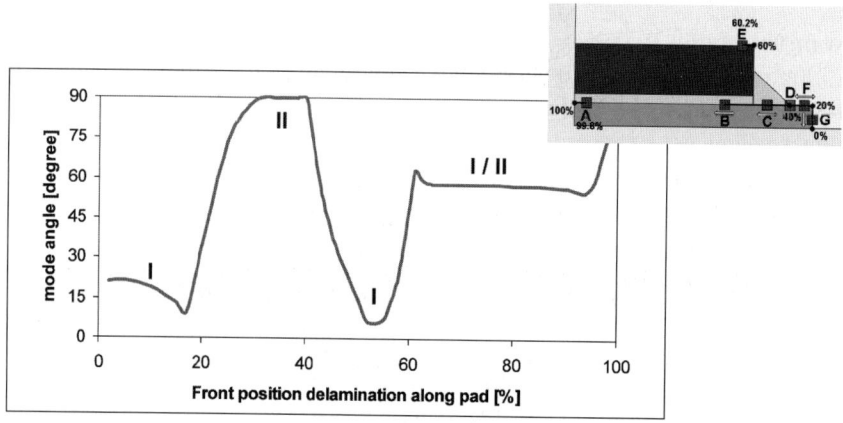

Figure 38. Mode mixity along the diepad interface

Cohesive Zone Results: Progression of Delamination

A FE simulation is performed using cohesive zone elements (or interface elements) along the diepad interface. In the cohesive zone elements the interface toughness values in the different modes are given as an input. For the nominal model only, a cohesive zone simulation is performed to predict the most likely pathway for interface delamination in exposed pad packages during processing. Figure 39 shows the results of this simulation, where the interface delamination is highlighted. The cohesive zone simulation confirmed the J-integral simulations with the place where the delamination will start: below the die corner in the die fillet region. As the loading increases, by thermo- or hygro-mechanics, the interface delamination progresses below the die. In a next stage, when the delamination below the die has reached a certain length, the forces at the side of the diepad are increased and exceed

5. Thermo-Mechanics of Integrated Circuits and Packages

the toughness values. Almost instantly the complete side and top of the diepad delaminate in this stage. Finally, total die-lift with total diepad delamination will occur, see the last picture in Figure 39. This is the thermo-hygro-mechanics based delamination pathway for exposed pad packages during processing and testing.

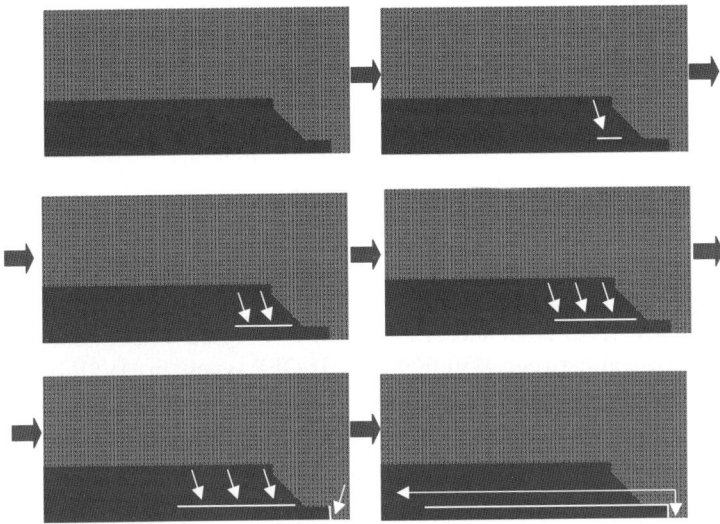

Figure 39. Delamination pathway driven by thermo-hygro-mechanics

J-values During Processing and Testing

Figure 40 shows the calculated J-values at the different locations as function of the loading conditions. In this case, J-values are calculated for the nominal package. The following can be concluded from this figure: At the side of the pad (location G and F), the J-values are below 5N/m during processing, indicating that this interface will not fail from a thermo-mechanical point of view. During TMCL, the J-values increase to 10N/m and are getting closer to the toughness values. During cool down from the moulding temperature the J-values at the locations B and C (interface die-attach with leadframe) increase dramatically, especially at location C, directly below the die corner. During MSL testing, the J-values drop. This is due to the expansion of the compound as a result of moisture uptake. When swelling the compound closes the interface and J-values decrease. The effect of the moisture is purely degrading the interface toughness with 20-40%.

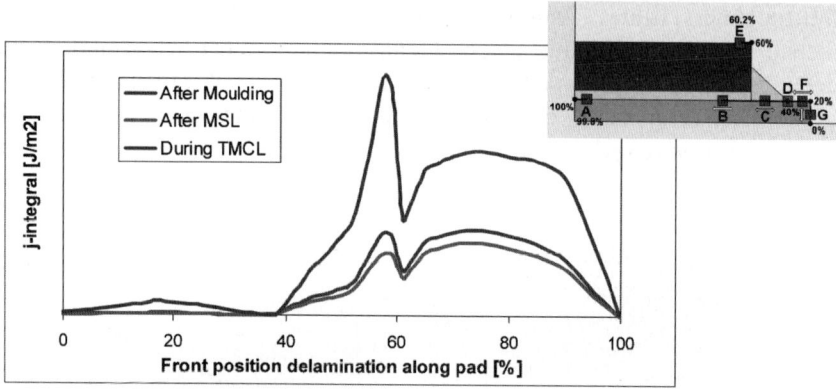

Figure 40. J-values at the different locations as function of the process conditions

Effect of Package Type

Exposed pad packages exist in different package types, all having a different body, leadframe, and/or die thickness. Figure 41 shows the J-values for a comparison between HVQFN and HLQFP types. In HVQFN, standard thickness equals 0.85 mm for the body, in HLQFP, this value is equal to 1.40 mm. The results clearly show that the HVQFN package type causes lower J values below the die, which is a combined effect of fillet height and vertical stiffness ratios.

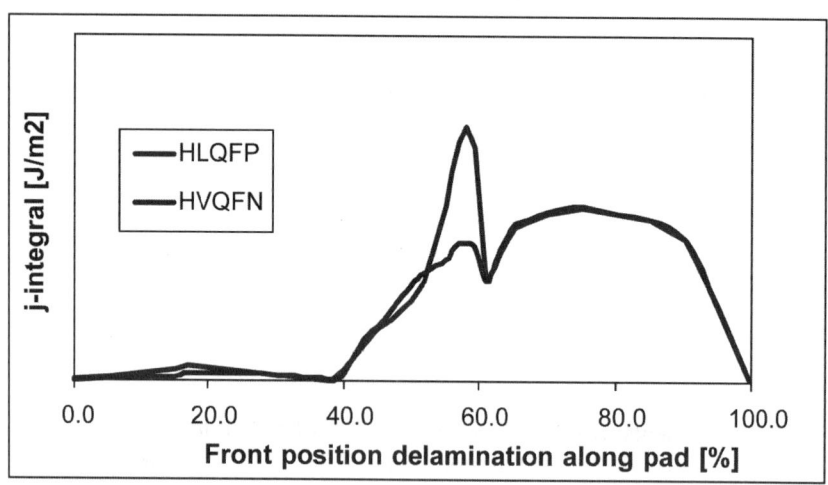

Figure 41. J-values along the diepad for HLQFP and HVQFN package type

Effect of Die-Attach Fillet Height

Figure 42 shows the effect of the die-attach fillet height for the J-values at the locations G (side of the pad) and D (below the IC corner). At the side of the pad, the J-values more or less remain constant when increasing the die-attach fillet height but below the IC, the J-values strongly increase when increasing the die-attach fillet height. This is a result of the increasing pulling forces the die-attach imposes to the leadframe.

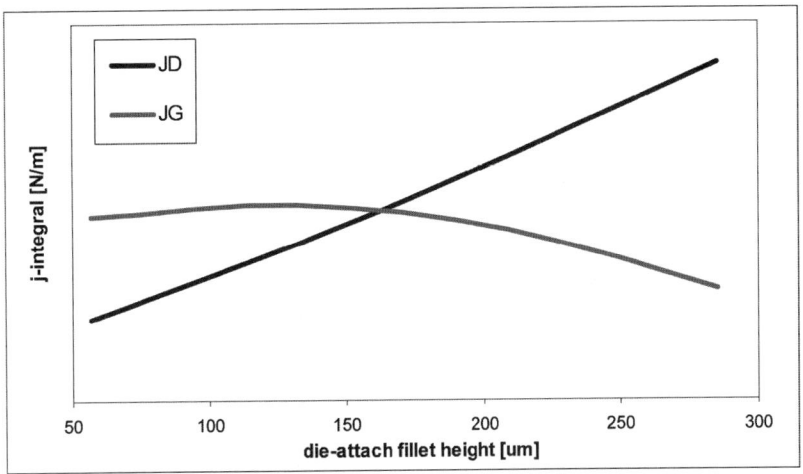

Figure 42. J-values as function of the die-attach fillet height for locations D and G

Combining J-integral and cohesive zone methods enables to predict initiation and progression of interface delamination. For fast calculation of interface(s) at risk, for design and material investigations, the J-integral method can be used. For the given interface(s) at risk, a cohesive zone simulation can be used to predict the delamination pathway. Although, current numerous numerical techniques are available and prove their use to predict interface delamination still: *Making the link between modelling results and reliability and/or field observations is the next challenge.*

4.7.4 Prediction of Curing Induced Warpage in Map Mould Systems

HVQFN (Heatspreader Very thin Quad Flat No lead) packages are packages composed of an Integrated Circuit (IC) attached to an exposed pad and in a later stage encapsulated with a moulding compound. It has been introduced into the semi-conductor market as a thin, cost effective, thermal and high frequency package solution [104]. The exposed pad is an exposed metal plate and is located on the bottom of the package. For reasons of cost reduction many HVQFN's are moulded in so called map moulds; i.e., an array of m × n devices created in one mould shot. After moulding and subsequent Post Mould Curing (PMC), the HVQFN's are separated by means of a sawing process. For a reliable sawing process the map mould must not exhibit a too large warpage at room temperature to enable vacuum chucking. However the cooling from Post Mould Curing temperature to room temperature inevitably introduces different thermal strains in the different materials. A certain amount of warpage will be the result. In this sections the mechanisms for warpage with the target to eventually reduce this warpage to an acceptable level is described. Figure 43 shows an example of a 5-map HVQFN system with a detail of the leadframe design of one map.

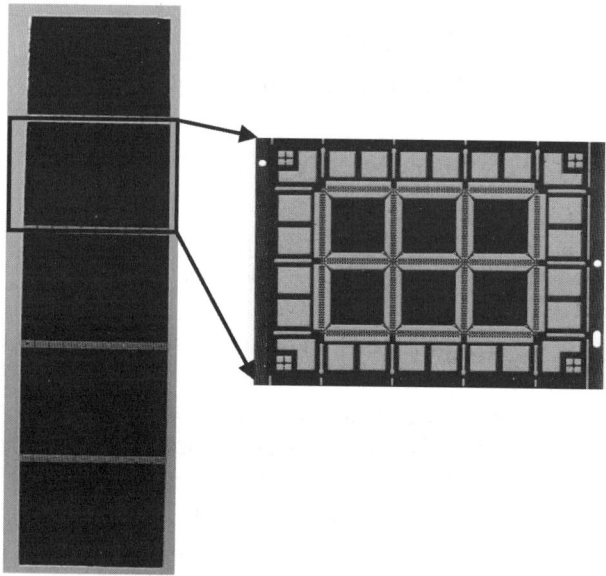

Figure 43. Typical 5-map HVQFN system (left) with a detail of the leadframe design for one map (right)

A parametric 3D FEM model of the map mould was created [105]. In this model the effect of a dead weight forcing the map mould to remain flat during Post Mould Curing was included. Figures 44 shows a top view and a cross section of the FEM model. Due to symmetry of the map mould only one quarter needs to be modelled. Symmetry boundary conditions are applied to the left and lower edge of the model. Also the vertical (z) displacement of one node was suppressed. Total package thickness equals 0.85 mm, die thickness 250 μm, and leadframe thickness 200 μm.

5. Thermo-Mechanics of Integrated Circuits and Packages

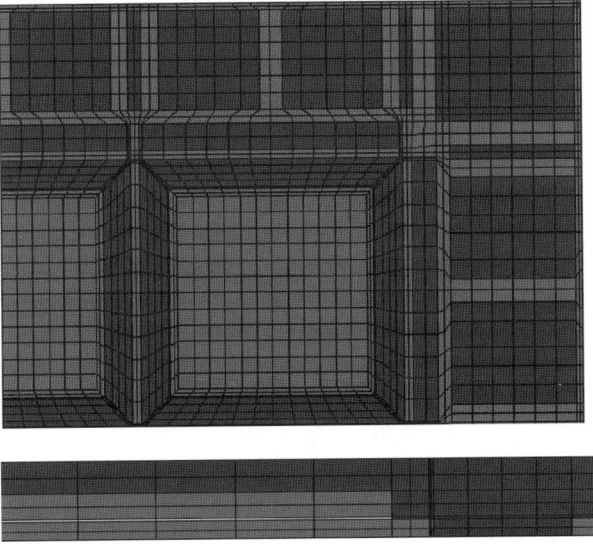

Figure 44. FEM model of ¼ HVQFN100 map mould, top view (top) and cross section (bottom)

The FEM simulation starts at the end of the moulding cycle, assuming that the HVQFN is perfectly flat after moulding at 175°C. During heating from room temperature to PMC temperature, PMC itself and cooling back to room temperature the map mould is placed on a flat surface and a dead weight of 9 kg is applied to 5 map moulds. The heating and cooling speed of the packages equals 150°C/313s at maximum when the product passes the glass transition temperature T_g during cooling. This speed is used in the simulations and was considered low enough to assume a constant temperature within the HVQFN. For the compound, a cure dependent viscoelastic model of the material was created [106-110]. The time-temperature superposition is described by the Arrhenius equation with the Arrhenius parameter $\Delta H/R$ determined earlier by fitting of full frequency data [108]. The silicon die is modelled as an anisotropic, the die-attach as a linear viscoelastic, and the leadframe as elasto-plastic material.

To correlate the numerical solutions, the warpage of the HVQFN100 map mould was measured using the Shadow Moiré method. This non-contact method is based on the geometric interference of a shadow grating projected on the sample surface and a real grating on a flat reference surface. The Akrometrix TherMoiré Model PS88+ device that was used for the measurements can obtain a resolution of approximately 2.5 µm. The map moulds were placed in a climate control box and the temperature was measured with

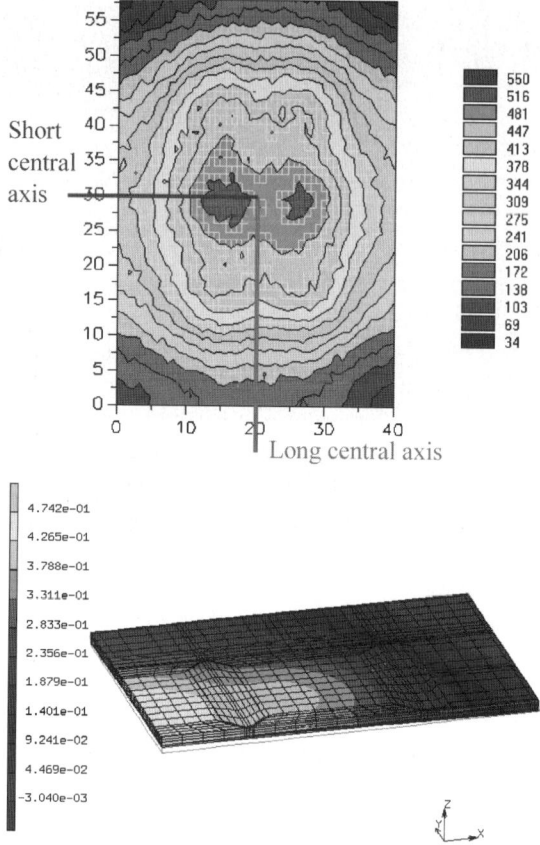

Figure 45. Z-displacement profile of a map mould measured at room temperature after 4 hours PMC at 175°C with 9 kg dead weight: measured (top) versus numerical (bottom) results

a thermocouple placed at the bottom of the map mould. Figure 45 shows the results of a warpage measurement together with the results of the FEM simulation.

Figure 46 compares the measured z-displacement data along the central axes with the FEM predictions. As can be observed the agreement is very good. A volume curing shrinkage of 0.25% was included. This shrinkage was assumed to progress with a constant speed during the 4 hours of PMC. The good agreement between experiment and FEM predictions seems to imply that in current situation (including a dead weight on top of the map) the effect of curing; i.e., stiffness change and curing shrinkage, can be simplified by a homogenous step shrinkage. Therefore it is correct to use the fully cured compound properties.

Simulations and measurements were also performed for the case that no dead weight was present. In that case the repeatability of the experiments was poor, whereas this was good for the dead weight case, and also the agreement between measurement and FEM prediction was poor. In figure 47 the FEM results using varying curing shrinkages are shown. Here the effect of the curing shrinkage on the end warpage is quite large. Therefore it is very likely that in this case the effect of curing must be included [106, 110]. The observations can be explained as follows. In the case with weight the map mould is forced to remain flat during PMC. The stresses that are caused by the curing are forced to relax to a (nearly) flat stress free situation since the temperature is above T_g. Curing effect can only reveal itself through the rubber modulus, which does not relax, and through the last part of curing shrinkage, no time to relax. Hence a small effect of the curing shrinkage on the end warpage is found when using a dead weight.

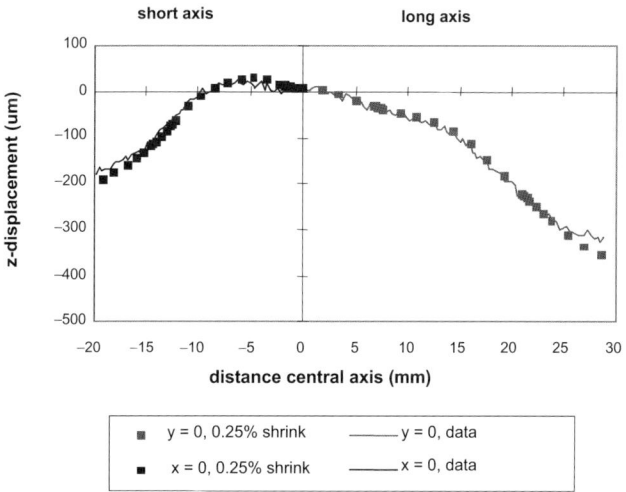

Figure 46. Measured (line) and FEM nodal (markers) z-displacement of the map mould at room temperature after 4 hours PMC at 175°C with 9 kg dead weight determined along central axes

In the no weight case the curing shrinkage during PMC can induce some warpage through the rubber modulus, especially in the no die areas. Upon cooling to room temperature the compound stiffens. This can be seen as the freezing of some Maxwell elements from liquid to solid. As they freeze, the zero-stress strain of each particular Maxwell element is fixed. In this case this freezing will occur in a curved situation and this curved situation depends on the curing shrinkage. Therefore, an effect of the curing shrinkage or of the curing in general can be expected. In case of a dead weight the freezing will always occur in a flat situation, not depending on the curing shrinkage.

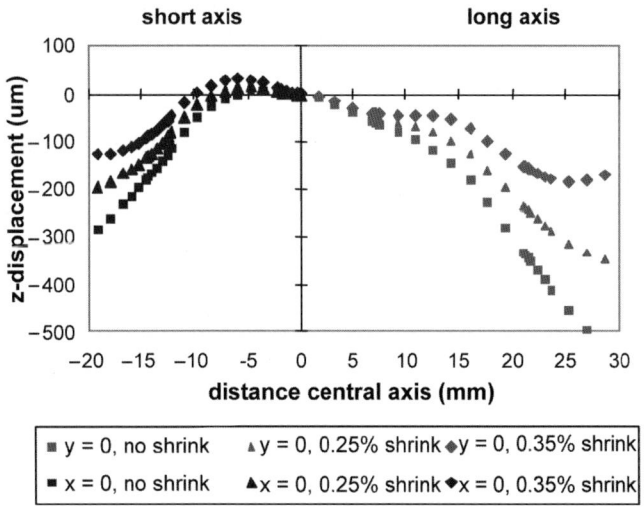

Figure 47. FEM nodal z-displacement of the map mould at room temperature after 4 hours PMC at 175°C with 9 kg dead weight and with no, 0.25 and 0.35 volume % curing shrinkage determined along central axes

Comparing the simulation with experimental results enables to tune and verify the constitutive behaviour of the moulding compound. In a next step, simulation-based optimisation can be used to optimise the compound behaviour to obtain minimal warpage of map mould systems. This will be described in Chapter 8. Going even further, the current 5-map systems can be replaced by a 1-map system when the warpage is understood and/or controlled [111].

5. THERMO-MECHANICS OF COUPLED IC BACKEND AND PACKAGING PROCESSES

In general an IC needs to be assembled before being put on an electronic board by the end customer. The materials of which the IC is made and the materials for the assembly normally have different thermo-mechanical behaviour, in particular thermal expansion. Temperature cycles during manufacturing, testing, and application will therefore cause stresses in the materials. These stresses, in extreme cases, may cause serious damages to the IC's, especially at the corners of the chip. Besides efforts of process improvements and material optimisation for both packaging and IC manufacturing, proper measures need to be taken in IC design in order to eliminate/suppress the problem sufficiently. The target of coupled IC package modelling is to obtain:

- The interaction between package processing and testing on the stress/strain response of the IC layers and its deeper structures.
- To deduct IC design rules to prevent possible failures after packaging, manufacturing, and testing. The package stress design rules provide guidelines in the design manual of those measures in the IC design to avoid the IC damages induced by package stresses, by:
  o A definition of the areas (the sizes and shapes) in different categories according to the severity of potential damages.
  o Regulation of the use of those areas. That is, the rules stipulate what kinds of elements of the IC are allowed/forbidden in each area.
  o The ways of using slots in wide metal lines are given (layout of bond pads is not regulated by these rules).
- To deduct package design rules to prevent possible failures after manufacturing and testing. The package design rules provide guidelines in the package manual to obtain optimal package quality and reliability, for example:
  o What IC thickness should be used?
  o What wire loop height and length should be used?
  o What Bill of Material (BOM) list should be used?
- It is evident that IC-package design rules will depend on the type of IC and the type of the package. For example, for ceramic packages in which no other materials are present above and next to the dies (except for the bonding wires) package stress rules can be much relaxed or even no layout rules are needed at all. However, at this moment the package design rules are independent of the package type. Integrated IC package modelling may help to deduce these rules in a very efficient way.

Given the large size-differences, reliability predictions on the coupled IC-package level require multi-level (global-local) modelling techniques. At present, these techniques are used to predict the behaviour of low-k debonding in flip chips [47, 114] and to predict the occurrence of passivation cracks and pattern shift [112, 113]. Multi-level modelling can be done by using 3D FE models, 2D FE models, and/or coupled 2D-3D FE models. Although the technique is very powerful, it has some major drawbacks:

- The eventual accuracy in the local FE model(s) is strongly connected with the mesh size (read: coarseness) in the global model;
- Displacements calculated in the global FE model serve as boundary conditions in the local model(s), with the risk of creating an over-constrained system;
- No coupling exists between the local FE model back to the global model, indicating that if delamination / cracks / etc. occur on a local level this will definitely have an effect on the global deformations. This is by far the major drawback of the currently available multi-level techniques.

Other options to predict the reliability of the integrated IC backend and package combination are by using 2D models including the thin IC layers [77] and/or 3D slice models [45]. This cannot be done in full 3D since it will lead to very large FE models. Two cases of the coupled approach are presented in the following sections.

5.1 Effect of IC Metal Design on Passivation Crack and Pattern Shift Occurrence

Passivation cracks and metal interconnect shifting are well-known failure modes inside IC packages. Metal shifting is often called pattern shift and cracking of the passivation layer always precedes this phenomena. Figure 48 shows an example of metal shift. Besides that, delamination at the interface between IC and compound is always observed. This section focuses on the effect of delamination on the occurrence of passivation cracks and pattern shift using a combined experimental and numerical approach. Multi-level FE models are developed in order to predict the crack energy levels in the passivation. Besides that, specially designed package samples are developed to observe passivation cracks and to verify crack predictions from simulations (crack energy). The samples contain IC's with various test structures in each corner.

Different 3D and 2D FE models are created to predict global and local stress and strain levels, including:

Figure 48. Intact (left) and shifted (right) metal lines

1. *3D: full package model.* In these models it is possible to include delamination and predict stresses as a consequence of this. In this way, critical locations for passivation cracking can be determined. It consists of the IC with moulding compound, leadframe, and die-attach to represent a typical QFN package.
2. *2D: IC layer model including the metal layout and a J-integral mesh.* The 2D models are used to predict stresses, deformations and crack energy (J-value) with and without delamination as a function of metal layout. These models contain the silicon substrate, interlayer dielectrics, top level of interconnect metal layout, and moulding compound. The metal layout includes the seal ring, bondpads and power line.

By combining the results of both the 2D and 3D models, a 3D J-integral distribution along the power-line can be obtained by multiplying the 2D J-integral values with the 3D distribution obtained from the 3D simulations. Both wafer and packaging processes are taken into account to include the thermal history, since significant stresses are already built in the backend processes. Other assumptions and simplifications are:

- Initial passivation cracks are located at the corner of the power-line or at the centre between metal lines (or power line and bond pad).
- An initial crack must be put in the models in order to extract crack energy. The initial crack length is 80 nm. The corner crack direction is 45° and the centre crack is vertical. The effect of different initial crack propagation was studied in separate models and 45° was found to be critical for a large variety of crack lengths. A length of 80 nm was applied to simulate the smallest possible initial crack.
- Delamination at the edge of the IC is taken into account in the models, during TMCL. The delaminated length is 0.5 mm from the edge.

Delamination growth is not modelled; delamination is used as an input by using contact bodies.
- Appropriate material behaviour is assumed, including time- and temperature-dependent properties.
- Backend is modelled as one cool down step from 450°C to RT. The individual build-up of metal, IMD and passivation layers are not taken into account.
- Isothermal conditions are assumed.
- Intrinsic stresses in the passivation layers are taken into account.

Figure 49 shows the different 2D and 3D FE models as used in the multi-level modelling. The 3D model aims at predicting the compound forces on the passivation in the IC corners. Links between the IC and the compound are added to simulate the interlocking effect of a metal line at that position. The distribution of the forces predicted by the 3D model is used to scale the numerical J-integral values predicted by the 2D simulations.

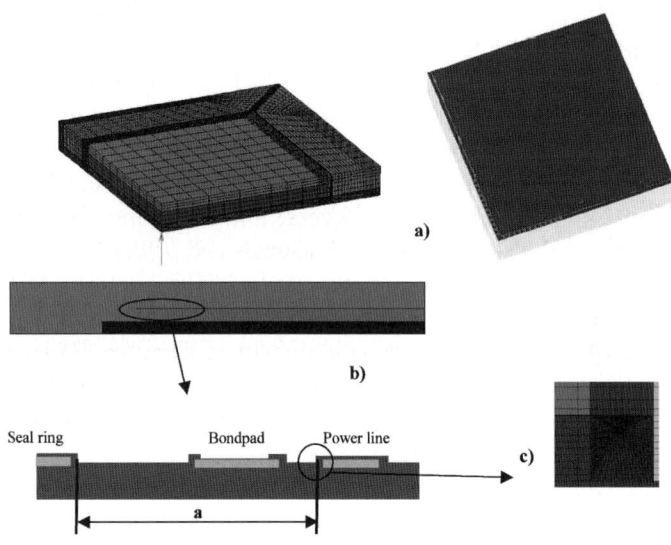

Figure 49. Multi-level modelling to predict the J-integral in passivation layers; a) 3D package model, b) 2D IC layer model, and c) J-integral mesh. Distance **a** is defined as the distance between the power line and the seal ring

Six different test structures, including a seal ring, bondpads, and power lines are constructed with varying values for the parameter *a*, which is defined as the distance between the power line and the seal ring. The structures are referred to as S1 till S6 with increasing distance *a*: for S1 a = 78 μm; for S2 a = 138 μm; for S3 a = 203 μm; for S4 a = 269 μm; for S5

5. Thermo-Mechanics of Integrated Circuits and Packages

a = 334 μm; for S6 = 400 μm. CMOS technology with a 0.18 μm line width is used to construct the test structures. The IC's are assembled in a 14 x 14 mm^2 plastic encapsulated packages (QFN). The packages are tested by using standard temperature cycle testing. At various stages in the assembly and testing processes, Scanning Acoustic Microscopy (SAM) analyses are performed to determine the amount of delamination at the IC-compound interface. After testing, packages are decapped and the following analyses are performed to observe different failures:

- Pinhole analyses to detect passivation cracks. A pinhole test is putting an etching fluid on top of the IC, which runs through passivation cracks (if present) and etches away the metal. This is visible by the changes in reflection/colour.
- Visual inspection to detect metal shift and passivation cracks by decapping (removal) of the moulding compound.
- Scanning Electron Microscopy (SEM) and Focused Ion Beam (FIB) analyses to detect both failures in cross-sections.

The experimental results are used to verify the numerical approach, which on its turn can be used to investigate the effect of IC-compound delamination, the effect of distance metal line to the IC edge, and the effect of metal design. The results are described below.

- *The effect of IC-compound delamination.* For some test structures, large amounts of delamination at the IC-compound interface is found before thermal cycling. In theses structures, after decapping, visual inspection clearly showed passivation cracks and metal shift, see for example Figure 48- right picture. The shifting of the metal line and bond pads runs from the corner towards the inside of the IC. The metal surface is wrinkled, showing that the metal is pushed towards the centre of the IC. Again, delamination at the IC - compound interface is always observed in failed samples, making it the major trigger causing passivation crack and metal shift. The physics behind the mechanism is determined by the simulations and agreed with the experimental observations. Figure 50 shows the maximum principal stress distribution during thermal cycling in the IC passivation layer for a non-delaminated (left) and a delaminated (right) IC-compound interface. In case of the delaminated interface, the stress level in the passivation material is increased with almost a factor 8 (595MPa versus 4500MPa).

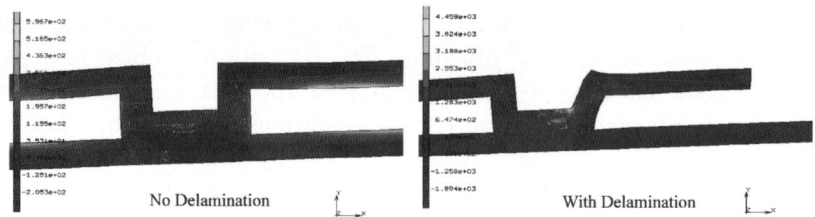

Figure 50. Maximum principal stress distributions during thermal cycling in the IC passivation layer: non-delaminated (left) versus delaminated (right) IC-compound interface

Without delamination high maximum principal stresses in the passivation layers occur close to the metal. They are mainly due to the high CTE of the metal compared to the surrounding materials. When delamination reaches the metal line and starts growing to the inner part of the IC, its effect starts to appear in the passivation layer. As such, the highest maximum principal stresses change progressively of location. Figure 51 shows the maximum principal stress distribution in the passivation near the metal for 3 delamination lengths:

- o A short delamination length: tensile stresses of 300MPa are maximum in the corner of the passivation near the metal feature.
- o A delamination length just until the metal line: the stress level and critical location are similar to the above, maximum is 300MPa.
- o The delamination length exceeds the metal line: the compound starts pushing against the passivation and stress levels increase to a peak of 1760MPa in this case.

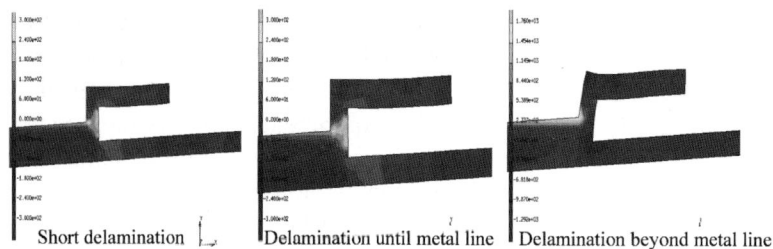

Figure 51. Maximum principal stress distribution during thermal cycling in the passivation layer with increasing delamination length (from left to right)

Figure 51 shows that the stress levels increase significantly and the critical location shifts to the corner of the passivation material. If delamination appears, the above part of the moulding is no more continuously constrained by the passivation, but only by the concentrated interlocking due to the topography. The interlocking causes higher stresses compared to continuous contact of compound and passivation. Figure 52 explains this mechanism. Actually, delamination and passivation cracking are part of one and the same failure mechanism.

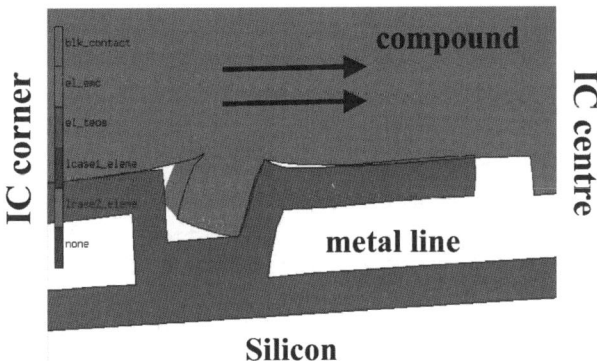

Figure 52. Mechanism for delamination related passivation crack and metal shift: compound pushes against the passivation leading to increased stress levels in the corner of the passivation layer

- *The effect of distance between metal line and IC edge.* The experimental results for the structures S1 to S6 showed that:
  o The structures S2, S1, S3, and S4 consistently showed failures to a different extent. For instance: structure S2 showed major cracking and metal shifting.
  o The structures S5 and S6 showed no failures.

  Figure 53 shows the predicted J-values for the different structures. For the calculations a 10% delamination between the IC and the moulding compound is assumed. Ranking the test structures from worst to best would give the following:
  o S2 -> S1 -> S3 -> S4 -> S5 -> S6.

  The calculated and experimental ranking agrees very well. The simulation results show that the distance from the metal line to the IC edge has a significant effect on the crack energy.

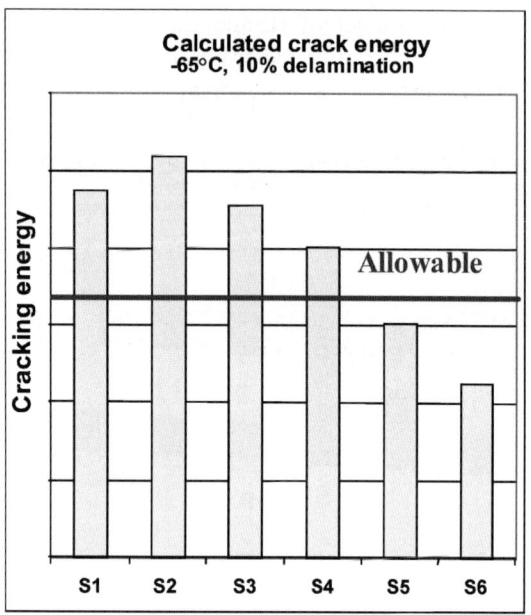

Figure 53. Calculated J-values at -65°C during thermal cycling for the different structures S1 to S6. A 10% delamination between the IC and the moulding compound is assumed

- *The effect of metal design.* Once delamination in the IC-compound interface is present, varying the metal design, in terms of width and distance from edge, influences the passivation stress levels. As such, it is possible to design the metal layout such that delamination will not lead to failures as passivation crack and metal shift. Figure 54 shows the maximum principle stress for a safe and a critical metal layout including 10% delamination. The maximum tensile stress reaches about 1700MPa for the safe metal layout and 4450MPa for the critical one. Figure 55 shows the influence of the metal width w and the distance to edge a on the J-value. The figure clearly shows that the effect of a is dominant, whereas the effect of w is negligible. This indicates that from a thermo-mechanical perspective the metal design rules should not focus on the width but rather on the placement of the structure.

5. Thermo-Mechanics of Integrated Circuits and Packages

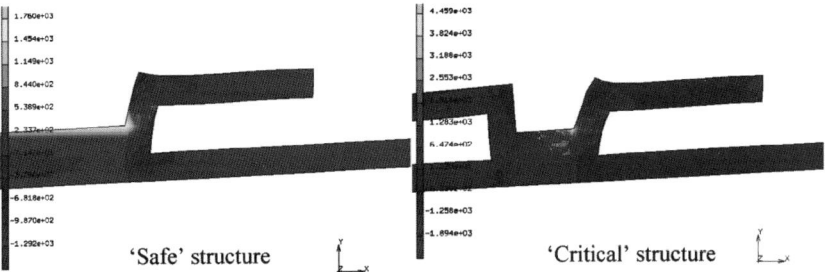

Figure 54. Maximum principal stress distributions during thermal cycling in the passivation layer of a delaminated safe (left) and critical structure (right)

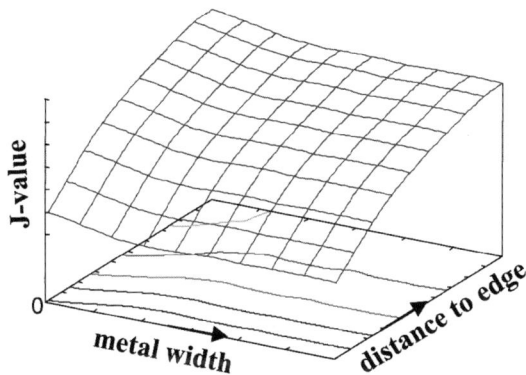

Figure 55. J-value as a function of metal width w and distance to edge a

Several FE models are built to determine the mechanisms of passivation cracking during wafer backend processes and package assembly. Based on the results, the simulation models are capable to predict the effects of metal layout on failure occurrence, critical locations, and root causes. Furthermore, a limited number of test samples can be used to predict failures prior to prototyping of real metal designs.

5.2 Effect of Package Structure on IC-Compound Interfacial Delamination

Interfacial delamination is a recognized failure mode in multi-layered micro-electronic components [77]. This failure mode is important both in determining the final failure of an electronic structure and in affecting the reliability of a complete micro-electronic package. Interfaces at risk are those between wire and lead, lead and compound, die-attach and diepad, IC and compound, etc. In other package types, such as substrate packages, the new materials, i.e., laminate and solder, introduce other interfaces, which might be at risk during processing, testing, and/or application. So far, interfacial delamination has not been sufficiently addressed and there is a strong desire from the micro-electronic industry to understand, control, and possibly predict these kinds of delamination failures.

There are different mechanisms from which delamination starts, i.e., mismatch in Poisson ratios, differences in CTE and in-plane shear stiffness mismatch. Interface delamination is known to be the starting point for many other types of failures [90, 112, 113]. Fracture mechanics can be applied to model this type of failure. Interfacial stresses occur between every two neighbouring layers of materials during the manufacture process, it's testing, and finally its application. Due to the large differences in the thermal-mechanical properties, some of the stresses could be so high that the material failure may occur at sites of high stresses, particularly near the interfaces. Or in another case, the stresses are so high that micro-cracks are initiated somewhere near an interface, while the global structure (the IC assembly) remains intact. Due to the time dependence of the material properties, such a micro crack may undergo further propagation during processing and application, and hence reduce the reliability. Because of the nature of a multi-layer structure, multiple sites of potential interfacial delamination will compete with each other in an assembly. Interface delamination, especially the initiation of it, is strongly driven by the moisture distribution in the IC package [116].

The interface between the top IC layer, i.e., the passivation layer, and the moulding compound is chosen as a carrier interface. The interface strength between both materials is characterised by combining button shear/tensile experiments with Finite Element (FE) simulations. For three possible package structures the failure sites for delamination at the interface between moulding compound and IC passivation are calculated. The following simplified package structures are used:
- A package having no leadframe and consisting of an IC surrounded by moulding compound.
- An exposed pad package, composed of an IC attached to a leadframe and, on one side surrounded by moulding compound. Typical exposed pad packages are HLQFP, HVQFN, and HTSSOP.
- A leadframe based package composed of an IC attached to a leadframe and, on both sides surrounded by moulding compound. Typical leadframe based packages are QFP, SSOP, and DIP.

The structures are schematically presented in Figure 56, without showing the details of the IC passivation structure.

5. Thermo-Mechanics of Integrated Circuits and Packages

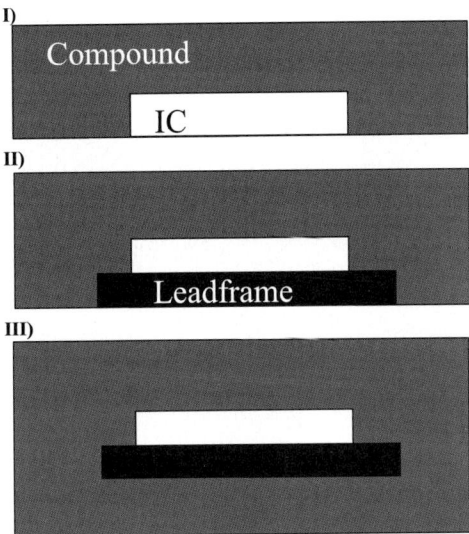

Figure 56. A simplified view of 3 package structures: I) without leadframe, II) exposed pad like structure and III) leadframe based like structure

2D generalized plane strain FE models are created to predict failure sites for delamination at the interface between moulding compound and IC passivation as function of the different package structures. The thermo-mechanical properties of the IC constituents like Si, SiN, and TEOS are listed in Table 4 and for the package constituents, like Cu alloys, are listed in Table 9. The modulus of the moulding compound strongly depends on the temperature. Accordingly, its CTE depends also on the temperature: α = 8ppm/K below the T_g and α = 32ppm/K above the T_g, which is 110°C. The load is applied through a global temperature change, 175°C to 25°C, which resembles the zero hour stress state of the package after moulding. For the top IC passivation layer, a topography consisting of a seal at the edge, a 70 µm wide bondpad, and a 60 µm wide metal line are taken into account. Figure 57 shows the 2D FE models of the three package structures and a zoom into the passivation level. Due to symmetry, only ½ of the structures is modelled. Contact bodies are used to overcome the mesh size discrepancies between the bulk package materials and the small IC bondpad/metal line.

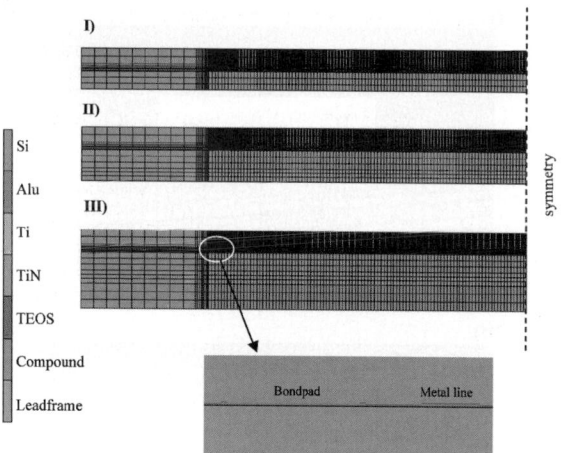

Figure 57. 2D FE models for the three package structures

The package structure has a strong influence on the global deformation of the IC package. This is indicated in Figure 58, which shows the package warpage for the three structures at 25°C. Package structure I, a package consisting of an IC surrounded by moulding compound, deforms in an upward direction (smiley or convex) which is an effect of the CTE-difference between the silicon IC (3ppm/°C) and the moulding compound (always > 3ppm/°C). Package structures II and III, however, which both consist of a leadframe, deform in a downward direction (grumpy or concave), which is an effect of the leadframe (17ppm/°C), which pulls the package downwards below the T_g of the compound.

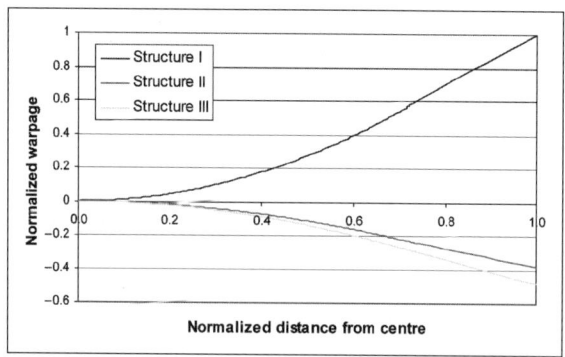

Figure 58. Package warpage as a function of distance from centre. The warpage is normalized to the maximum value of structure I

5. Thermo-Mechanics of Integrated Circuits and Packages

Figure 59 shows the maximum principal stress distribution in the passivation around the metal line at 25°C for the three structures. There is no significant difference between structure II and III for these stress levels. In case of structure I, the stress level in the passivation material is much lower as for structures II and III. This is an effect of the leadframe, which has changed the shape of the deformed structure by pulling the IC downwards. Remind that the stress levels in this case will be dramatically increased when delamination occurs between moulding compound and passivation (see the previous section).

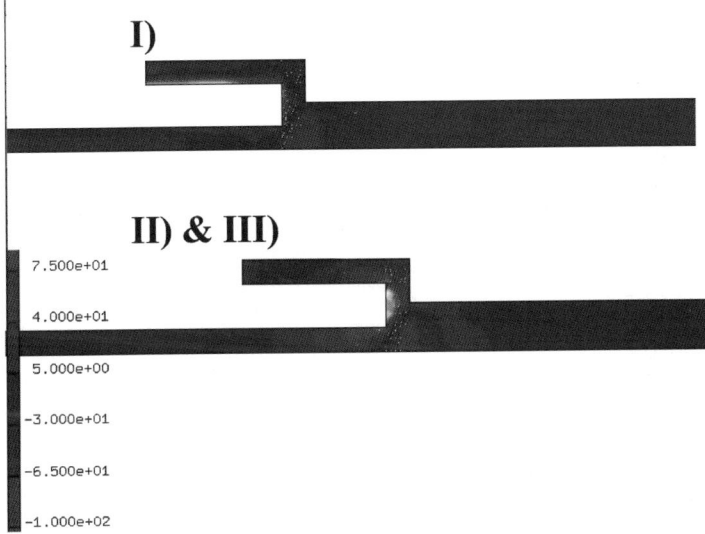

Figure 59. Maximum principal stress distribution at 25°C in the passivation layer around the metal line for the three structures. No delamination is assumed

Figure 60 presents the simulated results with the potential delamination failure sites (in grey) at 25°C for all three structures. For all structures the interfacial delamination tends to occur at the IC corner between the moulding compound and the passivation layer. For structure I, first failures occur at a temperature of 100°C and no other local failure sites are observed. Also, the delamination area at 25°C for structure I remains quite small, see the grey area in the corner of IC and compound. The previous results indicated that the leadframe has a strong influence on the global deformation of an IC package, i.e., it even changes the curvature direction of the package. When a leadframe is added to the IC package (structures II and III), additional effects occur, see Figure 60:

- It delays the occurrence of delamination to a temperature of 50°C, compared to the case without leadframe.
- It increases the size of the interface delamination at 25°C, especially for structures II where a somewhat larger delaminated area is found as for structure III.
- Failure tends to occur at more sites with the temperature drop. These sites are located at the corners of the topography (due to metal structures such as bondpads and metal lines) at the top passivation layer.

The simulations results for the three different structures may help to understand the interfacial delamination observed in IC packages. Further, the influence of geometrical and material variations, such as the thickness and type of moulding compound can be investigated.

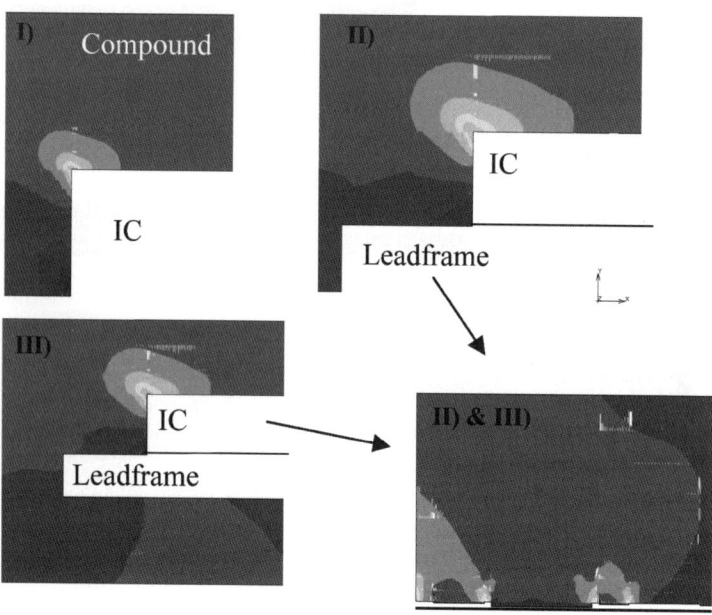

Figure 60. Simulated potential failure sites between the IC and moulding compound for the three structures at 25°C. The gray areas are the areas where the failure criteria are exceeded and delamination will occur

6. CASE STUDIES

In the following, two case studies are presented for reliability predictions of the damascene process and for material selections for a substrate based package.

6.1 Reliability Predictions of Thermo-Mechanical Integrity of the Damascene Process

The implementation of new low-k dielectric materials has serious consequences for the structural integrity of Cu/low-k interconnects. The new materials have substantially different thermo-mechanical properties than silicon oxide (TEOS), which is being used up to now as the dielectric material. Particularly, the elastic modulus of the new materials is two orders of magnitude smaller, whereas the coefficient of thermal expansion is 20 to 60 times larger. Moreover, the use of copper requires a dedicated processing technique, which is called the damascene process. The main process steps involved in making this structure are the following:
1. Deposition of dielectric layers,
2. Patterning the dielectric (etching),
3. Deposition of a diffusion barrier layer (e.g., tantalum nitride (TaN),
4. Copper electroplating,
5. Annealing,
6. Planarisation (chemical-mechanical polishing, CMP).

The different process steps are performed at different temperatures, and therefore the stack experiences many temperature cycles during the processing. This leads to high thermal stresses/strains, which can lead to delamination or fracture of the layers. We have studied the mechanical stresses that are introduced during the main process steps in the damascene process. Prior to our FE study, we have determined [42 - 45]:
- The visco-elastic properties of the low-k dielectric material by using Nano-indentation. This study revealed that the behaviour of the low-k film could be described with a linear visco-elastic model up to stress and strain levels of 200MPa and 3% respectively.
- Adhesion strength of low-k dielectrics with various other materials by using the four-point bending adhesion test. This study revealed that the adhesion between low-k with adjacent materials is poor.

A typical damascene structure is used, consisting of a two-level interconnect stack on a silicon wafer, see Figure 61. To make use of the symmetry and the periodic layout, we meshed the half pitch of the structure. It was assumed that the global deformations of this small part are determined by the much thicker silicon underneath. Initially a 2D model was made using generalized plane strain elements. Various material configurations were studied with this simple model, considering that stresses are overestimated by using plane strain conditions. Subsequently the model was expanded into the 3rd dimension to have a one-element thick model. The size of the meshed part is approximately 1.5 x 3 µm^2, it is shown in Figure 61.

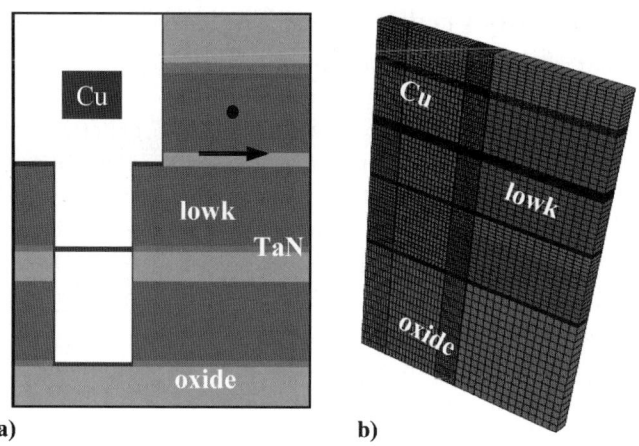

Figure 61. Repetitive Cu/low-k damascene structure (a) and 3D slide FE model (b)

All process steps, which are thermal loadings during the main process, are modelled by using activating and deactivating of the elements. The boundary conditions are set up to fulfil the symmetry and silicon driven global deformation conditions. The bending of the substrate and the internal force acting on the substrate can be ignored [117]. Symmetry conditions are applied at the left hand side of the model, as the vertical degree of freedom was constrained. The bottom nodes are supported vertically, but can move horizontally. Coupled boundary conditions are applied at the right side of the module, where the applied deformations are calculated from the thermal expansion/shrinkage of the silicon. The front plane was constrained against out-of-plane deformations, and the back plane was loaded with a thermal deformation.

Material properties are listed in Table 4, for the copper a temperature dependent yield is assumed of 300MPa at low temperature, and 100MPa at 400°C. Different configurations are analysed:
- The traditional Al - Teos configuration. Both the Al metal line and the Teos dielectric are modelled as elastic materials. No process modelling is used.
- The traditional Al - Teos configuration. The Al metal line is modelled as an elastic-plastic material. No process modelling is used.

5. Thermo-Mechanics of Integrated Circuits and Packages

- The damascene configuration. The Cu metal line is modelled as an elastic material. No process modelling is used.
- The damascene configuration. The Cu metal line is modelled as an elastic-plastic material. No process modelling is used.
- The damascene configuration. The Cu metal line is modelled as an elastic-plastic material. Process modelling is used.

Note that in all configurations the low-k material, being a polymeric material, is modelled as a linear visco-elastic material.

Peel stresses along the dielectric (Teos/Low-k) - oxide interface (indicated in Figure 61), are depicted in Figure 62. By simulating the cooling down process peel stresses are tensile and about 100MPa in magnitude in the configurations where the dielectric is the low-k material. Using Teos as a dielectric results into compressive stresses along the path. In terms of delamination tensile peel stresses can be related to the opening mode (mode I), and this mode is the most critical. Therefore configurations including low-k materials can be critical compared to configurations with Teos. The evolution of tensile stresses can be explained by the thermal expansion coefficient of the low-k compared to these coefficients of the other materials:

- The CTE of low-k is much larger than that of copper, and hence it tends to shrink more. Or: ΔCTE (Cu - low-k) < 0 -> during cooling down tension on path.
- In the Al/Teos configuration, the CTE of Teos is very small so it is pulled on by the other materials, which causes compressive stresses. Or: ΔCTE (Al - oxide) > 0 -> during cooling down compression on path.

Thus, In the Al/Teos configuration these peel stresses are not critical (compressive), in configurations that includes low-k peel stress becomes tensile. Implementing the process steps changes the simulation results considerably. Figure 63 depicts the film stress history during the process steps in the low-k material at the location indicated in Figure 61. This figure shows the stress evolution in the low-k during processing. First the dielectric layers are deposited at high temperature, the layers are assumed to be stress free. During the cooling down step the film stress rises to about 70MPa tensile. The etching step releases the stress to 45MPa. Then it is further released to 20MPa by the heating up process prior to barrier deposition. The stack has to be cooled down to room temperature again before the copper plating, the stresses are raised to 60MPa, somewhat lower then after the first cooling down step. The annealing step causes compressive stresses, about 30MPa of magnitude during the hold time, and for this reason the residual stress in the low-k is about the same as after the first cooling down. The visco-elastic character of the low-k material can also be observed in Figure 63: especially during the annealing step relaxation of stresses can be seen.

Peel stress levels along the low-k-oxide interface path including the process modelling is depicted in Figure 62. Clearly, including elastic-plastic copper and linear visco-elastic low-k behaviour has a decreasing influence on the stress response along the path. However, the stresses remain tensile and are, thus, critical for the occurrence of cracks.

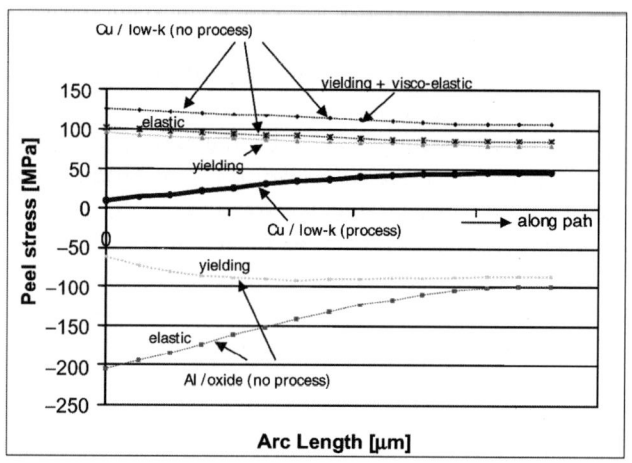

Figure 62. Peel stresses along the dielectric (Teos/Low-k) - Oxide interface

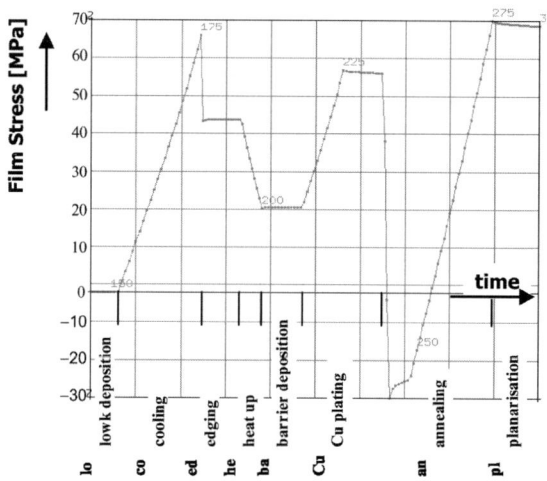

Figure 63. Stress history in the low-k dielectric during processing

In this case study the thermo-mechanical integrity of Cu/low-k interconnect structures is studied using finite element simulations. The evolution of stresses due to processing is computed using a finite element model in which the material configurations as well as the material models are varied. The main conclusions are as follows.
- The Cu/low-k configuration, leading to tensile peel stresses at interfaces, is more critical to layer delamination than the conventional Al/Teos combination, which gives compressive peel stresses.
- The process temperatures are below the T_g of low-k, hence its modulus is more influenced by temperature than by time.
- The dual damascene process steps should be taken into account to obtain a reliable prediction of stress and strain levels. In particular, simulation of the etching steps is important, since these lead to a decrease in the residual stresses in the stack. Detailed process simulation avoids an overestimation of the stress levels.

6.2 Simulation-based Material Selection for a TBGA Package

Material choices are of major importance for the reliability of micro-electronic packages. Moreover, the differences in CTE of the packaging constituents are one of the determining factors for its thermo-mechanical behaviour. Making the wrong material choice may result in various kinds of failures, such as passivation crack, wire shift, and/or wire break and are closely related with the occurrence of delamination at specific interfaces.

IC packages subjected to thermal loads and/or moisture during processing and testing are vulnerable to delamination at all possible interfaces [118-120]. Prediction of the initiation and propagation of interface delamination, as well as the response of the package to the delamination is vital for the micro-electronic industry [121]. Studies have been performed to calculate interface delamination, based on Linear Fracture Mechanics [122] and Micro-Mechanics [116, 123] approaches. These studies have found that differences in Coefficients of Thermal and Moisture Expansion are the driving factors for interface delamination in micro-electronic packages.

Cavity down Tape Ball Grid Array (TBGA) is developed and patented by VLSI in 1995 [124]. In comparison with a normal BGA, this package's substrate is a 1 Cu layer flexible substrate laminated to a thick heat slug so as to enhance the thermal performance. Potting material is used to encapsulate the die area instead of over-mould it. The typical structure of this package is shown in Figure 64. Based on the experience of this specific package with regular materials, assembly houses and material suppliers continue their cooperation to develop green material solutions for this package to fulfil the thermal and thermo-mechanical demands. Since no solution is available at the time, qualifications tests are performed for different compound and die-attach material combinations. A parametric 3D Finite Element Model (FEM) is constructed to calculate the driving stresses thereby predicting the reliability of these different material combinations. The FE modelling results will enable an improved interpretation of the process and product qualifications needed to qualify manufacturing processes, and to improve the lifetime and successful operation of the product (from ITRS 2001 roadmap).

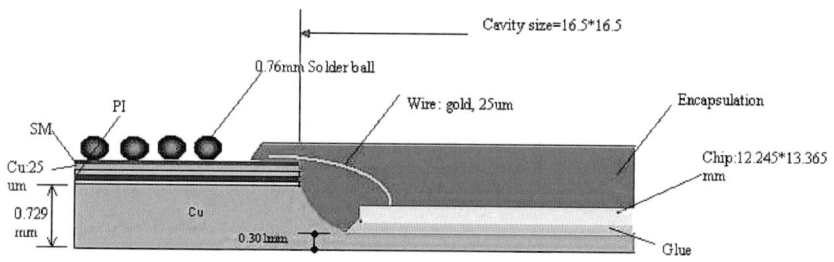

Figure 64. Cross-section of a cavity down TBGA package

Six different material combinations for the cavity down TBGA package are selected for qualification tests. Qualification data from subcontractor companies for this package are also available [124]. The bill of materials (BOM) is indicated in Table 10. The package features a cavity size of 16.5 x 16.5 mm^2, with an IC size of 12.2 x 13.4 mm^2. Standard tests based on General Quality Specifications (GQS) for micro-electronic packages are performed, including Moisture Sensitivity Level tests (MSL), Temperature Cycling Tests (TMCL), etc. Notice that for the TMCL test, a precondition step is preceded inducing moisture to the product, to test the effect of moisture driven delamination on other types of failures. At several steps during the process, Scanning Acoustic Microscope measurements (SAM) are performed. After manufacturing of the package, construction analyses are performed to indicate any geometrical mismatch. For all BOM's, these construction analyses showed no abnormalities. For all BOM's, samples are taken after the GQS testing and inspected for failure occurrence by:
- Visual inspection to observe global cracks,
- SAM measurement for delamination,
- Decapping to observe i) passivation cracks, ii) wireball shift and/or iii) wirebond break.

The results of the inspections are listed in Table 11. The following can be concluded from the reliability data:

5. Thermo-Mechanics of Integrated Circuits and Packages

- For all BOM combinations, except for BOME, different failure modes are detected.
- There is evidence that passivation cracks are associated with the occurrence of delamination. All cracked IC's showed delamination, and in only BOMF delamination is found without any cracks.
- Ranking from worse to best turns out to:
  BOMA (worst) – C/D – B – F – E (best)

Table 10. Indication for Bill Of Materials (BOM)

	Compound Material	
Die-attach Material	Type 1	Type 2
Type a	BOMA	Not done
Type b	BOMB	Not done
Type c	BOMC	BOME
Type d	BOMD	BOMF

Table 11. Observed failures after testing

BOM	Observed Failures	Picture
A	Failures are detected after 500 and 1000 TMCL cycles. The failure modes are serious corner delamination, passivation crack and wire ball shift.	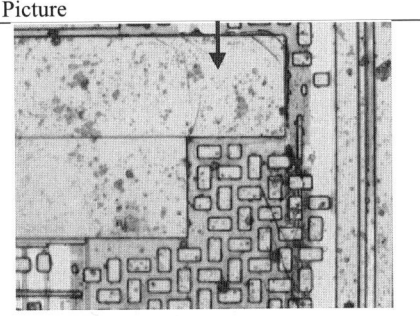
B	The result showed that similar defect modes, including corner delamination, ball lift, as found in the BOMA package, are found. Delamination is detected in the two lower corners at the IC – compound interface. The delaminated area is correlated with the occurrence of passivation cracks in the IC.	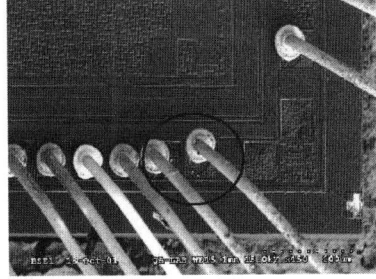

BOM	Observed Failures	Picture
C	After precondition, no matter MSL level 3 or 5, a common failure mode happened for both BOMC and BOMD. Through SAM, serious delamination, not only at corners, but also extending at the complete package, is detected for almost all samples. Cross sectioning is used to further identify the failure modes after subsequent 1000 TMCL cycles. Some failure modes, including i) glue/die delamination, ii) glue bulk, iii) encapsulants / heatslug delamination, and iv) substrate layers delamination, are detected by cross sectioning.	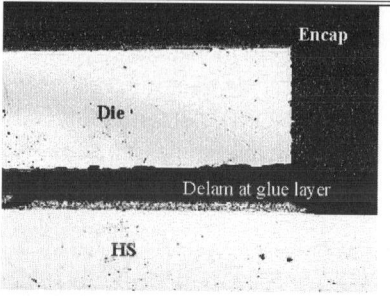
D		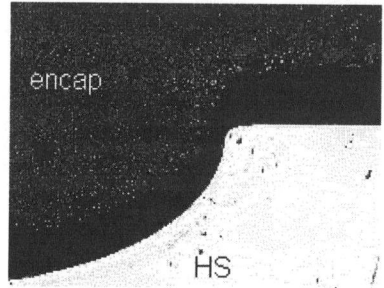
E	All sub-groups of BOME passed preconditioning and 1000 TMCL cycles. No delamination is found at all interfaces. In all parts of the package, including glue, compound, no defects are detected. One remarkable point is that after TMCL cycling, there is no passivation crack or ball bond shift. The typical die surface condition is shown.	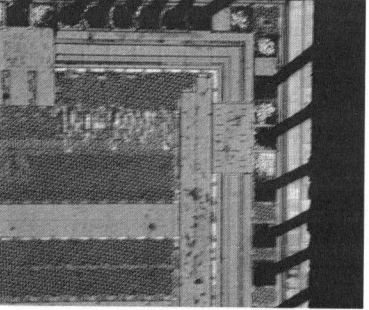
F	For BOMF defects (delamination, circuits open) are detected after testing (preconditioning and 1000 TMCL cycles) and wire bond break is found by decapping failed samples. In addition, there is no passivation crack or ball bond shift at the samples after 1000 TMCL cycles.	

5. Thermo-Mechanics of Integrated Circuits and Packages

Parametric 3D FEM models representing the cavity down TBGA package are developed. Because of symmetry, only one quarter of the package is modelled. The effect of element sensitivity is explored by using several distributions and/or discretizations. In total the model consists of approximately 20,000 8-noded elements, see Figure 65. Appropriate boundary conditions are used along the symmetry axis. The single crystal silicon die is modelled as temperature independent anisotropic [65]. For the compound and die-attach constituent, supplier info and/or temperature dependent properties are used [65, 68]. For the copper heatspreader material ideally plastic model is assumed. For the Upilex substrate, temperature dependent properties are used. Table 12 lists some typical material properties for the different die-attach and compound combinations. Based on the data listed in the table it is hard to estimate which combination would perform the best.

For the loading in the model, the complete time and temperature profile during manufacturing is used in the simulations, followed by the GQS conditions.

To develop reliable and efficient thermo-mechanical prediction models for the product/process designs of electronic packaging, various justified simplifications and assumptions are needed. In this paper, the following simplification and assumptions are used:

- The warpage/stress free state is assumed for the silicon die, diepad, glue, heatspreader, and substrate at 150°C. The choice for this temperature is based on curing measurements for the die-attach. For the compound a warpage/stress free temperature of 170°C is assumed, which is the processing temperature for the compound. For the solder balls, a warpage/stress free temperature of 183°C is assumed, which is the temperature of eutectic solder to solidify.
- Isothermal loading conditions are used for the modelling of both the packaging processes and GQS testing conditions. This assumption is valid since thermal cycle dwell times are 15 minutes, times where steady state thermal distributions are well reached.
- The initial die warpage/stress is neglected. Although these warpage/stress values can be significant, they are neglected in this study.
- Perfect adhesion is assumed between all constituents.
- The curing process induced stresses are neglected. Curing effects in compounds may lead to very significant stress levels, but this effect is neglected here since the failures are observed after testing and not after mould curing.
- Time effects such as material degradation are not taken into account.

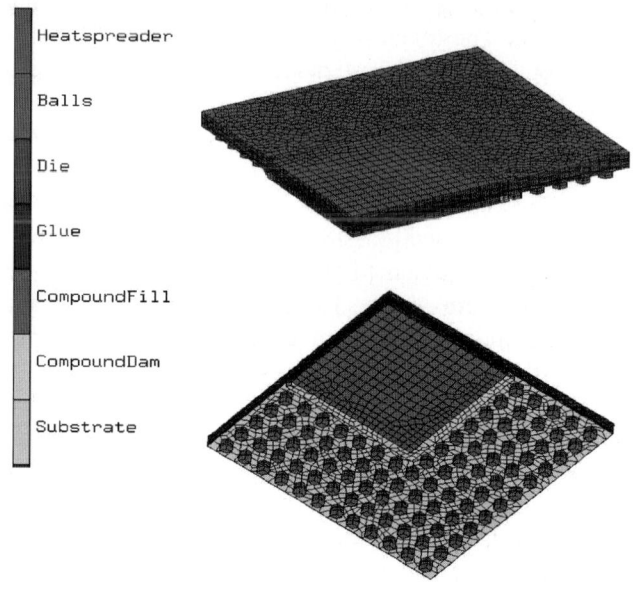

Figure 65. ¼ FE model for the cavity down TBGA

Table 12. Typical material properties for the die-attach and compound materials

Material	T_g [°C]	CTE_1 [ppm/°C]	CTE_2 [ppm/°C]	Young's Modulus [MPa]
Die-attach				
Type a	−10	72	170	860 @ 25°C
Type b	60	70	180	500 @ 25°C
Type c	15	60	180	2500 @ 25°C
Type d	33	65	114	3060 @ 25°C
Compound				
Type 1	169	17	72	10000 @ 25°C
Type 2	145	17	67	13000 @ 25°C

- *On Delamination and Passivation Crack.* After the GQS tests, such as temperature cycling, cracks in the thin passivation layers of the IC are found in BOMA-D. In all BOMA-D, it is found that the passivation cracks are associated with delamination between IC and moulding compound. We have studied the influence of interfacial delamination on the passivation crack systematically by using J-integral methods, see [112, 113]. The mechanism of the effect of delamination on passivation cracking is explained in section 5.1: the delaminated compound will push the passivation layer in such a way that it initiates and propagates cracks in this layer. Delamination and passivation cracking are one and the same failure mechanism. Or in other words, the occurrence of delamination during manufacturing and/or testing (reflow) may predict the occurrence of passivation cracking in a later stage (temperature cycling).

- *On Delamination and interface strength.* Interfacial delamination inside an IC package occurs because of the large differences in material behaviour (read differences in CTE) between materials such as polymer and metal and polymer and passivation. However, no effective methodologies, models, and tools are available for the prediction of the interfacial strength of these material combinations. Techniques to study the interfacial strength of IC and packaging materials are described in sections 3.5 and 4.5. In this case, two types of tests are investigated, i.e. the button shear/pull test and 4-point bending with pre-notch crack [100]. By tensile loading the strength of the compound – passivation interface is much lower as by shear loading. Combining these test results with dedicated FE models, it is possible to quantify the strength of interfaces.
- *On Reliable FEM modelling.* To check the reliability of the FE model for the cavity down TBGA package, warpage measurements are performed. For this, 35 samples are built by BOME and used to check their warpage and coplanarity. Figure 66 shows an example of the measured together with the calculated deformations at 25°C. In the measurement, an average warpage of 37.8 ± 13.7 µm is found, in the FE model, for this BOM, a warpage of 31.0 µm is calculated. In both cases, a smiley-faced deformation is found.

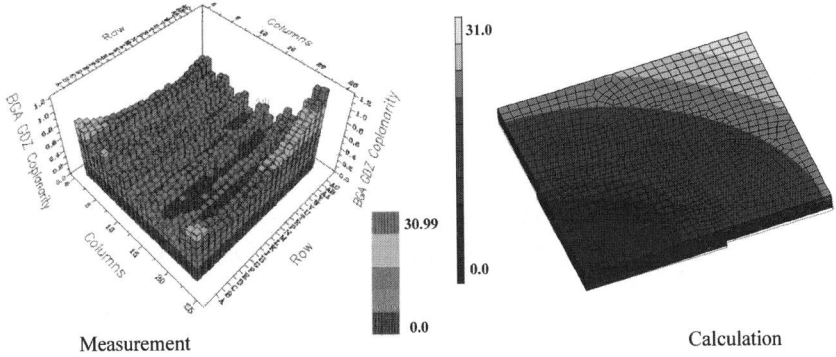

Figure 66. Example warpage at 25°C of measurement (left) and FE model (right)

- *On Material Choices*
  Tensile interface stress levels (maximum principal stress) should not exceed threshold values to prevent interface delamination. Figure 67 shows the interface stress distribution after cool down to –55°C at the IC compound and IC die-attach interface for three material combinations:
  o BOMA, a package with severe reliability problems,
  o BOMF, a package with less reliability problems,
  o BOME, a package with no reliability problems.
  Much lower stresses are found at the IC-compound interface for BOME, which in reality also behaves much better. Since the delamination provokes passivation cracking, for BOMA a much worse reliability is anticipated from this modelling result.

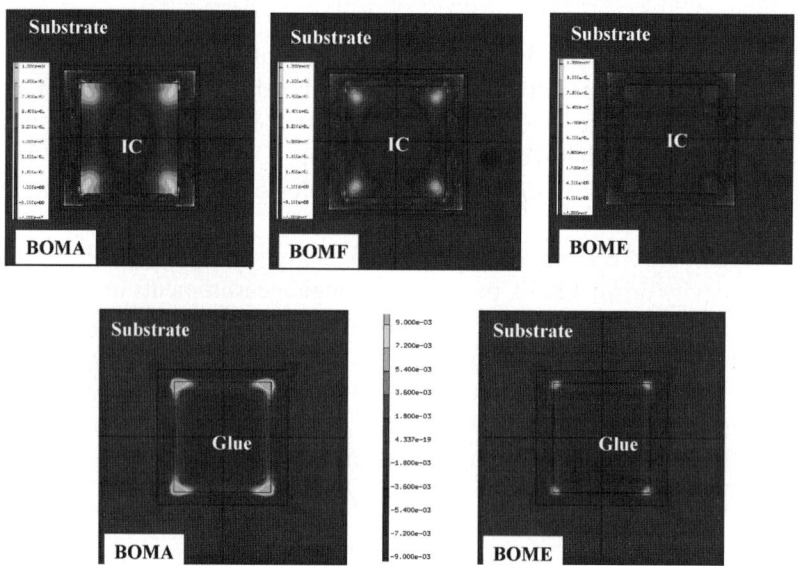

Figure 67. Interfaces stress (maximum principal stress) for BOMA, BOMF and BOME at the IC-compound (top) and the IC-die-attach (bottom) interface

Figure 68 shows the ranking for different BOM's in terms of an accumulated risk to failure. Again, very clear is the extreme low risk for BOME, which is about 50% lower that for the worst-case combination BOMA.

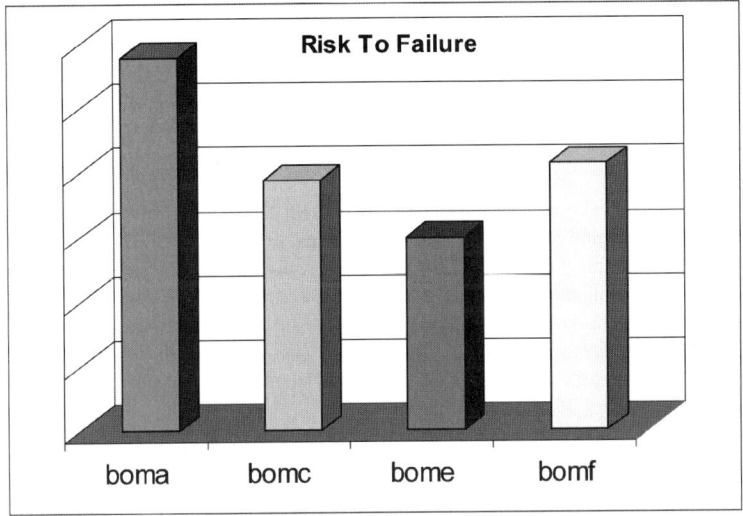

Figure 68. Ranking for different BOM's in terms of accumulated risk to failure

In this case study, we have examined the occurrence of different failure types depending on the compound and die-attach material choice in a cavity down TBGA package. Qualification of the different products is combined with 3D Finite Element calculations to predict the delamination driving stresses for the different material combinations in the cavity down TBGA package. In some combinations of compound and die-attach the package performed poor, with only one combination without failures. The results of the Finite Element calculations indicated the possibility of interfacial delamination for different combinations. The results show that the developed simulation models can be used to assess the possibility of reliability problems as the consequence of using different material combinations. Using sophisticated simulation techniques cost- and time-expensive reliability tests can be reduced to a minimum.

7. REFERENCES

[1] G. Kelly, The simulation of thermo-mechanically induced stress in plastic encapsulated EC packages, Dordrecht: Kluwer Academic Publishers, 2000.
[2] Microchip fabrication – a practical guide to semiconductor processing, edited by P. van Zant, McGraw-Hill, 4th edition, 2000.
[3] Joint Electron Device Engineering Council (JEDEC), www.jedec.org
[4] Institute for Interconnection and Packaging Electronic Circuits (IPC), www.ipc.org
[5] International Electro technical Commission (IEC), www.iec.ch
[6] G.G. Stoney, Proc R. Soc. London, Ser. A, 82 172, 1909.

[7] C.A. Klein, How accurate are Stoney's equation and recent modifications, J. Appl. Phys. 88 (9), 2000.
[8] A. Atkinson, Proc Br. Ceram., 54 (1), 1995.
[9] L.B. Freund, J.A. Floro, E. Chason, Extensions of the Stoney formula for substrate curvature to configurations with thin substrates or large deformations, Appl. Phys. Lett. 74 (14), 1999.
[10] M. Stadtmueller, Mechanical Stress of Chemical Vapour Deposition Dielectrics, Journal of the Electrochemical Society 139 (12) pp. 3669-3674, 1992.
[11] T.H. Wu and R.S. Rosler, Solid State Technology 35 (4), pp. 65-72, 1992.
[12] K. Ramkumar and A.N. Saxena, Stress in SiO2 Films Deposited by Plasma and Ozone Tetraethylorthosilicate Chemical Vapour Deposition Processes, Journal of the Electrochemical Society 139 (5), pp. 1437-1442, 1992.
[13] J.-H. Zhao, T. Ryan, P.S. Ho, A.J. McKerrow and W.-Y. Shih, On-wafer characterization of thermomechanical properties of dielectric thin films by a bending beam technique, J. Appl. Phys., 88 (5), 2000.
[14] S.H. Lee, Y.C. Bae, Thermal stress analysis for polyimide thin film and a substrate layer system, Macromol. Chem. Phys. 201, pp. 1286-1291, 2000.
[15] J.S. Kim, K.W. Paik, S.H. Oh, The multilayer-modified Stoney's formula for laminated polymer composites on a silicon substrate, Journal of Applied Physics 86 (10), pp. 5474-5479, 1999.
[16] C.A. Klein, Comment on The multilayer-modified Stoney's formula for laminated polymer composites on a silicon substrate (J. Appl. Phys. 86, pp. 5474, 1999), J. Appl. Phys. 88 (9), 2000.
[17] H.J. Kook, D. Kim, In situ measurements and analysis of imidization extent, thickness, and stress during the curing of polyimide films, J. of Mat. Sci. 35, pp. 2949-2954, 2000.
[18] J.-H. Zhao, M. Kiene, C. Hu, and P.S. Ho, Thermal stress and glass transition of ultrathin polystyrene films, Appl. Phys. Lett. 77 (18), 2000.
[19] C. Pelletier, H.J.L. Bressers, G.Q. Zhang, K.M.B. Jansen, V. Gonda, L.J. Ernst, The state-of-the-art of thermo-mechanical characterization of thin polymer films, Proc EuroSimE 2002, pp. 293-300.
[20] A.J. Kalkman, A.H. Verbruggen, G.C.A.M. Janssen, Young's modulus measurements and grain boundary sliding in free-standing thin metal films, Applied Physics Letters 78 (18), pp. 2673-2675, 2001.
[21] H-J. Lee et al., Applied Physics Letters 76 (23), pp. 3415-3417, 2000.
[22] H. Huang, F. Spaepen, Tensile testing of free-standing Cu, Ag and Al thin films and Ag/Cu multilayers, Acta Mater. 48, pp. 3261-3269, 2000.
[23] M.M. de Lima, R.G. Lacerda, J. Vilcarromero, F.C. Marques, Coefficient of thermal expansion and elastic modulus of thin films, J. Appl. Phys. 86 (9), pp. 4936-4942, 1999.
[24] A.J. Kalkman et al., Rev. Sci. Instrum. 70 (10), pp. 4026-4031, 1999.
[25] J. Malzbender, J.M.J. den Toonder, A.R. Balkenende, G. de With, Measuring mechanical properties of coatings: a methodology applied to nano-particle filled sol-gel coatings on glass, J. of Mat. Science and Engineering 36, pp. 47-103, 2002.
[26] W.C. Oliver, G.M. Pharr, An improved technique for determining hardness and elastic modulus using load and displacement sensing indentation experiments, J. Mater. Res. 7 (6), 1992.
[27] A. Bolshakov, G.M. Pharr, Influences of pileup on the measurement of mechanical properties by load and depth sensing indentation techniques, J. Mater Res. 13 (4), 1998.

5. Thermo-Mechanics of Integrated Circuits and Packages

[28] J.C. Hay, A. Bolshakov, G.M. Pharr, A critical examination of the fundamental relations used in the analysis of nano indentation data, J. Mater. Res. 14 (6), 1999.
[29] Y.T. Cheng, C.M. Cheng, Int. J. Solids Structures 36, pp. 1231, 1999.
[30] R.B. King, Int. J. Solids Structures 23 (12), pp. 1657, 1987.
[31] J. den Toonder, A.R. van Dijken, V. Gonda, J. Beijer, G.Q. Zhang, J Waeterloos, L.J. Ernst, Mechanical characterization and modelling of low-dielectric-constant silk film using nano-indentation: time- and temperature-effects, Proc ECTC 2003, pp. 708-711.
[32] V. Gonda, L.J. Ernst, J. den Toonder, J. Beijer, G.Q. Zhang, Finite thickness influence on spherical and conical indentation on visco-elastic thin polymer film, Proc. Polytronic 2003, pp. 51-55.
[33] V. Gonda, K.M.B. Jansen, L.J. Ernst, J. den Toonder, G.Q. Zhang, Mechanical characterisation of silk by nano indentation and substrate curvature techniques, Proc. EuroSimE 2004, pp. 373-376.
[34] V. Gonda, J.M.J. Den Toonder, J. Beijer, G.Q.Zhang, W.D. van Driel, R.J.O.M. Hoofman, L.J. Ernst, Prediction of thermo-mechanical integrity of wafer backend processes, J. Microelectronics Reliability 44(12), 2004, pp. 2003-2009.
[35] P.G. Charambalides, A test specimen for determining the fracture resistance of bimaterial interfaces, J. of applied mechanics 56, pp. 77-82, 1989.
[36] G, Wang, C. Merrill, J-H. Zhao, S.K. Groothuis, P.S. Ho, Packaging effects on reliability of Cu/low-k interconnects, IEEE Transactions on Device and Materials Reliability 3 (4) , pp. 119-128, 2003.
[37] I. Hofinger, Modified four-point bending specimen for determining the interface fracture energy for thin, brittle layers, Intern. J. of Fracture 92, pp. 213-220, 1998.
[38] J.W. Beams, Structure and properties of thin films, edited by C.A. Neugebauer, J.B. Newkirk, and D.A. Vermilyea, Wiley, New York, pp. 183, 1959.
[39] T. Tsakalakos, The bulge test: comparison of theory and experiment for isotropic and anisotropic films, Thin Solid Films, Vol. 75, pp. 293, 1981.
[40] D.W. Zheng, Y.H. Xu, Y.P. Tsai, K.N. Tu, P. Patterson, Bin Zhao, Q.-Z. Liu, Maureen Brongo, Mechanical property measurement of thin polymeric-low dielectric-constant films using bulge testing method, Appl. Phys. Lett. 76 (15), 2000.
[41] T.-T. WU, Elastic Wave Propagation and non destructive evaluation of materials, Proc. Natl. Sci. Counc. ROC(A) 23 (6), pp. 703-715, 1999.
[42] T.-T. Wu, Y.-H. Liu, Inverse determinations of thickness and elastic properties of a bonding layer using laser-generated surface waves, Ultrasonics 37 pp. 23-30, 1999.
[43] A.R. Duggal, J.A. Rogers, K.A. Nelson, Mordechai Rothschild, Real-time characterization of acoustic modes of polyimide thin-film coatings using impulsive stimulated thermal scattering, Appl. Phys. Lett. 60 (6), 1992.
[44] J.K. Cocson, C.S. Hau, P.M. Lee, C.C. Poon, A.H. Zhong, J.A. Rogers, K.A. Nelson, Transverse isotropic elastic moduli and in-plane thermal diffusivity in silicon-supported thin films of a photosensitive polyimide measured using impulsive stimulated thermal scattering, Polymer 36 (21), pp. 4069-4075, 1995.
[45] J.K. Cocson, C.S. Hau, P.M. Lee, C.C. Poon, A.H. Zhong, J.A. Rogers, K.A. Nelson, Characterization of 6FDA-APBP polyimide films through impulsive stimulated thermal scattering, Journal of Materials Science 30, pp. 5960-5966, 1995.
[46] J.A. Rogers, Y. Yang, K.A. Nelson, Elastic modulus and in-plane thermal diffusivity measurements in thin polyimide films using symmetry-selective real-time impulsive stimulated thermal scattering, Appl. Phys. A 58, pp. 523-534, 1994.
[47] M.J. Banet, M. Fuchs, J.A. Rogers, J.H. Reinold, Jr., J.M. Knecht, M. Rothschild, R. Logan, A.A. Maznev, K.A. Nelson, High-precision film thickness determination using a laser-based ultrasonic technique, Appl. Phys. Lett., 73 (2), 1998.

[48] T. Nishino, M. Kotera, N. Inayoshi, N. Miki, K. Nakame, Residual stress and microstructures of aromatic polyimide with different imidization processes, Polymer 41, pp. 6913-6918, 2000.
[49] D. Temple et.al, Journal of the Electrochemical Society 140 (2), pp. 564-567, 1993.
[50] K. Ramkumar, S.K. Ghosh and A.N. Saxena, J. of the Electrochemical Society 140 (9), pp. 2669-2672, 1993.
[51] P. Scafidi, P., M. Ignat, Journal of Adhesion Science and Technology 12 (11), pp. 1219-1242, 1998.
[52] T. Tsuchiya et al., Sensors and Actuators A, 82 (1-3), pp. 286-296, 2000.
[53] Y.L. Shen, S. Suresh, I.A. Blech, Stresses, warpage and shape changes arising from patterned lines on silicon wafers, J. Appl. Phys. 80, pp. 1388-1398, 1996.
[54] J. Kasthurirangan, P.S. Ho, Thermal stress & microstructure in Al(Cu) and Cu interconnects for advanced USLI applications, PhD thesis, 1998.
[55] R.B.R. van Silfhout, Y. Li, W.D. van Driel, J.H.J. Janssen, J. Bisschop, F. Kuper, R. Schravendeel, G.Q. Zhang, M. van Gils, M. Jansen, The state-of-the-art on thermal-mechanical modelling of IC Backend processes, Proc. EuroSimE 2001, pp. 277-289.
[56] G.Q. Zhang, R. van Silfhout, Y. Li, W.D. van Driel, M. van Gils, J. Beijers, L.J. Ernst, Prediction of backend process-induced wafer warpage and experimental verification, Proc. ECTC 2002, pp. 1182-1187.
[57] J. Bisschop, Failure mechanisms in plastic package IC's, Proc. EurosimE 2002, pp. 328-331.
[58] Y.T. He, M.A.J. van Gils, W.D. van Driel, G.Q. Zhang, R.B.R. van Silfhout, L.J. Ernst, Prediction of crack growth in IC passivation layers, J. of Microelectronics & Reliability 44(12), 2004, pp. 2011-2017.
[59] X.H. Liu et al., Developing design rules to avert cracking and debonding in integrated circuit structures, Engineering Fracture Mechanics 66, pp. 387-402, 2000.
[60] J.R. Lloyd, M.R. Lane, X.-H. Liu, E.Liniger, T.M. Shaw, C.-K. Hu, R. Rosenberg, Reliability Challenges with Ultra-low k Interlevel Dielectrics, Microelectronics Reliability 44 (11), 2004, pp. 1835-1841.
[61] M.A.J. van Gils, G.Q. Zhang, J.H.J. Janssen, O. van der Sluis, R.M.J. Voncken, Analysis of Cu/low-k bond pad delamination by using a novel failure index, Proc. EurosimE 2005, pp. 190-196.
[62] G. Wang, S. Groothuis and P. S. Ho, Effect of packaging on interfacial cracking in cu/low-k damascene structures, Proc. ECTC, pp. 727-732, 2003.
[63] C.D. Hartfield, E.T. Ogawa, Y-J. Park, Tz-C. Chiu, Interface reliability assessments for copper/low-k products, IEEE Transactions on Device and Materials Reliability, Vol. 4, No. 2, pp. 129-141, 2004.
[64] G.A.O. Davies, Benchmarks for composite delamination, NAFEMS publication R0084, 2002.
[65] W.D. van Driel, J.H.J. Janssen, G.Q. Zhang, D.G. Yang, L.J. Ernst, Packaging induced die stresses - effect of chip anisotropy and time-dependent behaviour of a moulding compound, Journal of Electronic Packaging 125 (4), 2003, pp. 520-526.
[66] MatWeb, free searchable database of material data sheets, www.matweb.com.
[67] H.J.L. Bressers, W.D. van Driel, G.Q. Zhang, The chemistry design rules of polymers for microelectronics reliability, Proc. EuroSimE 2004, pp. 621-625.
[68] K.M.B. Jansen, L. Wang, D.G. Yang, C. van't Hof, L.J. Ernst, H.J.L. Bressers, G.Q. Zhang, Constitutive modelling of moulding compounds, Proc. ECTC 2004, pp. 890-894.

5. Thermo-Mechanics of Integrated Circuits and Packages

[69] T.S. Yeung, M.M F. Yuen, Visco-elastic analysis of IC package warpage, sensing, modelling and simulation in emerging electronic packaging, Journal of Electronic Packaging 17, pp. 101-107, 1996.

[70] S. Yi, K.Y. Sze, Cooling rate effect on post cure stresses in molded plastic IC packages, Journal of Electronic Packaging 120, pp. 385-390, 1998.

[71] D.G. Yang, K.M.B. Jansen, L.J. Ernst, G.Q. Zhang, W.D. van Driel, H.J.L. Bressers, X.J. Fan, Prediction of process-induced warpage of IC packages encapsulated with thermosetting polymers, Proc. ECTC 2004, pp. 98-105.

[72] J. Chen, M. Chan, I. De Wolf, Local stress measurements in packaging by Raman spectroscopy, Proc. EPTC 2000, pp. 159-162.

[73] J. Chen, I. De Wolf, Raman spectroscopy as a stress sensor in packaging: correct formulae for different sample surfaces, Proc. ECTC 2002, pp. 1310-1317.

[74] X.Q. Shi, John H.L. Pang, X.R. Zhang, Q.J. Liu, and M. Ying, In-situ micro-digital image speckle correlation technique for characterization of materials properties and verification of numerical models, IEEE Transactions on Components & Packaging 27 (4), pp. 659-667, 2004.

[75] M.M.F. Yuen, H. Fan, Delamination of electronic packages, Proc. EuroSimE 2003, pp. 5-21.

[76] W.K. Szeto, M.Y. Xie, J.K. Kim, M.M.F. Yuen, P. Tong, S. Yi, Interface failure criterion of button shear test as a means of interface adhesion measurement in plastic packages, Proc. Electronic Materials and Packaging Conference, pp. 263-268, 2000.

[77] W.D. van Driel, Liu C.J., Zhang G.Q., J.H.J. Janssen, van Silfhout R.B.R., van Gils M.A.J., Ernst L.J, Prediction of interfacial delamination in stacked IC structures using combined experimental and simulation methods, J. of Microelectronics & Reliability 44(12), 2004, pp. 2019-2027.

[78] M.A.J. van Gils, P.J.J.H.A. Habets, G.Q. Zhang, W.D. van Driel, P.J.G. Schreurs, Characterization and modelling of moisture driven interface failure, J. of Microelectronics & Reliability 44(11), 2004, pp. 1317-1322.

[79] X. Dai, M.V. Brillhard, P.S. Ho, Adhesion measurement for electronic packaging applications using double cantilever beam method, IEEE Transactions on Components & Packaging 23 (1), 2000, pp. 101-116.

[80] J. Taweeplengsangsuke, R.A. Pearson, Processing adhesion relations for die-attach adhesives and underfill resins, Proc. ECTC 1998, pp. 160-167.

[81] R.D. Adams, Engineered Materials Handbook, Adhesives and Sealants 3, Materials Park, OH: ASM International, 1995.

[82] S. Xu, D.A. Dillard, Determining the impact resistance of electrically conductive adhesives using a falling wedge test, IEEE Transactions on Components & Packaging 26 (3), 2003, pp. 554-562.

[83] A.A.O. Tay, Y.Y. Ma, S.H. Ong, T. Nakamura, Measurement of interface toughness as a function of temperature, moisture concentration and mode mixity, Advances in Electronic Packaging 26 (2), 1999, pp. 1129-1136.

[84] D.-Y. Shih, J. Kim, P. Buchwalter, P. Lauro, H. Clearfield, K-W. Lee, J. Paraszczak, S. Purushothaman, A. Viebeck, S. Kamath, C. Lund, H.M. Tong, M. Anschel, R. Lacombe, Adhesion test standardization for multichip module packages, Proc. of 3rd International Conference on Adhesive Joining and Coating Technology in Electronics Manufacturing, 1995, pp. 884-888.

[85] N. Tanaka, M. Kitano, T. Kumazawa, A. Nishimura, Evaluating IC-package interface delamination by considering moisture-induced moulding-compound swelling, IEEE Transactions on Components and Packaging Technologies 22 (3), 1999, pp. 426-432.

[86] C.C. Merrill, P.S. Ho, Effect of mode-mixity and porosity on interfacial fracture of low-k dielectrics, Proc. Mat. Res. Soc. Symp. Vol. 812, 2004.
[87] J. Thijsse, W.D. van Driel, M.A.J. van Gils, O. van der Sluis, Interfacial adhesion method for semiconductor applications covering the full mode mixity, Proc. EurosimE, 2005.
[88] W.D. van Driel, G.Q. Zhang, J.H.J. Janssen, L.J. Ernst, F. Su, K.S. Chian, S. Yi, Prediction and verification of process induced warpage of electronic packages, J. of Microelectronics & Reliability 43 (5), 2003, pp. 765-774.
[89] S. Yi, A Novel Portable 3D Moiré Systems, Nanyang Technological University, Singapore, Patent Pending, 2000.
[90] W.D. van Driel, J.H.J. Janssen, R.B.R. van Silfhout, M.A.J. van Gils, G.Q. Zhang, L.J. Ernst, On wire failures in microelectronic packages, Proc. EuroSimE 2004, pp. 53-58.
[91] D. Degryse, B. Vandevelde, E. Beyne, Mechanical FEM simulation of bonding process on Cu low-k wafers, IEEE Transactions on Components & Packaging 27 (4), 2004, pp. 643-650.
[92] Y. Liu, S. Irving, T. Luk, Development of thermosonic wire bonding process simulation and bond pad over active stress analysis, Proc. ECTC 2004, pp. 383-391.
[93] W.D. van Driel et al., Driving mechanisms of delamination related reliability problems in exposed pad packages, Proc. EuroSimE 2005, pp. 183-189.
[94] W.D. van Driel et al., Prediction of interfacial delamination in stacked IC structures using combined experimental and simulation methods, Microelectronics Reliability, 44 (12), 2004, pp. 2019-2027.
[95] F.G. Bucholz, R. Sistla, T. Krishnamurthy, 2D and 3D applications of the improved and generalized modified crack closure integral method, in Computational Mechanics, S.N. Atluri and G. Yagawa, Eds. New York: Springer Verlag, 1988.
[96] R. Krueger, The Virtual Crack Closure Technique: History, Approach and Applications, NASA/CR-2002-211628, ICASE Report No. 2002-10.
[97] C.P. Cherepanov, Crack propagation in continuous media, Appl. Math. Mech. 31 (1967), 476-488.
[98] J.R. Rice, A path independent integral and the approximate analysis of strain concentrations by notches and cracks, J. Appl. Mech. 35 (1968), 379-386.
[99] W.K. Wilson, J.R. Osias, A comparison of finite element solutions for an elasto-plastic crack problem, Int. J. Fracture 14 (1978), R95.
[100] M.A.J. van Gils, et al., Characterization and modelling of moistures driven interface failures, Microelectronics Reliability 44 (11), 2004, pp. 1317-1322.
[101] M.A.J. van Gils, et al., Virtual qualification of moisture induced failures of advanced packages, Proc. EuroSimE 2004, pp. 157-162.
[102] E.H. Wong et al, Moisture Diffusion and Vapour Pressure Modelling of IC Packaging, Proc. ECTC 1998, pp. 1372-1378.
[103] R. Dudek et al., Studies on Moisture Diffusion and Popcorn Cracking, Proc. EuroSimE 2002, pp. 225-232.
[104] HVQFN™, Amkor Technology Inc., Chandler, USA.
[105] J.G.J. Beijer, J.H.J.Janssen, H.J.L. Bressers, W.D. van Driel, K.M.B. Janse3, D.G. Yang, G.Q. Zhang, Warpage minimization of the HVQFN map mould, Proc. EuroSimE, 2005, pp. 168-174.
[106] D.G. Yang, K.M.B. Jansen, L.J. Ernst, G.Q. Zhang, W.D. van Driel, H.J.L. Bressers, X.J. Fan, Prediction of Process-Induced Warpage of IC Packages Encapsulated with Thermosetting Polymers, Proc. ECTC 2004, pp. 98-105.

5. Thermo-Mechanics of Integrated Circuits and Packages

[107] D.G. Yang, K.M.B. Jansen, L.J. Ernst, G.Q. Zhang, W.D. van Driel, H.J.L. Bressers, Modelling of Cure-Induced Warpage of Plastic IC Packages, Proc. EuroSimE 2004, pp. 33-40.

[108] K.M.B. Jansen, L. Wang, C. van 't Hof, L.J. Ernst, H.J.L. Bressers, G.Q. Zhang, Cure, temperature and time dependent constitutive modelling of moulding compounds, Proc. EuroSimE 2004, p. 581-585.

[109] K.M.B. Jansen, L. Wang, D.G. Yang, C. van't Hof, L.J. Ernst, H.J.L. Bressers, G.Q. Zhang, Constitutive modelling of moulding compounds, Proc. ECTC 04, pp. 890-894.

[110] C. van 't Hof, G. Wisse, L.J. Ernst, K.M.B. Jansen, D.G. Yang, G.Q. Zhang, H.J.L. Bressers, A novel tool for cure dependent visco-elastic characterization of packaging polymers, Proc. EuroSimE 2004, pp. 385-390.

[111] D.G. Yang, K.M.B. Jansen, L.J. Ernst, G.Q. Zhang, J.G.J. Beijer, J.H.J. Janssen, Experimental and Numerical Investigation on Warpage of QFN Packages Induced during the Array Moulding Process, Proc. ICEPT 2005, pp. 94-98.

[112] R.B.R. van Silfhout, W.D. van Driel, Y.Li, M.A.J. van Gils, J.H.J. Janssen, G.Q. Zhang, G. Tao, J. Bisschop, Effect of metal layout design on passivation crack occurrence using both experimental and simulation techniques, Proc. EuroSimE 2004, pp. 69-74, 2004.

[113] R.B.R van Silfhout, J.D. Roustant, W.D. van Driel, Y. Li, G.Q. Zhang, Effect of delamination of IC/compound interface on passivation cracking, Proc. EuroSimE 2003, pp. 353-358.

[114] L.L. Mercado, C. Goldberg, S.-M. Kuo, T. Yu, T. Lee, S.K. Pozder, Analysis of flip-chip packaging challenges on copper/low-k interconnects, IEEE Transactions on Device and Materials Reliability 3 (4), 2003, pp. 111-118.

[115] L.L. Mercado, S.-M. Kuo, C. Goldberg, D. Frear, Impact of flip-chip packaging on copper/low-k structures, IEEE Transactions on Advanced Packaging 26 (4), 2003, pp. 433-440.

[116] X. Fan, G.Q. Zhang, W.D. van Driel, L.J. Ernst, Analytical solution for moisture-induced interface delamination in electronic packaging, Proc. ECTC 2003, pp. 733-738.

[117] Y. Du, G. Wang, C. Merill, P.S. Ho, Thermal stress and debonding in Cu/Low-k damascene line structures, Proc. ECTC 2002, pp. 859-864.

[118] A.A.O. Tay, T.Y. Lin, Influence of temperature, humidity and defect location on delamination in plastic IC packages, Thermal and Thermomechanical Phenomena in Electronic Systems, 1998, pp. 179-184.

[119] Y. Liu, S. Irving, M. Rioux, A.J. Schoenberg, D. Chong, Die-attach delamination characterization modelling for SOIC package, Proc. ECTC 2002, pp. 839-846.

[120] S. Liu, Y. Mei, Behaviour of delaminated plastic IC packages subjected to encapsulation cooling, moisture absorption, and wave soldering, IEEE Transactions on Components, Packaging, and Manufacturing Technology, 18 (3), 1995, pp. 634-645.

[121] W.H. Wong, L. Cheng, Initiation and propagation of interface delamination in plastic IC packages, Proc. EPTC 2000, pp. 277-282.

[122] T. Saitoh, H. Matsuyama, M. Toya, Linear fracture mechanics analysis on growth of interfacial delamination in LSI plastic packages under temperature cyclic loading-Part II: Material properties and package geometry factors, IEEE Transactions on Advanced Packaging, 23 (3), 2000, pp. 554-560.

[123] P. Liu, L. Cheng, Y.-W. Zhang, Interface delamination in plastic IC packages induced by thermal loading and vapour pressure - a micromechanics model, IEEE Transactions on Advanced Packaging, 26 (1), 2003, pp. 1-9.

[124] W.D. van Driel, G. Wisse, A.Y.L. Chang, J.H.J. Janssen, G.Q. Zhang, L.J. Ernst, Influence of material combinations on delamination failures in a cavity down TBGA package, IEEE Transactions on Components & Packaging 27 (4), 2004, pp. 651-658.

8. EXERCISES

1. What are the steps to manufacture an IC package?
2. What are the functions of a package? Describe them.
3. In what classes can packages be divided?
4. What is reliability? Give 2 examples of reliability tests.
5. Consider a 600 nm SION layer deposited on a 725 μm Silicon wafer. The warpage of this wafer at RT is measured at –175m, the warpage before deposition at –3500m. The warpage as a function of temperature is measured, revealing a slope $d\sigma/dt$ of 0.015 MPa/°C.
   - What is the total mechanical stress in the deposited layer at RT?
   - What is the CTE value of the deposited material?
6. Derive the equation to calculate the intrinsic stress level directly from measured room temperature warpage.
   - Hint: use equations (**1**), (**2**), and (**3**).
7. Which techniques are available to measure the properties of thin layers of IC materials:
   - To measure the stiffness?
   - To measure the CTE?
8. What is delamination? Describe the phenomenon? What are the driving factors for this failure mechanism?
9. Which techniques are available:
   - To measure the properties of packaging materials?
   - To measure the interface strength between two packaging materials?
   - To measure deformations in IC packages?
10. What is warpage? What is the effect of mould compound curing on warpage? Explain. How can the curing effect be controlled?
11. Consider an exposed pad package with the following constituents:
    Size 10 x 10 x 0.85 mm^3.
    Cu alloy based leadframe 10 x 10 x 0.2 mm^3.
    Moulding compound with: E(25°C) = 10000MPa; E(100°C) = 8000MPa; E(125°C) = 1000MPa; E(175°C) = 800MPa, α = 8ppm/K below the T$_g$ and α = 32ppm/K above the T$_g$, which is 110°C.
    Stress free temperature for both constituents is 175°C.
    Construct a parametric 2D FE model using generalized plain strain element type and calculate the warpage of the product as function of a temperature decrease to 25°C. Answer the following questions:

5. Thermo-Mechanics of Integrated Circuits and Packages

- What is the warpage of the product as a function of temperature? Explain.
- What is the shape of the product at 25°C? Explain.
- What is the shape of the product at 125°C? Explain.
- What is the influence of a decreased package size (6 x 6 mm2) on the warpage?
- What is the influence of an increased package size (14 x 14 mm2) on the warpage?

12. Adapt the parametric 2D FE model of exercise 10 by adding an IC with: size 6 x 6 x 0.28 mm^3, E = 169000MPa, v = 0.23, α = 3.0ppm/°C.
    - What is the warpage of the product with the IC as a function of temperature? Explain the difference.
    - What is the shape of the product with the IC at 25°C? Explain the difference.
    - What is the influence of a decreased IC size (2 x 2 mm2) on the warpage?
    - What is the influence of an increased IC size (8 x 8 mm2) on the warpage?

Chapter 6

CHARACTERIZATION AND MODELLING OF MOISTURE BEHAVIOUR

X.J. Fan[1], G.Q. Zhang[2,4], W.D. van Driel[3,4], and J. Zhou[5]
[1]*Intel Corporation, CH5-263, 5000 W. Chandler Blvd., Chandler, AZ 85226 USA*
[2]*Philips Semiconductors, HTC 60, 5656AG Eindhoven, The Netherlands*
[3]*Philips Semiconductors, P.O. Box 30008, 6503HK Nijmegen, The Netherlands*
[4]*Delft University of Technology, Mekelweg 2, 2628CD Delft, The Netherlands*
[5]*Lamar University, Department of Mechanical Engineering, Beaumont TX 77710 USA*

Abstract: Two major types of failures related to moisture absorption dominate in microelectronic packaging. One of them is the delamination and cracking, which occurs at reflow when the package is subjected to moisture before soldering. Another failure is the delamination and cracking during highly accelerated stress conditions (HAST) under isothermal conditions. This chapter first covers moisture diffusion modelling (in particular, moisture diffusion modelling in multi-material system), characterization of moisture diffusivity, solubility and saturated moisture concentration at various temperatures and humidity conditions. A vapour pressure model based on a micromechanics approach, failure mechanism at reflow and the degradation of adhesion with moisture and temperature are introduced. The in-depth understanding of hygroscopic swelling, including the modelling and characterization, and the effect of non-uniform moisture distribution and transient hygroscopic swelling stresses are then presented. The impact of hygroscopic swelling on copper/low K structures as well as under bump metallurgy (UBM) failures is discussed. Void behaviour on interface or in bulk subjected to moisture and thermal stresses is studied. The characterization and modelling of interface strength as function of moisture and temperature is introduced. The interface delamination & fracture modelling methodologies and applications are presented through several case studies in designing substrate-based and leadframe-based packages, such as chip scale packages (CSP), ball grid array package (BGA) and quad flat non-lead QFN packages.

Key words: Moisture, diffusion, vapour pressure, hygroscopic swelling, moisture sensitivity test, reflow, delamination, interfacial fracture toughness, void, characterization, diffusivity, saturated moisture concentration, coefficient of hygroscopic swelling, cracking, instability.

1. INTRODUCTION

Moisture plays an important role in the integrity and reliability of plastic electronic packages. Ever since the discovery of the "popcorn" failure of plastic-encapsulated IC packages in the 1980s [1], numerous studies in the literature were presented to understand the moisture-induced failure mechanism. The presence of moisture in plastic packaging alters thermal stress through alteration of thermo-mechanical properties; induces hygroscopic stress through differential swelling; induces vapour pressure that is responsible for the eventual popcorn cracking; reduces interfacial adhesion strength; induces corrosion; and alters dielectric properties of materials.

Moisture resistance sensitivity tests were introduced as one of the major reliability tests for package qualification. When subjected to moisture, the package generally undergoes four different stages, which are schematically depicted in Figure 1. In stage 1 (preconditioning), the package absorbs moisture from the environment, which condenses in micropores in polymer materials such as substrate, die-attach, moulding compound and various adhesives along interfaces. Preconditioning is a time-consuming process, and usually takes a few days or even months in controlled or un-controlled humid environment. In stage 2, the package is mounted onto the printed circuit board by soldering. The entire package is exposed to temperatures as high as 220°C when eutectic solder is applied and may go up to 260°C for lead-free solders. As a result, the condensed moisture vaporizes under the sudden temperature rise associated with the soldering process. The vapour pressure and the reduction of interface strength at high temperature due to the previous moisture intake will cause local interfacial delamination. In stage 3, the vapour pressure exerts compressive loading on the delaminated surfaces, eventually causing the package to bulge. In the final stage, a package crack forms and propagates laterally outwards. When the crack reaches the package exterior, high-vapour pressure is suddenly released, producing an audible sound like popcorning. Figure 2 shows an actual package bulging during reflow due to the moisture vaporization. In chip scale packages and flip chip packages, package bulge in stage 3 and package (mould compound) cracking in stage 4 may not happen. Delamination or cohesive failures of adhesives at stage 2 are dominant failure modes.

Pressure Cooker Test (PCT) under 121°C/100%RH or Highly Accelerated Stress Test (HAST), is another stringent accelerated test for moisture-sensitive packages. The moisture weakens the interfacial adhesion strength, generates internal vapour pressure during reflow, and induces tensile hygroscopic swelling stress on under bump metallization (UBM) during PCT. Figures 2 and 3 show various failure modes during PCT due to hygroscopic swelling and thermal stresses.

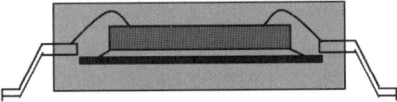

Stage 1: moisture absorption

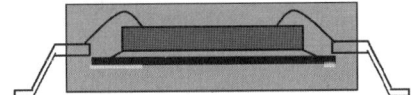

Stage 2: Initiation of delamination at reflow

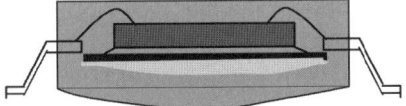

Stage 3: Delamination propagation and package bulge

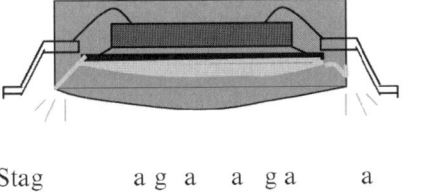

Stag   a g a   a ga   a   a

Figure 1. Schematic description of four stages of moisture-induced failures in a plastic electronic package

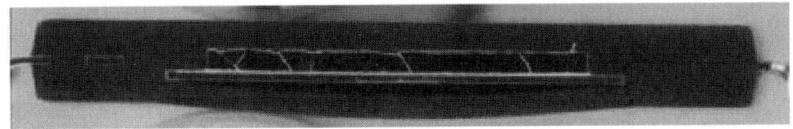

Figure 2. LQFP package die paddle belly bulge during reflow

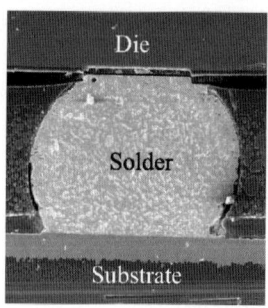

Figure 3. Hygroswelling induced failure in flip chip assembly during PCT

Despite the pivotal role of moisture, research activities in moisture-induced failure remain relatively low, compared to the thermally induced failure. This is partly due to the lack of material data and aggravated by the lack of material characterization techniques and procedures, and the near absence of such properties from material vendors.

Fukuzawa et al. [1] first reported and studied popcorn problems in 1985. Kitano et al. [2], Tay and Lin [3-6] conducted a series of work on the moisture diffusion and heat transfer in plastic IC packages, and studied the dynamics of moisture diffusion, hygrothermal stresses and delamination using an interface fracture mechanics approach. Liu and Mei [7-8] also published a series of work on behaviour of delaminated plastic IC packages subjected to encapsulation cooling, moisture absorption and wave soldering. Nguyen et al. [9] did pioneering work to understand the failure mechanism associated with the solder reflow process. Galloway et al. [10] made an excellent contribution to the moisture diffusion modelling and characterization for various kinds of plastic materials. Wong et al. [11] followed Galloway's approach to propose an alternative variable for moisture diffusion modelling. Shook et al. [12] studied diffusion models to derate moisture sensitive surface mount IC's for factory use conditions. Tee et al. [13-18] developed a fully integrated modelling approach to investigate the moisture behaviour. Dudek et al. [19, 20] presented parametric studies on moisture diffusion and popcorn cracking. The authors [21-25, 51-53, 55-57] recently introduced a micromechanics approach to study the moisture behaviour and delamination initiation, and studied the impact of non-uniform moisture distribution on charactering hygroscopic material behaviours of materials.

This chapter presents a systematic knowledge and methodology to deal with the moisture-induced failures in electronic packaging. The moisture diffusion through multi-material system will be introduced first. The measurement techniques on the diffusivity and solubility as function of temperatures are described. The evolution of vapour pressure during the

6. Characterization and Modelling of Moisture Behaviour

temperature rise is derived based on a micromechanics approach. The modelling and characterization of hygroscopic swelling are introduced. Single void behaviour either on interface or in bulk subjected to moisture and thermal stresses is studied. The characterization and modelling of interface strength as function of moisture and temperature is introduced. The interface delamination & fracture modelling methodologies and applications are presented through several case studies in designing substrate-based and leadframe-based packages.

2. MOISTURE DIFFUSION MODELLING

2.1 Diffusion in Multi-Material System

Polymeric packaging materials transport moisture primarily by diffusion, although secondary effects such as surface tension and pressure driven flows may also contribute. Moisture transport strictly by diffusion is modelled using the standard transient diffusion as following

$$\frac{\partial^2 C^2}{\partial x^2} + \frac{\partial^2 C^2}{\partial y^2} + \frac{\partial^2 C^2}{\partial z^2} = \frac{1}{\alpha_D} \frac{\partial C}{\partial t} \tag{1}$$

where C is the local concentration (g/cm^3, weight of water or water vapour per unit volume in bulk material), x, y, z are coordinates (cm), α_D is the moisture diffusivity (cm^2/s), and t is the time (s).

Secondary moisture transport mechanisms are ignored in developing equation (1). Later, this assumption will be verified. Since equation (1) follows the same governing differential equation as the diffusion of heat, but the dependent variable, temperature, is replaced by moisture concentration and the thermal diffusivity is replaced by moisture diffusivity, commercially available finite element analysis (FEA) software can be used to solve transient moisture diffusion problems. However, a unique problem arises in the diffusion of moisture. An interfacial concentration discontinuity results where two materials having different saturated concentrations are joined, as shown in Figure 4. Standard FEA software will not easily accommodate a step discontinuity at an interface.

The interfacial discontinuity can be removed by normalizing the field variable. Galloway et al. [10] defines a new variable as following

$$\varphi = C / S \tag{2}$$

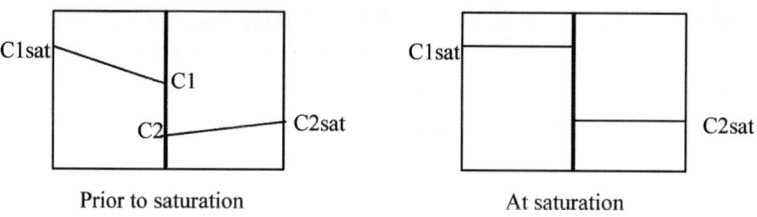

Figure 4. Moisture concentration across bi-material interface

which is continuous across the bimaterial interface. The S is the solubility and related to the saturated concentration C_{sat} by

$$S = C_{sat} / p_{ext} \qquad (3)$$

where p_{ext} is the ambient vapour pressure under the given humid conditions. Galloway interpreted the new variable φ as the partial vapour pressure of moisture across the interface, which must be continuous.

As an alternative variable w, Wong et al. [11] introduced the so-called 'wetness' defined by

$$w = C / C_{sat} \qquad (4)$$

which is also continuous along the interface.

It is pointed out by Fan et al. [24] that the two approaches are equivalent with the use of equation (3). When φ is used, the diffusion equation can be re-written as

$$\frac{\partial^2 \varphi}{\partial x^2} + \frac{\partial^2 \varphi}{\partial y^2} + \frac{\partial^2 \varphi}{\partial z^2} = \frac{1}{\alpha_D} \frac{\partial \varphi}{\partial t} \qquad (5)$$

with

$$\varphi_1 = \varphi_2 \qquad (6)$$

$$\alpha_{D1} S_1 \frac{\partial \varphi_1}{\partial n} = \alpha_{D2} S_2 \frac{\partial \varphi_2}{\partial n} \qquad (7)$$

6. Characterization and Modelling of Moisture Behaviour

When w is used, the diffusion equation can be re-written as

$$\frac{\partial^2 w}{\partial x^2} + \frac{\partial^2 w}{\partial y^2} + \frac{\partial^2 w}{\partial z^2} = \frac{1}{\alpha_D}\frac{\partial w}{\partial t} \tag{8}$$

with

$$w_1 = w_2 \tag{9}$$

$$\alpha_{D1} C_{sat1} \frac{\partial w_1}{\partial n} = \alpha_{D2} C_{sat2} \frac{\partial w_2}{\partial n} \tag{10}$$

where the subscripts 1 and 2 represent different materials respectively. Equations (7) and (10) are the continuity of flux of moisture across the interface. With the established thermal-moisture analogy, the commercial thermal FEA software can be used to model moisture diffusion in a multi-material system. The implementation procedure is described in details in [57].

In order to illustrate the nature of moisture diffusion in plastic materials, we consider an example of a simple transient moisture diffusion problem in a slab, with the boundary conditions as shown in Figure 5. The thickness of 2 mm is taken as a typical value. The material properties (e.g., a typical mould compound), together with the ambient humid conditions for three types of moisture preconditioning are listed in Table 1. The solution can be represented as

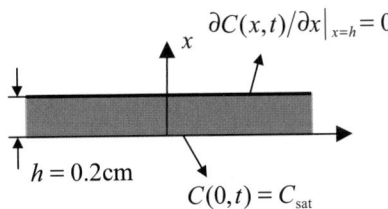

Figure 5. One-dimensional moisture diffusion problem

$$C(x,t) = C_{sat}[1 - \sum_{n=0}^{\infty} \frac{[2(-1)^n]e^{-\lambda_n^2 \alpha_D t/h^2}}{\lambda_n} \cos(\lambda_n (h-x)/h)] \tag{11}$$

where

$$\lambda_n = (\frac{2n+1}{2})\pi \qquad (12)$$

Figure 6 shows the results for the local moisture concentration at $x = h$ as the function of time. It takes hundreds of hours (or even longer) to the saturated moisture state. Assume the thermal diffusivity for this slab as 4.22×10^{-3} cm²/s, Figure 7 shows the temperature rise at the side $x = h$ as function of time, when the side $x = 0$ is suddenly subjected to a step-temperature-rise ΔT. Figures 6 and 7 show that the heat transfer is much faster than moisture diffusion. This implies that the package temperature can be approximately assumed as uniform at reflow temperature.

Table 1. Three types of moisture preconditioning

Moisture pre-conditioning	Saturated vapour density ρ_g (g/cm³)	Saturated vapour pressure p_g (MPa)	Ambient vapour density ρ_{ext} (g/cm³)	Ambient vapour pressure p_{ext} (MPa)	Moisture diffusivity α_D (cm²/s)	Saturated concentration C_{sat} (g/cm³)
30°C/60%RH	3.04×10^{-5}	4.24×10^{-3}	$0.6 \rho_g$	$0.6 p_g$	3.13×10^{-9}	7.86×10^{-3}
85°C/60%RH	3.58×10^{-4}	5.87×10^{-2}	$0.6 \rho_g$	$0.6 p_g$	2.85×10^{-8}	8.84×10^{-3}
85°C/85%RH	3.58×10^{-4}	5.87×10^{-2}	$0.85 \rho_g$	$0.85 p_g$	2.85×10^{-8}	1.25×10^{-2}

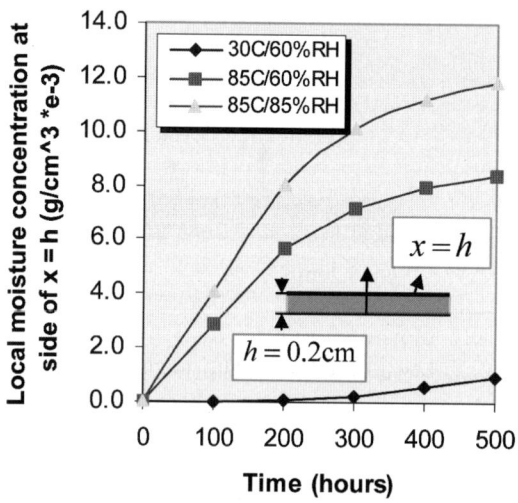

Figure 6. Local moisture concentration at $x = h$ as function of time

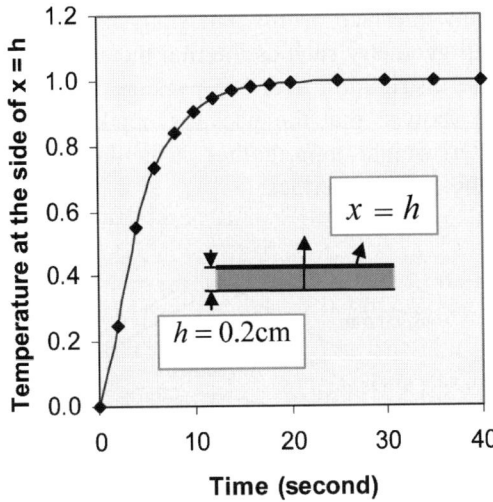

Figure 7. Dimensionless temperature at x = h as function of time

2.2 Application to PBGA Package

Figure 8 shows the geometry and dimensions of the 68 I/O PBGA package [10]. Realistic geometry effects, e.g., as internal via, metallization on the die pad, and metallization at the solder joint interconnect are included in the model. Moisture absorption processes are modelled with a standard FEA code using a 3-D eight-noded transient diffusion element. The FEA model predicts φ (see equation (5)) for each element from which the local concentration may be determined using (2). The volume and concentration are known for each element; thus, the mass of each element can be calculated. The total package weight gain is calculated by summing the individual mass contributions for each element for a given time step, which can be compared with the weight gain test results.

The moisture properties of diffusivity (D) and C_{sat} used in JEDEC level 1, 85°C/85%RH, are listed in Table 2. Die, copper, gold plate, and solder bump do not absorb moisture, and are assigned with very small values of diffusivity and solubility in analysis (The next section will discuss how to measure diffusivity and saturated concentration).

Transient weight gain predictions at the JEDEC Standard level 1 and level 3 conditions (i.e., 85°C/85%RH and 30°C/60%RH, respectively) are compared to experimental data in Figure 9. The agreement between data and simulations is within 10% for absorption process. Although a more simplified one-dimensional analysis may be used to estimate the local concentration at any interface, a fully 3-D analysis is required to account for realistic packaging geometry such as internal metallization or via. Figure 10 gives the moisture distribution over the package length cross section [14]. Kitano et al. [2] showed that the package cracking is not controlled by the absolute water weight gain, rather it is due to the local moisture concentration at the critical interface.

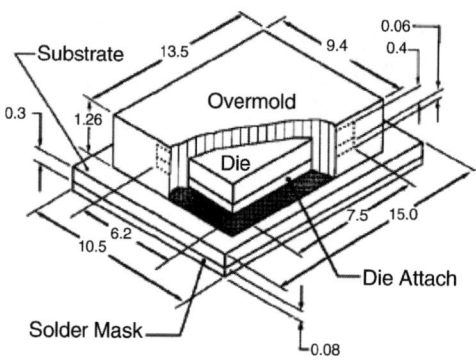

Figure 8. PBGA package geometry

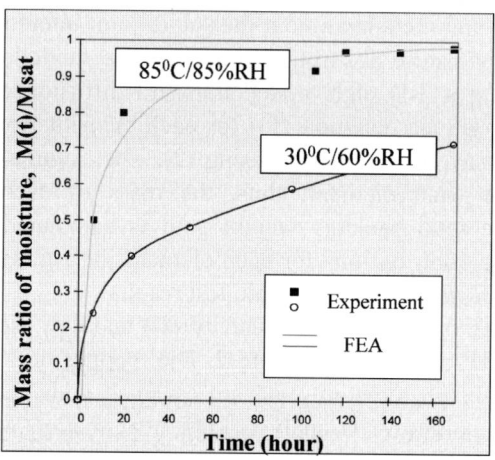

Figure 9. Comparison of PBGA experimental weight gain to finite element analysis simulation

6. Characterization and Modelling of Moisture Behaviour

Table 2. Diffusivity and C_{sat} used in the moisture diffusion modelling

Material	Diffusivity D (cm^2/s)	C_{sat} (g/cm^3)
BT	8.55e-9	2.40e-2
Die Attach	1.68e-7	5.30e-3
Mould Compound	5.40e-8	4.00e-3
Solder Resist	2.47e-8	3.88e-2
Underfill	5.60e-9	2.47e-2

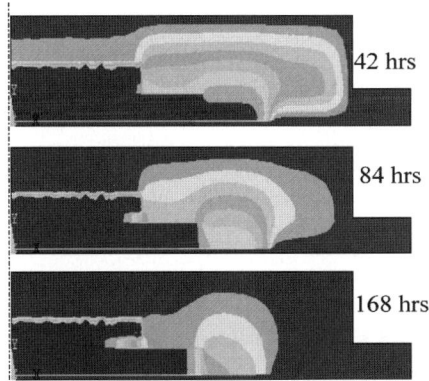

Figure 10. The moisture distribution contour in a PBGA package

2.3 Moisture Desorption

Moisture desorption is considered because moisture content in the package is lost during reflow, and this may affect the moisture distributions. For moisture desorption, the initial condition for the package is the moisture distribution after moisture preconditioning. The boundary condition is that the moisture concentration is zero at the exposed interfaces, since during reflow the oven is 100% dry. The C_{sat} data used are the same as those in preconditioning, but the diffusivity values are assumed to be a few order higher than in the moisture absorption at preconditioning, such as 85°C/85%RH. Figure 11 shows the moisture diffusion distributions of FCBGA before and after 2 minutes of desorption at 220°C [14]. The initial condition is level 1 (85°C/85%RH, 168 hours). It is found that the moisture desorption during reflow affects the moisture distribution greatly, though the time duration is only 2 minutes. However, even though there is significant amount of moisture loss in the package during reflow, the location moisture concentration at underfill/substrate interface may remain unchanged.

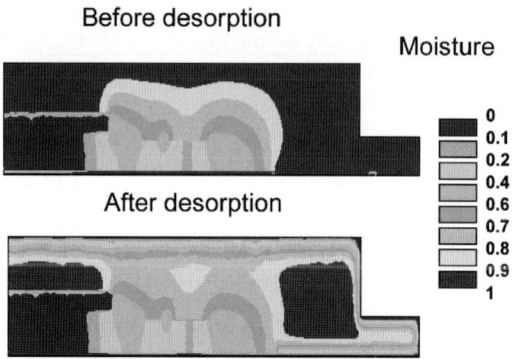

Figure 11. Moisture distribution before and after moisture desorption

3. CHARACTERIZATION OF MOISTURE DIFFUSIVITY AND SATURATION CONCENTRATION

Moisture absorption and desorption properties of packaging materials are determined by measuring the weight gain or loss as a function of time, and fitting the response to an analytical solution. A 3-D diffusion model is presented in the following to provide physical insight into the moisture absorption and desorption processes. Experimental data are presented to validate the proposed diffusion model and to provide numerical values for the diffusivity and solubility as a function of temperature.

3.1 Diffusivity Measurement

For a 3-D rectangular bar as shown in Figure 12, the equation (1) can be solved using standard separation of variable techniques, yielding an expression for the local moisture concentration as a function of time, position, and diffusivity. Local concentration data are unavailable but the total weight gain is easily measured. Hence, an analytical expression for the total weight gain as a function of time is obtained by integrating the local concentration over volume of the rectangular block sample

6. *Characterization and Modelling of Moisture Behaviour*   293

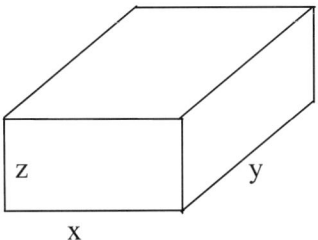

Figure 12. A 3-D rectangular bar

$$\frac{M_t}{M_\infty} = 1 - \frac{512}{\pi^6} \sum_{l=0}^{\infty} \sum_{m=0}^{\infty} \sum_{n=0}^{\infty} \frac{e^{\frac{-Dt}{L_{eqv}}}}{[(2l+1)\pi]^2 [(2m+1)\pi]^2 [(2n+1)\pi]^2}$$

where
$$L_{eqv} = \left\{ \left[\frac{(2l+1)\pi}{x}\right]^2 + \left[\frac{(2m+1)\pi}{y}\right]^2 + \left[\frac{(2n+1)\pi}{z}\right]^2 \right\}^{-1} \quad (13)$$

where M_t is the instantaneous mass of the sample, M_∞ is the saturated mass, L_{eqv} is an equivalent length scale, x, y and z are the length, width, and thickness dimensions, respectively. Equation (13) converges within 1 to 2% for the sample dimensions reported here when four terms are included in each dimension.

Moisture diffusivity and solubility data were gathered by periodically weighing samples soaked in a humidity chamber controlled to an accuracy of ±1°C and ±1% relative humidity (RH). An analytical balance having an accuracy of ±0.01 mg was used to weigh samples. The samples were first dried in an oven above 100°C for at least 24 hours, and weighed to determine the dry weight. The samples were then placed in vented containers and inserted into the humidity chamber through a side port, rather than distributing the test conditions by opening and closing the chamber door. Samples were periodically removed from the humidity chamber and weighed. The ambient moisture conditions were set by specifying the dry bulb temperature and the relative humidity. Desorption studies were conducted using data gathered with a thermogravimetric analysis (TGA) instrument.

Diffusivity constants were determined by minimizing the difference between the model predictions given by equation (13) and the experimental weight gain versus time data using nonlinear regression method with least square method [57]. Figure 13 gives a typical test data with the calculated results. The diffusivity and the saturated moisture concentration at each measured temperature can then be determined.

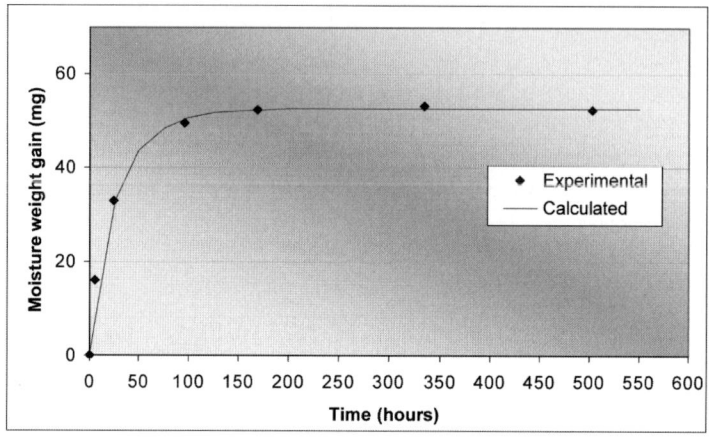

Figure 13. Moisture weight gain curve and analytical fit

The next step is the second regression analysis to fit the Arrhenius equation given by equation (14) as following

$$D = D_o e^{\frac{Q}{RT}} \tag{14}$$

where D_o is the diffusivity coefficient, Q is the activation energy, R is the Boltzmann constant (8.83e-5 ev/K), and T is the absolute temperature (K). If the moisture transport mechanism is controlled by diffusion, then the rate of absorption or desorption should be accurately predicted by the Arrhenius equation, generally within a 30% error. Numerical values for the diffusivity coefficient and activation energy are given in Table 3. The Arrhenius equation fits the experimental absorption data shown by solid symbols in Figure 14, within 15% over the range of temperatures [10].

Belton et al. [26] states that the absorption and desorption processes are controlled by diffusion when the following three conditions are met:
1. The weight gain curve appears linear at low non-dimensional times.
2. The curve becomes concave downward at later times.
3. The weight gain versus non-dimensional time for different samples thickness all collapses on a single curve.

All of the aforementioned conditions are met as shown in Figure 15[10].

6. Characterization and Modelling of Moisture Behaviour

Table 3. Absorption and desorption diffusivity constants

Material	Absorption		Desorption	
	α_o (cm²/s)	Q_α (ev/K)	α_o (cm²/s)	Q_α (ev/K)
BT Epoxy	1.2×10^{-4}	-0.295	6.0×10^{-2}	-0.465
Die Attach	7.0×10^{-2}	-0.474	2.0×10^{-2}	-0.392
Over-mold	5.5×10^{-3}	-0.376	1.1×10^{-2}	-0.366
Solder Mask	5.5×10^{-6}	-0.167	2.7×10^{-2}	-0.366

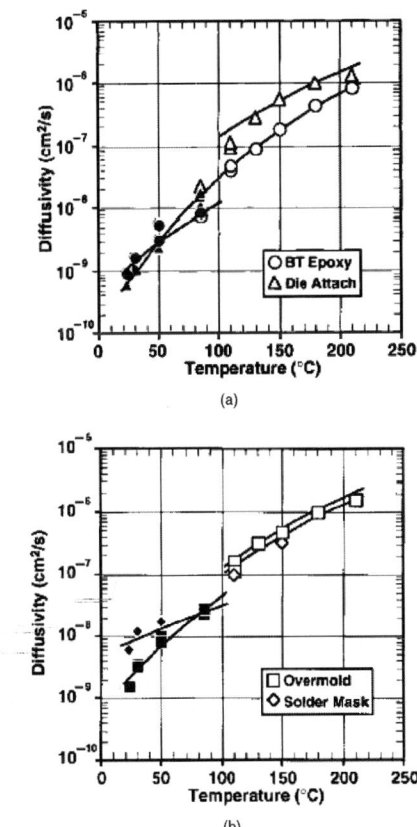

Figure 14. Moisture absorption and desorption diffusivity as a function of temperature

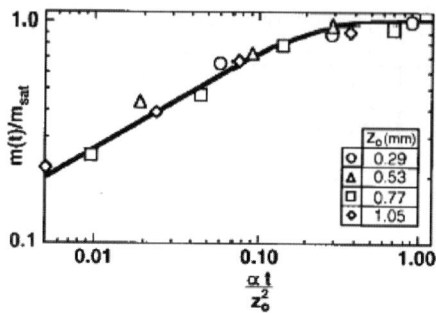

Figure 15. Experimental weight gain compared to a 3-D analytical model predictions for overmould at 50°C/85%RH

3.2 Saturated Moisture Concentration

Saturation concentration data were measured to determine the maximum possible weight gain per sample volume for a particular ambient condition. The saturation concentration, C_{sat}, is a function of temperature, humidity, and material. When C_{sat} is normalized by the ambient water vapour density ρ_g, the concentration ratio, ψ, becomes a function of temperature and does not depend on the relative humidity. The concentration ratio dependency on temperature is modelled using the form of the Arrhenius equation given by

$$\psi = \frac{C_{sat}}{\rho_g} = \psi_0 \exp(Q_\psi / RT) \tag{15}$$

The agreement between correlations based on (15) and experimental concentration ratio data is within 10% as shown in Figure 16. Empirical constants used to model the concentration ratio dependency on temperature are given in Table 4, where T is measured in degrees Kelvin.

The physical state of moisture, either liquid or vapour, residing in polymeric materials may be inferred from the data shown in Figure 16. Because the concentration of moisture, based on the volume of the sample, is several orders of magnitude greater than the ambient water vapour density at the same temperature, the majority of moisture residing in the package by mass must reside in the liquid state. This conclusion has enormous consequences when considering the moisture-induced stresses during reflow. Vapourizing moisture from liquid deposits in micro-pores from which high pressure steam is generated during reflow, which will be discussed in the next section.

6. Characterization and Modelling of Moisture Behaviour

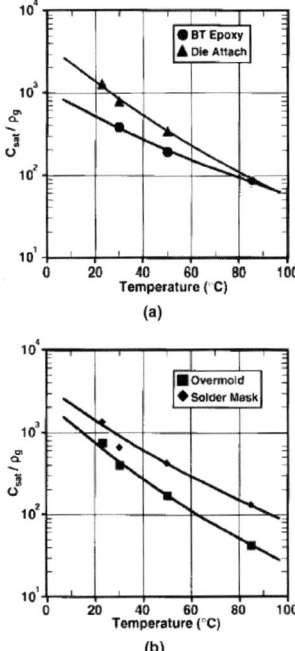

Figure 16. Concentration ratio versus temperature

Table 4. Concentration constants

Material	ψ_o (-)	Q_ψ (ev/K)
BT Epoxy	2.1×10^{-2}	0.255
Die Attach	4.8×10^{-4}	0.375
Overmold	9.8×10^{-5}	0.400
Solder Mask	2.8×10^{-3}	0.332

4. VAPOUR PRESSURE MODELLING

The modelling of ensuing vapour pressure within the package during the reflow is the key element in understanding the failure mechanism. Previous researchers assumed that the delamination exists before the reflow, and considered the vapour pressure as traction loading subjected to the delaminated interfaces. There were some studies done and a few methods were proposed to estimate the vapour pressure acting on the delaminated interface. Since the vapour pressure is generated anywhere in the package, it is necessary to investigate the whole field vapour pressure distribution before the package delamination.

4.1 Micromechanics-based Vapour Pressure Model

One of the critical issues in developing a vapour pressure model is to find out the moisture density in *voids*, denoted as ρ,

$$\rho = \frac{dm}{dV_f} \tag{16}$$

where dm is the mass of moisture per unit volume of free spaces in material, dV_f. The moisture concentration C is defined as

$$C = \frac{dm}{dV} \tag{17}$$

where dV is the element volume of the porous material, which contains free spaces dV_f. It should be noted that, because of the inhomogeneous character of a porous material, the element should be established over a (finite) representative volume, RVE [27].

Introduce the void volume fraction f according to

$$f = \frac{dV_f}{dV} \quad 0 < f \leq 1 \tag{18}$$

The following relation between ρ and C can be obtained

$$\rho = \frac{dm}{dV_f} = \frac{dm}{dV}\frac{dV}{dV_f} = C/f \tag{19}$$

6. Characterization and Modelling of Moisture Behaviour

Let's make a comparison of the magnitude of ρ with the ambient moisture density at 85°C/85RH condition, i.e., $\rho_{ext} = 0.85 \cdot \rho_g = 3.04 \cdot 10^{-4}$ g/cm^3. ρ_g is the saturated moisture density. Assuming that the void volume fraction f is 0.05, and the saturated moisture concentration C is $1.25 \cdot 10^{-2}$ g/cm^3 for a typical epoxy-based polymer according to the measurement by Galloway et al. [10], equation (19) gives $\rho = 820 \rho_{ext}$. This number clearly shows how much moisture a typical polymer material could absorb. Such an amount of moisture must condense into the mixed liquid/vapour phase in the material.

The following condition is used to determine the moisture state in voids at preconditioning of temperature T_0,

$$\begin{cases} \rho \leq \rho_g(T_0) & \text{for vapor phase at } T_0 \\ \rho > \rho_g(T_0) & \text{for mixed liquid/vapor phase at } T_0 \end{cases} \quad (20)$$

where ρ_g is the saturated vapour density, which can be obtained from the steam table as function of temperature.

When the moisture is at mixed liquid/vapour phase, it is necessary to know at which temperature the moisture can be *fully* vaporized. This temperature is called the *phase transition temperature*, denoted by T_1, which can be determined by

$$\rho(T_1) = \rho_g(T_1) \quad (21)$$

Now the vapour pressure in voids can be determined by the moisture state analysed above. When the moisture is in the mixed liquid/vapour phase, the vapour pressure maintains the saturated vapour pressure p_g as function of temperature (from steam table), i.e.,

$$p(T) = p_g(T) \quad \text{for mixed liquid/vapor phase} \quad (22)$$

When the moisture is in single vapour phase, the ideal gas law can be followed to calculate the vapour pressure as following,

$$p dV_f = dmRT \quad \text{or} \quad p = \rho RT \quad (23)$$

Dividing both sides by dV, we obtain

$$pf = CRT \quad (24)$$

where R is the universal gas constant ($= 8.314$ J/mol).

The two vapour-phase states (p, f, T, C) and (p_r, f_r, T_r, C_r) are then related by

$$\frac{p}{p_r} = \frac{Tf_r C}{T_r f C_r} \quad \text{for single vapor phase from } T_r \text{ to } T \tag{25}$$

Assuming that the material is incompressible, the change of volume element due to the temperature change is related by

$$\frac{dV}{dV_0} \approx 1 + 3\alpha \Delta T \tag{26}$$

where $\Delta T = T - T_0$ and α the coefficient of thermal expansion, respectively. Thus

$$C = \frac{dm}{dV} = \frac{dm}{dV_0} \frac{dV_0}{dV} = C_0 (1 - 3\alpha \Delta T) \tag{27}$$

from which, it is noted that although the moisture mass is assumed conserved during the temperature rise (the desorption effect is neglected), the moisture concentration may change due to the change of the bulk volume by thermal expansion.

Three distinct cases for the vapour pressure evolution have been identified [23], and are shown in Figure 17. In the following, the detailed description and derivation for the vapour pressure evolution for each case are presented, and some corrections on the errors in previous publications are made.

In case 1, the moisture in a void is in single vapour phase at T_0 after the moisture absorption. The condition for this case can be mathematically expressed as

$$p(T_0) \leq \rho_g(T_0) \tag{28}$$

or equivalently, according to equation (19)

$$C_0 / f_0 \leq \rho_g(T_0) \tag{29}$$

6. Characterization and Modelling of Moisture Behaviour

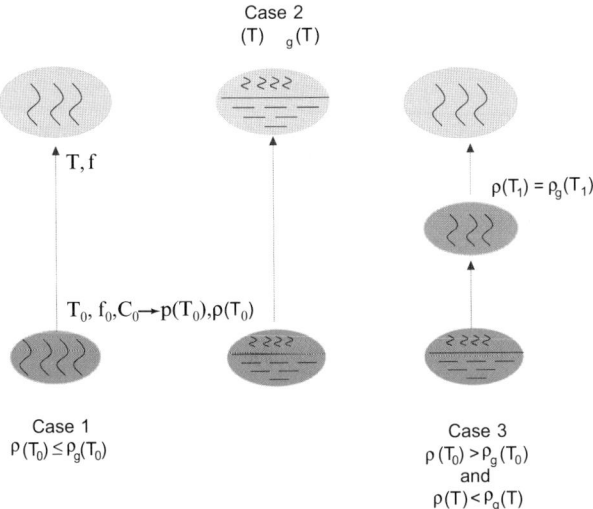

Figure 17. Three distinct cases for the vapour pressure evolution from the preconditioning temperature T_0 to the current temperature T

The initial vapour pressure at T_0 can be determined from equation (23), by relating with a fully saturated state at temperature T_0, as following

$$p_0 = p(T_0) = \rho(T_0) RT_0$$
$$p_g(T_0) = \rho_g(T_0) RT_0$$
(30)

thus

$$p_0 = \frac{\rho(T_0)}{\rho_g(T_0)} p_g(T_0) = \frac{C_0 p_g(T_0)}{f_0 \rho_g(T_0)}$$
(31)

by which the vapour pressure at temperature T_0 can be calculated when the moisture concentration and the initial void volume fraction are known.

The vapour pressure at temperature T for case 1 can then be obtained from equation (25) with the reference state (p_r, f_r, T_r, C_r) to be substituted by (p_0, f_0, T_0, C_0), as following

$$p(T) = \frac{Tf_0 C}{T_0 f C_0} p_0 = \frac{C_0 p_g(T_0)}{\rho_g(T_0) f} \frac{T}{T_0}[1 - 3\alpha(T - T_0)]$$
(32)

In case 2, the moisture in the voids is in the mixed liquid/vapour phase at *current* temperature T. Therefore, the moisture must also be in the mixed liquid/vapour phase at initial T_0. The condition for case 2 is thus as following

$$p(T) \geq p_g(T) \tag{33}$$

or by equation (21)

$$\frac{C_0}{f}[1 - 3\alpha(T - T_0)] \geq p_g(T) \tag{34}$$

In this case the vapour pressure maintains the saturated vapour pressure during the course of the temperature rise. Thus the vapour pressure at temperature T follows equation (21) as follows

$$p(T) = p_g(T) \tag{35}$$

Case 3 is an intermediate case between case 1 and 2, where the moisture is in the mixed liquid/vapour phase at initial T_0, but in the single vapour phase at current T. The condition for this case can be written as

$$p(T_0) > p_g(T_0) \quad \text{and} \quad p(T) < p_g(T) \tag{36}$$

or using equation (21)

$$C_0 / f_0 > p_g(T_0) \quad \text{and} \quad \frac{C_0}{f}[1 - 3\alpha(T - T_0)] < p_g(T) \tag{37}$$

The phase transition temperature T_1 where the moisture is just fully vaporized should be determined first according to equation (21), which can be rewritten as

$$\frac{C_0}{f(T_1)}[1 - 3\alpha(T_1 - T_0)] = p_g(T_1) \tag{38}$$

Then from T_1 to temperature T the equation (25) can be used with the reference state (p_r, f_r, T_r, C_r) to be substituted by ($p(T_1)$, $f(T_1)$, T_1, $C(T_1)$), as follows

6. Characterization and Modelling of Moisture Behaviour

$$p(T) = p(T_1)\frac{T}{T_1}\frac{f(T_1)C}{fC(T_1)} = p_g(T_1)\frac{T}{T_1}\frac{f(T_1)C}{fC(T_1)}$$

$$= p_g(T_1)\frac{T}{T_1}\frac{f(T_1)}{f}\frac{1-3\alpha(T-T_0)}{1-3\alpha(T_1-T_0)} \tag{39}$$

Now the equations for calculating the vapour pressure are complete, which can be summarized as follows

Case 1: when $C_0/f_0 \leq p_g(T_0)$,

$$p(T) = \frac{C_0 p_g(T_0)}{p_g(T_0)f}\frac{T}{T_0}[1 - 3\alpha(T-T_0)] \tag{40}$$

Case 2: when $\dfrac{C_0}{f}[1 - 3\alpha(T - T_0)] \geq p_g(T)$

$$p(T) = p_g(T) \tag{41}$$

Case 3: when $C_0/f_0 > p_g(T_0)$, and $\dfrac{C_0}{f}[1 - 3\alpha(T - T_0)] < p_g(T)$

$$p(T) = p_g(T_1)\frac{T}{T_1}\frac{f(T_1)}{f}\frac{1-3\alpha(T-T_0)}{1-3\alpha(T_1-T_0)} \tag{42}$$

where T_1 is determined by equation (38).

The above model includes an unknown f, the current void volume fraction. Obviously, the vapour pressure is dependent on the void deformation behaviours, and should be solved together with the governing equations of deformation.

Let's investigate the magnitude of vapour pressure for case 1, where the moisture is in single-vapour phase at preconditioning. Assuming that the preconditioning temperature T_0 is 85°C, the *maximum* vapour pressure allowed in voids at T_0 is the saturated vapour pressure $p_g(T_0=85°C) = 5.27\text{e-}2\text{Mpa}$.

The vapour pressure at reflow temperature $T = 220°C$ is plotted as function of the current void volume fraction f in Figure 18, by using equation (40) ($\alpha = 200$ppm/°C, $f_0 = 0.03$). The vapour pressure decreases with the current void volume fraction. The pressure may be lower than the initial vapour pressure of T_0 when the void becomes large. The maximum vapour pressure developed at 220°C is 7.92e-2 MPa, when the void does not grow ($f = f_0$). The results imply that the vapour pressure for *case 1* is substantially low such that it has almost negligible effect on the void growth.

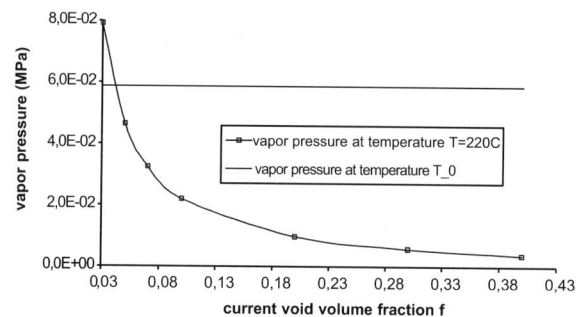

Figure 18. Vapour pressure p at 220°C versus the current void volume fraction f by equation (23) for case 1, with $f_0 = 0.03$, $\alpha = 200$ppm/°C, $T_0 = 85°C$, and $p_0 = p_{0max} = p_g(85°C) = 5.27$e-2 MPa

Consider the case 2 where the moisture is not fully vaporized at reflow temperature T. In this case the vapour pressure is the saturated vapour pressure, i.e., $p = p_g(T = 220°C) = 2.32$ MPa.

Questions remain that how to measure the initial void volume fraction f_0. An approximate method in estimating the initial void volume fraction was proposed by using the moisture absorption test. From equation (19), when moisture absorption is saturated, the initial void volume fraction is given by

$$f_0 = \frac{C_{sat}}{\rho} \tag{43}$$

Given the fact that the moisture condenses mostly into the liquid form and the water liquid density is 1.0 g/cm³, f_0 can be estimated from

$$f_0 \approx C_{sat}\big|_{100°C/100RH} \tag{44}$$

Equation (44) provides a simple way to predict the approximate magnitude of the voids fraction existing in polymer materials using the moisture property data given by Galloway et al. [10]. It shows that the initial void volume fraction is usually between 0.01 and 0.05.

Equation (44) can be simplified as $C \approx C_0$ when the thermal expansion is much smaller than 1. Equations (40)-(42) can then be simplified as following,

Case 1: when $C_0/f_0 \leq \rho_g(T_0)$,

$$p(T) = \frac{C_0 p_g(T_0)}{\rho_g(T_0) f} \frac{T}{T_0} \tag{45}$$

Case 2: when $\dfrac{C_0}{f} \geq \rho_g(T)$

$$p(T) = p_g(T) \tag{46}$$

Case 3: when $C_0/f_0 > \rho_g(T_0)$, and $\dfrac{C_0}{f} < \rho_g(T)$

$$p(T) = p_g(T_1) \frac{T}{T_1} \frac{f(T_1)}{f} \tag{47}$$

4.2 Vapour Pressure as External Loading in Delaminated Areas

The internal vapour pressure in voids at the interfaces instantaneously becomes an external pressure subjected to the delaminated interfaces when delamination is formed at reflow. By equation (18), the delamination is defined when $f = 1$, which implies that the material element is totally voided. The initial vapour pressure subjected to the delaminated surfaces thus can be obtained from the above analysis with a special case $f = 1$. This makes the problem simpler since the final state of the void growth is known already. By substituting $f = 1$ into equations (45) to (47), we obtain

Case 1: when $C_0/f_0 \leq \rho_g(T_0)$,

$$p(T) = \frac{C_0 \rho_g(T_0)}{\rho_g(T_0)} \frac{T}{T_0} \tag{48}$$

Case 2: when $C_0 \geq \rho_g(T)$

$$p(T) = p_g(T) \tag{49}$$

Case 3: when $C_0/f_0 > \rho_g(T_0)$, and $C_0 < \rho_g(T)$

$$p(T) = p_g(T_1) f(T_1) \frac{T}{T_1} \tag{50}$$

where T_1 is determined by equation (38), and $T_0 < T < T_1$.

The vapour pressure at the delaminated area will be immediately uniform when the delamination is complete, although the local moisture concentrations at interfaces are different with the location. Thus, the average local concentration can be defined as following

$$C^{ave} = \frac{\int_A C dA}{A} \tag{51}$$

where A is the delaminated area. In calculating the vapour pressure after delamination by above equations, the concentration C should be replaced by C^{ave}.

One of main differences between equations (45) to (47) and equations (48) to (50) is that the initial vapour pressure subjected to the delaminated surfaces can be computed directly, without knowing the deformation history of the voids, except for equation (51). For example, assuming that the void volume fraction is 0.03, if the local moisture concentration $C_0 \leq f_0$ $\rho_g(T_0) = 0.03 \times 3.58\text{e-}4 = 0.11 \times 10^{-4}$ g/cm^3 (under 85°C), then the vapour pressure can be calculated directly from equation (48). As indicated in Figure 18, the vapour pressure in this case is substantially small such that it has negligible impact on the void growth. On the other hands, from equation (49), when the local concentration $C_0 \geq \rho_g(T) = 1.16 \times 10^{-2}$ g/cm^3 (under 220°C), the initial vapour pressure will remain the saturated vapour pressure 2.32MPa. Equation (49) provides a very simple criterion for the vapour pressure estimation on the delaminated surfaces. When the moisture diffusion modelling is performed, the averaged moisture concentration can be calculated according to equation (51). When

$$C^{ave} \geq \rho_g(T) \tag{52}$$

the initial vapour pressure on the delaminated interface will remain the saturated vapour pressure at current temperature T according to equation (49). Otherwise, the vapour pressure will be dependent on the deformation history of the voids by equation (50).

4.3 Vapour Pressure-Induced Expansion

Our previous finite element modelling of whole field vapour pressure according to the model developed above concludes that the vapour pressure saturated much faster than the moisture diffusion. This implies that the vapour pressure may be uniformly distributed in the plastic material regardless of moisture saturation. The Young's modulus of plastic material drops a few orders at the reflow temperature, thus the vapour pressure-induced expansion may become as important as thermal expansion. For instance, assume the Young's modulus of a typical underfill at 220°C is 500 MPa, and Poisson ratio is 0.3. Therefore, the volume change caused by vapour pressure, $P_g(220°C)$ of 2.32 MPa, can be estimated as

$$\frac{\Delta V}{V} = \frac{3(1-2v)}{E} p = 5.568e-3 \tag{53}$$

which, is equivalent to the coefficient of thermal expansion (CTE) of 5.568e-3/3/(220-175) = 41ppm/°C under the 175-220°C temperature loading. The magnitude of this vapour pressure-induced strain is in the same order as thermal strain. It is obvious that the vapour pressure-induced expansion introduces additional mismatch. It must also be pointed out that such an expansion is directly related to the vapour pressure distribution, rather than the moisture distribution.

4.4 Whole-Field Vapour Pressure Modelling

The finite element results of vapour pressure for FCBGA at level 1, reflow temperature of 220°C, are shown in Figure 19 (a), with different times of moisture absorption. The corresponding moisture diffusion distributions are shown in Figure 19 (b) [14].

It is found that moisture diffusion and vapour pressure have different distributions. The vapour pressure in the package saturated much faster than the moisture diffusion. At level 1 condition (168 hours), the package is almost fully saturated with vapour pressure of 2.32 MPa, which corresponds to p_g at 220°C. Equation (49) shows that the vapour pressure will remain its saturated pressure at 220°C when the transition temperature, T_1, is larger than 220°C. According to equation (38), T_1 will be above 220°C as long as $\rho_m(T_0) \geq \rho_g(220°C) = 0.0116$ g/cm^3. Such a condition can be easily satisfied even though less moisture is absorbed. Similarly, the vapour pressure distributions at 220°C for wire bond PBGA at level 3 are given in Figure 20 (a), with different times of moisture absorption. The corresponding moisture diffusion distributions are shown in Figure 20 (b).

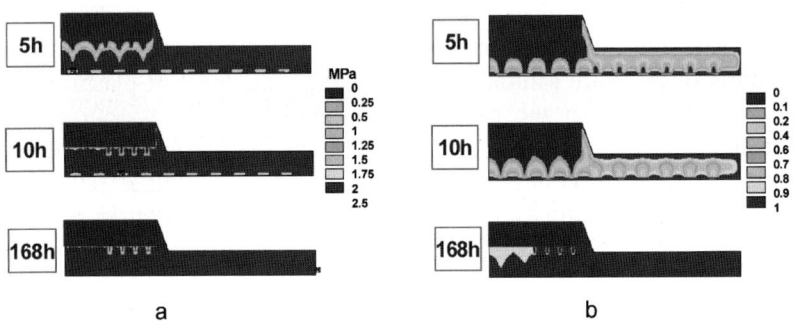

Figure 19. (a) Transient vapour pressure distribution in FCBGA at level 1,220°C; (b) Transient moisture distribution in FCBGA at level 1,220°C

6. *Characterization and Modelling of Moisture Behaviour* 309

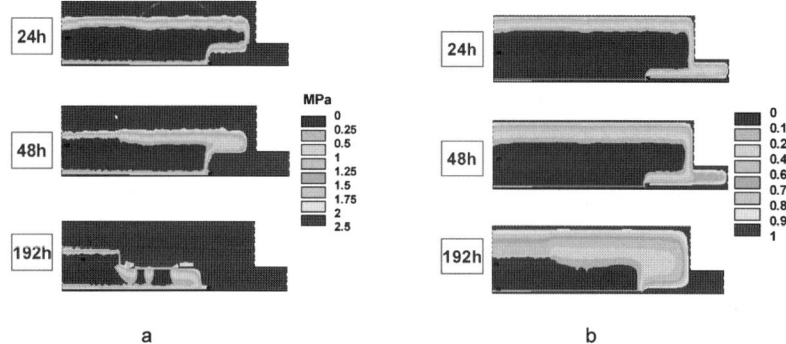

Figure 20. Transient vapour pressure distribution in FCBGA at level 3, 220°C; (b) Transient moisture distribution in FCBGA at level 3, 220°C

4.5 Failure Mechanism

The moisture affects the package reliability at reflow from two aspects: generation of vapour pressure and degradation of interfacial adhesion. Figure 21 shows the relative effect of moisture absorption on the interfacial adhesion and vapour pressure in the package. The interfacial adhesion is weakened with higher moisture level.

Previous results of vapour pressure estimation show that the saturated vapour pressure can be reached even with less moisture absorbed. For FCBGA and wire bond PBGA modelled here, the saturated pressure can be reached at the critical interface, even at the level 3 condition. However, the interfacial adhesion will be significantly decreased with more moisture absorption. When the adhesion strength is reduced to the level below the vapour pressure, delamination will occur. Therefore, the knowledge of material interfacial adhesion strength with moisture effect at high temperature condition is important in determining the failure criteria.

4.6 Underfill Selection for Flip Chip BGA Package for Moisture Performance

Figure 22 shows the weight gain curves of six underfills. Underfill *A* has the least moisture absorption among six materials. Moisture sensitivity test was performed with the FCBGA package using underfill A. Four different configurations of test vehicles were tested under JEDEC level 3. These four configurations include two different ball layouts with and without mould. Table 5 summarizes the test results. Unfortunately all samples failed at level 3 with the delamination between underfill and polyimide at chip side.

Additional tests were performed using underfill *C* and *E*. The results in Table 6 showed the packages with underfill C and E passed JEDEC level 3, and packages with underfill C even passed level 2. Further adhesion test showed that the underfill C has the strongest adhesion with the moisture at high temperature. These results support the failure mechanism described in the previous section. It is noted that it is important to select a material with high adhesion with moisture at reflow temperature, regardless of moisture absorption.

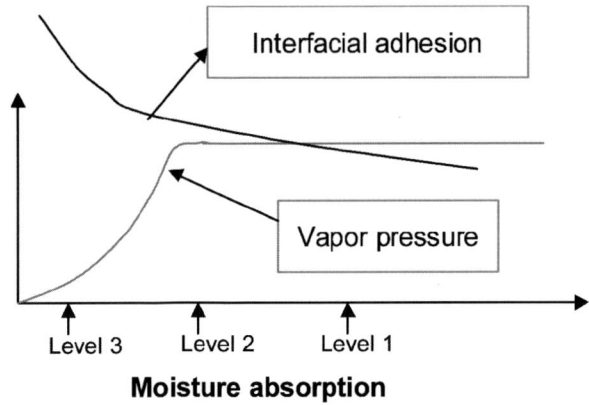

Figure 21. Relative effect of moisture absorption on interfacial adhesion and vapour pressure

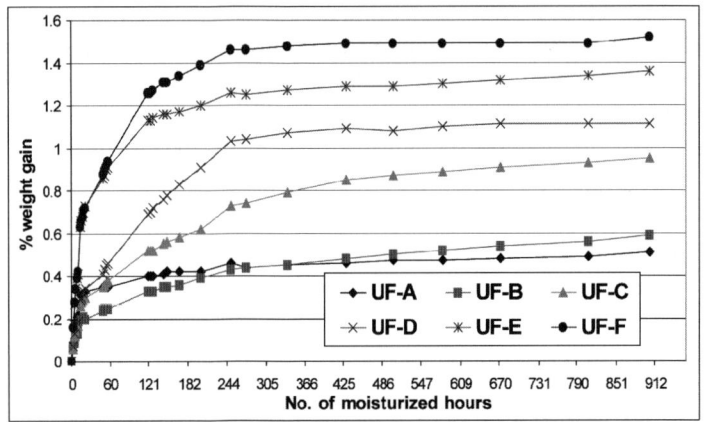

Figure 22. Moisture weight gain data for six underfills

6. Characterization and Modelling of Moisture Behaviour

Table 5. Controlled samples tested in Level 3 (30C/60 %RH) with underfill A

Sample ID	Configuration	# unit with this failure mode			Total number of failure
		Electrical failure	Delamination or pop-corn	Solder joint failure	
C12	peripheral, molded	0/24	3/24	0/24	3/24
C22	peripheral, non-molded	0/24	5/24	5/24	5/24
C11	full array, molded	4/31	10/31	15/31	15/31
C21	full array, non-molded	4/16	4/16	4/16	4/16

Table 6. Test results with underfill C and E

Underfill	Total number of failure units		
	Level 3	Level 2	Level 1
UF-C	0/24	0/18	10/11 *
UF-E	0/18	Not Available	Not Available

* failure mode is solder joint open / short induced by PI/UF interfacial delamination

5. HYGROSCOPIC SWELLING CHARACTERIZATION & MODELLING

Polymeric materials swell when moisture is absorbed. Differential swelling occurs between the polymeric and non-polymeric materials as well as among the polymeric materials constituting the electronic packages. This differential swelling induces hygroscopic swelling stresses in the package, which adds to the thermal stress at elevated temperature.

5.1 Hygroscopic Swelling Characterization

The change in dimension and weight can be related by the coefficient of moisture expansion β for the strain versus moisture concentration as following,

$$\varepsilon_h = \beta C \tag{54}$$

where ε_h is the hygro strain, and C is the moisture concentration. The hygroscopic swelling-induced stresses are analogous to the CTE mismatch-induced thermo-mechanical stresses.

There are several methods in characterizing the hygroscopic swelling properties of polymer materials, such as TMA/TGA method [17, 48], and more recently, the Moiré interferometry method [50]. Two identical material samples are preconditioned with moisture under certain temperature and humidity (e.g., 85°C/85%RH) for about 2 weeks until saturation. Then the specimens are monitored for desorption under thermo-gravimetric analyser (TGA) and thermo-mechanical analyser (TMA) respectively at the same time under 85°C constant temperature until the moisture is fully dried out. Both the reads of TMA and TGA show time dependent curves, as in Figure 23. The change in thickness dimension direction can be calculated from the TMA readings. Then, average strain in thickness direction is obtained as the change in dimension divided by the original length. The change in weight or moisture loss of the specimen during desorption can be obtained from the readings of TGA. Then the average moisture concentration can be calculated from the moisture loss divided by the specimen volume. The average strain and moisture concentration C can be defined as following respectively [60],

$$\varepsilon_{ave}^{I} = \frac{L_t - L_{dry}}{L_{dry}} \tag{55}$$

$$C_{ave}^{I} = \frac{M_t - M_{dry}}{M_{dry}} \tag{56}$$

in which, the average strain is defined with the reference point in dry conditions. L_t and M_t are the thickness and the total mass of specimen at time t during desorption, and L_{dry} and M_{dry} are the thickness and the total mass of specimen at 'infinite time' when all moisture is dried out.

Alternatively, the average strain and moisture concentration C can also defined as following, respectively,

$$\varepsilon_{ave}^{II} = \frac{L_{sat} - L_t}{L_{sat}} \tag{57}$$

$$C_{ave}^{II} = \frac{M_{sat} - M_t}{M_{sat}} \tag{58}$$

in which, the average strain is defined with the reference point in saturated conditions. L_t and M_t are the thickness and the total mass of specimen at time t during desorption, and L_{sat} and M_{sat} are the thickness and the total

6. Characterization and Modelling of Moisture Behaviour

mass of specimen at 'time zero' when the sample is fully saturated. If the second method is used to analyse the test data in Figure 23, the average strain is thus plotted as a function of average moisture concentration, as shown in Figure 24. Linear regression is then performed to obtain the slope of the data set, which is the coefficient of hygroscopic swelling.

The values of coefficient of hygroscopic swelling measured for various packaging polymeric materials are shown in Table 7, based on the above method [17]. It shows that the coefficient of hygroscopic swelling doesn't have direct correlation with the amount of moisture absorbed. Some materials absorb more moisture, but have less ability to swell (smaller coefficient of hygroscopic swelling). Some materials absorb less moisture, but swell more because of greater coefficient of hygroscopic swelling.

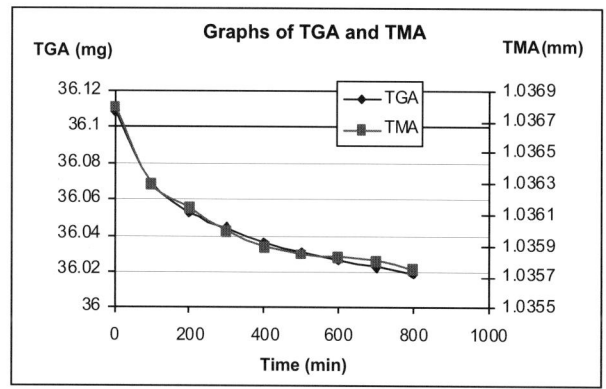

Figure 23. TGA and TMA curves for Underfill C

5.1.1 Effect of Non-Uniform Moisture Distribution

In the above methods in measuring hygroscopic swelling properties, moisture distribution is not uniform across the test specimen during the measurement. The coefficient of hygroscopic swelling is obtained based on the averaged moisture concentration over the total volume as function of specimen deformation. It is necessary to understand how much error might be introduced based on such an averaged approach. Our recent studies have shown that the current averaged approach might over-estimate the coefficient of hygroscopic swelling as much as 250% [51-53, 60]. In the following, the detailed analysis is described.

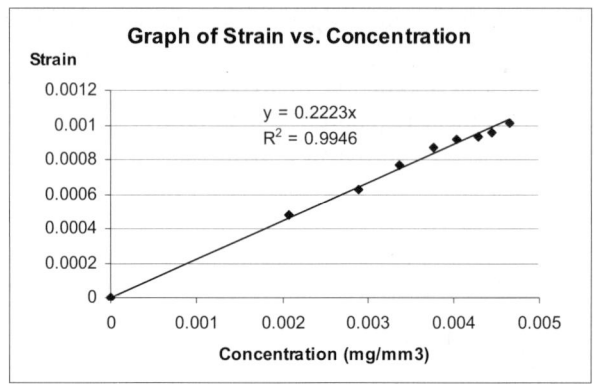

Figure 24. Computation of coefficient of hygroscopic swelling for Underfill C

Table 7. Moisture and hygroscopic swelling material properties

Materials	D (mm^2/s)	C_{sat} (mg/mm^3)	CME (mm^3/mg)	Total Hygro Strain (C_{sat} x CME)
Underfill A	9.02e6	0.0152	0.18	0.0027
Underfill B	1.55e6	0.0329	0.22	0.0072
Underfill C	1.14e5	0.0112	0.31	0.0035
Mould Compound	2.79e6	0.0043	0.4	0.0017
Solder Mask	4.83e5	0.0143	0.2	0.0029
BT Substrate	2.13e6	0.0075	0.4	0.0030

If a plate, as shown in Figure 25, is taken to be infinitely long in y and z directions the moisture content inside the plate varies only in the x-direction. Fick's law as following can describe this one dimensional moisture diffusion problem

$$\frac{\partial C}{\partial t} = D \frac{\partial^2 C^2}{\partial x^2} \qquad (59)$$

Assume that the moisture concentration inside the plate is uniformly saturated at time zero, that is $C = C_{sat}$, where C_{sat} is the saturated moisture concentration of the material. Then the plate is suddenly exposed to the absolutely dry environment and the moisture concentration on exposed faces instantaneously reaches zero. Mathematically, the initial condition and boundary conditions are

6. Characterization and Modelling of Moisture Behaviour

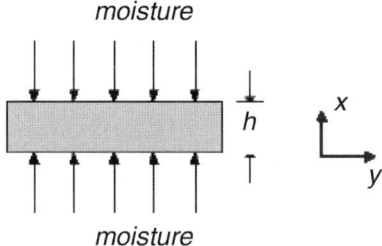

Figure 25. Graphical representation of one-dimensional diffusion problem

I.C.:

$$C(x,0) = C_{sat} \qquad (60a)$$

B.C.s:

$$C(0,t) = C(h,t) = 0. \qquad (60b)$$

Using the separation variable method, the solution of the described one-dimensional problem can be derived as

$$C(x,t) = \frac{4C_{sat}}{\pi} \sum_{n=0}^{\infty} \frac{1}{(2n+1)} \sin\frac{(2n+1)\pi x}{h} \exp\left[-\frac{Dt}{h^2}(2n+1)^2 \pi^2\right] \qquad (61)$$

If the plate possesses finite length in three dimensions as h_x, h_y, and h_z, it becomes a general three-dimensional moisture diffusion problem. Assuming that the initial condition and boundary condition are applied on each boundary same as the above one-dimensional problem, and then governing equations for the moisture distribution are presented by

$$\frac{\partial C}{\partial t} = D\left(\frac{\partial^2 C^2}{\partial x^2} + \frac{\partial^2 C^2}{\partial y^2} + \frac{\partial^2 C^2}{\partial z^2}\right) \qquad (62)$$

I.C.:

$$C(x,y,z,0) = C_{sat} \qquad (63a)$$

B.C.s:

$$C(0,y,z,t) = C(h_x,y,z,t) = 0$$
$$C(x,0,z,t) = C(x,h_y,z,t) = 0 \quad (63b)$$
$$C(x,y,0,t) = C(x,y,h_z,t) = 0$$

By introducing:

$$C(x,y,z,t) = C_x(x,t)C_y(y,t)C_z(z,t) \quad (64)$$

this three-dimensional moisture distribution problem becomes three independent one-dimensional problems about C_x, C_y, and C_z, respectively. As an example, the equations about C_x are given below.

$$\frac{\partial C_x}{\partial t} = D \frac{\partial^2 C_x^2}{\partial x^2} \quad (65)$$

$$C_x(x,0) = C_{sat}^{1/3} \quad (66a)$$

$$C_x(0,t) = C_x(h_x,t) = 0. \quad (66b)$$

Similarly, one can write the other two sets of equations about C_y and C_z, respectively. Hence the solution of this general three-dimensional moisture distribution problem is basically the multiplication of the results of three 1-D problems.

$$C(x,y,z,t) = \frac{4^3 C_{sat}}{\pi^3} \left\{ \sum_{n=0}^{\infty} \frac{1}{(2n+1)} \sin \frac{(2n+1)\pi x}{h_x} \exp\left[-\frac{Dt}{h_x^2}(2n+1)^2 \pi^2\right] \right\}$$
$$\left\{ \sum_{n=0}^{\infty} \frac{1}{(2n+1)} \sin \frac{(2n+1)\pi y}{h_y} \exp\left[-\frac{Dt}{h_y^2}(2n+1)^2 \pi^2\right] \right\} \quad (67)$$
$$\left\{ \sum_{n=0}^{\infty} \frac{1}{(2n+1)} \sin \frac{(2n+1)\pi z}{h_z} \exp\left[-\frac{Dt}{h_z^2}(2n+1)^2 \pi^2\right] \right\}$$

As discussed before, the *averaged* coefficient of hygroscopic swelling is obtained using the hygroscopic strain ε_{ave} versus *averaged* moisture content in weight over total volume (C_{ave}) above. Mathematically, such an averaged approach can be written as

6. Characterization and Modelling of Moisture Behaviour

$$\varepsilon_{ave} = \beta_{ave} C_{ave} \qquad (68)$$

where β_{ave} is the averaged coefficient of hygroscopic swelling. We know that $\varepsilon_{ave} = \Delta h_x/h_x$, where Δh_x is the displacement accrossing x direction, and $C_{ave} = \int_V C(x,y,z,t)dxdydz / h_x h_y h_z$ by its definition, thus

$$\beta_{ave} = \Delta h_x h_y h_z / \int_V C(x,y,z,t)dxdydz \qquad (69)$$

The integral in the above equation actually is the specimen weight change caused by moisture desorption. It can be obtained based on the solution of 3-D moisture diffusion shown in Equation (67), as following

$$\int_V Cdxdydz = \frac{8^3 C_{sat} h_x h_y h_z}{\pi^6} \left\{ \sum_{n=0}^{\infty} \frac{1}{(2n+1)^2} \exp\left[-\frac{Dt}{h_x^2}(2n+1)^2 \pi^2\right] \right\}$$
$$\left\{ \sum_{n=0}^{\infty} \frac{1}{(2n+1)^2} \exp\left[-\frac{Dt}{h_y^2}(2n+1)^2 \pi^2\right] \right\}$$
$$\left\{ \sum_{n=0}^{\infty} \frac{1}{(2n+1)^2} \exp\left[-\frac{Dt}{h_z^2}(2n+1)^2 \pi^2\right] \right\} \qquad (70)$$

On the other hands, for any point of the moisture field in specimen, the constitutive model of hygroscopic swelling is

$$\varepsilon = \beta C \qquad (71)$$

where ε and C are the hygroscopic stain and moisture concentration, respectively, and β is the coefficient of hygroscopic swelling of the material.

Let u be the displacement in x direction, then $du = \beta C dx$, where β is constant and $C = C(x,y,z,t)$. Take integral to both side of the equation to x along the measurement line of $y = h_y/2$, and $z = h_z/2$ (the centre point of specimen), that is

$$\beta = \Delta h_x / \int_0^{hx} C(x, \frac{h_y}{2}, \frac{h_z}{2}, t)dx \qquad (72)$$

Based on the solution of 3-D moisture diffusion in Equation (67), the integral in the above equation is

$$\int_0^{h_x} C(x, \frac{h_y}{2}, \frac{h_z}{2}) dx = \frac{128 C_{sat} h_x}{\pi^4} \left\{ \sum_{n=0}^{\infty} \frac{1}{(2n+1)^2} \exp\left[-\frac{Dt}{h_x^2}(2n+1)^2 \pi^2\right] \right.$$
$$\left\{ \sum_{n=0}^{\infty} \frac{1}{(2n+1)} \exp\left[-\frac{Dt}{h_y^2}(2n+1)^2 \pi^2\right] \right\}$$
$$\left. \left\{ \sum_{n=0}^{\infty} \frac{1}{(2n+1)} \exp\left[-\frac{Dt}{h_z^2}(2n+1)^2 \pi^2\right] \right\} \right\}$$
(73)

Using Equations (69) to (73), we can find that

$$\frac{\beta}{\beta_{Ave}} = \frac{4}{\pi^2} \frac{\sum_{n=0}^{\infty} \frac{1}{(2n+1)^2} \exp\left[-\frac{Dt}{h_y^2}(2n+1)^2 \pi^2\right] \sum_{n=0}^{\infty} \frac{1}{(2n+1)^2} \exp\left[-\frac{Dt}{h_z^2}(2n+1)^2 \pi^2\right]}{\sum_{n=0}^{\infty} \frac{(-1)^n}{(2n+1)} \exp\left[-\frac{Dt}{h_y^2}(2n+1)^2 \pi^2\right] \sum_{n=0}^{\infty} \frac{(-1)^n}{(2n+1)} \exp\left[-\frac{Dt}{h_z^2}(2n+1)^2 \pi^2\right]}$$
(74)

Equation (74) is the analytical expression of the correlation between the locally defined hygroscopic swelling coefficient and the averaged hygroscopic swelling coefficient.

It is interesting to note that Equation (74) is independent of the saturated moisture concentration C_{sat}. The ratio of the accurate β to the average β_{ave} depends on the specimen dimension, measuring time, and the diffusivity of material.

For a specific type of underfill material tested [51], the saturated concentration C_{sat} is 4.50e-2 mg/mm^3, and moisture diffusivity of the material is 1.2e-6 mm^2/s. The theoretical β/β_{ave} versus measuring time are plotted in Figure 26 according to equation (74). The ratio β/β_{ave} ranges from 0.4 to 1. The ratio increases with the increase of measuring time. Initially, the specimens are moisture saturated, then desorption process starts. The moisture concentration distributes uniformly initially, which explains why β/β_{Ave} is closer to 1 in the beginning as shown in Figure 26. When the measuring time increases, the ratio approaches 0.4. It implies that the error by the averaged approach can be as high as 250%. The corresponding experimental data are also plotted in the figure. The predication shows a remarkable agreement with the experiment results.

6. Characterization and Modelling of Moisture Behaviour

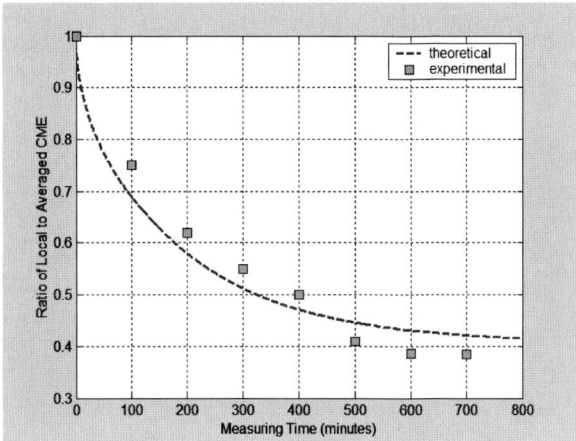

Figure 26. Ratio of accurate coefficient of hygroscopic swelling to averaged coefficient of hygroscopic swelling as function measuring time during desorption

Assume the specimen has the same length in y and z direction, that is $h_y = h_z$. An aspect ratio α is defined as $\alpha = h_y/h_x = h_z/h_x$. When the size of the specimen is very large in y and z direction, which means aspect ratio α goes to infinite, the ratio of the accurate β to β_{ave} will be unity. It implies that as long as the in-plane specimen size is much larger than the thickness, the averaged approach will give the accurate coefficient of hygroscopic swelling, even through the moisture in thickness direction is not uniformly distributed. The aspect ratio is usually around 4 when TMA/TGA is used. Figure 27 shows the ratio of accurate coefficient of hygroscopic swelling to averaged coefficient of moisture expansion as a function of aspect ratio α, where α changes in a range from 2 to 50. The measuring time is given as four different values as 100 minutes, 500 minutes, 1000 minutes, and 2000 minutes, respectively. Based on Figure 27, one can also conclude that with the increase of α, the ratio β/β_{ave} increases, and will be closer to 1, which means the larger specimen introduces less error in characterizing the hygroscopic swelling properties.

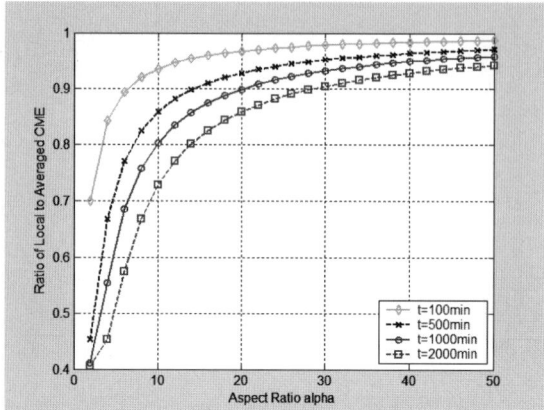

Figure 27. Ratio of local coefficient of hygroscopic swelling to averaged coefficient of hygroscopic swelling versus aspect ratio

The above analysis doesn't consider the effect of hygroscopic stresses induced during desorption. Since the moisture distribution is not uniform, the hygroscopic mechanical deformation is developed. In the following section, a sequentially coupled moisture diffusion and hygroscopic stress analysis is described to investigate the impact of hygroscopic stresses on the accurate determination of hygroscopic swelling characterization.

5.1.2 Effect of Hygroscopic Stresses

Considering the hygroscopic swelling strain and thermal strain, the Hook's law for the isotopic material can be expressed as,

$$\{\varepsilon\} = [D]^{-1}\{\sigma\} + \{\varepsilon^{hygro}\} + \{\varepsilon^{th}\} \qquad (75)$$

where $\{\sigma\}$ is the stress vector,

$$\{\sigma\} = [\sigma_x \quad \sigma_y \quad \sigma_z \quad \sigma_{xy} \quad \sigma_{yz} \quad \sigma_{xz}]^T;$$

$\{\varepsilon\}$ is the strain vector,

$$\{\varepsilon\} = [\varepsilon_x \quad \varepsilon_y \quad \varepsilon_z \quad \varepsilon_{xy} \quad \varepsilon_{yz} \quad \varepsilon_{xz}]^T.$$

$[D]^{-1}$ is the flexibility or compliance matrix. Assume E is the Young's modulus, G is the shear modulus, and ν is the Poisson ratio, then

6. Characterization and Modelling of Moisture Behaviour

$$[D]^{-1} = \begin{bmatrix} 1/E & -v/E & -v/E & 0 & 0 & 0 \\ -v/E & 1/E & -v/E & 0 & 0 & 0 \\ -v/E & -v/E & 1/E & 0 & 0 & 0 \\ 0 & 0 & 0 & 1/G & 0 & 0 \\ 0 & 0 & 0 & 0 & 1/G & 0 \\ 0 & 0 & 0 & 0 & 0 & 1/G \end{bmatrix};$$

$\{\varepsilon^{th}\}$ is the thermal strain vector. Assume α is the coefficient of thermal expansion, ΔT is the temperature difference, then

$$\{\varepsilon^{th}\} = \Delta T [\alpha \quad \alpha \quad \alpha \quad 0 \quad 0 \quad 0]^T; \tag{76}$$

$\{\varepsilon^{hygro}\}$ is the hygroscopic swelling strain vector. Assume β is the coefficient of hygroscopic swelling, and C is the local moisture concentration, or weight of the moisture per unit volume in bulk material, then

$$\{\varepsilon^{hygro}\} = C[\beta \quad \beta \quad \beta \quad 0 \quad 0 \quad 0]^T. \tag{77}$$

Without considering the temperature change, the total strain contains hygroscopic strain and elastic strain. Hygroscopic strain is due to the change of moisture concentration, while the elastic strain is caused by the hygroscopic stress. In the experiment of the characterization of the coefficient of hygroscopic swelling, moisture is non-uniformly distributed during the test, which induces hygroscopic stress. This hygroscopic stress in turn causes the elastic strain.

In the previous analysis, the elastic strain by hygroscopic stress, which is induced by the non-uniform moisture distribution, has been ignored. In the following, a sequentially coupled diffusion and stress analysis is used. In a sequentially coupled physics analysis, one can couple the two fields by applying results from moisture analysis as loads in structural analysis.

The specimen described in the previous section is used. Structural material properties are Young's modulus $E = 10$ GPa, and Poisson's ratio $v = 0.4$.

Only one eighth of the geometry is simulated due to symmetry. Three symmetric surface conditions are applied. One can see the geometry in finite element analysis in Figure 28. Point O in the figure is the centre of the whole specimen, and point A is the centre on the outer surface. In the characterization experiment, the dimension change is actually measured along the line AO.

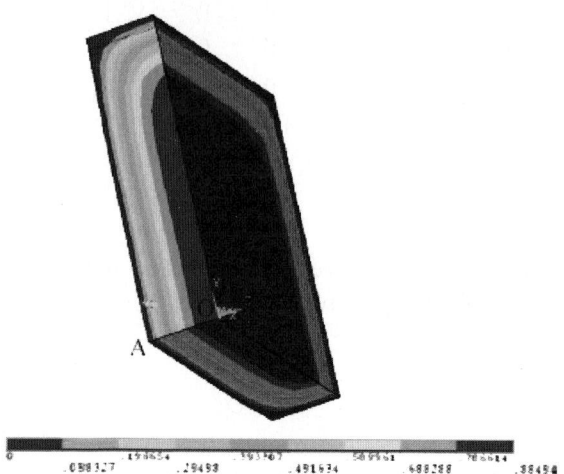

Figure 28. Finite element model of the one-eighth of the specimen and moisture distribution in FEA analysis

For the moisture diffusion analysis, initial condition is moisture fully saturated; and boundary conditions are that all outer surfaces are absolute dry with zero moisture concentration, i.e.,

I.C.:

$$C(x,y,z,0) = C_{sat}$$

B.C.s:

$$\begin{aligned} C(0,y,z,t) &= C(h_x,y,z,t) = 0 \\ C(x,0,z,t) &= C(x,h_y,z,t) = 0 \\ C(x,y,0,t) &= C(x,y,h_z,t) = 0 \end{aligned} \qquad (78)$$

Figure 28 also shows the non-uniform moisture distribution inside the specimen at time 180 minutes during desorption. On the surface of the specimen it is absolute dry and the moisture concentration is zero; while in the centre O, the moisture concentration is about 88% saturated moisture concentration.

Both theoretical hygro-strain from equation (71) in previous analysis and the hygro-strain from finite element analysis at the centre of the specimen (point O) are plotted as function of time in Figure 29. As expected, the two results are in excellent agreement, which verifies our finite element modelling.

6. Characterization and Modelling of Moisture Behaviour

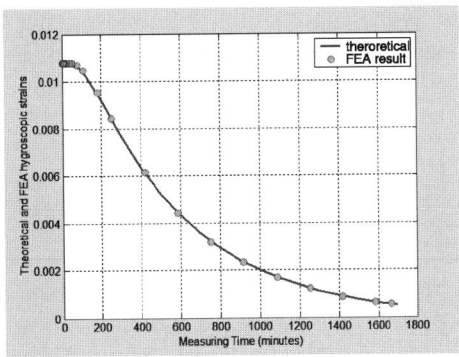

Figure 29. Comparison of the theoretical hygroscopic-strain versus FEA hygroscopic-strain at the centre of specimen (point O)

Figure 30 shows the FEA results of elastic strain and hygroscopic strain at the centre of specimen (point O) versus time. Hygroscopic strain decreases with time. The elastic strain, however, starts from zero and increases to its maximum at ~180 minutes, then drops gradually. Both hygroscopic strain and elastic strain will eventually become zero when all moisture is dried out. It is noted that the elastic strain accounts about the half of hygroscopic strain at peak, which implies that one-third of total deformation comes from the hygroscopic mechanical stress.

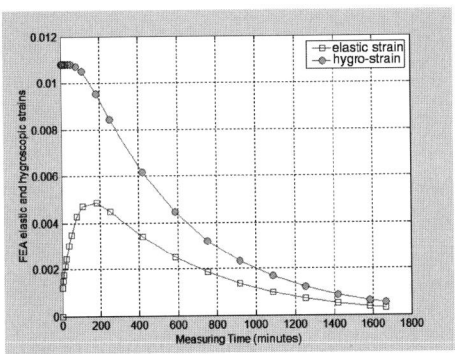

Figure 30. Elastic strain and hygroscopic strain at the centre of specimen (point O) versus time

Elastic strain and hygroscopic strain along the measure line of the specimen (line OA) are plotted in Figure 31 at time t = 180 minutes during desorption. The horizontal axis is the distance to the centre point (point O). One can see that the sign of elastic strain changes from positive to negative along the measuring path, and its average is more than one order lower than that of the average hygroscopic strain. This means that the average elastic strain along the measure path is close to zero compared to the average hygroscopic strain. Hygroscopic strain is positive along the measure path. The total displacement of specimen centre is calculated from:

$$\Delta h_x = 2\int_0^A \varepsilon_x^{total} dx = 2\int_0^A (\varepsilon_x^{hygro} + \varepsilon_x^{elastic}) dx = 2(\int_0^A \varepsilon_x^{hygro} dx + \int_0^A \varepsilon_x^{elastic} dx) \qquad (79)$$

from which it can be seen that the integral of the elastic strain portion (i.e., the second term of the right side of equation (79)) will have negligible contribution to the total deformation according to the above analysis, even though the elastic strain is significantly comparable to hygroscopic strain at each location. This explains why the prediction purely based on the hygroscopic deformation analysis in previous section gives very good agreement with experimental results in Figure 26.

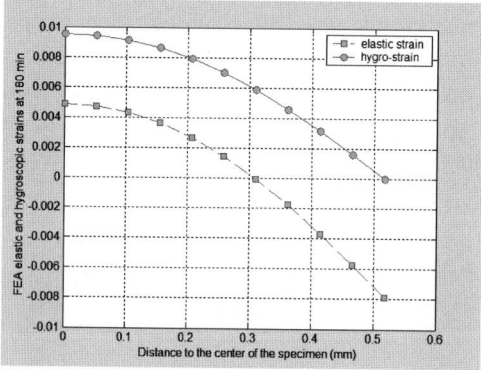

Figure 31. Elastic strain and hygroscopic strain along the measure line of the specimen (line OA) at time = 180 minutes

Figure 32 describes the comparison of total displacement to the displacement caused by the hygroscopic strain only at the centre on the surface (point A). The total displacement was obtained from the finite element analysis, and the hygroscopic displacement was derived by the theoretical work in previous section by equation (72). When the time is less than 80 minutes, the total displacement is very close to the hygroscopic displacement. When the time ranges from 300 to 600 minutes, the relative error becomes higher, but within 10%.

6. Characterization and Modelling of Moisture Behaviour

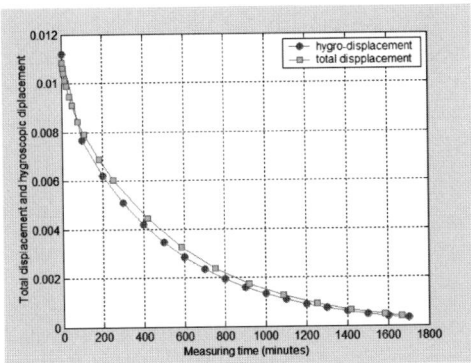

Figure 32. Comparison of the total displacement to the displacement caused by the hygroscopic strain only at the centre on the surface (point A)

Von Mises stress distribution in the specimen is shown in Figure 33. The stress pattern is very complicated. Figure 34 shows the Von Mises stress distribution along the measure line of the specimen (line OA) with time at 180 minutes. The horizontal axis is the distance to the centre of the specimen. The surface of the specimen is the place to have the highest Von Mises stress, with the value of ~100 MPa. The lowest stress zone is the layer somewhere ~0.31 mm off the centre of the specimen. The Von Mises stress is about 60 MPa in the centre O.

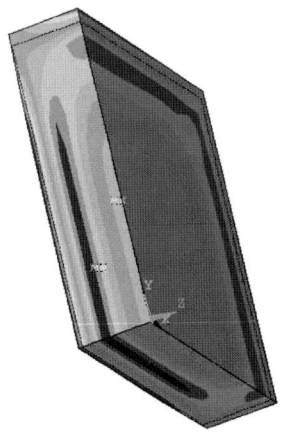

Figure 33. Von Mises stress distribution in the specimen at time 180 minutes

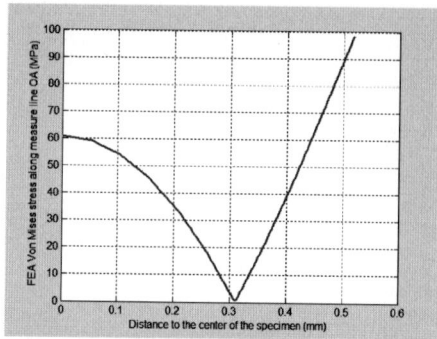

Figure 34. Von Mises stress distribution along the measure line of the specimen (line OA) at time 180 minutes

Figure 35 shows the Von Mises stress at the centre O and the surface centre A as function of time. Initially, the stress at centre surface A has a value of ~160 MPa, and it then drops gradually with time. The stress at the centre O starts form zero, and reaches its maximum which is ~60 MPa at 180 minutes, and then decreases with time afterwards.

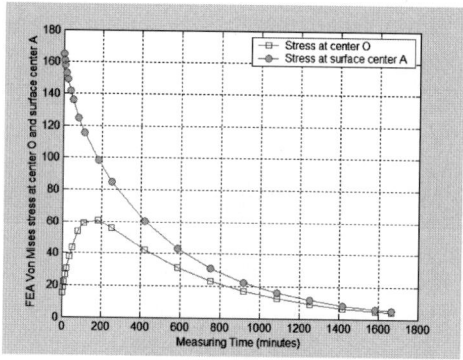

Figure 35. Von Mises stress at the points O and A as function of time

It is interesting to notice that even through the elastic strain caused by hygroscopic stress takes up to more than 30% of the total strain at each location, and the hygroscopic stress is as high as 160 MPa, the displacement induced by the elastic deformation at the surface centre is negligible compared to the displacement by hygroscopic strain. The modelling results also show that the surface centre displacement is not sensitive to Young's modulus. The reason behind these phenomena can be explained from the results that the average elastic strain along the measuring line is very small, as discussed in Figure 31 and equation (79).

5.1.3 General Guidelines for Characterizing Hygroscopic Swelling Properties

From the above analysis, a general guideline in determining the accurate coefficient of hygroscopic swelling can be suggested, with following procedures,
1. Bake the specimen until it is absolutely dry;
2. Let the specimen absorb moisture at certain temperature and certain relative humidity until full saturation;
3. Measure the length h_x and weight M of specimen initially and keep the probe the same location;
4. Let the sample dry out completely and measure the length h_x and weight M of specimen, denoted by h_{x0} and weight M_0
5. The coefficient of hygroscopic swelling β is calculated by

$$\beta = \frac{(h_x - h_{x0})/h_{x0}}{(M - M_0)/(h_{x0}h_{y0}h_{z0})} \qquad (80)$$

The coefficient of hygroscopic swelling for polymer materials determined in this way is easier and accurate. The errors caused by selecting the data sets for the linear regression can be avoided.

5.2 Hygroscopic Swelling Modelling for FCBGA Package

In order to perform the hydroscopic swelling induced stress analysis, the moisture diffusion modelling must be conducted first to obtain the local concentration. Then from equation (54) the hygroscopic strain can be computed. Then the hygro-mechanical problem can be solved using the same procedure as a typical thermo-mechanical solution.

In the following study [17], FCBGA (8 x 8 mm, 80 I/O), both unmoulded and moulded packages (see Figure 36), are used as the test vehicles for modelling studies. The substrate is similar as used in TFBGA package, which only has 0.22 mm thickness. The control case chosen is an un-moulded FCBGA-8 x 8, with no-flow underfill type A, and die thickness of 0.525 mm. Parametric studies are performed to study the effects of no-flow underfill materials (types A, B & C), package type (moulded vs. un-moulded), die thickness (0.525 vs. 0.325 mm), and substrate size (8 x 8 vs. 6 x 6 mm^2) on the stresses of the UBM during reflow and PCT. For each case, only one design parameter is varied with respect to the control case. The moisture diffusion modelling (85°C/85%RH) is first performed to study the relative moisture distribution of various underfill materials. Then, the results of moisture diffusion are used in the hygroswelling modelling. Independent thermo-mechanical stress modelling is also applied to calculate the stress during reflow. The failure criteria applied are the maximum normal peeling stress (S_y) and shear (S_{xy}) stresses on UBM, which contribute to the UBM-opening failure during PCT.

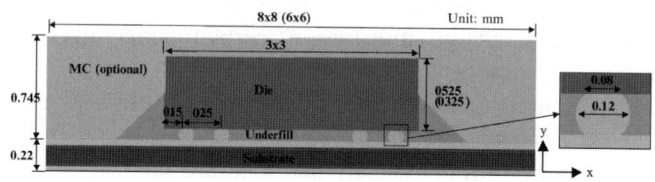

Figure 36. Schematic of FCBGA-8 x 8mm^2 package

Figure 37 shows that during PCT, the normal hygroswelling stress (S_y) acting on UBM and solder bump is mostly tensile (peeling mode), causing the UBM-opening failure. Shear stress of lower magnitude also contributes to the interfacial delamination. The swelling of underfill induces compressive normal stress on the die/underfill interface, but the UBM (die/bump) is under tensile normal stress, because the solder bump doesn't absorb, nor swell with moisture.

From the parametric studies, un-moulded FCBGA with Underfill-A has the lowest hygroswelling stress. Previous reliability test results also show that un-moulded FCBGA with Underfill-A has the best performance. Underfill-A has the least hygroswelling strain (see Table 7), due to relatively low C_{sat} and CME. Interfacial adhesion strength of underfill under high temperature and high moisture conditions is also an important consideration. For better UBM reliability during PCT, material suppliers may need to compromise in the formulation of no-flow underfill materials, to achieve low values of both C_{sat} and CME.

Un-moulded FCBGA with smaller die thickness of 0.325 mm has 23% lower stress than FCBGA with die thickness of 0.525 mm, but substrate size has little effect on UBM stress. Moulded package has 38% larger stress than un-moulded package (for Underfill A), due to additional CME mismatch with mould compound material. This correlates well with reliability test results that the un-moulded package generally performs better than moulded package during PCT.

Since the test vehicles experience several tests (moisture preconditioning at Level 3, followed by reflow and 168 hours of PCT), the actual failure mechanism is more complex, and could be a mixed mode. Combined effects of process defects, interfacial adhesion strength, moisture, vapour pressure,

6. Characterization and Modelling of Moisture Behaviour

thermal stress, and hygroswelling stress could contribute to the moisture-induced failures.

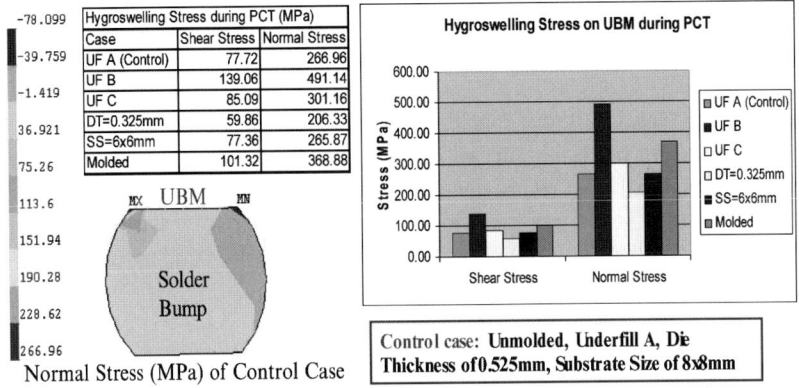

Figure 37. Design analysis on hygroswelling stress induced during PCT

At PCT condition of 121°C/100%RH, there is a combination of both hygroswelling stress and thermal stress. During temperature cooling, from underfill curing temperature to 121°C of PCT, the thermal stress induced on UBM and solder bump is compressive, acting against the tensile hygroswelling stress. For the un-moulded package with Underfill-A, the compressive stress is 262MPa, close to the tensile hygroswelling stress of 267MPa. Under high CTE, the thermal stress induced is even more compressive, 278MPa, greater than the hygroswelling stress. Underfill with high CTE helps to generate compressive stress, against the tensile hygroswelling stress.

In order for failure to occur during PCT, the hygroswelling stress must be much greater than the thermal stress. Previous hygroswelling stress modelling was performed based on the coefficient of hygroscopic swelling at 85°C. At 121°C, the coefficients of hygroscopic swelling of materials can be about 2 times higher than values measured at 85°C. When the coefficient of hygroscopic swelling is doubled, the hygroswelling stress is also doubled accordingly. For the case with Underfill-A, the magnitude of hygroswelling stress can be as high as 534MPa, much larger than the thermal stress of 278MPa. Therefore, the hygroswelling stress is the dominant stress. For relative comparison among design variations, the previous results based on assumption of the coefficient of hygroscopic swelling at 85°C are still valid.

Figure 38 shows that during reflow, the normal thermal stress (S_y) acting on UBM and solder bump is also mostly tensile, and therefore has a negative effect on UBM reliability. The magnitude of normal and shear thermal stresses induced during reflow is much lower than the hygroswelling stresses during PCT. More failures are observed during PCT than during reflow. The UBM and underfill/die adhesion strength may be affected during reflow, but the package still manages to pass the electrical test (solder bump resistance measurement). The subsequent hygroswelling stress during PCT is the main factor causing the underfill/die interfacial delamination and UBM opening failure.

Parametric studies show that the thermal stresses are close for these three types of underfill materials, because there is no large variation in the underfill thermo-mechanical material properties. Un-moulded FCBGA with smaller die thickness of 0.325 mm has 24% lower stress than FCBGA with die thickness of 0.525 mm. Substrate size and moulded package have little effect on thermal stress of UBM during reflow.

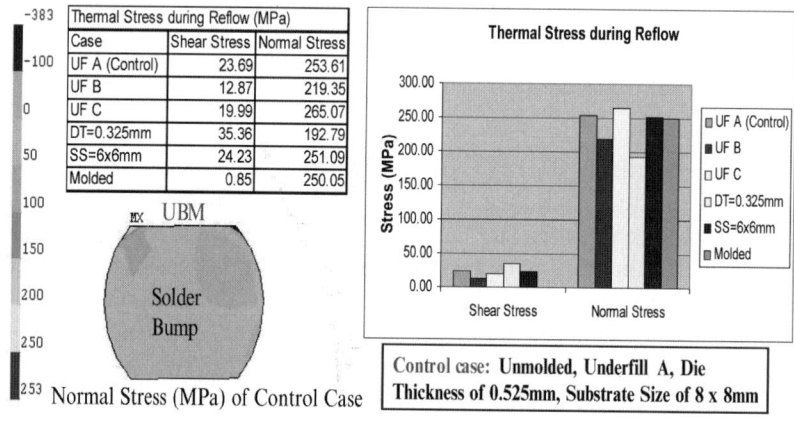

Figure 38. Design analysis on thermal stress induced during reflow

6. SINGLE VOID INSTABILITY BEHAVIOUR SUBJECTED TO VAPOUR PRESSURE AND THERMAL STRESS

The interface delamination is considered as the consequences of the micro-voids growth, nucleation and coalescence. Therefore, it is important to understand the deformation behaviours in micro-void level. A micro-mechanics analysis of a single void is very useful to reveal some salient features and fundamental failure mechanisms associated with the initiation of interfacial delamination. The use of single void model is also helpful to investigate the role of vapour pressure on void behaviours.

6.1 Void Behaviour at Bulk

Many researches have studied the effect of the mean stress on the void growth [28-30]. An exponential dependence of void growth rate on the tri-axial stress was found. However, since the void growth is finite-deformation, Huang [31] pointed out that the consideration of the finite-deformation would lead to an *unstable* void growth. This implies that the void cell will 'burst' suddenly when the applied stress reaches its critical value.

For the purpose of analysis, we consider a spherical volume of material containing a microvoid of spherical shape, as shown in Figure 39. The material is incompressible. The inner radial surface is subjected to internal vapour pressure, induced by the moisture inside. The vapour pressure p follows the rules we discussed in section 4. A radial stress σ^T is applied to outer radius to represent the thermal stress as function of temperature rise. The radial equilibrium solution of a spherically symmetric cell in current configuration is the found to be

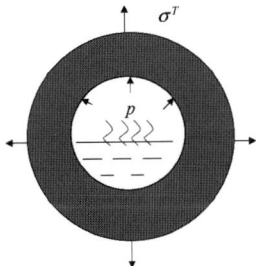

Figure 39. A single sphere shape void

$$\frac{\sigma^T(T) + p(T, f, T_0, f_0, C_0)}{\sigma_0} = \int_{\varepsilon_1}^{\varepsilon_2} \frac{H(\varepsilon)d\varepsilon}{1 - \exp(-\frac{3}{2}\varepsilon)} \qquad (81)$$

where ε_1 and ε_2 are the radial strains at the two-end points of the cell, which can be determined by the current and initial void volume fractions:

$$\varepsilon_1 = \frac{2}{3}\ln(\frac{f_0}{f}\frac{1-f}{1-f_0})$$
$$\varepsilon_2 = \frac{2}{3}\ln(\frac{1-f}{1-f_0}) \qquad (82)$$

$H(\varepsilon)$ in equation (81) denotes the true stress- logarithmic strain relation as following

$$\frac{\sigma}{\sigma_0} = H(\varepsilon) = \begin{cases} \varepsilon/\varepsilon_0 & \text{if } |\varepsilon| < \varepsilon_0 \\ (|\varepsilon|/\varepsilon_0)^N \text{sign}(\varepsilon) & \text{if } |\varepsilon| \geq \varepsilon_0 \end{cases} \tag{83}$$

where σ_0 is the yield stress.

Equation (81) displays a nonlinear and non-monotonic relation between the current volume fraction f and the sum of the internal vapour pressure p and the externally imposed radial thermal stress σ^T. Figure 40 presents the results of the void volume increase with respect to the applied traction $\sigma^T + p$, when the initial void volume fraction f_0 takes 0.01 and 0.05 respectively. The applied traction increases continuously from zero to its peak value and then decreases, which implies that the cell will collapse when the peak valued is reached. From Figure 40 it can be seen that the critical stress σ_{cr} for the unstable void growth is about 2-3 times of the yielding stress σ_0, when the initial void volume fraction is in the range 0.01-0.05.

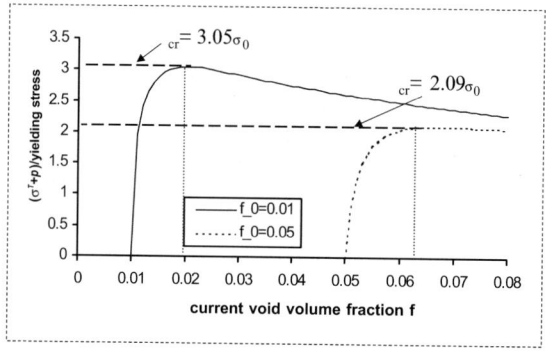

Figure 40. Vapour pressure p at 220°C versus the current void volume fraction f by equation (56) for case 1, with $f_0 = 0.03$, $\alpha = 200$ppm/°C, $T_0 = 85$°C, and $p_0 = p_{0\max} = p_g(85°C) = 5.27\text{e-}2$ MPa

Given the initial moisture concentration C_0, the thermal expansion (to obtain σ^T) and the initial void volume fraction f_0, equation (81) can be solved together with the vapour equations (45)–(47) to determine the current void volume fraction as well as the vapour pressure as function of temperature. A critical temperature can then be found when the sum of applied thermal stress and vapour pressure reaches the peak value. The growth of the void cell at this temperature becomes unstable, and failure takes place.

6. Characterization and Modelling of Moisture Behaviour

Assume that the initial void volume fraction f_0 is 0.05, and the thermal stress can be estimated by $\sigma^T = \alpha E(T-T_{stress-free})$, in which $\alpha = 200$ ppm/°C and $T_{stress-free} = 150$°C. The critical temperature where the void unstable growth occurs is usually well above the glass transition temperature. The Young's modulus is decreased dramatically after the glass transition temperature. At reflow temperature 220°C, a typical Young's modulus may be in the range of 500 MPa. Though the yielding strength at high temperature is not available to our best knowledge, an estimation can be made based on the Young's modulus by $\sigma_0 = E\varepsilon_0$. Assume that $\varepsilon_0 = 0.01$, then the yield stress $\sigma_0 = 5$ MPa.

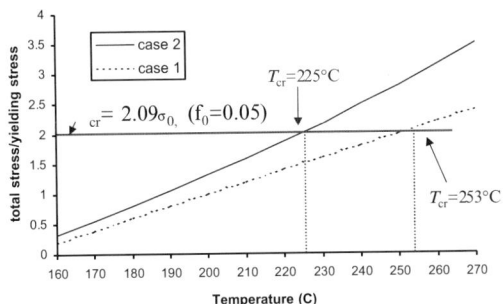

Figure 41. The applied stress (sum of thermal stress and vapour pressure) as function of temperature rise for case 1 and 2 respectively ($\alpha = 200$ ppm/°C, N = 0.1, $\varepsilon_0 = 0.01$, $f_0 = 0.05$, E = 500 MPa, $\sigma_0 = 5$ MPa). When the total stress reaches its critical stress during temperature rise, the void growth becomes unstable. The critical temperature T_{cr} for case 2 is 225°C, and 255°C for case 1. In case 1, the vapour pressure is so low that it has negligible effect on the void growth

Let's consider the vapour pressure in two extreme cases: case 1 and case 2 (case 3 is in between case 1 and case 2 as illustrated in section 4). The vapour pressure in case 1 is extremely low from the analysis in section 4 (see Figure 18), thus on calculating the left-hand side of equation (81), the vapour pressure can be neglected. For the case 2, the vapour pressure maintains the saturated state throughout the course of temperature rise. It is noted that the saturated vapour pressure increases with the temperature rise significantly. In Figure 41, the applied stress is plotted as function of temperature rise for a void with vapour pressure of case 1 and 2, respectively. The void-growth becomes unstable when temperature reaches 225°C for the case 2. However, in case 1 where the vapour is almost negligible, the failure takes place at a temperature as high as 255°C, well above the soldering temperature when happen at 220°C.

As we may recall, the saturated vapour pressure at 220°C is 2.32 MPa. The critical stress σ_{cr} is about 2-3 times of σ_0 (see Figure 40), i.e.,

σ_{cr} = 10 –15 MPa, when the initial void volume fraction is in the range 0.01-0.05. It shows that the vapour pressure takes a significant portion over the critical stress. When the lead-free solder material is used, the reflow temperature can be as high as 260-270°C. Therefore, the saturated vapour pressure can reach 5.51 MPa. Besides, the yield stress σ_0 can be even lower. The impact of the vapour pressure on the void unstable behaviours will become even more significant.

Although the predicted critical stress from the elastic-plastic model shows the important role of the vapour pressure on void deformation, the model is not able to make differences between the void behaviours at interface and in bulk. As we know, the unstable void growth is present at interfaces only during the reflow. In addition, the model is not able to explain some other phenomenon associated with the moisture-induced failure as well. For example, Figure 41 tells us that the cell may collapse when the temperature is reached to the level where the thermal stress alone reaches the critical stress for a given initial void volume fraction (case 1). However, a package without moisture intake will very unlikely fail even though the package is heated up well above the reflow temperature.

Many of the polymer materials used in electronic packages are thermoset materials, which display a rubble-like state above the glass transition temperature. The stress-strain relation expressed by the equation (83) may not be appropriate for a rubber-like material. Guo and Cheng [32] introduce the neo-Hookean model to describe the deformation behaviours of the rubber-like materials. The stored energy function can be written as following

$$W = \frac{\mu}{2}(\lambda_1^2 + \lambda_2^2 + \lambda_3^2 - 3), \quad \lambda_1 \lambda_2 \lambda_3 = 1 \tag{84}$$

where μ is the shear modulus and λ_I are the principal stretches. It can be seen that the shear modulus is the only material property introduced in this stress-strain relation. The equilibrium solution of a spherically symmetric cell in current configuration, which is similar to equation (81), can then be expressed explicitly in terms of the initial and current void volume fractions f_0 and f as following

$$\frac{\sigma^T(T) + p(f_0, f, C_0, T, T_0)}{\mu} = 2(\frac{1-f}{1-f_0})^{1/3}$$
$$+ \frac{1}{2}(\frac{f_0}{f}\frac{1-f}{1-f_0})^{4/3} - 2(\frac{f_0}{f}\frac{1-f}{1-f_0})^{1/3} \tag{85}$$
$$- \frac{1}{2}(\frac{f_0}{f}\frac{1-f}{1-f_0})^{4/3}$$

6. Characterization and Modelling of Moisture Behaviour

Again, equation (85) displays a nonlinear and non-monotonic relation between the applied stress and the void volume fraction f, as shown in Figure 42. The unstable void growth takes place when the peak value of the sum of the thermal stress and vapour pressure is reached.

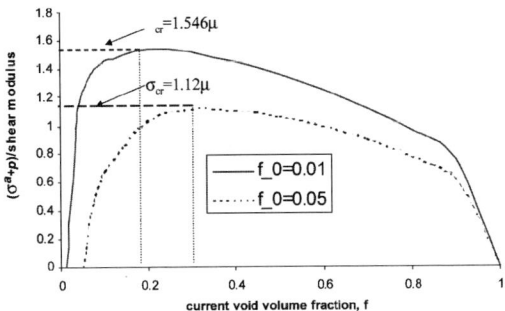

Figure 42. The sum of thermal stress and vapour pressure applied to a cavity in a finite neo-Hookean rubber-like matrix versus the evolution of the void volume fraction f (f_0 = 0.01 and 0.05)

It is noted that using the neo-Hookean model by equation (84), the critical stress is of the order of the shear modulus μ. Assume that the shear modulus $\mu = E/2(1+\nu) = E/3$ in incompressible case, where E is the Young's modulus, μ is about 168 MPa when $E = 500$ MPa. This implies that the critical stress is in the order of 168MPa. This value is about one order higher than the predicted critical stress of 10–15 MPa, when the elastic-plastic stress-strain relation is used. The saturated vapour pressure 2.32 MPa at 220°C thus is very small compared to this critical stress 168 MPa. This seems to suggest that the model study using the neo-Hookean relation does not explain well the impact of the moisture (vapour) on the material failures in reflow soldering.

On the other hands, the results obtained from the rubber-like material assumption indeed explain that the void unstable-growth can never happen in bulk material, rather than at the interfaces only. After the moisture absorption, the moisture exists anywhere in polymer materials. The moisture concentration close to the outside surface is higher that the moisture at the interface if the moisture absorption is not saturated. During the soldering, the entire package is exposed to the reflow temperature at 220°C. the moisture is evaporated anywhere in material. However, the rupture of the bulk material due to the moisture prior to the interface delamination has never been observed. Failure always starts only from the interfaces with delamination. Afterwards the vapour pressure exerts traction loading on the delamination area, eventually causing the package bulge, and cracking inside bulk. This implies that even though the vapour pressure is built up anywhere in the polymer material, it has insignificant effect on the void growth in *bulk* since the portion over the critical stress is very small.

One of main reasons that the above models do not fully explain the moisture-induced failures in electronic packages is that the models are limited to homogeneous materials only. The effect of interface, in particular, the strength reduction with moisture absorption, is not accounted for. An alternative way to take the interface into consideration is to treat the interface as a special material layer that is completely different from the bulk material. This special material layer has very low Young's modulus and higher porosity. In this case the magnitude of the critical stress may be significantly reduced and the impact of the vapour pressure becomes prominent at the interface layer only. However such an approach does not incorporate the interface mechanism associated with the moisture absorption. A best approach is to incorporate the interface mechanism into the base cell model, in which the void growth is not only controlled by the applied stress, but also additionally controlled by the interface characteristic as function of moisture absorption.

6.2 Void Behaviour at Interface

The void behaviour at the interface is different from that within the bulk. The void growth at the interface is not only controlled by the total stress but also the interface strength. Figure 43 sketches the void behaviours at the interface, in which three stages are involved. At the beginning, the void at the interface has an initial void volume fraction f_0, in which a certain amount of moisture is condensed into liquid. With the increase of the temperature, thermal stress and vapour pressure are developed and subjected to the void. The void will reach the equilibrium at the void volume fraction f_1. Stage 1 shows no difference with behaviours of voids in bulk and the stress-level is much less than the critical stress at which the void will 'burst'. However, due to the fact that the interfacial strength is weaken by the moisture intake at high temperature, the void will continue to grow, as shown in stage 2 of Figure 43. The new equilibrium will be reached at the void volume fraction f_2. At this new equilibrium position, the problem can be treated as the equilibrium for a void with the initial volume void fraction $(f_0+f_2-f_1)$. If the applied stress reaches the critical stress with $(f_0+f_2-f_1)$, the void growth becomes unstable (3rd stage). Otherwise, the void growth will stop here and no further delamination is formed.

6. Characterization and Modelling of Moisture Behaviour

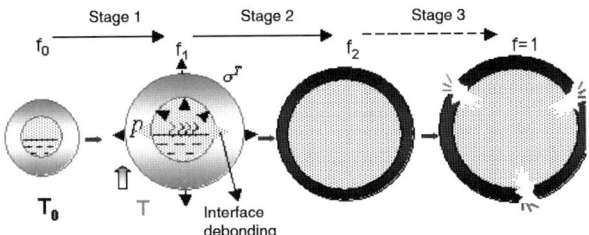

Figure 43. Schematic representation of the void behaviour at the interface. Three stages are involved. Stage 1 describes the deformation during the temperature rise due to the vapour and thermal stresses. Stage 2 is the void-growth stage due to the debonding of the interface under certain stresses with the effects of the moisture intake at certain temperature. Stage 3 describes the unstable void-growth

The void behaviour at stage 2 is related to the interface properties. A general relationship between the void-growth and the moisture contents and the temperature may be postulated as following form

$$\dot{f} = kCe^{-\frac{Q}{RT}}\dot{T} \tag{86}$$

where C is the moisture concentration and T the *temperature*, R is the universal gas constant. K and Q are interface constants to be determined. It can be seen that the rate of void-growth is proportional to moisture concentration C and temperature T. Some materials exhibit excellent resistance to moisture absorption with low C, while other materials show interface strengths being quite sensitive to moisture absorption with high k. There is no direct correlation between the absorbed moisture amount and the void-growth at interface since different materials have different k. Some failures may occur for materials with minor moisture absorption (low C), but the material has very high k. Some other materials do not fail, even with major moisture absorption (very high C), due to the excellent resistance of interface strength at high moisture concentration and high temperature (very low k). However, for *same* material the correlation between the delamination and the moisture absorption is direct and obvious. Therefore, equation (86) is a general form to describe the void behaviour on interface.

The exact determination of the material properties such as k is challenging. Instead of determining equation (86), in the following, equation (85) will be applied to see how much void growth in stage 2 will lead the void grow unstably. Let assume that the initial void volume fraction is 0.01. At the beginning the void will deform along the solid line shown in Figure 44. Then void growth enters second stage, in which the stress-level does not change but void grows following equation (86). This stage is shown in Figure 44 as dotted line, which will intersect with another equilibrium curve. The void will not grow further if stress-level is still below the critical stress. Otherwise, like case 2 shown in Figure 18, the delamination will take place.

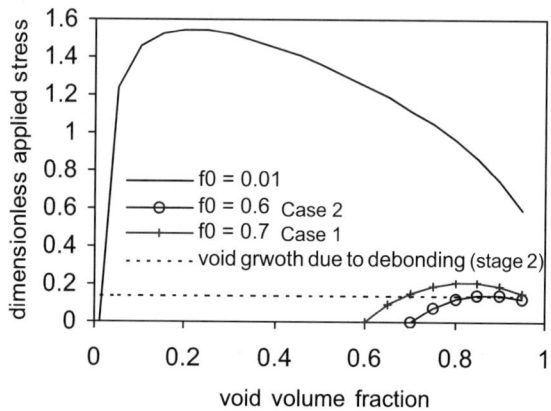

Figure 44. Void growth at interfaces

7. INTERFACE STRENGTH CHARACTERIZATION AND MODELLING

Moisture absorption has been long recognized as the root cause for the delamination failure at reflow. However, one of the important conclusions from the vapour pressure modelling is that the maximum vapour pressure at reflow is not always proportional to the moisture absorption. The vapour pressure maintains the saturated value (e.g. = 2.32 MPa at 220°C) no matter how much moisture is absorbed, as long as the moisture is not fully vapourized. In fact, the interface delamination not only depends on the vapour pressure, but also on the interface strength as well. When the vapour pressure maintains its saturated value, the interface strength becomes a key factor for the delamination.

7.1 Mechanics of Interfacial Delamination

A brief summary of interfacial fracture mechanics is provided below. Theoretical details can be found in Chapter 4 and in Hutchinson and Suo [33]. Consider a semi-infinite traction-free crack lying along the interface

6. Characterization and Modelling of Moisture Behaviour

between two homogeneous isotropic half planes, with material 1 above the interface and material 2 below (see Figure 45). Also depicted in the figure are two coordinate systems (x_1, x_2) and (r, θ). It can be shown that the stress field depends on only two Dundurs's parameters

$$\alpha_D = \frac{\mu_1(\kappa_2+1) - \mu_2(\kappa_1+1)}{\mu_1(\kappa_2+1) + \mu_2(\kappa_1+1)} \tag{87}$$

$$\beta_D = \frac{\mu_1(\kappa_2-1) - \mu_2(\kappa_1-1)}{\mu_1(\kappa_2+1) + \mu_2(\kappa_1+1)} \tag{88}$$

where μ_i, E_i, and v_i ($i = 1, 2$) are the shear modulus, Young's modulus, and Poisson ratio of the respective materials. $\kappa_i = 3 - 4v_i$ for plane strain and $\kappa_i = (3-v_i)/(1+v_i)$ for plane stress.

The tractions ahead of the crack tip are given as

$$\sigma_{22} + i\sigma_{12} = K(2\pi r)^{-1/2} r^{i\varepsilon} \tag{89}$$

$$\text{where } \boldsymbol{K} = K_I + iK_{II} \tag{90}$$

r is the distance from the crack tip and ε is the oscillatory index,

$$\varepsilon = \frac{1}{2\pi} \ln\left(\frac{1-\beta}{1+\beta}\right) \tag{91}$$

Since both K and $r^{i\varepsilon}$ are complex quantities, it is clear that the ratio of the shear stress to normal stress depends on r. For lack of a better choice and due to its similarity to the classical stress intensity factor, the complex product $Kr^{i\varepsilon}$ can be characterized by its modulus $|Kr^{i\varepsilon}|$ and its argument or phase angle $\hat{\psi}$ at an arbitrary reference length $\hat{r}$. It is noted that $|Kr^{i\varepsilon}| = |K|$ numerically although the units are different. Also, $\hat{\psi}$ is known in the literature as the mode mixity of the stresses and is given by

$$\tan\hat{\psi} = \left(\frac{\sigma_{12}}{\sigma_{22}}\right)_{r=\hat{r}} \tag{92}$$

Thus, in reporting values of the complex stress intensity factor for bimaterial interface cracks, it is usual to specify $|K|$ and $\hat{\psi}$ at $r = \hat{r}$ (usually taken within the range 1 to 100 μm, see Rice [34] and Shih [35]). Clearly, it is important to ensure that a common $\hat{r}$ has been used when comparing experimental or numerical results, and to take note of the fact that phase angles at different values of r are related by

$$\psi_{r_2} - \psi_{r_1} = \varepsilon \ln\left(\frac{r_2}{r_1}\right) \tag{93}$$

The associated crack flank displacements at a distance r behind the tip, $\delta_i = u_i(r,\pi) - u_i(r,-\pi)$ are given by

$$\delta_2 + i\delta_1 = \frac{8}{(1+2i\varepsilon)\cosh(\pi\varepsilon)} \frac{(K_I + iK_{II})}{E^*} \left(\frac{r}{2\pi}\right)^{-1/2} r^{i\varepsilon} \tag{94}$$

where
$$\frac{1}{E^*} = \frac{1}{2}\left(\frac{1}{\overline{E}_1} + \frac{1}{\overline{E}_2}\right) \tag{95}$$

where $\overline{E}_i \equiv E_i/(1-v_i^2)$ in plane strain and $\overline{E}_i \equiv E_i$ in plane stress. Since $r^{i\varepsilon} = \cos(\varepsilon \ln r) + i\sin(\varepsilon \ln r)$ the displacements δ_1, δ_2 have an oscillatory singularity resulting in overlapping of the cracked faces for small values of r. The crack driving force is the strain energy release rate G given by [36]

$$G = \frac{(1-\beta_D^2)}{E^*} K^2 \tag{96}$$

where $K^2 = K_I^2 + K_{II}^2$ (97)

6. Characterization and Modelling of Moisture Behaviour

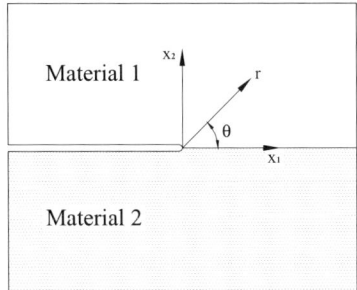

Figure 45. A crack along a bimaterial interface

7.2 Interfacial Fracture Toughness

Interface toughness is defined as the critical value of G required for crack propagation along the interface. This critical value G_c has to be measured. However, its measurement is rather complicated by the fact that it depends strongly on the mode mixity ψ. A typical variation of G_c with ψ is illustrated in Figure 46.

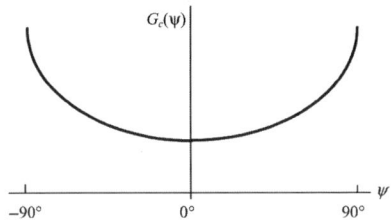

Figure 46. Typical variation of $G_c(\psi)$ with ψ

It can be seen that G_c has a minimum at $\psi = 0°$ (pure mode I) and a maximum at $\psi = 90°$ (pure mode II). For interfaces between mould compounds and leadframe materials, there is the further complication that the interface toughness is also a function of temperature and moisture content.

In order to predict the onset of delamination at pad-encapsulant interfaces, data on the variation of interface toughness G_c with ψ, temperature T and moisture concentration C must be available. Unfortunately, such data are scarce in the literature, but are presently being accumulated. Liechti and Chai [37]

[37] measured the variation of G_c with ψ for an epoxy-glass interface. Liu et al. [7] measured the variation of G_c with ψ at room temperature for a copper-prepreg interface. Tanaka and Nishimura [38] conducted three-point-bending tests on ENF (end-notched flexure) specimens to measure the interface toughness of interfaces between Alloy-42 leadframes and two mould compounds at various temperatures. Tay and Lin [3] also used the same method to measure the interface toughness of interfaces between copper leadframes and a mould compound at various temperatures and humidity levels. They found that interfacial fracture toughness decreases with both temperature T and moisture concentration. Later Tay et al. [39] carried out some comprehensive measurement of the toughness of the interface between copper leadframe and mould compound as a function of temperature, moisture concentration and mode mixity. Their results are shown in Figure 47. Van Driel et al. [61, 62] combined the four point bending with pre-crack notch with the shaft-loaded blister test to measure the interface toughness between epoxy systems and leadframe as function of temperature and moisture conditions. Their results are in agreement with those shown in Figure 47.

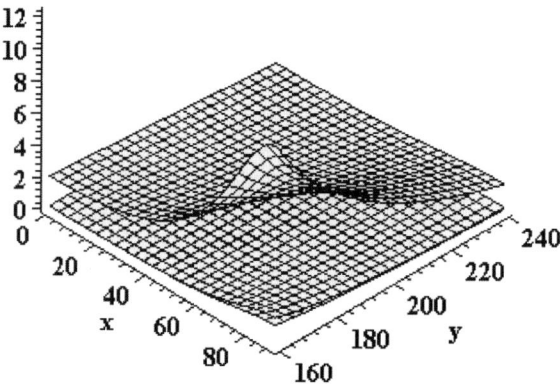

Figure 47. Variation of G_c with ψ, T and C, from [39]

7.3 Interface Modelling using Cell Element

The micromechanics analysis based on the single-void model study, as discussed above, reveals some fundamental features associated with the failure mechanism for porous material such as the unstable growth of voids. How to link the results of the single void behaviour to descriptions of material behaviour in a macroscopic sense is one of critical issues.

6. Characterization and Modelling of Moisture Behaviour

Homogenization processes can be applied for this purpose. There are several theories to establish the relationships between the microscopic and macroscopic variables [e.g., 27, 40] for porous material, in which the void volume fraction f is treated as a *field* variable: a damage parameter to represent the local material behaviour. $f = 1$ at a particular (continuum) point implies that delamination takes place at this 'point'. The evolution equation is required for the void volume fraction f. For homogeneous material in bulk, the growth rate can be written as

$$\dot{f} = \dot{f}_{growth} + \dot{f}_{nucleation} \tag{98}$$

$$\dot{f}_{growth} = (1-f)\dot{E}_{kk} \tag{99}$$

$$\dot{f}_{nucleation} = A\dot{\sigma}_e + B\dot{\Sigma}_m \tag{100}$$

At the interface, the impact of interface on void growth should be included

$$\dot{f} = \dot{f}_{growth} + \dot{f}_{nucleation} + \dot{f}_{debonding} \tag{101}$$

$$\dot{f}_{growth} = (1-f)\dot{E}_{kk} \tag{102}$$

$$\dot{f}_{nucleation} = A\dot{\sigma}_e + B\dot{\Sigma}_m \tag{103}$$

$$\dot{f}_{debonding} = kCe^{-\frac{Q}{RT}}\dot{T} \tag{104}$$

Gurson [27] assumes that the matrix material follows the classical elastic-plastic flow rule with Von-Mises yielding criterion. He established the relationship between the macroscopic and microscopic variables by the averaging method over a cell containing a single void. Finally the macroscopic plastic potential that represents the yielding condition has the form

$$\Phi = (\frac{\Sigma_e}{\sigma_e})^2 + 2f \cosh(\frac{3\Sigma_m}{2\sigma_e}) - (1+f^2) = 0 \tag{105}$$

in which Σ_e denotes Mises equivalent macroscopic stress, Σ_m the mean macroscopic stress, σ_e the current matrix flow strength of matrix and f the current void volume fraction. Equation (80) shows the effect of mean stress and void volume fraction on the material's yielding. Tvergaard [41] improved the model predictions for periodic arrays of cylindrical and spherical voids by introducing two factors q_1 and q_2, as following

$$\Phi = (\frac{\Sigma_e}{\sigma_e})^2 + 2q_1 f \cosh(\frac{3q_2\Sigma_m}{2\sigma_e}) - (1+q_1 f^2) = 0 \tag{106}$$

The original Gurson-Tvergaard model requires each finite element to be modelled as porous material anywhere over a given structure. This usually gives rise to the difficulties in numerical implementation. Also, the model is very sensitive to the size of the element to be taken. The concept of *cell model* was first introduced by Xia and Shih [42] to tackle this problem. In the cell model, only a material layer of characteristic thickness D is modelled by Gurson-Tvergaard relation (see Figure 48). Beyond this region, the conventional material without voids is applied. Therefore, it can be assumed that voids are present only in the material layer from the very beginning. This model has advantage that each cell behaves as a basic material unit containing a void and can be considered as a representative volume element pertaining to the specific material considered. The discrete, three-dimensional nature of a cell enables it to capture the important features from the crack formation to the propagation of a macro-crack. Since the Gurson-Tvergaard model is not able to capture the coalescence phase, the cell model will use a linear traction-separation law to supersede the Gurson model when the void volume fraction reaches a critical value.

The cell model has recently been extended to model the interface delamination in plastic IC packages [43]. The vapour pressure effect is investigated by the comparison between the baked and unbaked packages. The vapour pressure model in section 4 for case 1 is used. However, the initial vapour pressure is assumed to be as high as 2.6 times of yielding strength. According to equation (40), the initial pressure is very small and can never exceed 5.27e-2 MPa.

One of concerns in using the Gurson-Tvergaard model is its validity for the polymer materials. Thermoset materials behave like the rubber-like at high temperatures. Thermoplastic materials behave more likely a viscous fluid or visco-elasto-plastic. Nevertheless, this mechanism-based approach

6. Characterization and Modelling of Moisture Behaviour

provides insights into the failure of plastic packages arising from thermal and vapour pressure effects in the initiation of micro-voids, void growth, and the coalescence of voids. A specific micro-mechanics model for porous polymer materials with moisture effect is the future of the study in this field.

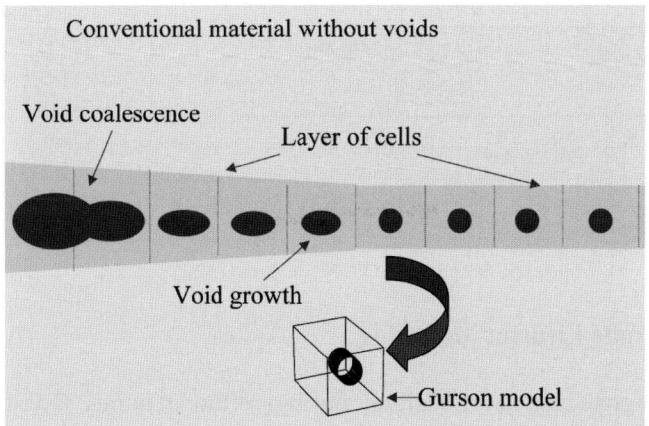

Figure 48. Schematic of the cell model that is used to model a special material layer only such as an interface layer. The cell model uses the Gurson model before the void enters the coalescence phase. The coalescence phase is modelled by a linear traction-separation law. The cell model behaves as a basic material unit containing a void and has a characteristic thickness D

8. CASE STUDIES

In the following, two example studies are presented for moisture induced failures in a leadframe based package, such as a QFN and a substrate based package such as a BGA.

8.1 Integrated Stress Study for QFN Package

QFN packages are getting popular as a low-cost solution for applications with low pin-count requirements. QFN is also known as MLF (Micro LeadFrame), MLP (Micro Lead Package), QON (Quad Outline Non-lead), or SON (Small Outline Non-lead) package. It is a type of CSP and can be assembled and moulded in matrix form for cost-saving. A schematic of QFN-8x8mm (52 leads) package used in this study is shown in Figure 49. The peripheral leads are exposed at the package bottom for soldering to the board. Extra soldering under the centre die pad and at diagonal corners is optional. Moisture induced failures during moisture preconditioning and reflow are main concerns for QFN packages.

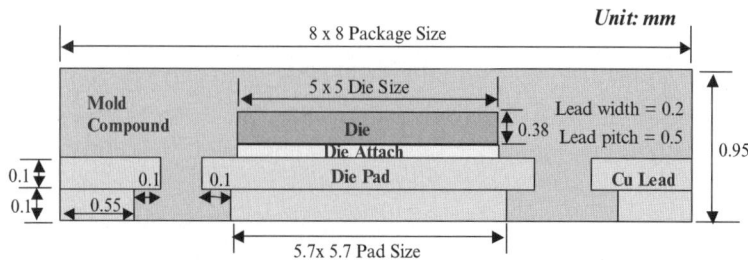

Figure 49. Schematic of QFN 8 x 8 mm (52L) package

8.1.1 Finite Element Modelling

The integrated stress modelling during reflow requires five types of modelling, i.e. moisture diffusion during moisture preconditioning and reflow, thermal modelling, hygro-mechanical modelling, thermo-mechanical modelling, and vapour pressure modelling. A 2D half-model is established for QFN. For all types of models, the same finite element geometry model is used, but different element types, boundary conditions, loadings, and solutions are applied. The theories and procedures for each type of modelling are briefly explained in the following sections.

1. Moisture Diffusion Modelling

The moisture properties, i.e. diffusivity and C_{sat}, characterized under 85°C/85%RH, are listed in Table 8. The moisture properties can be obtained by curve-fitting the moisture weight gain data of a thin disc specimen with the analytical solution of 1D moisture diffusion. Die and copper leadframe are assumed to be impermeable to moisture, i.e. do not absorb moisture, and therefore they have zero diffusivity and C_{sat}.

The moisture weight gain curve of a mould compound under 85°C/85%RH is shown in Figure 50. This is a typical curve with a high moisture absorption rate at the beginning (first few days), and then reaching steady state or saturation at longer time. It was found that the weight gain curve-fitted by Fick's Law solution overestimates the moisture absorption rate at longer time. The non-Fickian behaviour of test sample could be that moisture diffusion close to saturation stage is dominant by a slow rate of water condensation from vapour to liquid phase. The C_{sat} of material is much higher than the moisture density in the humidity chamber. Therefore, some

6. Characterization and Modelling of Moisture Behaviour

of the vapour-phase water must be condensed into liquid water, to make more room for moisture diffusion at vapour phase from ambient into the microvoids in the material. At the beginning of weight gain test, moisture density in the microvoid is low, therefore there is no condensation, and the curve can be fitted well with Fick's Law. At longer time, the slower condensation process is dominant. Therefore, the moisture diffusivity at vapour phase is reduced at higher moisture concentration, or the so-called non-Fickian behaviour. The effect of this condensation process shall be studied in detail in future.

Table 8. Moisture and hygroswelling material properties

Material	D (mm^2/s)	C_{sat} (mg/mm^3)	β, coefficient of hygroscopic swelling (mm^3/mg)	Hygro Strain ($\beta \times C_{sat}$)
Mould Compound	7.43e-7	7.06e-3	0.222	1.57e-3
Die Attach	1.25e-5	6.20e-3	0.520	3.22e-3

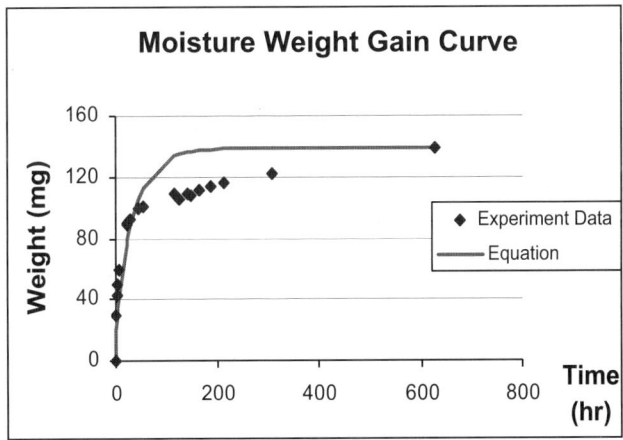

Figure 50. Moisture weight gain curve for mould compound

Moisture desorption diffusivity is also an important material property, especially during reflow, to determine the moisture weight loss. Moisture diffusivity during desorption at reflow is larger than absorption at preconditioning temperature. Figure 51 shows the effect of temperature (30°C to 220°C) on moisture diffusivity during desorption for mould compound, D (MC), and die attach, D (DA). The D_o and Q of mould compound and die attach materials during desorption are listed in Table 9.

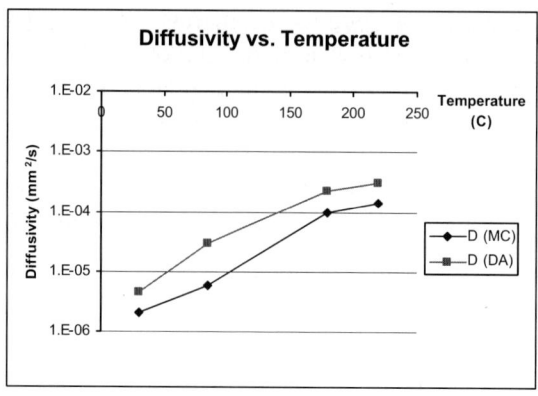

Figure 51. Diffusivity as a function of temperature

Table 9. Diffusivity constants of moisture desorption

Material	D_o (mm²/s)	Q (ev)
Mould Compound	0.18	−0.304
Die Attach	0.35	−0.293

2. Thermal Modelling

Similar to mass diffusion, the following transient heat conduction equation can be solved to obtain the temperature distribution in the IC package during reflow. The thermal material properties, specific heat (C_p), thermal conductivity (k), and density (ρ) are listed in Table 10. They are related to thermal diffusivity (α_T) by:

$$\alpha_T = \frac{k}{\rho \cdot C_p} \tag{107}$$

Table 10. Thermal material properties

Material	Cp (J/kg.K)	k (W/m.K)	ρ (kg/m³)
Copper	385	390	8950
Die	712	108	2330
Mould Compound	900	0.67	2088
Die Attach	800	1.2	2400

6. Characterization and Modelling of Moisture Behaviour

The initial condition is interior of package set at room temperature (see Figure 52). The boundary condition used is fixed external surface temperature, according to the reflow temperature profile measured, within the specification of JEDEC test standard (e.g. maximum temperature of 220°C, +5/–0°C). The effects of convection coefficient (h) and ambient temperature in the multi-zone oven are considered indirectly by measuring the surface temperature of the package. The centreline of package is symmetric (heat flux = 0).

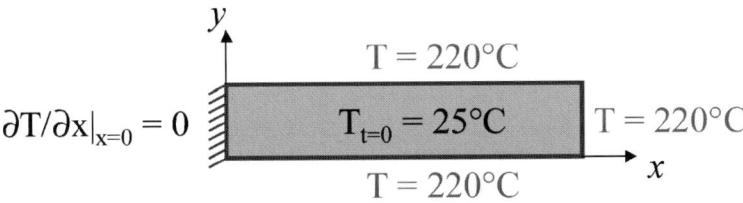

Figure 52. Initial and boundary conditions of heat transfer model

3. Hygro-mechanical Modelling

The theory of hygro-mechanical modelling and material characterization procedure of CME were explained in previous section. For mould compound samples tested, both TMA and TGA show very close time-dependent curves (see Figure 53) because the initial moisture content and desorption condition are also close. According to Equation (54), the coefficient of hygroscopic swelling can be computed from the slope of the graph (see Figure 54), strain vs. concentration. The values of the coefficient of hygroscopic swelling measured for mould compound and die attach are shown in Table 8. Die and copper leadframe materials are assumed to have zero coefficient of hygroscopic swelling, i.e., no hygroswelling. All materials are considered as linear elastic.

For this case, the hygro strain is as high as the thermal strain. For example, the mould compound hygro strain is equivalent to thermal strain with mould compound CTE of 34.9 ppm/°C under the same temperature loading of 175–220°C.

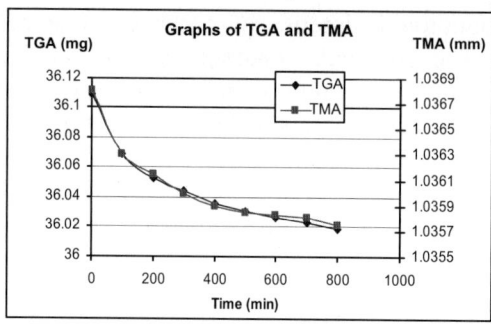

Figure 53. Hygroswelling material characterization

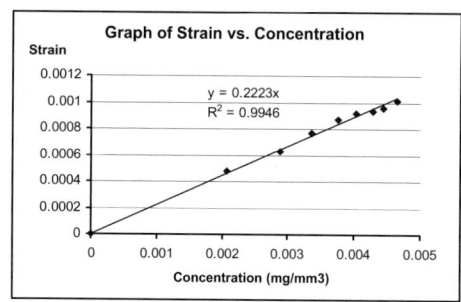

Figure 54. Computation of mould compound CME

4. Thermo-mechanical Modelling

Linear-elastic thermo-mechanical stress model is applied. The temperature loading applied is from the post mould cure temperature of mould compound, 175°C, to the reflow temperature of 220°C. The thermo-mechanical material properties used in the modelling are shown in Table 11. The boundary conditions applied are similar to previous thermo-mechanical model.

Table 11. Thermo-mechanical material properties

Material	Modulus at 220°C (GPa)	Mean CTE (ppm/°C)
Copper	127.4	17.4
Die	131	2.8
Mould Compound	1.1	34
Die Attach	0.043	170

5. Vapour Pressure Modelling

As discussed in section 4.3, the linear vapour pressure induced strain (ε_p) may be estimated by

$$\varepsilon_p = \frac{1-2v}{E} p \qquad (108)$$

where v is the Poisson's ratio, E is the modulus, and p is the average vapour pressure. The modulus of mould compound drops a few orders at the reflow temperature, and thus the vapour pressure strain may become as important as thermal or hygro strain. For example, for mould compound with saturated vapour pressure of 2.32 MPa, the strain induced is equivalent to thermal strain with CTE of 18 ppm/°C under the same 175–220°C temperature loading. The magnitude of this vapour pressure induced strain is in the same order as hygro and thermal strains. The vapour pressure strain induces additional mismatch to the package, in addition to the CTE (thermo-mechanical strain) and CME (hygro-mechanical) mismatch. It must also be pointed out that such an expansion is directly related to the vapour pressure distribution, rather than the moisture distribution.

8.1.2 Integrated Stress Modelling

Figure 55 shows the methodology of integrated stress modelling to calculate the package stress induced during reflow. The five models mentioned earlier are related to one another. Result of moisture distribution from the moisture diffusion model is used as input for both the vapour pressure model and the hygro-mechanical model. On the other hand, the temperature distribution from the thermal model is applied in both vapour pressure model and thermo-mechanical model.

The stress and strain induced by vapour pressure, thermo-mechanical, and hygro-mechanical models are combined into an equivalent linear elastic stress model to compute the package stress and strain induced during reflow. The hygro strain and vapour pressure induced strain are converted to an equivalent thermal strain, defined by CTE under the same temperature loading (175–220°C).

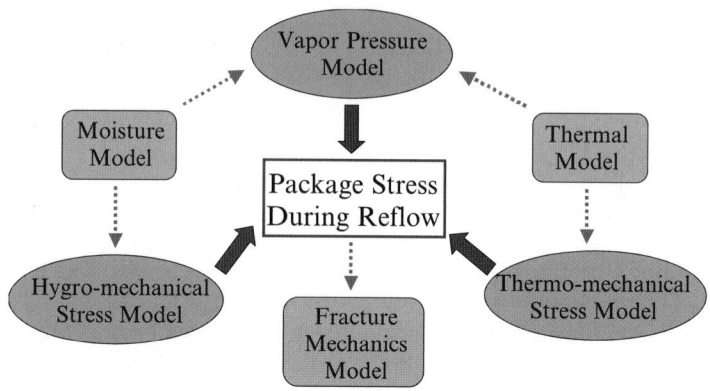

Figure 55. Integrated package stress model during reflow

8.1.3 Interfacial Fracture Mechanics Modelling

Subsequently, the integrated package stress is analysed with linear interfacial fracture mechanics. There are several possible ways used by researchers for fracture mechanics analyses, e.g., J-integral method, Virtual Crack Closure (VCC) method, and Crack Flank Displacement Extrapolation Method (CFDEM), depending on the type of application. Here, VCC method with non-singular crack-tip element (see Figure 56) is applied to calculate the strain energy release rate, G, of a crack, assumed along the die/mould compound and die/die attach corner interfaces, which are the potential delamination layers. The work required to extend the crack by Δa is equivalent to the work needed to close the crack tip by Δa. The mode I (G_I) and mode II (G_{II}) components of G and mode mixity are described by

$$G_I = \frac{1}{2\Delta a}(F_{y1}\Delta v_j + F_{y2}\Delta v_i) \tag{109}$$

$$G_{II} = \frac{1}{2\Delta a}(F_{x1}\Delta u_j + F_{x2}\Delta u_i) \tag{110}$$

$$G = G_I + G_{II} \tag{111}$$

6. Characterization and Modelling of Moisture Behaviour

$$\psi = \tan^{-1} \sqrt{\frac{G_{II}}{G_{I}}} \tag{112}$$

where Δa is the crack extension, F_{x1}, F_{x2}, F_{y1}, and F_{y2} are the x and y components of nodal forces per unit thickness at nodes 1 and 2, Δu_i, Δu_j, Δv_i, and Δv_j are the x and y displacements at nodes i and j (see Figure 56). From Equation (111) it is clear that when mode I is dominant, $G_I > G_{II}$, mode mixity is less than 45° (vice versa for mode II).

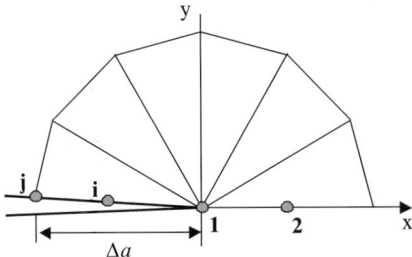

Figure 56. Non-singular crack-tip elements

8.1.4   Results

1. Moisture Diffusion Modelling

The transient moisture wetness distribution in QFN (symmetric half model) is shown in Figure 57, ranging from 0 to 100% saturation. The moisture diffuses into the package through mould compound, and gradually spreads into die attach layer. At the end of 168 hours of moisture pre-conditioning under 85°C/85%RH, the package is almost fully saturated with moisture. During the 5-minute solder reflow, external package surface loses a significant amount of moisture due to high moisture desorption rate. The moisture diffusivity is a few orders higher at reflow temperature than moisture preconditioning temperature. However, the moisture concentration in the interior of the package, including critical die attach and die/mould compound interfaces, still remains relatively unchanged. The local moisture concentration in these critical interfaces determines the strength of interfacial adhesion and magnitude of internal vapour pressure induced, which partially lead to moisture induced failures, e.g., delamination and popcorning, in the package.

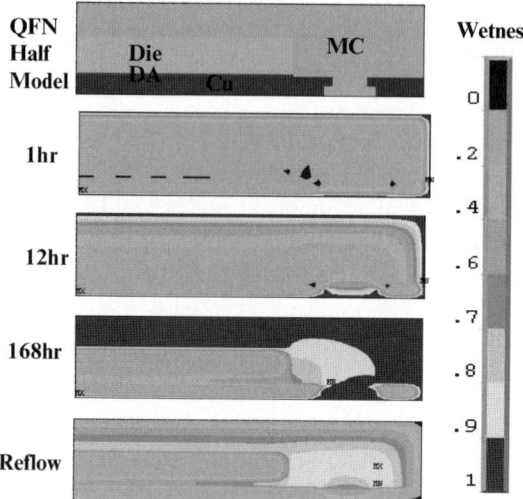

Figure 57. Transient moisture wetness distribution

2. Thermal Modelling

The thermal solution is analogous to moisture diffusion. However, the thermal diffusivity is much larger than the moisture diffusivity. Figure 58 shows the temperature distribution in the package during 5-minute reflow (from 25°C to peak temperature of 220°C). Heat is conducted faster in die and copper leadframe than in mould compound. When the external surface is heated to 220°C after 5 minutes, the internal package reaches this uniform temperature within a few seconds. Therefore, in the subsequent thermo-mechanical and vapour pressure models, temperature distribution during reflow (5 minutes) can be assumed to be uniform throughout the package.

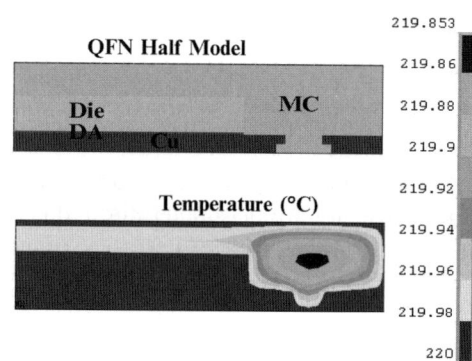

Figure 58. Package temperature distribution during reflow

6. Characterization and Modelling of Moisture Behaviour

3. Vapour Pressure Modelling

From the transient moisture distribution (see Figure 57), the corresponding vapour pressure distribution at reflow temperature can be calculated (see Figure 59) based on Equations (45)-(47). Case 3 of saturated pressure is mostly encountered here.

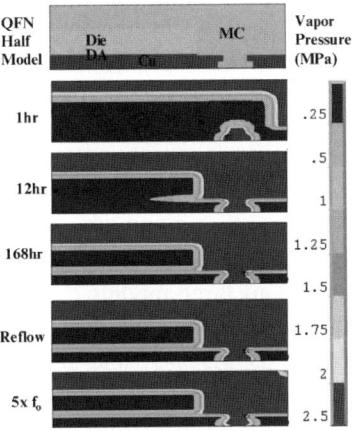

Figure 59. Vapour pressure distribution at reflow temperature

It is interesting to note that moisture diffusion and vapour pressure models have very different rates of saturation. For moisture diffusion, the distribution gradually reaches near-saturation at 168 hours of moisture preconditioning. However, for vapour pressure, the distribution reaches saturated pressure at reflow temperature with only 12 hours of moisture preconditioning under 85°C/85%RH. This implies that at JEDEC Level 1, the void moisture density exceeds its critical density value after about 12 hours. Subsequent addition of moisture density has no effect on the saturated pressure. Therefore, the vapour pressure induced strain is related directly to vapour pressure distribution, rather than to moisture distribution.

At 1 hour of moisture preconditioning (see Figure 57), the void moisture density is below the critical value. Therefore, the vapour pressure has closer distribution pattern as moisture diffusion, with lower vapour pressure inside the package, and gradually larger towards the external package surface. This corresponds to Cases 1 and 2 (Equations (45)-(46)), which assume ideal gas behaviour for non-saturated pressure conditions. Comparison is made with vapour pressure obtained from thermodynamics table, and the ideal gas assumption only has less than 5% of error. Anyway, for JEDEC Level 1 with 168 hours of moisture preconditioning, saturated pressure (Case 3, Equation

(47)) is always encountered, and its magnitude is much larger than the vapour pressure in Cases 1 and 2.

The effect of initial void volume fraction (f_o) is also studied. When f_o is increased 5 times, the vapour pressure distribution is almost unchanged, because the void moisture density is still above the critical value. It is important to point out that although f_o (until 5 times) has little effect on vapour pressure, it may weaken the interfacial adhesion. Exact measurement of f_o is not critical for Level 1, but it can be important for low moisture conditions, e.g., Level 3 or below, due to low void moisture density.

4. Integrated Stress Modelling

The total strains induced by thermo-mechanical, hygro-mechanical, and vapour pressure loadings in mould compound and die attach are listed in Table 12. For hygro-mechanical and vapour pressure induced strains, they are converted into equivalent mean CTEs, under the same 175–220°C temperature loading, so that all three models can be integrated in a thermo-mechanical model with equivalent strains. For example, hygro-mechanical strain of mould compound is 1.57e-3, and the equivalent mean CTE is computed as 1.57e-3/(220°C–175°C) = 34.9 ppm/°C. The total equivalent CTE is much larger than only considering equivalent CTE of individual model. Therefore, the thermal stress and strain due to mismatch of these three models are much higher than values of individual model.

Table 12. Total strains and equivalent CTEs in integrated stress model

	Mould Compound		Die Attach	
	Total Strain [-]	Equivalent mean CTE (ppm/°C)	Total Strain [-]	Equivalent mean CTE (ppm/°C)
Thermo-mechanical	1.53e-3	34	7.65e-3	170
Hygro-mechanical	1.57e-3	34.9	3.22e-3	71.6
Vapour Pressure	8.14e-4	18.1	2.16e-2	479.6
Integrated (total)	3.91e-3	87	3.25e-2	721.2

The relative package warpage or vertical displacement distributions during reflow among various models (not the same scale) are shown in Figure 60. Thermo-mechanical model has upward warpage, opposite in direction as compared to hygro-mechanical and vapour pressure induced stress models. This is because for hygro-mechanical and vapour pressure induced stress models, the die and copper leadframe do not absorb moisture, and therefore have zero coefficient of hygroscopic swelling and vapour pressure. For thermo-mechanical model, all the materials have different non-zero CTEs. As for the integrated model, the warpage distribution is more complex, a mixture of all three models. The integrated model is very different from the individual models.

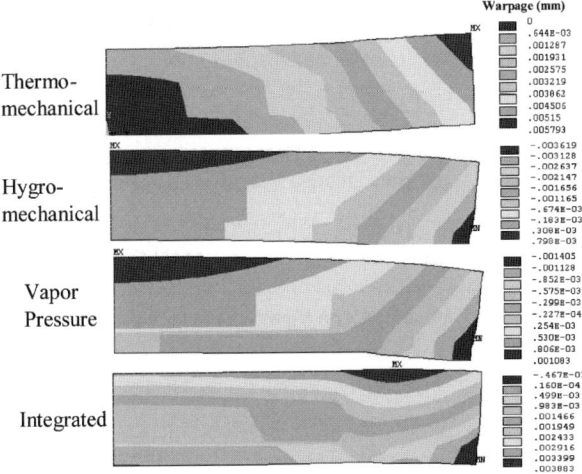

Figure 60. Comparison of warpage distributions

The Von Mises stress distribution in mould compound for the integrated model is shown in Figure 61. The maximum stress is located at the junction of die, die attach, and mould compound. This location is the same for all other models, despite they have different warpage distribution. However, the magnitude of the maximum stress is different (see Figure 62). For Von Mises stress, the principle of superposition is not applicable here, because it is an equivalent stress, linearity is no longer valid. For other stress components, the stress computed by integrated stress model equals exactly to the sum of stresses given by three basic stress models mentioned above. Among the three models, thermo-mechanical stress is the highest. The integrated stress model has about 50% larger maximum stress in mould compound than the thermo-mechanical stress model. Without considering the hygro-mechanical and vapour pressure induced stresses, the package stress during reflow would have been underestimated.

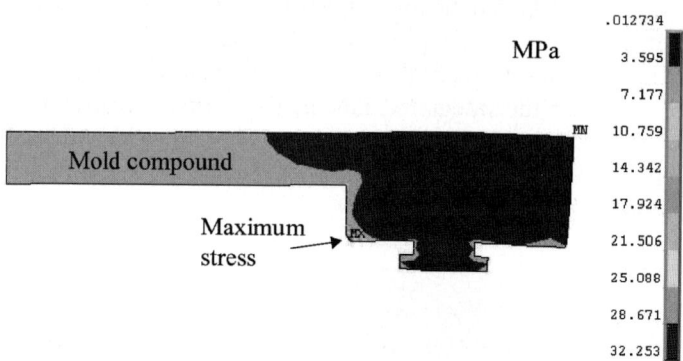

Figure 61. Integrated model: Von Mises stress distribution

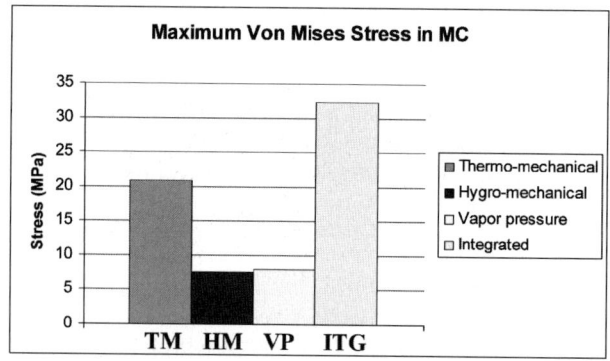

Figure 62. Comparison of maximum Von Mises stresses

Besides the difference in stresses, the considerations of moisture diffusion and vapour pressure also have impact on interfacial adhesion strength. The interfacial adhesion should be characterized under reflow temperature with the same level of moisture preconditioning.

5. Interfacial Fracture Mechanics Modelling

Two potential failure interfaces, i.e. die/mould compound and die/die attach, are further analysed for delamination. An initial crack of 0.1mm is assumed at the die corner (both top and bottom interfaces), which is induced

6. Characterization and Modelling of Moisture Behaviour

by the integrated stress under the thermo-mechanical, hygro-mechanical, and vapour pressure loadings. Once the delamination is initiated, the vapour pressure fills up the cavity quickly and saturated vapour pressure of 2.32 MPa is generated to open-up and extend the crack length. The initial delamination will propagate when G is larger than G_c, critical interfacial toughness of the bi-material. G_c depends strongly on mode mixity, and generally G_c increases with larger mode mixity. It is the strongest under the pure mode II (mode mixity = 90°). The values of G_c have to be determined experimentally.

Figure 63 shows that G (calculated by Equations (109)–(111)) increases with crack length at both interfaces studied. The effect of crack length is very significant for die/mould compound interface. If this interface has an initial crack, the crack may be extended by the vapour pressure until the whole interface is fully delaminated.

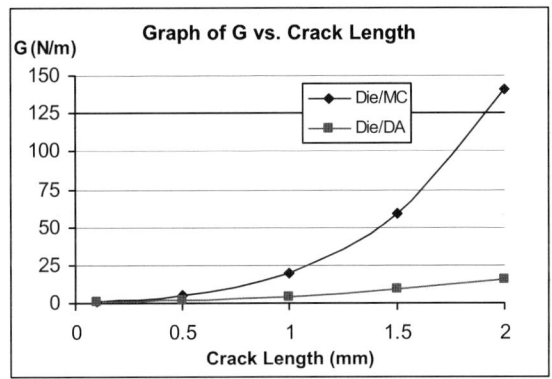

Figure 63. Effect of crack length on strain energy release rate (G)

Figure 64 shows that the mode mixity (calculated by Equation (112)) decreases with longer crack length, gradually changed from in-plane shear mode (mode II) to opening mode (mode I). This is because the contribution of vapour pressure loading becomes greater with longer crack length, and naturally it acts to open-up the interfacial gap. For die/mould compound interface, mode mixity becomes stable when crack length exceeds 1mm. However, for die/die attach interface, mode mixity continues to decrease with crack length. Therefore, the interfacial fracture toughness is also reduced, and the crack is more susceptible to further delamination.

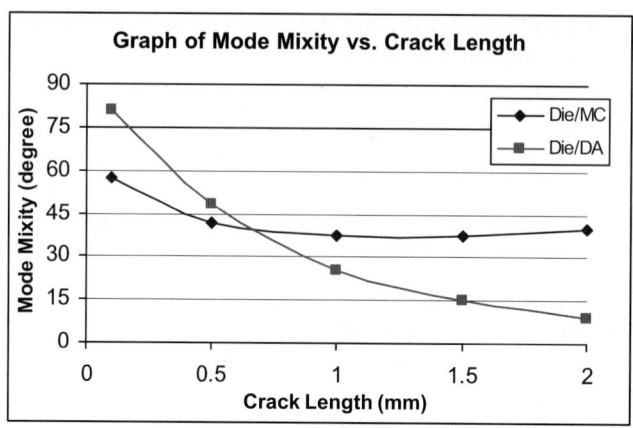

Figure 64. Effect of crack length on mode mixity

8.1.5 Summary

The actual package failure mechanism during reflow is complex, contributed by combined effects of process defects, interfacial adhesion strength, moisture, vapour pressure, thermo-mechanical stress, and hygro-mechanical stress. Therefore, there is a need for comprehensive studies on moisture diffusion, thermal, hygro-mechanical stress, thermo-mechanical stress, and vapour pressure modelling. An integrated stress model with interfacial fracture mechanics is established to study the QFN package stress, and compute the strain energy release rate when delamination is initiated. It is an useful tool to analyse and enhance the reliability of package, minimizing the delamination and popcorn failures encountered during reflow.

The package vapour pressure distribution during reflow is the key factor in understanding the failure mechanism. Moisture diffusion model is applied to predict the local moisture concentration at the critical interfaces, which can be used for subsequent vapour pressure calculations. High moisture concentration weakens the critical interfacial adhesion, generates vapour pressure during reflow, and induces hygro-mechanical stress in the package. The vapour pressure induces additional mismatch to the package, which is of the same order as the CTE and CME mismatch. When the interfacial

6. Characterization and Modelling of Moisture Behaviour 361

adhesion is reduced to the level below the stress predicted by the integrated stress model, delamination will occur.

For both die/mould compound and die/die attach interfaces, G increases with a longer crack length, and mode mixity decreases with a longer crack length. The propagation of an initial crack depends on the relative values of G and G_c. The contribution of vapour pressure to interfacial delamination is significant, especially when there is a defect or crack along the interface.

8.2 BGA Moulding Compound Selection with Optimal Resistance to Moisture Induced Failures

This section presents a combined numerical and experimental methodology for predicting and preventing moisture induced failures in encapsulated packages. Prevention of such failures will enable a pre-selection of materials and geometry combinations with the highest resistance to moisture. This virtual qualification methodology is illustrated for a specific BGA package which showed 50% failures (broken stitch-bonds) during HAST testing due to excessive warpage and/or delamination of different interfaces. For three different material combinations the moisture diffusion during the HAST is predicted and subsequently thermo-mechanical-moisture simulations are performed where the effects of hygroscopic swelling, vapour pressure, thermal expansion and delamination on the failure mechanisms is predicted. The comparison of the simulation results for different moulding compounds are compared to the experimental observations during HAST qualification testing. Application of the presented methodology will result in a shorter time-to-market and a significant reduction of costs due to reduced trial-and-error design and material runs.

8.2.1 Description of the Carrier

As carrier a specific BGA package is chosen. This BGA package with low alpha radiation compound showed 50% failures (broken stitch-bonds) during HAST (Highly Accelerated Stress Testing) qualification testing due to excessive warpage and/or delamination of different interfaces. The global geometry of the BGA package is plotted in Figures 65 and 66. The package is very thin in nature with relatively large size. The wire bonds are used and the substrate has four copper layers.

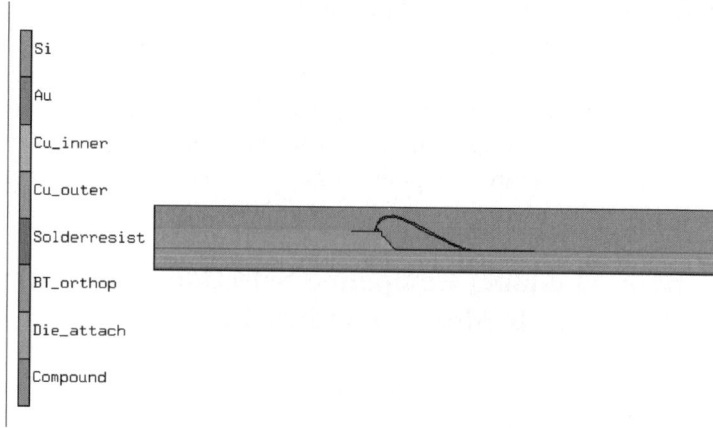

Figure 65. Cross section of BGA package

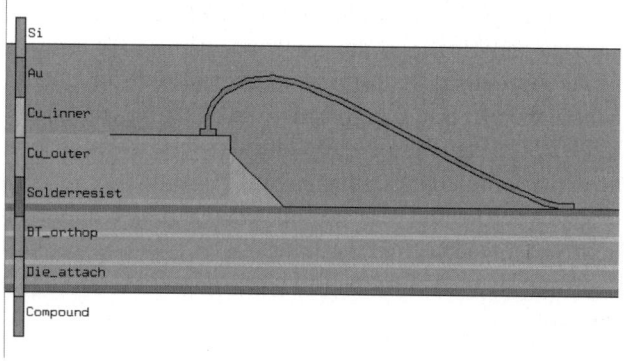

Figure 66. Detail of wire geometry

The observed failures are shown in Figures 67 and 68, indicating a broken stitch bond and delamination at the compound/substrate interface.

Three different moulding compounds were experimentally evaluated resulting in significantly different behaviours in the HAST qualification test as presented in Table 13.

6. Characterization and Modelling of Moisture Behaviour 363

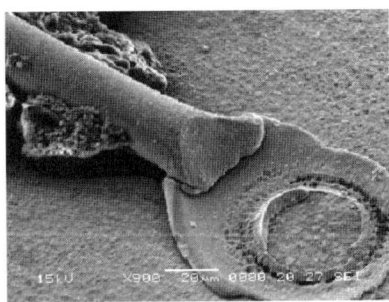

Figure 67. Observed stitch failure after HAST qualification test (MCA compound)

Figure 68. SCAT pictures showing serious delamination at the compound/substrate interface for MCA compound

Table 13. Observed failures during HAST qualification

Compound	Observations
MCA	Failed qualification
	Severe delamination at compound/substrate interface
	Many stitch breaks
MCB	Just within qualification limits
	Some delamination at compound
MCC	Qualified
	Minor delamination at compound

The present work of virtual qualification methodology was initiated in order to understand the observed failure modes and differences between the moulding compounds used. The following sections will introduce the present methodology and the obtained results.

8.2.2 Material Characterization

1. Moisture related properties

All the polymer materials in the BGA package have been characterized with regard to their moisture and thermo-mechanical behaviour. The diffusivity D and the saturated moisture concentration C_{sat} have been determined under 85°C/85RH conditions. In Figure 69 the normalized moisture uptake at specific time intervals is plotted for the three different moulding compounds and the substrate (both measured and fitted response are visualized). The moisture uptake is normalized with the saturated moisture level C_{sat} of moulding compound B (MCB). It should be noted that the dimensions of the substrate specimen differ from the other specimens resulting in a faster moisture uptake.

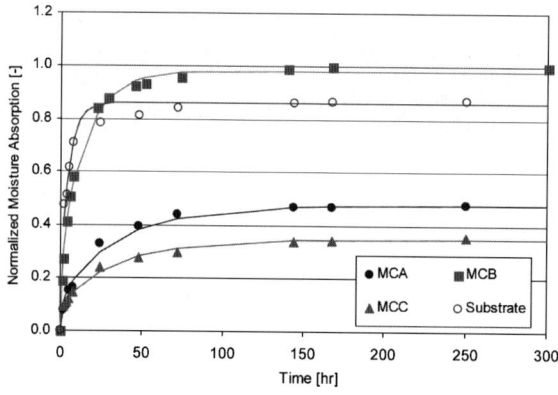

Figure 69. Normalized moisture uptake for the three different moulding compounds and the substrate

Using this Fickian behaviour, the diffusivities and saturated moisture concentrations are extracted using the method described in section 3. The resulting Fickian fit is shown as the lines in Figure 69 with the associated normalized values listed in Table 14.

For the determination of the moisture expansion coefficient, combined TMA/TGA experiments have been performed at 130°C on saturated samples. Combining the obtained results of moisture desorption and shrinkage as function of time can estimate the coefficient of hygroscopic

6. Characterization and Modelling of Moisture Behaviour

swelling. In Figure 70 the combined results of the TGA/TMA measurements are plotted together with the linear fit. The curves don't cross the origin due to the non-uniform moisture distributions at long measurement times, as discussed in section 5. The coefficient of hygroscopic swelling of the substrate differs significantly in x,y and z-direction due to its orthotropic nature.

Table 14. Fitted values for D and Csat (normalized)

Material	D [-]	Csat [-]
MCA	0.33	0.48
MCB	1	1
MCC	0.35	0.35
Substrate	0.246	0.87

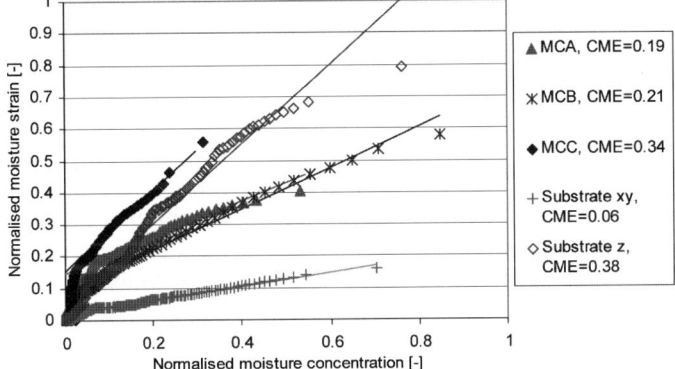

Figure 70. CME fits of all the materials

2. Thermo-mechanical properties

The thermomechanical behaviour of the polymers has been measured using TMA and DMTA techniques. With the DMTA a full viscoelastic characterisation of the materials is determined. In Table 15 the CTE values are listed according to the TMA measurements.

The following material models are used for the different materials, which are present in the package:

Table 15. Measured CTE values

Material	CTE < Tg [1e-6/°C]	Tg [°C]	CTE > Tg [1e-6/°C]
MCA	11	155	35
MCB	17	189	62
MCC	6	108	35
Substrate xy	16		16
Substrate z	51	178	239

- The single crystal silicon is modelled as a linear elastic material.
- The substrate is modelled as an orthotropic layered material taking into account the thickness and density of the copper layers. These copper layers are modelled as an ideal elasto-plastic material.
- The gold wire is also modelled as an ideal elasto-plastic material.
- All polymeric materials are modelled using visoelastic or temperature dependent properties [45, 46].

8.2.3 Finite Element Modelling

The temperature and humidity conditions during the HAST qualification test (130°C, 85RH, 96 hrs) have been simulated using commercially available finite element software combined with additional user subroutines. These simulations consist of two consecutive steps: first a combined temperature and moisture diffusion simulation followed by a thermo-mechanical simulation with uses the predicted temperature and moisture fields as input. The initial stress/warpage free temperature is the process temperature of 175°C. The characteristics of temperature and moisture simulations are:

- Temperature is assumed homogeneous in the package due to the large time scales compared to the moisture effects.
- At the start of the HAST test the moisture boundary conditions (W = 1) are applied at the outer edges of the package.
- The predicted moisture profile after the HAST loading step is further maintained during the last cooling down step in order to simulate the effect of removing the package from the test chamber.
- The presence of copper layers in the substrate is taken into account by reducing the diffusivity and moisture concentration relative to the percentage copper.
- The characteristics of the thermo-mechanical simulations are:
- The effect of delamination between moulding compound and substrate/die on resulting strains/forces on the wire is investigated. The delamination is modelled using contact bodies.

6. Characterization and Modelling of Moisture Behaviour

- In order to predict strains/forces on the wire, the wire is modelled as a frictionless contact body within the compound, allowing the wire to move within the compound (tunneling effect). This simulates the lack of adhesion due to the inert properties of the gold with respect to the moulding compound.
- The wire remains attached to the die and leadframe regardless whether delamination between compound and leadframe/die exists at that place.
- The simulations take into account the viscoelastic properties of the different materials, in particular the moulding compound.

8.2.4 Results

1. Moisture diffusion

The predicted weighted moisture concentration profiles after HAST testing for the three different moulding compounds are visualized in Figure 71.

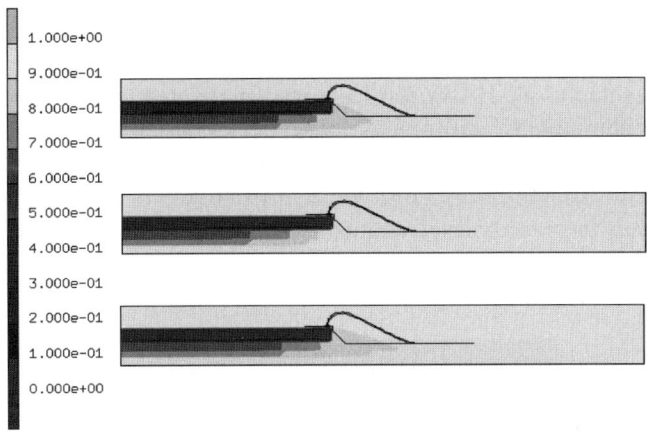

Figure 71. Predicted weighted moisture concentration W after HAST test for the three different moulding compounds (from top to bottom MCA, MCB and MCC respectively)

The predicted weighted moisture concentrations for the three moulding compounds show a large resemblance with an almost saturated package. The absolute values of moisture concentration are however different as is visualized in Figure 72.

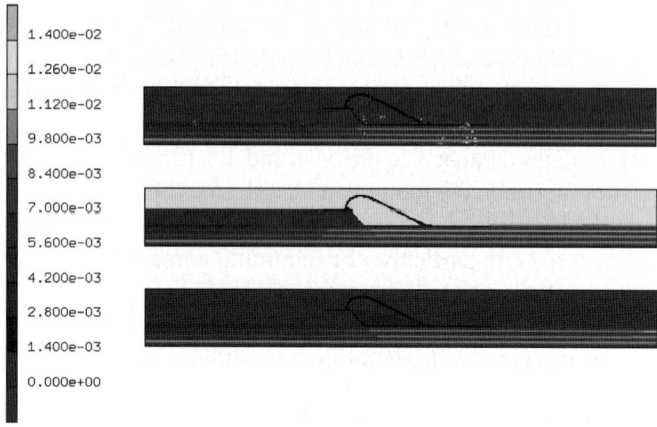

Figure 72. Predicted moisture concentration C after HAST test for the three different moulding compounds (from top to bottom MCA, MCB and MCC respectively)

2. Thermo-mechanical results

Using the predicted moisture diffusion distributions of the previous paragraph combined with the thermo-mechanical loading, the displacements and stresses due to the HAST test have been simulated for the three different moulding compounds with or without delamination between moulding compound and substrate/die. In Figures 73 and 74 the effect of delamination on the plastic strains in the gold wire is visualised. It is clear that delamination causes increased plastic strains near the heel of the stitch bond, which is in agreement with the observed stitch failures.

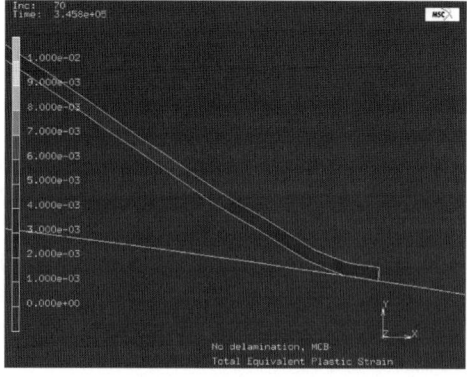

Figure 73. Predicted plastic strain in gold wire after HAST test for moulding compound MCB without delamination

6. Characterization and Modelling of Moisture Behaviour

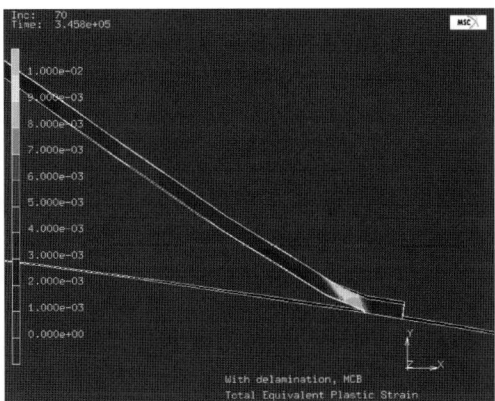

Figure 74. Predicted plastic strain in gold wire after HAST test for moulding compound MCB with delamination

The predicted forces on the stitch bond at the end of the HAST test are visualized in Figure 75. It appears that delamination has a large effect on the forces for MCA and MCC but a limited effect for MCB. This is consistent with the observations, which indicated delamination for MCA and MCB but only wire break for MCA. MCC exhibits no delamination in HAST testing but the simulations indicate that if delamination should occur this will result in a large increase in forces on the wire and potential problems. The normalized warpage of the package after HAST testing at 130°C and after the following cool down to room temperature is visualized in the Figure 76.

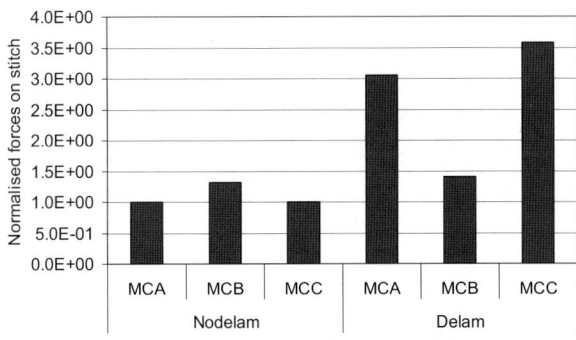

Figure 75. Predicted normalized forces on the stitch bond at the end of the HAST test

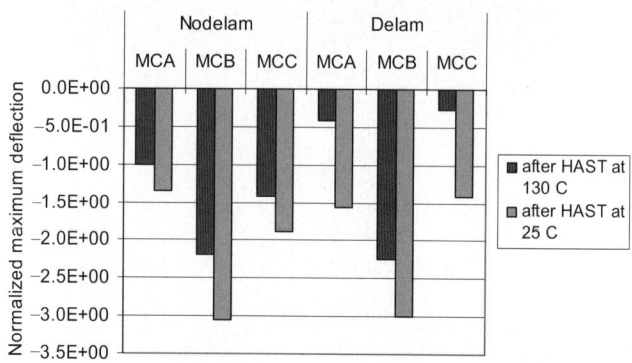

Figure 76. Normalized maximum deflection of the package

This figure shows that delamination also has an effect on the global deflection of the package. The difference between the deflection at 130°C and the deflection at 25°C further gives an indication on the relative influence of temperature strains versus the influence of moisture strains.

8.2.5 Summary

Moisture induced failures are one of the most important mechanisms within microelectronic products. Combined with the ongoing trend of miniaturisation (thinner packages) and the increasing number of material interfaces, virtual prototyping and virtual qualification techniques can contribute to shorter time-to-market and cost reduction. The example presented focuses on the effect of moisture and delamination on the observed wire failures and warpage of BGA packages. The results indicate that the occurence of delamination is the key factor for the observed failures. If delamination can be controlled, Moulding Compound C would be the best choice, but when this is not possible, Moulding Compound B is an alternative option. Future work will focus on the ability to predict the initiation and propagation of delamination due to moisture and temperature effects for the three Moulding Compounds.

9. REFERENCES

[1] I. Fukuzawa, S. Ishiguro, and S. Nanbu, Moisture Resistance Degradation of Plastic LST's by Reflow Soldering, Proceedings, 23rd International Reliability Physics Symposium, pp. 192-197, (1985).

[2] M., Kitano, A. Nishimura, and S. Kawai, Analysis of Package Cracking during Reflow Soldering Process, Proc. IRPS, pp. 90-95, (1988).
[3] A.A.O. Tay and T.Y. Lin, Influence of temperature, humidity and defect location on delamination in plastics packages, IEEE Transactions on Components, Packaging and Manufacturing Technology, Part A, 22, no. 4, pp. 512-518, (1999).
[4] T.Y. Lin and A.A.O. Tay, Dynamics of Moisture Diffusion, Hygrothermal Stresses and delamination in Plastic IC Packages, ASME EEP-Vol. 19-1, Advances in Electronic Packaging, pp. 1429-1436, (1997).
[5] A.A.O. Tay and T.Y. Lin, Moisture diffusion and heat transfer in plastic IC packages, IEEE Transactions on Components, Packaging and Manufacturing Technology, Part-A, vol. 19, no. 2, pp. 186-193, (1996).
[6] A.A.O. Tay and T.Y. Lin, The impact of moisture diffusion during solder reflow on package reliability. Proc. of the 49th Electronic Components and Technology Conference, pp. 830-836, (1999).
[7] S. Liu, Y. Mei and T. Y. Wu, Bimaterial Interfacial Crack Growth as a Function of Mode-Mixity, IEEE Transactions on Components, Packaging, and Manufacturing Technology – Part A, Vol. 18, No. 3, pp. 618-626, (1995).
[8] S. Liu and Y. H. Mei, Behaviour of delaminated plastic IC packages subjected to encapsulation cooling, moisture absorption, and wave soldering, IEEE Transactions on Components, Packaging, and Manufacturing Technology – Part A, Vol. 18, No. 3, (1995).
[9] L. T. Nguyen, Lo, R.H.Y., Belani, J.G., Moulding compound trends in a denser packaging world, IEEE Int. Electron. Manufact. Techol. Symp., Kanazawa, Japan, June, 9-11, (1993).
[10] J.E. Galloway and B.M. Miles, Moisture Absorption and Desorption Predictions for Plastic Ball Grid Array Packages, IEEE Transactions on Components, Packaging and Manufacturing Technology, Part-A, 20(3), pp. 274-279, (1997).
[11] E.H. Wong, Y.C. Teo, and T.B. Lim, Moisture Diffusion and Vapour Pressure Modelling of IC Packaging, 48th Electronic Components and Technology Conference, pp. 1372-1378, (1998).
[12] Shook R, Vaccaro BT, Gerlach DL. Method for equivalent acceleration of JEDEC/IPC moisture sensitivity levels, Proc. 36th IRPS, pp. 214-9, (1998).
[13] T. Y. Tee and H.S. Ng, Whole Field Vapour Pressure Modelling of QFN during Reflow with Coupled Hygro-mechanical and Thermo-mechanical Stresses, 52nd ECTC Conference, USA, pp. 1552-1559, (2002).
[14] T.Y. Tee, X.-J. Fan & T.B. Lim, Modelling of whole field vapour pressure during reflow for flip chip and wire-bond PGBA packages, 1st International Workshop on Electronic Materials & Packaging, Singapore, September 29-October 1, (1999).
[15] T. Y. Tee, C.L. Kho, D. Yap, X. Baraton, and K. Sivakumar, "Comprehensive Moisture Diffusion, Hygroswelling, and Thermo-Mechanical Modelling of FCBGA Package with No-flow Underfill", APACK 2001 Conference, Singapore, pp. 210-216, (1999).
[16] C.L. Kho, T. Y. Tee, D. Yap, C. Toh, and X. Baraton, Flip Chip in Package Assembly and Reliability with No-Flow Underfill Materials, SEMICON Advanced Packaging Technology Symposium, Singapore, Aug 2003, Singapore, pp. 81-88, (2002).
[17] T. Y. Tee, C.L. Kho, D. Yap, C. Toh, X. Baraton, and Z.W. Zhong, Reliability Assessment and Hygroswelling Modelling of FCBGA with No-flow Underfill, Microelectronics Reliability Journal, Vol. 43(5), pp. 741-749, (2003).
[18] T. Y. Tee and Z.W. Zhong, Integrated Vapour Pressure, Hygroswelling, and Thermomechanical Stress Modelling of QFN Package during Reflow with Interfacial Fracture

Mechanics Analysis, Microelectronics Reliability Journal, Vol. 44(1), pp. 105-114, (2004).

[19] R. Dudek, J. Walter; B. Michel, P. Alpern, R. Schmidt, R. Tilgner, Studies on Parameters for Popcorn Cracking, Proc. Polytronic 2001, Potsdam, Germany, October 21-24, pp. 140-148, (2001).

[20] R. Dudek, H. Walter, B. Michel, Studies on Moisture Diffusion and Popcorn Cracking, Conference EuroSimE 2002, Paris, France, April 2002, Proc. pp. 225-232, (2002).

[21] X.J. Fan, G.Q. Zhang and L.J. Ernst, A Micromechanics based vapour pressure model in electronic packages, ASME Journal of Electronic Packaging, September Issue, (2005).

[22] X.J. Fan, G.Q. Zhang and L.J. Ernst, Analytical solution for moisture-induced delamination in electronic packaging, Proc. of 53rd IEEE Electronic Component and Technology Conference, New Orleans, Louisiana, 733-738, (2003).

[23] X.J. Fan, G.Q. Zhang and L.J. Ernst, A Micro-mechanics approach in polymeric material failures in microelectronic packaging, Proc. of 3rd International Conference on Thermal & Mechanical Simulation in Micro-Electronics, 154-164, Paris, France, April, (2002).

[24] X.J. Fan and T.B. Lim, Mechanism analysis for moisture-induced failures in IC packages ASME 1999 International Mechanical Engineering Congress, 11th Symposium on Mechanics of Surface Mount Assemblies, Nashville, Tennessee, 14-19 November, IMECE/EPE-14, (1999).

[25] X.J. Fan and G.Q. Zhang, Moisture related issues in microelectronic packaging, short course note, 4th International Conference in Thermal and ThermoMechaincal Simulation and Experiment in Microelectronics and Microsystem, Aix-Provence, France, April, (2003).

[26] D. J. Belton, E. A. Sullivan, and M. J. Molter, Moisture sorption and its effects upon the microstructure of epoxy moulding compounds, Proc. 3rd Int. Electron. Manufact. Technol. Symp., pp. 158–169, (1987).

[27] A.L. Gurson, Continuum theory in ductile rupture by void nucleation and growth: part I- yield criteria and flow rules for porous ductile media, Journal of Engineering Materials and Technology, 99, pp. 2-15, (1977).

[28] X.J. Fan and S.Y. Zhang, Void Behaviour Due to Internal Vapour Pressure Induced by Temperature Rise, Journal of Materials Science, 30, 3483-3489, (1997).

[29] F.A. Mclintock, a criterion for ductile fracture by growth of holes, Journal of Applied Mechanics, 35, 363-371, (1968).

[30] J.R.Rice and D.M. Tracey, On the ductile enlargement of voids in triaxial stress fields, Journal of Mechanics and Physics of Solids, 17, 201-217, (1969).

[31] Y. Huang, J.W. Hutchinson & V. Tvergaard Cavitation instabilities in elastic plastic solids, Journal of Mechanics and Physics of Solids, 39, 223, (1991).

[32] T.F. Guo and L. Cheng, Unstable Void Growth in Plastic IC Packaging Material, Scripta Materialia, 2001, submitted.

[33] J.W. Hutchinson and Z. Suo, Mixed Mode Cracking in Layered Materials, Advances in Applied Mechanics, vol. 29, pp. 63-191, (1992).

[34] J.R. Rice, Elastic Fracture Mechanics Concepts for Interfacial Cracks, J. Appl. Mech., Vol. 55, pp. 98-103, (1988).

[35] C.F. Shih, Cracks on Bimaterial Interfaces: Elasticity and plasticity Aspects, Material Sci. Engng., A143, pp.77-90, (1991).

[36] B.M. Malyshev and R.L. Salganik, The strength of adhesive joints using the theory of cracks, International Journal of Fracture Mechanics, v. 1, pp. 114-128, (1965).

[37] K.M. Liechti and Y.S. Chai, Biaxial Loading Experiments for Determining Interfacial Toughness, ASME Journal of Applied Mechanics, vol. 58, pp. 680-687, (1991).

6. Characterization and Modelling of Moisture Behaviour 373

[38] N. Tanaka and A. Nishimura, Measurement of IC moulding Compound adhesion strength and prediction of interface delamination within package, ASME EEP-Vol, 10-2. Advances in Electronic Packaging, pp. 765-773, (1995).

[39] A.A.O. Tay, Y. Ma, S.H. Ong and T. Nakamura, Measurement of interface toughness as a function of temperature, moisture concentration and mode mixity, in Advances in Electronic Packaging, Vol. No. EEP-Vol. 26-2, pp. 1129-1136, (1999).

[40] B.S. Bakhvalov and G.P. Panasenko, Homogenization: Averaging Processes in Perioidc Media, Kluwer Academic Publisher, (1989).

[41] V. Tvergaard and J.W. Hutchinson, The relation between crack growth resistance and fracture process parameters in elastic-plastic solids, Journal of Mechanics and Physics of Solids, 40, pp. 1377-1397, (1992).

[42] L. Xia and C.F. Shih, Ductile crack growth – I, a numerical study using computational cells with microstructurally-based length scales, Journal of Mechanics and Physics of Solids, 43, pp. 233-259, (1995).

[43] H.B. Chew, T.F. Guo and L. Cheng, Modelling interface delamination in plastic IC packages, APACK Conference on Advances in Packaging, Singapore. (2001).

[44] M.A.J. van Gils, W.D. van Driel, G.Q. Zhang, H.J.L. Bressers, R.B.R. van Silfhout, X.J. Fan, J.H.J. Janssen, Virtual qualification of moisture induced failures of advanced packages, Proceedings of the 5th international conference on thermal & mechanical simulation and experiments in micro-electronics and micro-systems, Brussel, Belgium, May 9-11, 2004, pp. 157-162, (2004).

[45] W.D. van Driel, G.Q. Zhang, J.H.J. Janssen, L. J. Ernst, F. Su, K.S. Chian, S. Yi, Prediction and verification of process induced warpage of electronic packages, Journal of Microelectronics & Reliability 43 (5), 2003, pp. 765-774, (2003).

[46] W.D. van Driel, J.H.J. Janssen, G.Q. Zhang, D.G. Yang, L.J. Ernst, Packaging Induced Die Stresses – Effect of Chip Anisotropy and Time-dependent Behaviour of a Moulding Compound, Journal of Electronic Packaging 125 (4), 2003, pp. 520-526, (2003).

[47] E.H. Wong, R. Rajoo, T.B. Lim, Moisture absorption and distribution characterisation of packaging materials-advanced treatment, Microelectronics Reliability, 2003 (43), pp. 2087-2096.

[48] E.H. Wong, K.C. Chan, R. Rajoo, T.B. Lim, The mechanics and impact of hygroscopic swelling of polymeric materials in electronic packaging, Proc. 50th Electron. Comp. Technol. Conf., Las Vegas, NV, 2000, pp. 576-580, 2000.

[49] T.Y. Tee, Z. Zhong, Integrated vapour pressure, hygroswelling, and thermo-mechanical stress modelling of QFN package during reflow with interfacial fracture mechanics analysis, Microelectronics Reliability, 2004, pp. 105-114.

[50] E. Stellrecht, B. Han, M.G. Pecht, Characterization of Hygroscopic swelling behaviour of mould compounds and plastic packages, IEEE transactions on Components and Packing Technologies, 2004, pp. 499-506.

[51] J. Zhou, S. Lahoti, and M. Sitlani, Investigation of inner-layer dielectric (ILD) failure by hygroscopic swelling, IEEE 55th Electronic Components and Technology Conference (ECTC), May 31-June 4, 2005, Orlando, Florida.

[52] J. Zhou, S. Lahoti, and M. Sitlani, Investigation of Non-Uniform Moisture Distribution on Determination of Hygroscopic Swelling Coefficient and Finite Element Modelling for a Flip Chip Package, IEEE International Conference on Thermal, Mechanical and Multiphysics Simulation and Experiments in Micro-Electronics and Micro-System (EuroSimE), Berlin, Germany, pp. 112-119, April 17-20, 2005.

[53] J. Zhou, T.Y. Tee, and X. Zhang, Transient analysis on hygroscopic swelling characterization using sequentially coupled moisture diffusion and hygroscopic stress modelling approach, Proceedings of IMECE2005, 2005 ASME International

Mechanical Engineering Congress and Exposition November 5-11, 2005, Orlando, Florida USA, IMECE2005-81847.
[54] H. Ardebelli, E.H. Wong, and M. Pecht, Hygroscopic swelling and sorption characteristics of epoxy moulding compounds used in electronic packaging, IEEE Trans. Comp. Packag. Technol., vol. 26, pp. 206-214, Mar. 2003.
[55] X. J. Fan and G.Q. Zhang, Overview: characterization and modelling of moisture behaviour of electronic packaging, Micromaterials and Nanomaterials, a special issue to Dr. Andreas Schubert in memoriam, 2004.
[56] X. J. Fan, J. Zhou, and G.Q. Zhang, Multi-physics modelling in virtual prototyping of electronic packages – combined thermal, thermo-mechanical and vapour pressure modelling, Journal of Microelectronics Reliability, 2003, 44, pp. 1967-1976.
[57] X. J. Fan, Moisture Related Reliability in Electronic Packging, ECTC 2005 Professional Development Course Note #6, Orlando, Florida, 2005.
[58] X. J. Fan, H.B. Wang and T.B. Lim, Investigation of the underfill delamination and cracking for flip chip module during thermal cyclic loading, IEEE Transaction of Component, Manufacturing and Packaging Technology, 24(1), 84-91, 2001.
[59] X. J. Fan, Modelling of vapour pressure during solder reflow for electronic packages, in Benifiting from Thermal and Mehaincal Simulation and Modelling in Microelectronics (ed. G.Q. Zhang), Kluwer Academic Publishers, Boston, 75-92, 2000.
[60] J. Zhou, Upper and Lower Bound Theoretical Analysis in Characterizing Hygroscopic Swelling of Polymeric Materials in Electronic Packaging, IMAPS 2005 – 38th International Symposium on Microelectronics, Philadelphia, Pennsylvania, Pennsylvania Convention Centre, September 25-29, 2005.
[61] W.D. van Driel, H.J.L. Bressers, M.A.J. van Gils, G.Q. Zhang, L.J. Ernst, Driving mechanisms of delamination related reliability problems in exposed pad packages, Proc. of 6th International Conference on thermal, mechanical & multi-physics simulation and experiments in micro-electronics and micro-systems, Berlin, Germany, April 17-20, 2005, pp. 183-189.
[62] W.D. van Driel, P.J.J.H.A. Habets, M.A.J. van Gils, G.Q. Zhang, Characterization of Interface Strength as function of Temperature and Moisture Conditions, Proc. of 6th International Conference on Electronics Packaging Technology, Shenzhen, China, Aug. 30 to Sep. 2, 2005, pp. 687-692.

10. EXERCISES

1. What is the moisture concentration? What are the material properties related to moisture diffusion?
2. How is the relative humidity (RH) defined? Is the saturated moisture concentration dependant on RH? What about the solubility?
3. Is the moisture concentration continuous at a bi-material interface? Why? List some other field variables describing the moisture distribution which are continuous at the interface.
4. Derive the solution given by equation (11) for the one-dimensional moisture diffusion from equation (1).
5. A material with rectangular parallelepiped shape (the lengths in x-, y-, and z-direction are Lx, Ly, and Lz respectively) is under controlled

6. Characterization and Modelling of Moisture Behaviour

humid chamber. Assume that the saturated moisture concentration and the diffusivity are known, obtain the 3-D analytical solution using one-dimensional solution given in equation (11).

6. In moisture weight gain test, the specimen has a rectangular shape with 1.5 mm thickness and 12 mm long in other two directions. The following are the measurement data (see the table given below), where time is in hours and M(t) is the weight gain (total moisture mass) as function of time. Determine the diffusivity and the saturated moisture concentration from those data. You need to use equation (11). You can use any software to process those data, or write your own code.

7. What equation can be used to describe the temperature dependency of diffusivity? Is the diffusivity dependant on the RH too? How to describe the temperature dependency of solubility?

8. Choose a plastic electronic package that you are familiar with, and use any finite element software to obtain the moisture distribution across the package.

9. What's the partial vapour pressure?

10. Will the moisture in material always be in vapour form at temperature above 100°C? Why?

11. What is the saturated vapour pressure? Where can you find the saturated vapour pressure as function of temperature?

12. What's the hygroscopic swelling? How does hygroscopic swelling affect the package behaviour?

13. How to characterize the coefficient of hygroscopic swelling? Is this coefficient temperature dependant?

14. How to characterize the void behaviour at the interface and bulk material? How different it would be for a void in bulk and at interface?

Table for Exercise No. 6

time, hr	M(t)exp, mg
25	1.5
50	2.25
100	3
175	3.9
325	5.5
500	6.5
700	7.5
1020	8.5
1350	9.1

Chapter 7

CHARACTERIZATION AND MODELLING OF SOLDER JOINT RELIABILITY

R. Dudek
Fraunhofer Institute Zuverlassigkeit und Mikrointegration, Gustav-Meyer-Allee 25, 13355 Berlin, Germany

Abstract: This chapter addresses finite-element analyses (FEA) of solder fatigue phenomena caused by low-cycle thermo-mechanical loading. To begin with, an introduction to board level solder joint fatigue, characteristic thermal loading situations, the effects of thermal mismatch and analytical lifetime estimates is provided. Subsequently, challenges to the FE-based methodologies are discussed, which are particularly related to the non-linear mechanical properties of soft solders. Material constitutive models and the implementation of time and temperature dependent behaviours of leaded and lead-free solders are described. Fatigue-life prediction modelling focuses on the strength of materials approaches, i.e., creep strain-based relations or energy-based relations. An overview on several fatigue-life prediction models from the literature is additionally provided. The FE-based methodology is applied to board-level solder joint reliability assessments for several components. Its wide applicability is illustrated by the choice of different types of components ranging from large ceramic surface mount to small flip-chip assemblies on different types of substrates. For some of the application cases, results of parametric studies are presented and comparisons between failure prediction and testing results are made.

Key words: Solder fatigue, Finite Element simulations, solder plasticity, solder creep, primary and secondary creep, reliability predictions, coffin manson, board level test validation.

1. INTRODUCTION

The computational design of reliable microsystems, electronic packages as well as their interconnects can minimize expensive prototype development and testing. Accordingly, finite element (FE-) modelling is widely used to perform parametric studies on the thermo-mechanical behaviour of components like e.g., silicon microstructures, plastic packages, chip size/wafer level packages, or flip chip assemblies.

From a mechanical point of view all these structures include constituents, which are subjected to different loading conditions. The theoretical analysis of stresses within these constituents induced by environmental conditions requires the characterization of loads and material properties, respectively, as well as the knowledge of the appropriate failure criteria. Figure 1 provides an overview on the characteristic tasks to be performed for those thermo-mechanical finite element analyses (FEA).

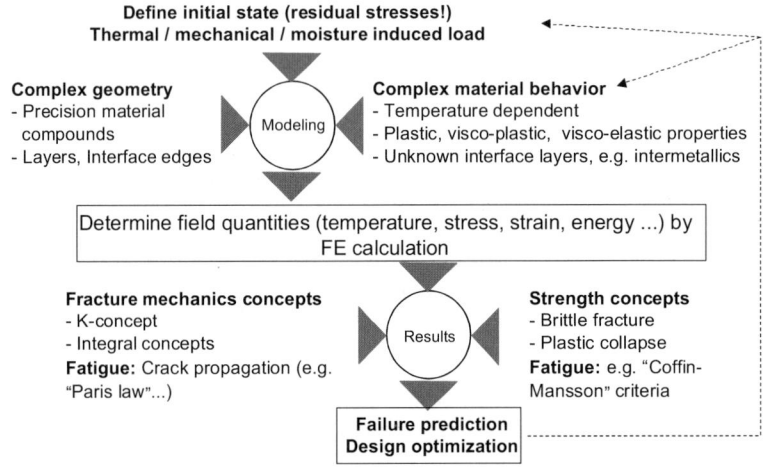

Figure 1. Finite element simulation flow chart for thermo-mechanical simulation

The difficulties of FEA for soldered assemblies are related to the definition of adequate geometric models for the constituents, exact specifications of loads and, particularly, the input of the strongly non-linear material properties and appropriate failure criteria to assess reliability from the non-linear analysis results. Because of the high complexity of the assemblies, e.g., packages mounted to printed circuit boards (PCBs), various modelling assumptions and simplifications are required. Neither the full geometry, nor the complete cyclic history can be considered in the calculations. Frequently, the definition of loading conditions is not straightforward, because stresses can be induced during the processing history. The constituents are usually stress-free at processing temperature, but chemical stress sources like e.g., migration, oxidation or polymerisation can occur even at this temperature. These residual stresses are typically overlaid by thermal mismatch induced stresses during cooling of the processing temperature or during environmental temperature changes. Within electronic polymers, which are typically parts of the packages, moisture swelling can induce subsequent stress states.

1.1 Low Cycle Fatigue Loading

Various sources of thermo-mechanical failure are possible at the board level. Mechanical loading caused by static PCB bending, high cyclic bending, drop down, and vibration can cause trace and PCB failures, failure at intermetallic layers, and sometimes also in bulk solder, especially if it is pre-damaged. Since one main failure source of the solder joints is thermal cyclic loading, induced by either environmental temperature changes or active power on/off cycling, the accelerated thermal cycling test is a standard procedure to evaluate product reliability. Accordingly, it is frequently used for FE-analyses of solder joint fatigue. In these analyses stress originates from testing conditions, mostly air-to-air thermal cycles. An example is shown in Figure 2. For simplicity the assumption of a homogeneous temperature change throughout the assembly is usually applied.

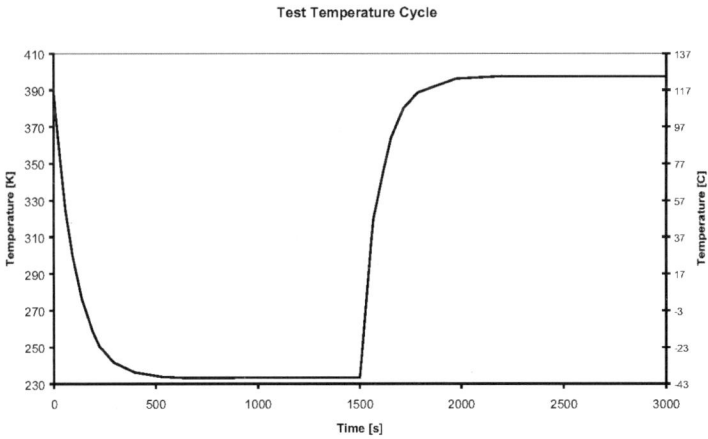

Figure 2. Definition of thermal boundary conditions for passive cycling (test cycle)

Transient thermal FE-calculations were performed to check this assumption for two packages on an FR-4 board, a Plastic Ball Grid Array 272 (PBGA 272) and a Chip Size Package (CSP) [1]. The temperatures in the PBGA at different locations with maximum temperature differences, the die centre and the encapsulant are shown in Figure 3 during cycling from

150°C to –40°C with 5 min ramps of the air temperature. The maximum temperature gradient between the die inner region and the outer edge of the encapsulant is approximately 5 K. This maximum temperature gradient is still lower for the CSP, where it does not exceed 1 K. It can be concluded that a homogeneous temperature change assumption is sufficient for the simulation of air-to-air thermal cycling at this kind of packages.

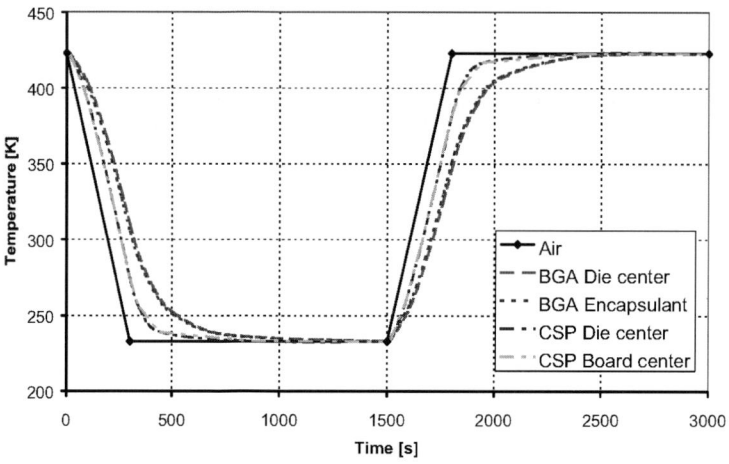

Figure 3. PBGA and CSP temperatures at different locations versus a linear air temperature cycle

In some cases the use conditions are dominated by passive heating, i.e., in automotive under-the-hood applications. In those investigations environmental temperature conditions have to be measured in dependence on time, the so-called "mission profile". In many of these slowly changing thermal cyclic loads local temperature gradients are negligible, too.

Thermal loading also can be dominated by active heating of the devices. The most complicated temperature loading conditions occur in cases when a cyclic temperature environment is overlaid by active heating, as depicted in Figure 4. These kinds of temperature loads as well as those where fast temperature changes or high thermal masses are involved require sequentially coupled thermal-mechanical analyses.

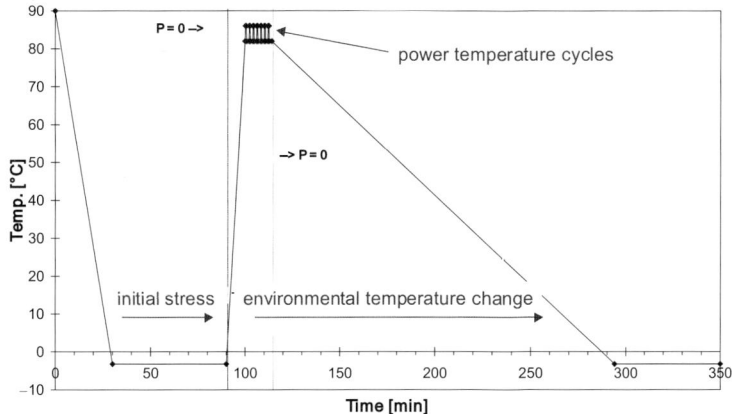

Figure 4. Definition of a service temperature cyclic environment

1.2 Thermally Induced Solder Joint Reliability

The prediction of solder joint reliability for low cycle fatigue loading has attracted attention for many years – compare the comprehensive publications [2-5]. Both, the more demanding environmental conditions, e.g., due to positioning of electronic devices under the hood in automotive applications, and the development of advanced interconnection technologies for surface mount (SM) and area array packages, e.g., BGAs and CSPs, have increased the interest in investigations of solder fatigue. Currently, various challenging questions are related to the introduction of lead-free solder materials.

The main source for thermal fatigue failure is the thermal expansion mismatch of the different materials soldered together, when the assemblies are subjected to a thermal cyclic environment already discussed above. The most damaging mismatch occurs usually between the component and the printed wiring board (PWB) the component is soldered to, also called "global mismatch". Another kind of mismatch is that between the component and the solder material itself, the so-called "local mismatch". Its importance depends on the size of the solder joints; for current joint sizes it is gaining importance only for cycle numbers greater than approximately

1000 of the characteristic temperature range –40°C to 125°C. The so-called "inner mismatch", occurring between different solder phases (i.e., tin rich and lead-rich phases), causes solder fatigue also within free-standing solder layers in the absence of any other material. However, for SnPb solders this kind of fatigue failure becomes important only for very high cycle numbers, i.e., higher than approximately 10,000. An example is depicted in Figure 5. Within tin based lead-free solders inner mismatch can be driven by the mismatch between the tin-rich matrix material and large rode-like or platelet-like intermetallic phases, which can gain importance at much lower cycle numbers, i.e., greater than approximately 1000 cycles. Additionally, Sn itself is anisotropic because of its crystal structure, which causes also different CTEs in the different spatial directions and, hence, additional inner mismatch.

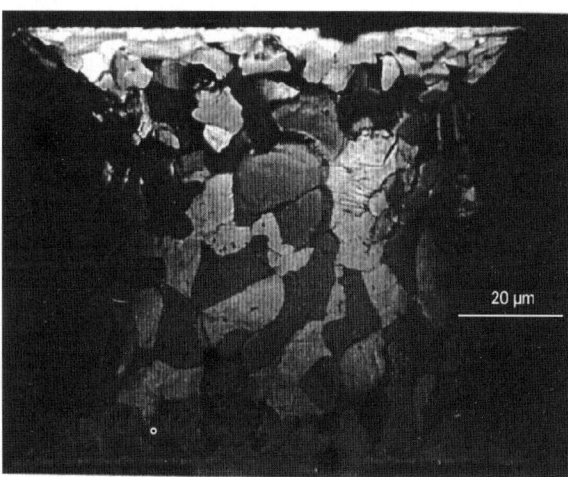

Figure 5. Solder fatigue in a free standing flip chip bump after 20,000 cycles –40 to 125°C (with permission S. Wiese, TU Dresden)

Two fundamental approaches are available to characterise the solder joint fatigue resistance. The first, which is applied to the majority of electronic products, is the thermal fatigue testing, i.e., the above mentioned thermal test cycling. To achieve comparable statements within reasonable time, various different test cycles have been defined. They operate usually with higher temperature cycling amplitudes and shorter cycling times compared to the

7. Characterization and Modelling of Solder Joint Reliability

use conditions, and are therefore called accelerated tests. It is important to note that cycling out of temperature range can engage inappropriate damage mechanisms. The occurrence of failure is detected either by electrical resistance measurements or by visual inspection of the solder joints with regard to solder cracking. Most often, electrical openings resulting from solder joint failures are intermittent and may be difficult to detect accurately. One example is broken flip chip solder bumps of underfilled assemblies, which are pressed together by the contracting underfill at room temperature and in this way pretend electrical contact. Whenever possible, on-line measurements should be preferred or off-line measurements should be made at elevated temperatures to avoid misleading results from the electrical measurements. Moreover, the latter method usually has to be combined with cross sectioning of the joints of interest.

The second approach for characterising solder fatigue resistance is based on a theoretical fatigue model. Two basic tasks have to be performed by the theoretical model: the first is to define a physical measure to indicate failure, e.g., inelastic strain or dissipated energy for appropriate strength theories, a damage parameter for damage theories, or a fracture toughness if a fracture mechanical approach is selected; the second task is to link this failure indicator to a critical cycle number. It should be noticed here that the failure indicator has not necessarily to be calculated by numerical means. Simplified analytical methods are given in the literature to estimate a critical cycles to failure number for simple solder joint geometries. In Figure 6 the different methodologies are compared to each other.

Experimental	Theoretical	
Temperature cycling	Approximative analytical models	Finite element calculation
- Real results (frequently micro-sectioning required)) - Soldering errors and component misalignment automatically included	- Low expense - Easy to handle - Limited explanations on fatigue influencing parameters	- Applicable for complex geometries - Solder material behavior approximatively included - Failure prediction increasingly certain - Well-suited for parametric studies and comparisons
- Time- and cost intensive - No results on failure sources and influencing parameters - Generalization difficult	- Strongly simplified geometry - Not applicable for complex geometries - Strong simplification of material behavior - Failure prediction frequently very uncertain	- High initial effort - Complicated Modeling - Currently still high calculation time - No simple algorithm for the evaluation of calculation results

Figure 6. Comparison of different means to evaluate solder fatigue

Building accuracy in a life prediction model is not an easy task. The acknowledged complexity, illustrated in Figure 7, arises from the following: The 3-dimensional package structures with solder joints are subjected to multi-axial non-linear material behaviours, complex joint shapes, and multi-axial loading. In a more specific sense the accuracy of life time predictions depends on the understanding and the consideration of the following problems:

- A solder joint is a multi-layered, non-homogeneous structure. For example, the standard eutectic SnPb solder consists of lead- and tin-rich phases with variations in composition.
- The initial microstructure can vary (e.g., grain size, intermetallics size and dispersion). The effects of this variation on the thermo-mechanical properties are not very well understood.
- Solder is not a stable material over time, since solder reacts with metallization to form interfacial intermetallics. Moreover, the Sn-Pb microstructure evolves in service. The microstructure coarsens due to thermally activated grain growth, a phenomenon that takes place under stress or at constant temperature.
- The effect of substrate finishes and component finishes on the reliability: soft solders react with metallizations to form interfacial intermetallics. Intermetallics grow with time and temperature. Metal consumption by intermetallic growth and intermetallics within the solder can be observed. Often the soldering process itself induces anomalies (e.g., changes in the reflow temperature profile, different cooling rates, excessive voids inside the solder, brittle phase formation, concentration gradients of elemental or metallurgical composition when the joint is formed). Smaller solder joints may impose increased interfacial effects and decreased fatigue-creep phenomena (joint volume versus interfacial effects).
- The mechanical behaviour of solder is non-linear and temperature dependent.
- Failure of solder joints is a complex sequence of possible failure mechanisms. For conventional SnPb solders, the sequence involves grain/phase coarsening, grain boundary sliding, matrix creep, micro-void formation and linking, and results in macro crack initiation and crack propagation. In the case of tin based lead free solder joints, the damage accumulation process leads to less well-known mechanisms with less coarsening of the microstructure.
- The failure criterion used for recording lifetime might vary, based on either mechanical cracks or electrical failure.

7. Characterization and Modelling of Solder Joint Reliability

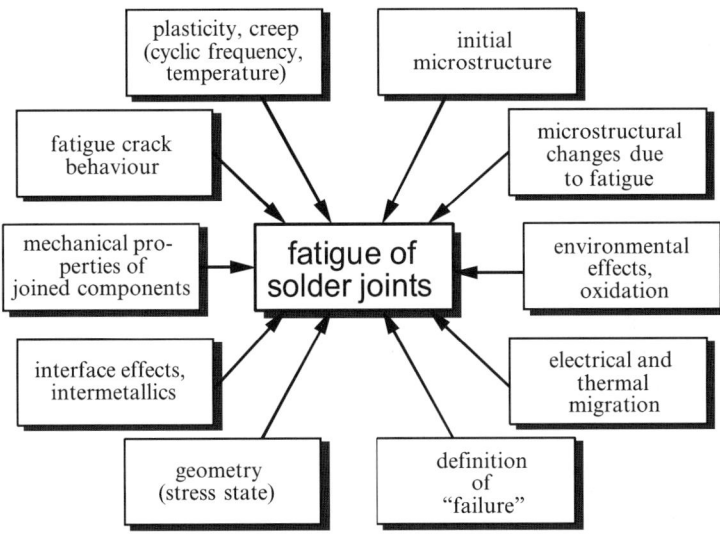

Figure 7. Fatigue of solders - complexity of the subject

2. ANALYTICAL-EMPIRICAL PROGNOSIS OF THE RELIABILITY

Analytical models are applicable in a limited number of cases. They have the advantage of linking important influential parameters with each other to allow easily the prediction of reliability trends. The easy use is, however, only possible due to the drastic simplifications with regard to geometry, loading and the material properties of the soldered joint.

Equation (1) shows an empirical model, confirmed by many investigations, for the prediction of a critical cycle number N_f of a material loaded with low-cycle fatigue: the Manson-Coffin relation [6] in simplified form, neglecting the elastic part of deformation.

$$(N_f)^c \Delta\varepsilon_{pl} = const. \quad or \quad N_f = \left(\frac{\Delta\varepsilon_{pl}}{\varepsilon_f}\right)^{\frac{1}{c}} \tag{1}$$

The analytical models for solder fatigue are generally based on this relation, i.e., they use the cyclic inelastic strain amplitude, or a similar equation for the strain energy dissipated in a cycle ("area of hysteresis loop"). Since it is not possible with elementary means to determine multi-axial

axial strain states or the inelastic parts of strain, the strain is calculated by assuming simplifications. In most cases the multiaxial strain state is simplified to a pure shear strain state with constant amplitude and all strain is assumed to be plastic strain. Regarding the rate and temperature dependencies of solder plasticity, experimentally determined correction parameters are introduced. The problem of the determination of the strain amplitude can only be solved for simplified geometrical relationships, as in Figure 7.

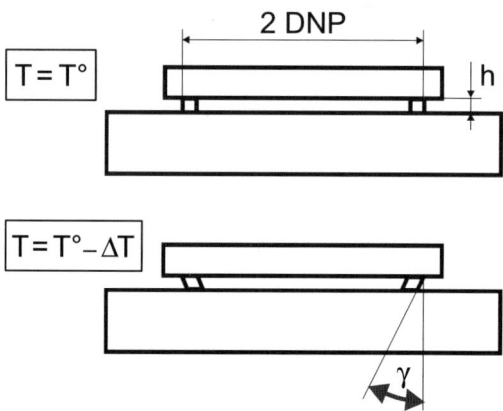

Figure 8. Schematic of a stiff leadless component on a substrate and its deformation due to cooling down

Neglecting any possible temperature gradients, as well as the three-dimensional character of the strain state (assumption of constant strain), the shear strain γ results in equation (2)

$$\gamma = \frac{DNP}{h}\Delta\alpha\Delta T, \quad \Delta\alpha = \alpha_S - \alpha_C, \quad \Delta T = T_{Max} - T_{Min} \qquad (2)$$

if $\Delta\alpha$ denotes the difference in the thermal coefficients of expansion of the substrate and component, ΔT the temperature amplitude of the cycle, DNP the distance to the neutral point, and h the joint height. It is obvious that a situation "stiff component on stiff board", where almost no local deformation or bending occurs, can be described reasonably well by the formula. In such a case, joint straining is determined only by the geometry and does not depend on the constitutive behavior of the joint material. The primary parameters for joint straining can be easily observed:
- Cyclic temperature difference
- Size of the component

- Thermal expansion difference
- Standoff height

These parameters are without doubt important ones, even in cases with more complicated geometries. Depending on the real geometry of the soldered component under investigation, the model geometry is a simplification to a greater or lesser degree. However, most real geometries cannot, not even approximately, be represented by the model. Many effects like deformation of board and component, multiaxial distribution of strain, the part of the strain resulting from local mismatch, and the complicated inelastic character of solder deformation can hardly be included in a similar model.

One model which is based on the described assumptions for the calculated strain is that of Engelmaier [7]. In the beginning the model was correlated to experimental data achieved by Wild [8] for eutectic SnPb solder, to which all the constants refer. The model reads in its basic form:

$$N_f(x\%) = \frac{1}{2}\left(\frac{F_k \Delta\gamma}{2\varepsilon^f}\right)^{\frac{1}{c}} \left(\frac{\ln(1-0.01x)}{\ln(0.5)}\right)^{\frac{1}{\beta}}$$

ε^f ...fatigue strain coefficient ($2\varepsilon^f \approx 0.65$ for Sn63Pb37)

$c = -0.442 - 6 \cdot 10^{-4} \overline{T}^S + 1.74 \cdot 10^{-2} \ln(1+f)$

$\overline{T}^S$ ...mean solder joint temperature in °C (3)

f ...cycle frequency in cycles/day

F_k ...correction factor ($0.7 < F_k < 1.5$)

β ...Weibull-slope ≈ 2 to 4

where the critical cycle number N_f is obtained with a probability of x% failure.

The model has been extended to many different applications by inclusion of corrective factors [9]. However, the predictive capabilities remain restricted to simple geometries and should be closely related to testing results.

An advanced analytical model for SM-components, especially BGAs, was proposed by Clech [10, 11], which runs on a PC. The model uses a strain-energy based fatigue law that was developed by correlating SMT fatigue data from 19 experiments. Inelastic strain energy is obtained as the area of solder joint stress/strain loops during thermal cycling. The solder constitutive model includes temperature-dependent creep and plastic flow.

During the last decade the rapid development of cheep and powerful computer technology has shifted the focus of reliability analysis of solder joints towards the use of FEA. However, the analytical models, two of them cited above, can be very useful in cases where a component with simple geometry varies in size etc. The model predictions should be complemented by testing results, at least for selected verification cases.

It was already mentioned that the thermal cyclic environment can vary, and that the so-called test cycles and field cycles are of special interest. The relation between the critical cycle numbers for both cycles, the so called "acceleration factor" can also be evaluated by analytical means. Again, the predictions are based in principle on the relation (1) and are developed by using thermal fatigue test data. It should be noted that these empirical relations are all related to the conventional Sn-Pb solder, and that no similar equations for lead-free materials are known yet.

In the most simplified case, when it is assumed that (1) holds true, all deformation converts to plastic joint strain, independent of the thermal cycle form, and c has the characteristic value –0.5. The relation between the critical number of field cycles N_{field} and test cycles N_{test} can then be calculated by the simple equation

$$N_{field} = N_{test} \left(\frac{\Delta T_{test}}{\Delta T_{field}} \right)^2 \tag{4}$$

which was given by de Kluizenaar [12].

An improved experimentally supported model for determining the acceleration factor N_{test}/N_{field} was developed by Norris and Landzberg [13]:

$$N_{field(x)} = N_{test(x)} \left(\frac{\Delta T_{test}}{\Delta T_{field}} \right)^{c1} \left(\frac{f_{field}}{f_{test}} \right)^{\frac{1}{3}} \exp\left[1414 \left(\frac{1}{T_{field}} - \frac{1}{T_{test}} \right) \right] \tag{5}$$

$N_{field(x)}$... Number of field cycles at which a percentage x of connections has failed, for example, $N_{field(50)}$ = mean cycles to failure at which 50% of connections have failed.

$N_{test(x)}$... analogous number of test cycles

c1 ... empirical constant, applicable is c1 = 1.9

f ... cycling frequency expressed in number of cycles/day

T ... maximum cycling temperature in K

The equation, which includes the dependence on the cyclic frequency as well as the maximum cyclic temperatures could be verified by experimental data for flip chip connections.

Finally, the relation of Engelmaier is given, which follows directly from the application of the relation (3) of test and field cyclic conditions:

$$\overline{N}_{field} = \overline{N}_{test} \frac{(\Delta\gamma_{field})^{\frac{1}{c_{field}}}}{(\Delta\gamma_{test})^{\frac{1}{c_{test}}}} \qquad (6)$$

3. THERMO-MECHANICAL CHARACTERISTICS OF SOFT SOLDERS

A variety of different solders are used in the electronic industry. The traditionally applied alloys typically contain lead, which is currently going to be replaced by other materials because of its toxic nature. But the solder still typical for electronic applications is the near-eutectic Sn60Pb40 alloy, frequently with 1–2% Ag. Worldwide, lead-free solders are implemented by the electronics industry. The properties-metallurgy dependencies of SnPb solders have been studied for decades and have turned out to be surprisingly complex. Much less is known about the lead-free alloys. In this area many research projects are still ongoing. However, all solder alloys under consideration are at or above 0.5 of their melting point at the typical minimum cyclic temperature of –40°C, see Table 1, so creep processes are expected to dominate the deformation kinetics.

Table 1. Basic characteristics of some solder alloys

Alloys based on	Composition (mass %)	Melting temperature (°C)
Sn-Pb	SnPb40	183
Sn-Ag-Cu	SnAg3.8Cu0.7	217
	SnAg3.8Cu0.7Sb0.25	
	SnAg3.8Cu0.7Sb0.5	
Sn-Ag	SnAg3.5	221
	SnAg3.5Sb0.25	
	SnAg3.5Sb0.5	
	SnAg3.5Ni0.05	
	SnAg5	
Sn-Cu	SnCu0.7	227
	SnCu0.7Sb0.25	
	SnCu0.7Sb0.5	
	SnCu0.7Ni0.05	
	SnCu0.7Bi0.5	
Sn-Bi-Ag	SnBi5Ag1Sb2	< 180–225

The preferred candidates for SnPb replacement are Sn based alloys. Currently, the ongoing efforts seem to be converging on SnAgCu ternary eutectic.

The mechanical behaviour of solder is non-linear and temperature dependent. Tensile tests on bulk specimen with 60 mm length and 3 mm diameter, shown in Figure 9, indicate also the dependence on the loading velocity. For measurements at room temperature this dependence is depicted in Figures 10 to 12, pointing to the creep properties of the investigated solder alloys.

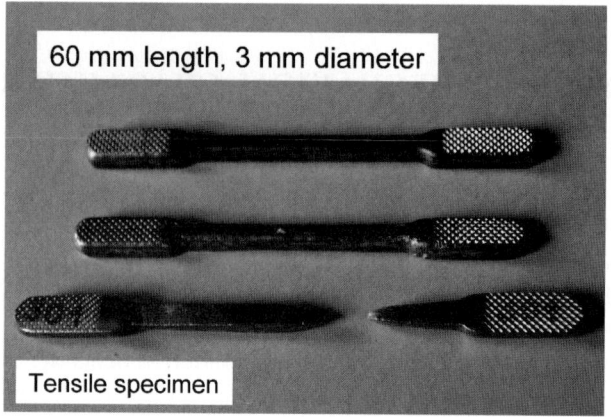

Figure 9. Tensile specimen made of bulk solder used for creep tests

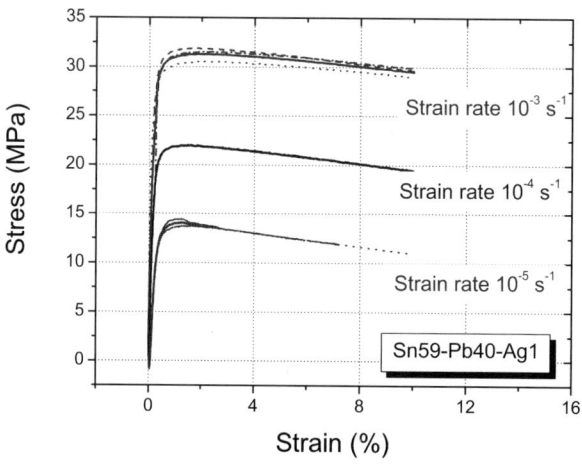

Figure 10. Stress-strain curves showing the effect of different strain rates for Sn59Pb40Ag1 solder at T = 25°C

7. Characterization and Modelling of Solder Joint Reliability

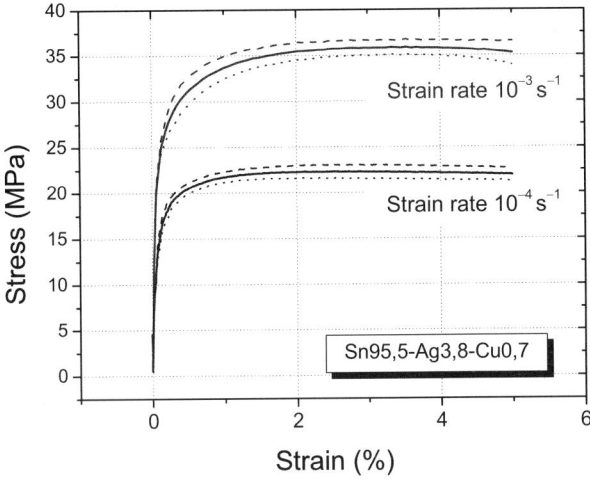

Figure 11. Stress-strain curves showing the effect of different strain rates for Sn95.5Ag3.8Cu0.7 solder at T = 25°C

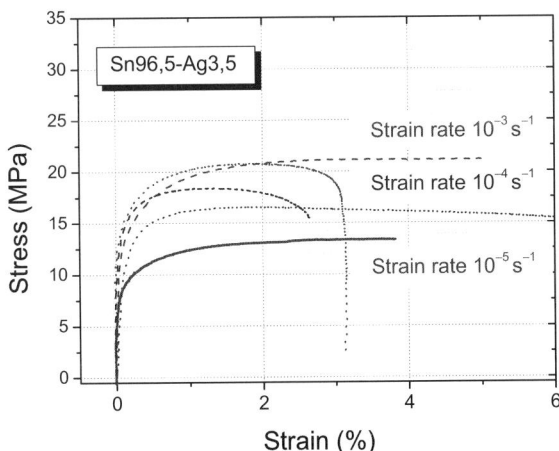

Figure 12. Stress-strain curves showing the effect of different strain rates for Sn96.5Ag3.5 solder at T = 25°C

Since the different thermal expansions are the main source for thermal fatigue failure, the CTEs of the solder alloys are of interest. Table 2 shows the data measured by TMA.

Table 2. Coefficients of thermal expansion (CTE) for different solder alloys

Solder	CTE [10^{-6} ppm/K^{-1}]		
	20 °C – 150 °C	20 °C – 100 °C	100 °C – 150 °C
Sn 59 Pb40 Ag1	25.1	24.0	26.9
Pb93.5 Sn5 Ag1.5	28.7	27.7	30.1
Sn 96.5 Ag 3.5	20.2–21.7	19.4–20.4	21.5–22.9
Sn 95.5 Ag3.8 Cu 0.7	17.6–21.0	16.7–20.4	18.8–22.0

When the constitutive model is considered, deformation can be split in elastic, plastic and creep parts, i.e.,

$$\frac{d\varepsilon}{dt} = \frac{d\varepsilon_{el}}{dt} + \frac{d\varepsilon_{cr}}{dt} + \frac{d\varepsilon_{pl}}{dt} \qquad (7)$$

where $\frac{d\varepsilon}{dt}$ denote the total strain rate and the other terms in (7) the elastic, creep, and plastic strain rates corresponding to the mentioned character of deformation.

These individual strain rates depend in different ways on the temperature and on the microstructural state of the material. A further, up to now little investigated dependency should also be on the loading trajectory, e.g., different properties in tension and compression, as were reported in [14]. Additionally, viscoelastic parts of the strain rate have been observed. The linear recoverable, elastic response can be described by Hooke's law. For the isotropic case there are two constants required for this law: the Young's modulus depicted in Figure 13 for different solders, determined by measurements and investigations of others [15–18], and Poisson's ratio, which is generally approximately 0.35. The measuring method is often different for the Young's modulus determination: tensile testing, DMA or ultrasonic pulse methods. To eliminate the effect of inelastic deformation and to get more accurate values, ultrasonic methods should be preferred.

The Young's moduli of the Sn based lead-free alloys seem to be substantially higher than that of eutectic Sn-Pb.

7. Characterization and Modelling of Solder Joint Reliability

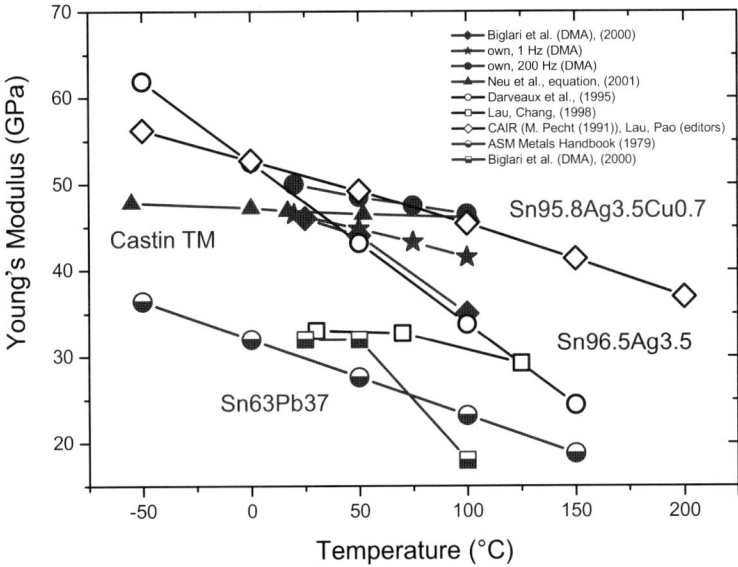

Figure 13. Young's modulus in dependence on temperature for different solder alloys

At high stresses, t/G > 10-3, there is also a time-independent plastic strain component to the deformation. A strain-hardening law can be used to describe high-stress deformation. Darveaux [17] proposed the following plastic flow or strain hardening law to include the plastic strain ε_{pl}:

$$\varepsilon_{pl} = C_t \left(\frac{\sigma}{G(T)} \right)^m \tag{8}$$

But it should be noted that, depending on the constitutive description used, all inelastic non-recoverable response could also be treated in a combined, time-dependent manner, at least for low cycle fatigue type loads.

Since the velocity dependent properties are the most critical parameters required for accurate stress analysis, which will be used in damage analysis to predict the fatigue life, the most challenging question is to describe solder creep.

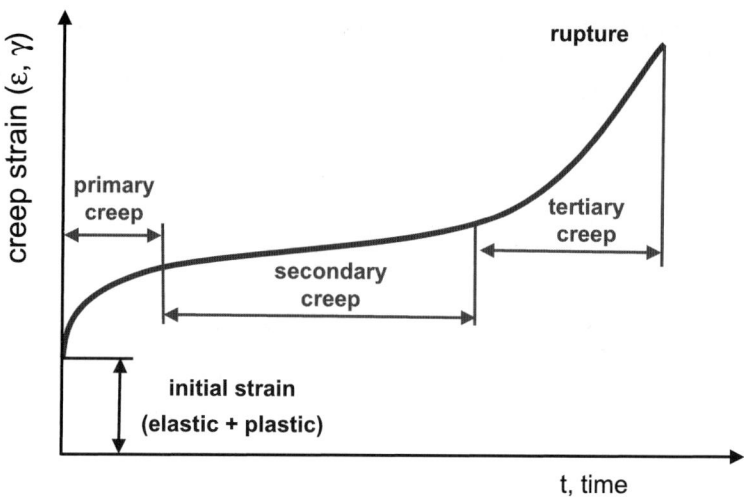

Figure 14. Creep Curve: strain vs. time under constant stress (or load) and temperature

Typical processes of high-temperature creep in metals under constant stress can be modelled in four phases, see Figure 14. In the first phase after application of the load, elastic and time-independent, non-elastic strains dominate. In the second phase, which starts immediately, creep occurs with decreasing creep rate (primary creep), which in the third phase transforms to a creep process with constant creep rate (secondary creep), while in the fourth phase the creep rate increases rapidly until fracture of the material. Despite complications due to the multiphase structure of soft solders, similar behaviour has also been shown for these materials. The creep response of solder materials relevant for low cycle fatigue loading is limited to primary and secondary creep. The creep strain rate in equation (7) is therefore split in a primary $\frac{d\varepsilon_p}{dt}$ and a secondary $\frac{d\varepsilon_s}{dt}$ part:

$$\frac{d\varepsilon_{cr}}{dt} = \frac{d\varepsilon_s}{dt} + \frac{d\varepsilon_p}{dt} \qquad (9)$$

However, during the usual thermal cycle regime with relatively slow temperature ramping rates discussed above, solder constitutive response is dominated by secondary creep. Accordingly, the basic non-linear constitutive model for solders is limited to a viscoplastic model, which includes an elastic and a secondary creep part. Due to its importance, various investigations on

7. Characterization and Modelling of Solder Joint Reliability

the determination of the secondary creep properties of solders, preliminarily for the Sn-Pb solders, have been performed.

The secondary creep deformation kinetics of soft solders can be divided into different stress dependent regimes, which are controlled by different dominating creep mechanisms [19], see Figure 15.

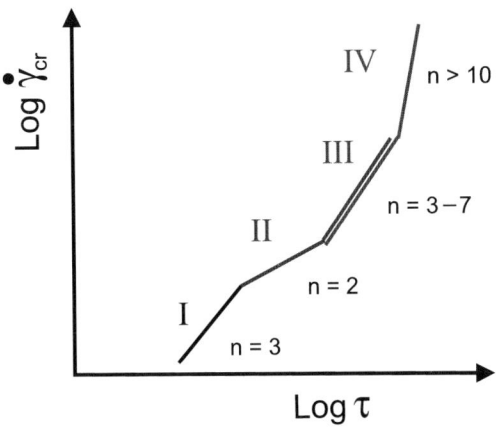

Figure 15. Schematic of phases of secondary creep deformation for eutectic Pb-Sn alloys at $T/T_M > 0.5$

The mathematical descriptions of the steady-state creep rate are based on the semi-empirical equation of Weertman-Dorn [20, 21]:

$$\dot{\gamma}_{cr} = A \frac{\mu b}{kT} \left(\frac{\tau}{\mu}\right)^n \left(\frac{b}{d}\right)^p D_0 \exp(-\frac{Q}{kT}) \qquad (10)$$

$\dot{\gamma}_{cr}$ – creep shear rate

D – appropriate diffusion coefficient
b – burgers vector
d – grain size (interphase spacing)
n – stress exponent
A – constant

μ – shear modulus
t – applied shear stress
k – Boltzmann's constant
p – grain size exponent
T – absolute temperature

3.1 Eutectic Sn-Pb-(Ag) Solder

Grivas et al. [19] suggested that a fine-grained (fine-phased) Sn-Pb-(Ag) eutectic alloy exhibited both superplastic deformation and matrix creep deformation. It was assumed that both deformation modes occur simultaneously via independent mechanisms. Superplastic deformation is rate controlling at "lower" stresses (region II), also referred to as grain boundary sliding deformation, and matrix creep deformation is rate controlling at "intermediate" stresses (region III). Combining the rates of the two deformation mechanisms, a semi-empirical equation was developed summing up the strain rates in both stress regions, so that

$$\frac{d\varepsilon_s}{dt} = \frac{A_{II}}{T} \sigma^{n_{II}} \exp\left(\frac{-Q_{a,II}}{kT}\right) + \frac{A_{III}}{T} \sigma^{n_{III}} \exp\left(\frac{-Q_{a,III}}{kT}\right) \qquad (11)$$

where $\frac{d\varepsilon_s}{dt}$ is the steady-state creep strain rate, k is the Boltzmann's constant, T is the absolute temperature, σ is the applied (tensile) stress, n_{II} and n_{III} are the stress exponents, and A_{II} and A_{III} are constants. Neglecting the dependence of Young's modulus on temperature, the activation energies are apparent ones.

These parameters were measured by tensile creep tests for Sn59-Pb40-Ag1 solder on bulk specimens, see Figure 8, as given in Table 3. Details of the testing procedure were reported by Schubert et al. [22].

Table 3. Deformation constants of the steady-state creep for Sn59Pb40Ag1, based on own measurements (equation (11))

Region	A (K s⁻¹)	n	Q_a (eV)
II	426.95	1.87	0.50
III	94.36	6.74	0.79

Many authors have reported analogous data, measured on a variety of different specimens ranging from bulk to joint size. A summary of some of the measurements, shown in Table 4, was given by Knecht et al. [23].

7. Characterization and Modelling of Solder Joint Reliability

Table 4. Sn-Pb solder alloy deformation constants given in different sources

	Q_{II} (eV)	n_{II}	Q_{III} (eV)	n_{III}
Grivas et al., 1979	0.5	2.0	0.84	7.1
Kashyap et al., 1981	0.46	1.7	1.0	11.1
Solomon, 1986/1987	–	3.1	–	7.3
Baker, 1979	0.69	3.4		
Lam et al., 1979	0.45	2.4	0.9	5.7
Wong et al., 1988	0.47	3.0	0.47	7.0

The description given by Grivas et al. [19] and Hacke et al. [24, 25]

$$\frac{d\gamma_s}{dt} = \frac{1.39 \; 10^{-6}}{T} \frac{\tau^{1.96}}{d^{1.8}} \exp\left(\frac{-48157}{RT}\right) + \frac{2.38 \; 10^3}{T} \tau^{7.1} \exp\left(\frac{-81239}{RT}\right) \tag{12}$$

contains an additional microstructural parameter d, which denotes the interphase spacing between tin and lead-rich phases. R = 8.314 Jmol^{-1}K^{-1} is the universal gas constant and T, as always, the absolute temperature.

A similar equation was given by Knecht, Fox [23, 26]:

$$\frac{d\gamma_s}{dt} = 8.31 \; \tau^2 \exp\left(\frac{-0.5}{kT}\right) + 1.12 \; \tau^{7.1} \exp\left(\frac{-0.84}{kT}\right) \tag{13}$$

Assuming isotropic material properties, the relations

$$\sigma = \tau\sqrt{3} \qquad \varepsilon = \frac{1}{\sqrt{3}}\gamma \tag{14}$$

can be applied to convert the shear and tensile (effective) stress and strain components.

The steady-state creep of solders can also be described by a hyperbolic sine stress function, the so-called "Garofalo law". At "intermediate" stresses (region III), the strain rate depends on stress to the power, n. At "high" stresses (region IV), the strain rate is an exponential function of stress. This power law break-down region can be described by the simple expression

$$\frac{d\varepsilon_s}{dt} = C_{III,IV} \left[\sinh(\alpha\sigma)\right]^n \exp\left(\frac{-Q_a}{kT}\right) \qquad (15)$$

where α prescribes the stress level at which the power law dependence breaks down. Q_a is the apparent activation energy and $C_{III,IV}$ is a constant.

The resulting description of the secondary creep strain rate dependent on the applied stress can be fitted to a very similar form as given in equation (11). However, for applications in FE codes this description is more susceptible to numerical instabilities because of the exponential growth of creep strain rate at high stress.

Darveaux et al. [17, 27] has used this kind of relation to describe various steady state creep rates for different solder alloys. Shear measurements were performed at solder joints of area-array ceramic chip carriers mounted on identical CCCs. The CCC on CCC assemblies had been aged for several months at RT or 100 hours at 100°C. For Sn62Pb36Ag2 the description given in the derived tensile form reads:

$$\frac{d\varepsilon_s}{dt} = 8.03*10^4 \left[\sinh(6.7*10^{-2}\sigma)\right]^{3.3} \exp\left(\frac{-0.7}{kT}\right) \qquad (16)$$

Figure 16 shows the graphs resulting from the different descriptions cited above for the temperature T = 293 K in a log-log plot. It is obvious that the creep rates differ. That means, even in the case of frequently investigated SnPb, the scatter among published data is on the order of several orders of magnitude. The lesson learned from these investigations is that there is no unique constitutive model for solder joints, hence the variety of models available throughout the literature. In the end, the applicability of a given model to real life assemblies and a reasonable agreement between the ensuing life predictions and test results determine whether a constitutive model will be of use to design engineers and reliability analysts. The variations in published data can be explained by differences in test methods, specimen preparation, and the problem in the nature of durability testing.

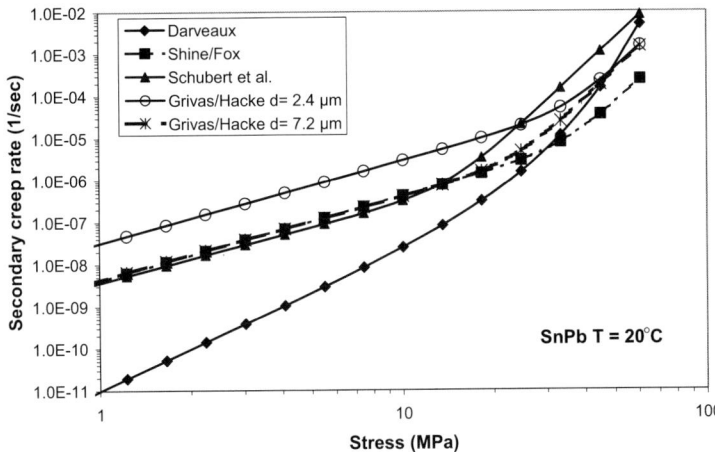

Figure 16. Secondary creep strain rate versus tensile stress for eutectic SnPb alloys at room temperature given by different researchers

3.2 Tin-Based Lead Free Solders

The alloy composition and microstructure of the tin based lead-free solders is dramatically different from the eutectic Sn-Pb-(Ag) solder. In accordance with these different microstructures differences are also expected in the creep behavior.

Tensile creep tests on bulk Sn-Ag and Sn-Ag-Cu solder specimens have been performed [22, 28] analogous to the Sn-Pb samples discussed above.

The measuring results for Sn95.5-Ag3.8-Cu0.7 were compared to similar results of other investigations [16, 29, 30] and a reasonable agreement was found. Averaging of all these data, shown in Figure 17, results in a hyperbolic sine stress function for the steady-state creep rate (15) with constants are given in Table 5.

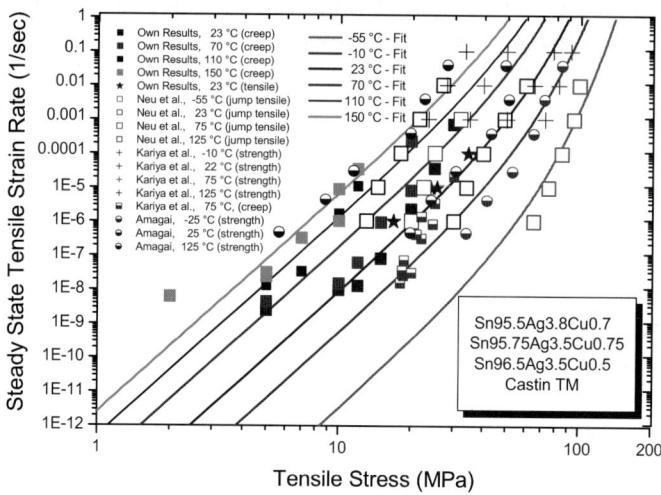

Figure 17. Curve-fitting of Sn-Ag-Cu creep data to hyperbolic sine model (equation (15)), based on 108 data points from own measurements and literature

Table 5. Deformation constants of the steady-state creep for SnAgCu-alloys, based on 108 data points from own measurements and literature (equation 15)

Sn95.5Ag3.8Cu0.7 Sn95.75Ag3.5Cu0.75 Sn96.5Ag3.5Cu0.5 Castin^{TM}	$C_{III,IV}$	277984 s^{-1}
	α	0.02447 MPa
	n	6.41
	Q_a	0.56 eV

A variety of other measuring results were reported in the literature for secondary creep of lead free solders. Wiese et al. [31] have investigated the microstructural dependence of secondary creep for tin based materials. Creep experiments for SnAgCu were performed for bulk material, for solder joints in a through-hole wire-PCB connection, and for flip chip bumps. A power law description analogous to equation (11) is proposed for the secondary creep rate, which for the solder joint is given by:

$$\frac{d\varepsilon_s}{dt} = 4*10^{-7} \sigma^3 \exp\left(\frac{-26.8*10^3}{RT}\right) + 1*10^{-12} \sigma^{12} \exp\left(\frac{-61.4*10^3}{RT}\right) \quad (17)$$

7. Characterization and Modelling of Solder Joint Reliability

This data agrees also with that measured for bulk material, only for the flip chip bump a much lower creep stain rate is reported.

The resulting graphs are shown in Figures 18-19 together with the data from from Lau et al. [32], based on measurements from Vianco and Rejent [33], and Zhang et al. [34].

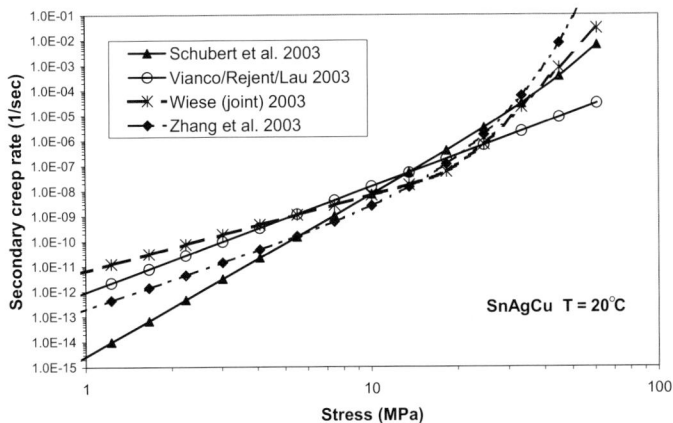

Figure 18. Secondary creep strain rate versus tensile stress for eutectic SnAgCu alloy at room temperature given by different researchers

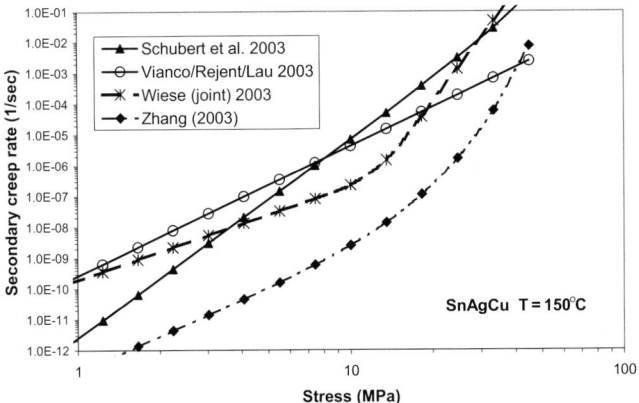

Figure 19. Secondary creep strain rate versus tensile stress for eutectic SnAgCu alloy at 150°C given by different researchers

Again, the scatter among the data is on the order of several orders of magnitude. When comparing different creep data of bulk and joint material it seems that the data scatter due to sample preparation, test methods, alloy composition and microstructure is more pronounced as due to real geometrical size effects.

The third solder treated here is eutectic tin-silver solder Sn96.5-Ag3.5.

A similar procedure of measuring bulk data and averaging with other published data, see Figure 20, has lead to the coefficients given in Table 6 for the hyperbolic sine model already described by equation (15).

Table 6. Deformation constants of the steady-state creep for SnAg-alloys, based on data points from own measurements and literature (equation (15))

	Sn96.5 Ag3.5
$C_{III,IV}$	4712.69
α	0.06354
n	4.704
Q_a	0.593 eV

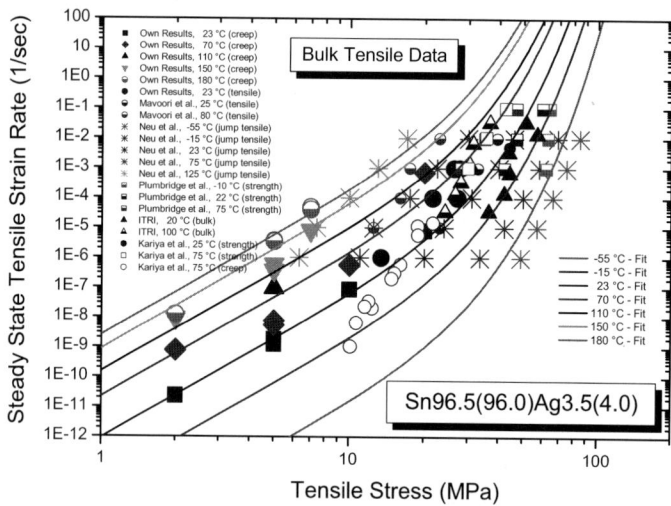

Figure 20. Curve-fitting of Sn-Ag creep data to hyperbolic sine model (equation (15)), based on 100 data points from own measurements and literature

A secondary creep law (18) was reported by Darveaux/Banerji [27], which again had been obtained by shear tests for solder joints of area-array

7. Characterization and Modelling of Solder Joint Reliability

ceramic chip carriers mounted on identical CCCs. The constants for equation (18) are given in Table 7.

$$\frac{d\gamma_s}{dt} = C_4 \frac{G(T)}{T} \left[\sinh\left(\alpha \frac{\tau}{G(T)}\right)\right]^n \exp\left(\frac{-Q}{kT}\right) \quad (18)$$

Table 7. Deformation constants of the steady-state creep for SnAg described by (18)

	Sn96.5 Ag3.5 (CCC solder joints)
C4	0.454 K/s/MPa
α	1500
n	5.5
Q	0.50 eV
G	17.437 GPa (300 K)
	13.780 GPa (353 K)
	10.192 GPa (405 K)

Clech and Handwerker report in [33] their results obtained by curve-fitting of Sn96.5Ag3.5 lap shear and ring & plug creep data to hyperbolic sine model data. The data, a total of 51 points measured at temperatures T = 20°C to 158°C, lead to the following result for the secondary creep shear rate:

$$\frac{d\gamma_s}{dt} = A \left[\sinh(\alpha\tau)\right]^n \exp\left(\frac{-Q_a}{kT}\right) \quad (19)$$

Table 8. Deformation constants of the steady-state creep for SnAg described by (19)

	Sn96.5 Ag3.5 (lap shear and ring & plug)
A	8.179 E11
α	0.0266
n	8.67
Q_a	0.80 eV

From bulk tests it turned out that Sn96.5Ag3.5 behaves like an uneven material with higher strength or more creep resistance in compression than in tension. This behaviour points to anisotropic properties of solder material, which is not yet studied in further detail.

The resulting graphs are shown in Figures 21–22 together with the data from Wiese et al. [31].

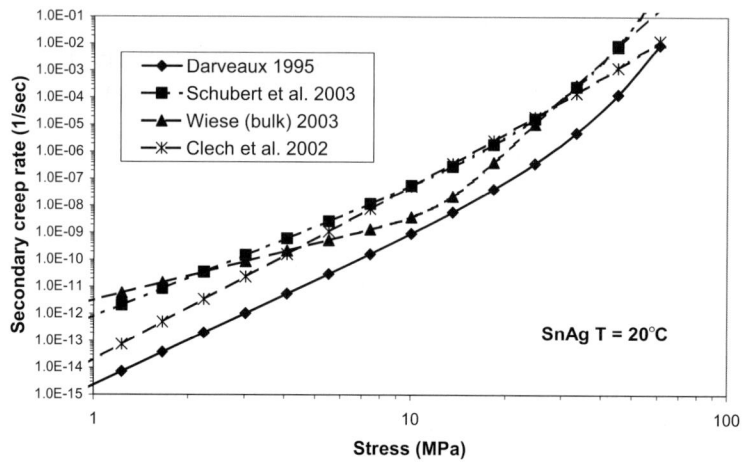

Figure 21. Secondary creep strain rate versus tensile stress for eutectic SnAg alloy at room temperature given by different researchers

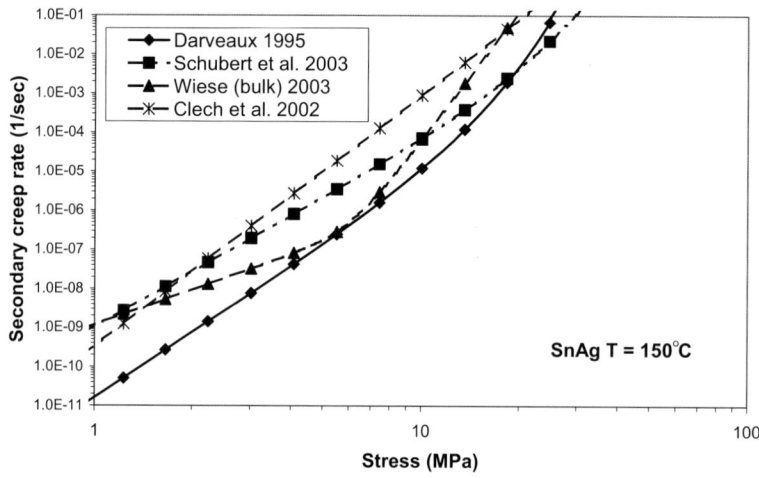

Figure 22. Secondary creep strain rate versus tensile stress for eutectic SnAg alloy at 150°C given by different researchers

7. Characterization and Modelling of Solder Joint Reliability 405

3.3 Discussion on the Solder Creep Characteristics

It can be seen from the data presented that the secondary creep rates given by different researchers can differ by several orders of magnitude. That means, even in the case of frequently investigated SnPb, no unique secondary creep law seems to exist. Some variations in published data can possibly be explained by differences in test methods, specimen preparation or measuring errors, but they point nonetheless to the phenomenon of the dependencies of solder creep properties on microstructure, which seem to be even stronger for lead-free solders than for SnPb solders.

In addition to the publications cited above, a variety of studies have been published concerning creep properties of solders. Since several projects on the behaviour of lead-free solders are still ongoing, these publications will continue and increase the understanding of open questions like the dependence of creep properties on the microstructure. Currently, comprehensive sources are papers of Clech (e.g., [14, 35]), Pang (e.g., [36], [37]), and the NIST database [38]. New obstacle-controlled creep models for various solders have been proposed by Clech [39].

Finally, the secondary creep properties of the three solder alloys discussed so far are compared to each other. The secondary creep laws used for this comparison, shown in Figures 23–25, are those of Schubert et al. [40]. It can be seen from the figures that the deformation of the three solders is a different function of stress, strain rate, and temperature. The SnAgCu solder shows generally a somewhat higher deformation resistance than the SnAg solder. This behaviour can be related to the strengthening mechanism of Sn-based alloys, which can be seen in the tiny dispersed intermetallic particles embedded in the inherently soft ß-Sn matrix, mainly composed of AgSn3 and Cu6Sn5. The two lead free materials are more creep resistant than SnPb at high temperatures, due to their higher melting point. They are also more creep resistant at low stresses. However, for high stresses, the superior creep resistance disappears, especially for temperatures at and below room temperature. A similar trend was reported by Zhang et al. [34]. He pointed out that the soft lead-rich phases at Sn boundaries facilitate grain boundary sliding (superplasticity) for low stresses in SnPb solder, whereas the tiny intermetallic Ag3Sn and Cu6Sn5 particles strengthen pure Sn grain boundaries and block grain boundary sliding in SnAgCu solder. In the high stress region matrix creep controls the deformation kinetics, which is not that much affected by the different microstructures. This could explain the observed trend.

It is concluded that in real joints the solder performance should be dependent on the temperature and induced stresses, i.e., solders perform differently for different package types. However, it is noticed that for the idealized assembly "soft joint between rigid plates" described above, the imposed cyclic strain does not depend on the creep behaviour of the joint material. Thus, the effect of the creep behaviours depends on whether there is a more "strain controlled" or "stress controlled" situation.

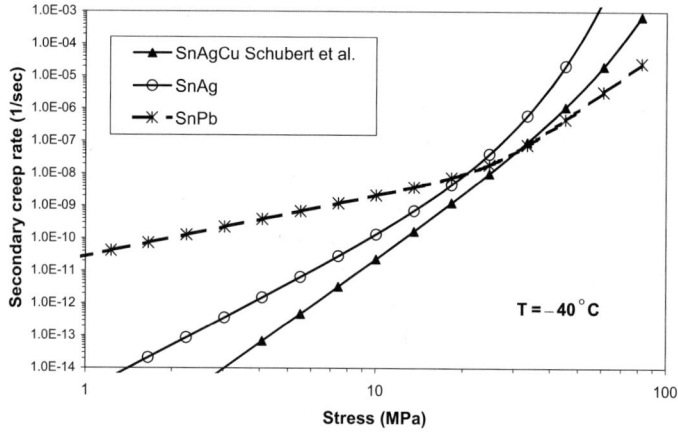

Figure 23. Secondary creep strain rate versus tensile stress for SnPb, SnAgCu, and SnAg at −40°C given by Schubert et al.

7. Characterization and Modelling of Solder Joint Reliability 407

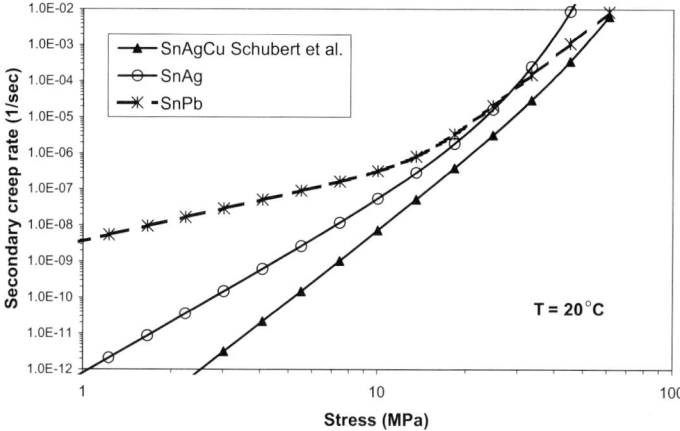

Figure 24. Secondary creep strain rate versus tensile stress for SnPb, SnAgCu, and SnAg at 20°C given by Schubert et al.

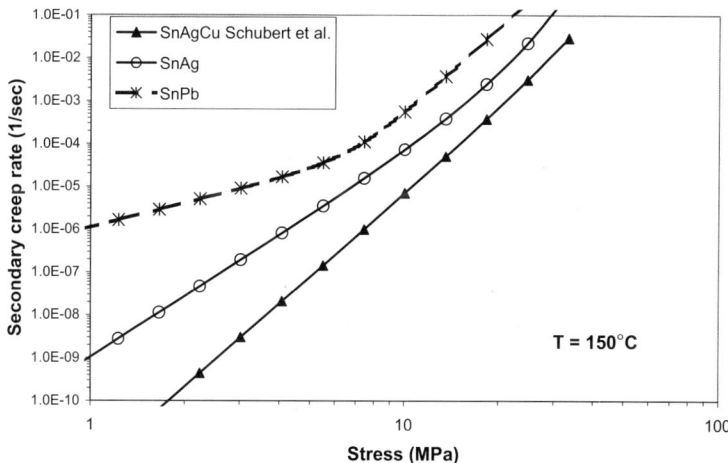

Figure 25. Secondary creep strain rate versus tensile stress for SnPb, SnAgCu, and SnAg at 150°C given by Schubert et al.

3.4 Primary Creep

It has been already mentioned that steady-state creep is not observed immediately when stress is applied (Figure 2). Transient (or primary) creep occurs before attaining steady state. Thus, primary creep may not be negligible in applications with high temperature ramp rate or under thermal cycling conditions with short dwell times. In general, there are two questions related to the relevance of primary creep: one is after the sensibility of the actual assembly towards the creep law, i.e., if there is a more stress-controlled or a more strain controlled situation. The second one is after the cyclic conditions, which should be evaluated numerically, i.e., do they activate a similar amount of primary creep or not? The latter case occurs if there are greatly differing temperature profiles, e.g., thermal shock test cycles and slow field cycles.

A simple implementation for primary solder creep was proposed in [40] and is referred to in this chapter. However, ongoing investigations have detected the saturated primary creep, which was found to be constant in the first investigations, to depend on the secondary creep rate [41]. Additionally, the adoption of the creep model measured at constant stress and temperature to the transient character of the cyclic stress built-up can be implemented, as also published in [41].

For normal decelerating transient creep, the strain rate starts high and decreases to the steady state value. The creep composed of a primary and secondary part at constant stress and temperature can be described by the equation

$$\varepsilon_{cr} = \varepsilon_p^{sat}\left\{1 - \exp\left[-K\left(\frac{d\varepsilon_s}{dt}\right)^n\right]\right\} + \frac{d\varepsilon_s}{dt}t \qquad (20)$$

where ε_p^{sat}, K, and n are constants [42, 43]. The saturated transient creep strain can be obtained by extrapolating the steady-state creep line back to the starting point.

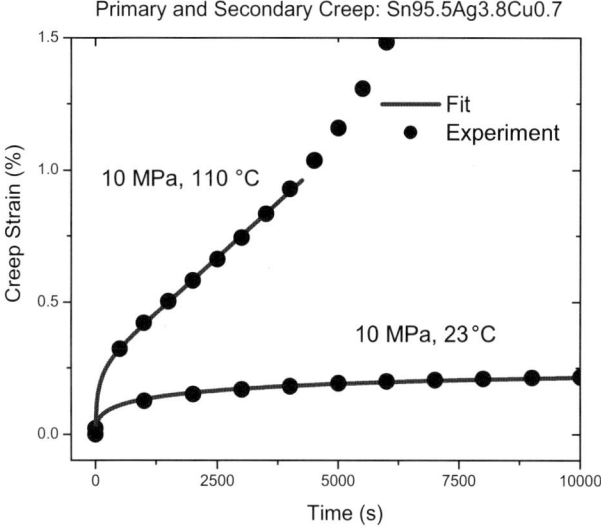

Figure 26. Creep of Sn95.5Ag3.8Cu0.7 composed of a primary and secondary part at constant stress and temperature, experimental curves, and fitted by equation (20)

Taking the time-derivative of both sides of equation (20) yields

$$\frac{\varepsilon_{cr}}{dt} = n \; K \; \varepsilon_p^{sat} \left(\frac{d\varepsilon_s}{dt}\right)^n (t)^{n-1} \exp\left[-K \left(\frac{d\varepsilon_s}{dt}\right)^n\right] + \frac{d\varepsilon_s}{dt} \quad (21)$$

where $\frac{\varepsilon_{cr}}{dt}$ is the total creep rate and $\frac{d\varepsilon_s}{dt}$ is the steady-state creep rate. For small times t, the instantaneous creep rate is greater than the steady-state creep rate (singular at t = 0), and after a long time, the instantaneous rate approaches zero, and the total creep strain rate is equal to the steady-state rate.

Table 9. Deformation constants of the primary (transient) creep for Sn59Pb40Ag1 and Sn95.5Ag3.8Cu0.7, based on own measurements (equation (21))

	ε_p^{sat} (%)	K (s^{-1})	n
Sn59Pb40Ag1	0.4735	145	0.66
Sn95.5Ag3.8Cu0.7	0.4778	69	0.55

The coefficients of equation (21) were fitted to measuring results. However, only few measuring data was available. Table 9 shows the results, slightly different for the two solder compositions. It should be mentioned that a simplified form of equation (21), which was given by Darveaux et al. [17], was not capable of fitting the stress strain response for both solders in a reasonable way. Another proposal to account for both primary and tertiary creep in solder constitutive models was recently made by Clech [44], which is capable of predicting the entire creep curves for Sn-based solders.

4. DATA EVALUATION AND LIFE-TIME ESTIMATION

Criteria based on the strength of materials are most commonly used because of their ease of application. These empirical laws, based on either the cyclic inelastic strain or dissipated energy, have already been introduced in conjunction with analytical reliability analysis. The main difference to these analytical solutions is the more precise determination of the failure strain or energy by FEA. It is obvious that the constitutive description used for the solder significantly affects the resulting field quantities. Thus, the application of these empirical hypotheses is connected to the constitutive relation, which has been applied in the FE-calculation. For the evaluation of the calculated creep strain amplitude, a Manson-Coffin criterion can be applied, i.e., the creep strain ε_{cr}^{acc} which has accumulated over one thermal cycle is expressed as an empirical relation and equated with N$_f$, the number of mean cycles to failure:

$$N_f = \Theta_1 (\varepsilon_{cr}^{acc})^{-c1} \tag{22}$$

As the crack initiation point and the point where joint failure occurs are usually numbers of cycles apart, the author proposed in [45] to use the amplitude distribution along the damage path to calculate different critical cycle numbers. This hypothesis is motivated by the observation of damage zones to coincide with the calculated bands of local maximum equivalent creep strain in many cases. The correspondence is expected to result from several factors like the relative rigidity of the components soldered, the displacement controlled character of the thermal mismatch-induced load and the perfect plastic flow described by the creep law at high strain rate levels. The perfect plastic response reduces the load carrying capacity of highly strained volume elements during a major part of the thermal cycle and, therefore, these volume elements react like weakened zones, a process which is similar to that occurring in the solder material during the cyclic history. Accordingly, it is assumed that the creep strain amplitude at a certain crack depth in direction towards the running crack would determine the number of cycles to macro crack initiation. An average value (or if approximately constant, the minimum value) over the remaining damage site is taken to determine the mean cycles to failure number. Figure 27 illustrates the procedure for a flip chip solder bump.

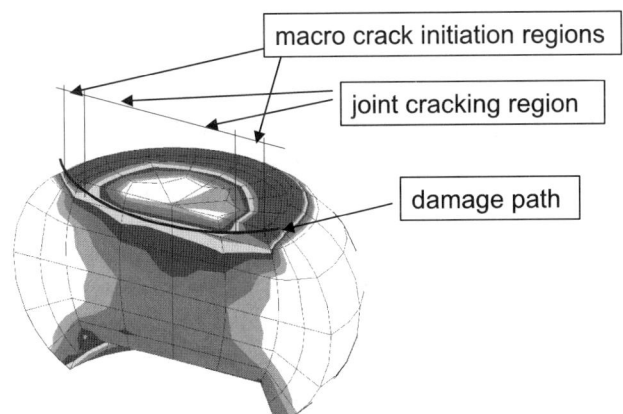

Figure 27. Damage path and averaging domains in a flip chip bump for the calculation of cycle numbers for "macro crack initiation" and "mean cycles to failure"

Besides creep strain criteria also creep strain energy-based methods have been applied to the fatigue of solder joints. The equation is to be read in terms of creep energy density:

$$N_f = \Theta_2 (\Delta W_{cr}^{acc})^{-c2} \tag{23}$$

where ΔW_{cr}^{acc} is the average value over the path of local maximum of creep strain energy density, dissipated per cycle. That means, an approach similar to that for the equivalent creep strain is chosen with regard to the distribution pattern of the creep energy density. Analogous to the evaluation method described above it is assumed that the creep energy density amplitude at a certain crack depth in direction towards the running crack would determine the number of cycles to macro crack initiation. The average (or minimum, if constant) of a band of maximum creep energy density crossing the volume of the solder joints is assumed to determine the critical cycle number.

However, this simple evaluation methodology, which can also be seen as a kind of local Manson-Coffin criterion, works well in cases of relatively smooth creep strain (or energy density) distribution. In these cases a certain part of the cross section, where the cyclic damage develops, is subjected to creep straining over the whole cycle time. Creep induced damage accumulates in these regions. In other geometric situations the stress and therefore the creep strain is strongly localized, mostly because of interface edge notch effects. In these cases cyclic damage does not develop relative uniformly over a cross section of the joint, but only very locally at the stress concentration location. It is noted that actually the stresses at these edges have a singular character [46, 47]. These stress singularities result also in singularities in the inelastic strains and strain energy densities, calculated at these interface edges. Mesh density dependent maximum values result from this singularity, which are therefore not suitable for evaluation. No simple way is currently available or implemented in standard codes to circumvent this difficulty, proposals of the use of constant element sizes and averaging techniques are given instead. Unfortunately, mostly macro-cracking starts at this location and the developing macro crack then progresses through the joint. In this crack, progress dominated kinds of failure-strength based concepts reach their limits. A conservative estimation is achieved by applying the criteria (22) or (23) to calculated creep strain or energy density amplitudes, averaged over an appropriate "damage zone", where the creep strain or energy density take their respective maxima. Advanced damage or fracture mechanics based fatigue models are required to improve this situation, which are not yet available in commercial FE packages.

However, because of the many idealizations inherent in modelling, comparisons of testing results are always recommended, at least for selected

7. Characterization and Modelling of Solder Joint Reliability

demonstrator cases. Coefficients for the empirical laws and the evaluation methodology described above were determined by correlation of joint experimentally determined characteristic lives versus cyclic accumulated equivalent creep strains or energy densities, respectively.

For a constitutive model where the SnPb solder is assumed to be described completely by an elastic-steady state creep constitutive law, the creep part following the description of Grivas/Hacke given in (12), the parameters in equation (22) were fitted to $\Theta_1 = 0.38$ and $c_1 = 1.96$, employing a comparison of calculation results with the damage zones observed in several solder joints after thermal cycling. These estimated parameters refer to test cycles with a temperature range of approximately −40 − 125°C and a cyclic period of about 1 hour. They differ in case of varying cycling conditions.

With reference to an elastic-creep model, where the creep part contains both primary and secondary creep, a quantitative comparison of the thermal fatigue lives of SnPb solder joints and SnAgCu joints was made for a number of assemblies [40]. The secondary creep part is described in Table 3 for SnPb solder and in Table 5 for SnAgCu solder, the primary part for both solders in Table 9. Figure 28 shows the correlation of fatigue life data from a number of accelerated tests on different package types with SnPb(Ag) and SnAgCu interconnects versus the determined accumulated creep strains by FEA in the solder interconnects. If equation (22) holds true, the curve fitting has to be a straight line in a log-log plot. It can be observed from the diagram that this behaviour is approximately true. The life prediction models for SnPb(Ag) and SnAgCu solder interconnects based on the best-fit line going through the data of 14 experiments.

The results suggest that SnAgCu thermo-mechanical failure data may correlate to accumulated creep strains as in the case of eutectic SnPb(Ag) soldered assemblies. In Figure 28, the slope of the best-fit line through the SnPb(Ag) data is −1.80 and for the SnAgCu data about −1.3. Furthermore, a crossover point between the two alloys is indicated at approximately 600 − 800 cycles.

A similar trend was found for the correlation of the thermo-mechanical failure data to the calculated viscoplastic strain energy density, as shown in Figure 29. In that case, a crossover point between the two alloys is indicated, at approximately 6000–8000 cycles. The determined slope of the best-fit line through the SnPb(Ag) data is −1.20 and for the SnAgCu data about −1.0.

The cycling conditions applied for these tests vary, but all were in the test range of 0.5 to 1 hour cycle time with temperature extremes between −55°C and 150°C. Inclusion of primary creep is expected to reduce the dependence of the empirical relations from the cycling regime, but no comparisons to field cycle conditions have been available up to now.

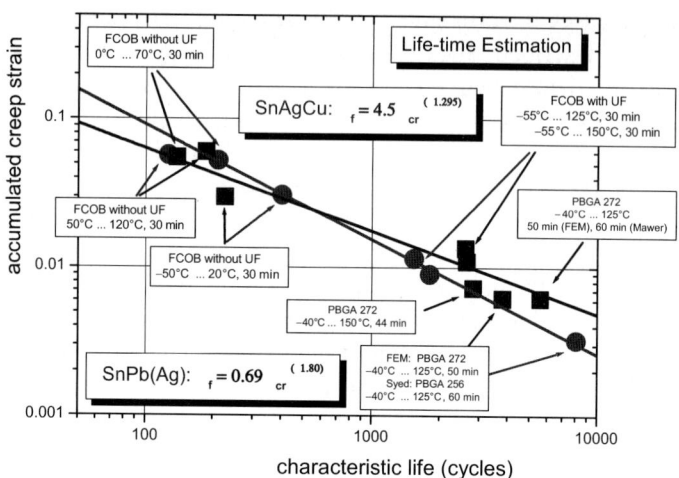

Figure 28. SnPb(Ag) and SnAgCu fatigue life versus accumulated creep strain during one thermal cycle

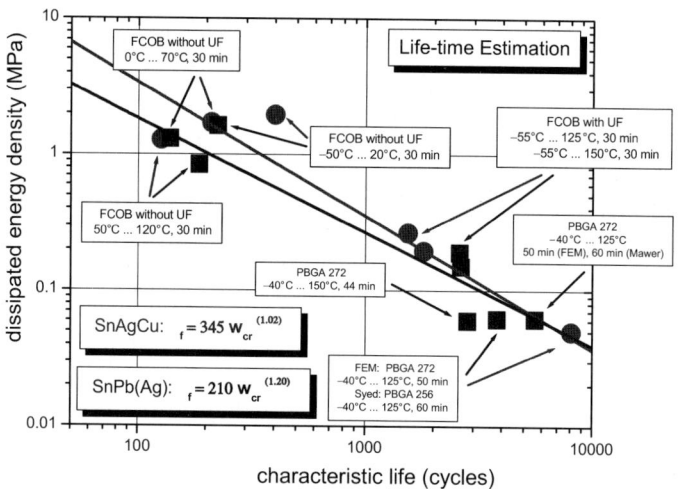

Figure 29. SnPb(Ag) and SnAgCu fatigue life versus dissipated energy density during one thermal cycle

Numerous similar empirical relations have been established for a variety of solders. Several investigations of Lau et al. [32, 48, 49], for leaded and lead-free solders should be mentioned. It is noted that, although the fatigue behaviour should be an invariant property of the solder materials, the different formulas are always related to the creep model used in the related calculations because of the various idealizations made.

Other hypotheses for fatigue life calculation based on strength of materials approaches were proposed in the literature. The most frequently cited fatigue hypothesis is that of Darveaux [17]. In his model the number of cycles to crack initiation is calculated by an energy-based equation analogous to equation (23). To predict the actual number of cycles to failure, a crack propagation term was introduced, which relates to the averaged cyclic creep strain energy ΔW_{cr}^{acc}

$$\frac{da}{dN} = \Theta_3 (\Delta W_{cr}^{acc})^{-c3}, \qquad a_W = N_0 + \frac{a}{da/dN} \qquad (24)$$

where da/dN is the cyclic crack extension rate. N_0 is the number of cycles until crack initiation, calculated by an equation analogous to (23), a is the joint diameter, a_W is the characteristic life of the joint, i.e., the sum of the number of cycles to crack initiation and the time the crack needs to grow through the whole joint. The constants Θ_2, Θ_3, c2 and c3 depend on both the simulated structure and the mesh density. In his first study on BGA balls, the view is taken that there are primary and secondary cracks, which propagate towards each other at different rates. However, no general rule is given for the estimation of the cyclic creep strain energy ΔW_{cr}^{acc}. The strain energy density is volume averaged across the elements along the solder joint interface where the crack propagates. For the finest data set from Darveaux used the average over a two element thick layer, each layer 1/2 mil (12.5 μm) thick. The main difference to equation (23) is the link of the creep energy density to a fatigue crack propagation rate instead of the critical cycle number itself. This proposal might be advantageous in cases where strongly localized creep straining is observed and crack progress is the dominant driver of fatigue, which is often observed for BGA and CSP ball type interconnects. However, the averaging technique along the whole interface seems questionable in this case.

One modified proposal is related to solders, for which the contribution of different major secondary creep mechanisms could be identified, which resulted in the formulation of double power laws analogous to equation (11). For SnPb, these mechanisms already mentioned were "grain boundary sliding" and "matrix creep" for low and high stresses, respectively. For this solder, Knecht and Fox [23, 26] proposed that exclusively the matrix creep strain part accumulated during the life of the solder joint can be related to the fatigue life. This matrix creep hypothesis predicts failure when the creep ductility of the solder has been exhausted during cycling. Linear accumulation of the matrix creep strain is assumed. The value of the creep ductility was determined to be 170% for pure shear loading.

Also for SnPb solder Syed has proposed a partitioned accumulated creep strain model based on damage mechanisms as described by the constitutive equation analogous to (11) [50, 51]. In this model again, the accumulation of secondary creep during thermal cycling is taken as damage measure, and split in the different actions of grain boundary sliding and matrix creep. An extension of the same model towards the evaluation of SnAgCu solder was proposed in [52]. As stated in equation (17), again two different creep mechanisms denoted with the superscript I and II seem to be dominating the secondary creep. If the contribution to cyclic creep damage is weighted individually, the corresponding equations become

$$N_f = (C_1 \varepsilon_{cr}^{acc\,I} + C_2 \varepsilon_{cr}^{acc\,II})^{-1} \qquad (25)$$

for the accumulated equivalent creep strain and

$$N_f = (C_3 \Delta W_{cr}^{acc\,I} + C_4 \Delta W_{cr}^{acc\,II})^{-1} \qquad (26)$$

for the creep energy density dissipated per cycle. In this model, both the strain exponent and energy density exponent are equal to 1. The constants given for SnPb, related to the mechanisms "grain boundary sliding" and "matrix creep", are C_1 = 0.02, C_2 = 0.063, C_3 = 0.0027, C_4 = 0.0019. For SnAgCu solder the contribution of the "low stress region I" creep was found to be so small to be negligible [51], hence the equations become analogous to equations (22) and (23) with both exponents c1 and c2 equal to 1.

A modified theoretical approach is the Energy Partitioning model. The constitutive model includes all the strain contributions of equation (7), i.e., elastic strain, primary and secondary creep strain and time independent plastic strain and assumes that the elastic, plastic and creep deformation contribute to the cyclic damage in a different manner. It predicts the cyclic creep-fatigue damage based on the elastic, plastic, and creep energy densities by calculating separate cycles-to-failure numbers for the three kinds of energy and the total numbers of cycles to failure by an appropriate average of the three numbers, see [34, 53].

In order to increase the accuracy of solder fatigue modelling, advanced theoretical features have been proposed. Since the degradation of the joint during the cycling history is accompanied by microstructural changes like voiding and microcracking causing local softening, the basic idea is to include these changes in the calculation. The inclusion of the evolution of the microstructure, i.e., the local softening, in the FE-analysis is attempted by so-called "damage models".

In continuum damage approaches, this is achieved by use of a special viscoplastic constitutive framework with damage evolution capabilities, which is not available in standard FE-packages. Examples for this methodology are given by Frear et al. [54], Qian et al. [55], Desai [56], Chandaroy and Barasan [57, 58], and Zhang et al. [59].

In cohesive zone approaches the material behavior is divided in a part without damage, within which continuum theory is applicable, and a part in which damage can evolve according to a cohesive zone law. These laws use a prescribed relationship between surface traction and corresponding crack opening displacement, specific for each material. A cohesive zone model for SnPb solder was proposed by Towashiraporn and Subbarayan [60]. A damage parameter is introduced which controls the integrity of the interface. This damage parameter is applied to simulate a propagating crack by a node connectivity release methodology on a pre-defined damage path. Another solder cohesive zone model is described in [61] by Abdul-Baqi et al., which accounts for details of the microstructure and uses a branched network of cohesive zone elements in-between the different microstructural constituents.

Going further into microstructural details is attempted by definition of constitutive properties on the micro-level and then introducing this behaviour in a computational model to describe the constitutive response of the solders on the macro-level. Sharma and Dasgupta describe in [62, 63] a damage model, particularly for SnPb solder, including void nucleation, void growth, and void coalescence based on micro-structural stress fields. Microstructural stress states are estimated under viscoplastic phenomena like grain boundary sliding, blocking at second-phase particles, and diffusional creep relaxation. This approach seems promising; however it is very difficult to describe the complex interactions within the evolving microstructure properly. Furthermore, material testing for the constitutive characterization is quite complicated on the micro-level.

Computational models to predict the evaluation of the microstructure, especially the diffusion controlled coarsening of SnPb solder, have also been proposed. The publications of Mueller et al., see e.g., [64], are to be mentioned here. An approach, which combines modelling of the microstructural evolution, limited to the diffusion-related phase region changes and a related constitutive characterization of each phase region by a separate hyperelasto-viscoplastic Perzyna model was proposed by Ubachs et al. [65].

Finally, all advanced damage models have the drawback in common that the cyclic history of solder fatigue has to be followed for many cycles, which requires high computer resources for complex assemblies and currently limits their practical application. But since these resources will certainly be available in the future and the theoretical developments are ongoing, such solutions might bring about improved predictive capabilities, which can be expected from the continuum mechanics point of view. However, it should also be kept in mind that a solder joint is a multi-material structure with process-related properties and process-induced defects, which cause scattering or sometimes conflicting testing results. Additionally, fatigue fracture is sometimes observed at other locations but not in the solder material. These uncertainties limit the predictive capabilities of any solder fatigue model.

5. CASE STUDIES

The four case studies selected to illustrate the proposed methods to evaluate thermal fatigue of solder joints by means of the finite element method use fatigue hypotheses based on the strength of materials. The wide applicability of the methodology is illustrated by the choice of different types of components ranging from large ceramic surface mount components mounted on FR-4 boards to small flip-chip assemblies on different types of substrates.

It is important to note that for FE analyses the precise description of the materials characteristics is a basic requirement. That means, not only the non-linear temperature dependent properties of solders must be precisely defined, but also other non-linear and temperature-dependent data. The viscoelastic behaviour of polymeric materials is mentioned here. As a rule of thumb, it has to be considered if the T_g of the respective polymer is within or slightly above the range of the temperature cycle loading. However, details of the definition of the temperature dependent viscoelasticity, used hereafter in the form of the mastercurve approach defined within the framework of linear viscoelasiciy, are not given here. Instead, reference is made to [66], [67] and other publications cited for each example.

5.1 Comparing Different Creep Laws for a Ceramic Capacitor and a PBGA on FR-4 Boards

The introductory examples are intended to illustrate the characteristic kind of the non-linear, temperature-dependent stress stain-response within solder joints. Effects of the choice of different creep laws on the calculated creep strains and energy densities are studied for standard components on FR-4 board, a ceramic chip capacitor of size 0402, a ceramic chip resistor of size 0805 and a plastic ball grid array with 225 I/O (PBGA 225). To relate calculation results to fatigue life, criteria of Manson-Coffin type are selected.

5.1.1 Finite Element Modelling

Different FE models of the standard components on 1.5 mm thick FR-4 boards were used: an idealized 3-D double symmetric quarter model with minimum solder gap height of 20 µm for the chip capacitor of size 0402, see Figure 30, a symmetric 2-D plane strain model for a cross section of the chip resistor with left solder gap height of 25 µm and right solder gap height of 15 µm, and a 3-D double-symmetric model containing each ball was chosen for the PBGA. The latter had a 27 mm body size, 4 rows of thermal balls and 4 perimeter rows with 1.27 mm ball pitch, and 0.55 mm stand off, see Figure 31. It contains a BT substrate with thickness 0.56mm, a 10×10 mm^2 die of 0.36 mm thickness and a die bonding adhesive layer of 50µm thickness.

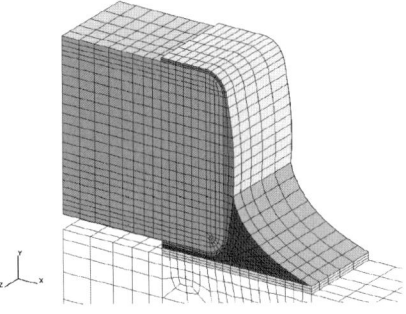

Figure 30. Quarter 3D-Finite element model of chip capacitor 0402 on FR-4 board

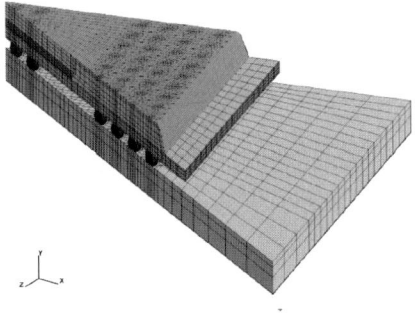

Figure 31. Octant 3D-Finite element model of PBGA 225 on FR-4 board

For SnPb two different secondary creep laws were compared, the one given by Grivas/Hacke, equation (12), and the one given by Schubert et al., based on in-house measurements and averaging with other published data, equation (11) and Table 3. For SnAgCu the hyperbolic sine stress law with coefficients given by Schubert et al., equation (15) with coefficients given in Table 5 was used. For both latter creep laws given by Schubert the primary creep terms were added as described by equation (21) with coefficients given in Table 9. Other selected materials data of the materials used in addition to the solders are given in Table 10.

Table 10. Selected materials data

	Constitutive Model	Young's Modulus [GPa]	CTE [1/°C]
Ceramic chip resistor	elastic	250.0	6.5E-6
Ceramic chip capacitor	elastic	130.0	10.5E-6
Board FR-4	elastic-orthotropic	230 K $E_{x,y}$ 21 E_z 8 398 K $E_{x,y}$ 18 E_z 5	$\alpha_{x,y}$ 17.E-6 α_z 58.E-6
Interposer (BT)	elastic-orthotropic	230 K $E_{x,y}$ 22.0 E_z 8.3 398 K $E_{x,y}$ 19.5 E_z 5.2	$\alpha_{x,y}$ 14.E-6 α_z 45.E-6
Die bonding adhesive	viscoelastic	230 K 5.5 340 K 1.4 423 K 0.1	230 K 63.E-6 420 K 75.E-6
Solder	elastic-viscoplastic	210 K 36.0 398 K 21.0	2.4E-5
Copper pad	elastic-plastic	90.0	16.5E-6
Encapsulant (BGA)	elastic	210 K 15.5 340 K 15.0 400 K 10.0	14.5E-6

For the study on the resistor and the PBGA a test temperature cycle – 40...125°C was considered for a 50min period with a ramp time of approximately 5 minutes, defined by decay functions, as shown in Figure 2. A 1-hour cycle – 40 to 150°C, analogously defined by decay functions, was used for the capacitor. It is noted that for these slow loading ramps secondary creep is expected to dominate the solders creep response.

5.1.2 Results

It was stated above that the temperature dependent creep of solder joints causes a highly non-linear stress strain behaviour when subjected to temperature cycling. The analysis results for the chip capacitor were chosen to illustrate the characteristic stress strain response of a solder joint. The assembly is assumed to be stress free after the end of the upper hold. Figure 32 shows the creep shear strain distribution pattern in the joint after the third ramp down, which obeys a maximum at the periphery of the solder gap because of the overlaid action of local and global mismatch.

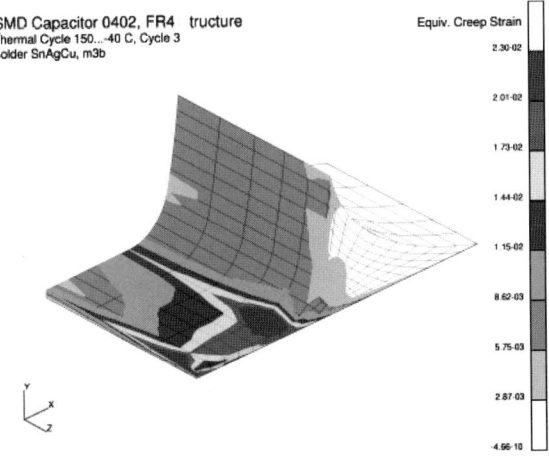

Figure 32. Acc. equivalent creep strain distribution pattern within a SnAgCu joint due to one thermal cycle 150°C to –40°C

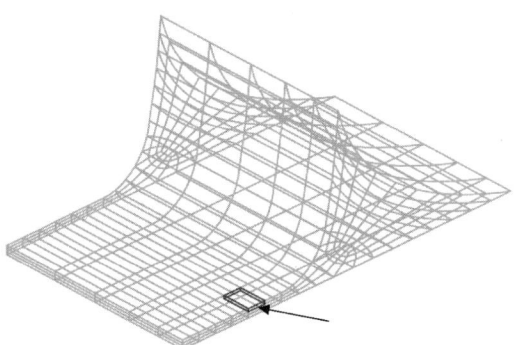

Figure 33. Element selected for hysteresis loop plots

A closer look is taken at the stress strain response within the area of maximum creep straining, shown in Figure 34. The hysteresis loop in Figure 35 depicts the maximum shear stress in dependence on shear strain for different creep laws after stabilization of the cyclic creep response

(3rd cycle). It is obvious that similarities as well as differences exist with regard to the solder and the creep law:
- At 150°C, where both solders are soft, shear stress is low. Because of the higher creep rates for SnPb solder at high temperatures the stress relaxes actually to a value close to zero for both creep laws applied, while for SnAgCu solder a small residual stress remains.
- During temperature ramp down, the different creep laws predict different rates of stress increase approaching similar maximum stress values at the end of the ramps. Because creep rates are rather small at –40°C, stresses decrease only slightly during the cold dwell. Significantly higher shear strain is calculated for SnPb with both of the creep laws when compared to SnAgCu.
- During temperature ramp up at first the shear stress is unloaded almost elastically. When the shear stress is close to zero, it changes sign and soon approaches its maximum in the opposite direction of shearing. The maximum is reached already during the ramp period: increasing softening, i.e., higher creep rates, of the solders lead to dominance of the stress relaxation part, the stress decreases during temperature increase and subsequent hold, and the loop closes at the end of the high dwell.
- For SnPb it is interesting to compare the hysteresis loop for solder joints of a LCCC on FR-4, experimentally determined by Hall [68]. Despite the differences in assembly and cycling conditions, the curves show a high qualitative similarity.

Another question is related to the stability of the hysteresis loops dependent on the number of cycles calculated. The stability depends on the creep strain constitutive description, it depends on the stress or strain controlled loading situation and on the residual stress, which is reached after the high temperature hold. From the hysteresis loops it is seen that a stress state, which is close to, the stress-free state is always reached at high temperature. Accordingly, this point in the loop is chosen as a starting point of the cyclic loading, because the FE-calculation assumes a stress-free initial state. While the shift of the loop is negligible for the SnPb joints due to the strain-controlled situation of the example and the return of the loop to zero stress, for SnAgCu a slight shift occurs, as can be observed from Figure 35. This ratcheting effect vanishes after the first cycle. However, the error in terms of cyclic equivalent creep strain is less than 10% also for the SnAgCu joint.

7. *Characterization and Modelling of Solder Joint Reliability* 423

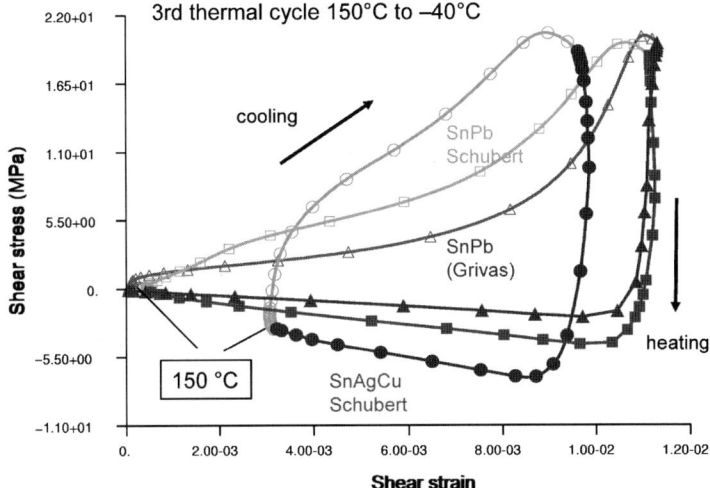

Figure 34. Comparison of stabilized creep hysteresis: shear strain versus shear stress for SnPb und SnAgCu solders

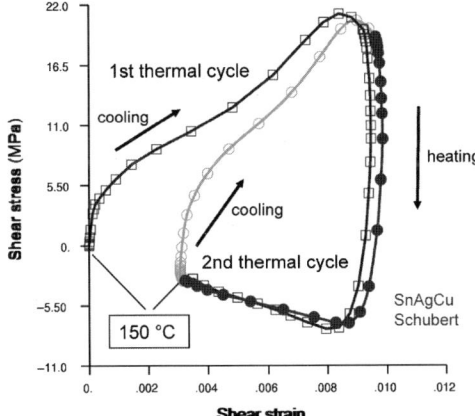

Figure 35. Comparison of creep hysteresis shear stress versus shear strain for the first and second cycle for SnAgCu solders

5.1.3 Failure Prediction at Chip Resistor and PBGA Solder Joints

After the description of the typical stress-strain response within solder joints, strength-based failure predictions are given. Effects of the choice of the creep laws on the failure prediction are discussed for a ceramic chip resistor with size 0805 and a PBGA 225, both mounted on FR-4 board. The components were chosen such that the deformation mode of the first assembly is more strain controlled and early failure is expected, while the second one is more stress-controlled and a high critical cycle number is known to occur.

The creep laws are those already explained for the investigations on the capacitor described above. The evaluation methodology follows the description given in paragraph 6. Table 11 summarizes the Coffin-Manson coefficients used for the three creep laws.

Table 11. Coefficients for the estimation of fatigue life

Solder Creep law	Type	Θ_1	c_1	Θ_2	c_2
SnPb- Grivas-Hacke	secondary	0.38	1.96		
SnPb- Schubert et al.	sec.+ primary	0.69	1.8	205	1.2
SnAgCu- Schubert et al.	sec.+ primary	4.5	1.295	345	1.02

For the ceramic resistor, the maximum creep strain and energy density values were calculated as averages in the solder gap, and the mean values within the solder fillets averaged along the metallization/solder interface, where local maximum straining occurs.

For the PBGA the creep strains or energies concentrate at the free interface edges of the balls to the package sites, which were taken as maximum values. The mean values were calculated as average along the ball-interposer pad interface. There was no strong dependence of the creep strain on the ball position, sometimes the outer thermal ball and some times the outer peripheral ball row exhibit a slightly higher creep strain amplitude than the others, dependent on the creep law chosen. The absolute maximum was taken for the evaluation.

The resulting creep strain and energy values are compared in Figures 36 to 39. They show that all values calculated for the PBGA are lower. It is also obvious from the figures that the choice of the creep law generally affects the

7. Characterization and Modelling of Solder Joint Reliability

PBGA results much more than the CC results, which can be attributed to the displacement controlled character of solder deformation for the CC assembly type and the plane strain assumption. For the PBGA the differences in calculated cyclic equiv. creep strain and energy density are in the range of approx. 30%, when either secondary or combined secondary/primary creep is considered. This applies analogous to both different solders, with a tendency to higher amplitude differences for SnAgCu solder. The differences are remarkably high for the slow temperature cycle considered. Even higher effects are expected for different cyclic regimes.

Estimations of fatigue lives are only possible for those results, for which the coefficients in equations (8) and (9) were determined, compare Table 11. The results are depicted in Figures 40 and 41. For the PBGA, higher critical cycle numbers are predicted than for the ceramic chip resistor, as has to be expected. The cycle numbers predicted by the creep strain criterion agree well with those predicted by the creep strain energy criterion. When the different solders are compared, the trends depend on the component. For the CC assembly only small differences in the critical cycle numbers are obtained, and they are below the expected prognosis precision. For the PBGA assembly, a higher fatigue life is predicted for SnAgCu solder balls.

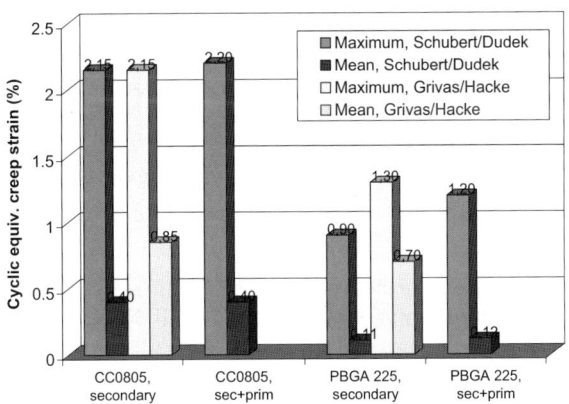

Figure 36. Calculated max. and mean creep strains for different secondary and combined secondary/primary creep laws: Solder SnPb

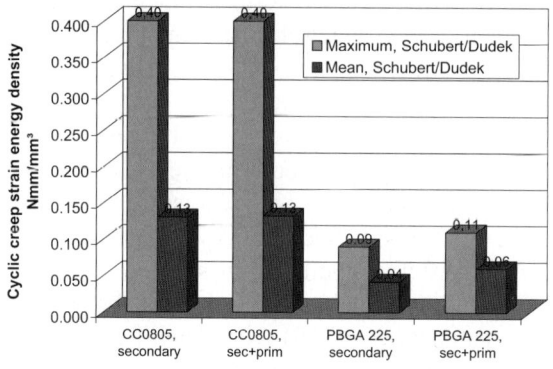

Figure 37. Calculated max. and mean creep strain energy densities for secondary and combined secondary/primary creep laws: solder SnPb

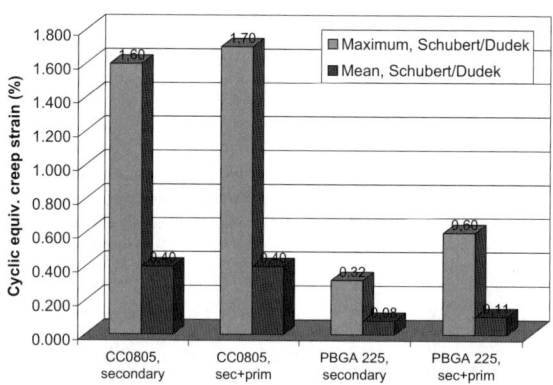

Figure 38. Calculated max. and mean creep strains for different secondary and combined secondary/primary creep laws: solder SnAgCu

7. Characterization and Modelling of Solder Joint Reliability

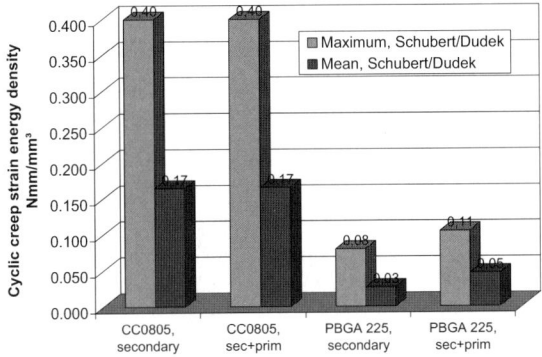

Figure 39. Calculated max. and mean creep strain energy densities for secondary and combined secondary/primary creep laws: solder SnAgCu

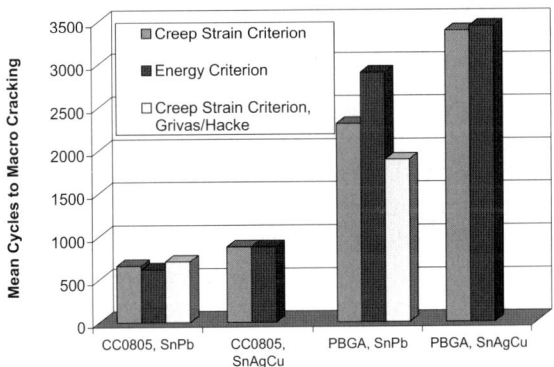

Figure 40. Estimated mean cycles to macro crack initiation for different creep laws and related Coffin/Manson coefficients: ceramic chip resistor (CC) versus PBGA 225

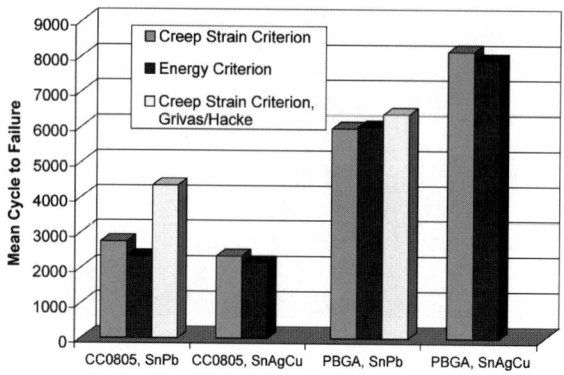

Figure 41. Estimated mean cycles to failure for different creep laws and related Coffin/Manson coefficients: ceramic chip resistor (CC) versus PBGA 225

5.2 Comparison of Predicted and Test Results for Surface Mount Quartz Components

A combined theoretical and experimental study on solder fatigue of surface mount quartz components was carried out to facilitate the reliable design of products [69, 70]. This study aimed at modelling of accumulated damage from mission profile, from customer specification and from accelerated testing on lead-free and lead-containing solders. Accordingly, different thermal test cyclic environments as well as different field cyclic environments were defined. For the test cycles the analysis results were compared to test results, which were performed in parallel.

Figure 42. Test assembly with surface mount ceramic quartz

7. Characterization and Modelling of Solder Joint Reliability

The test vehicle was a standard assembly on single-sided FR-4 board, see Figure 42. The board size was $333 \times 228 \times 1$ mm^3 with 55 microns thick copper circuit and Au ENIG surface finish. The lead-free solder chosen for comparison to the eutectic Sn59Pb40Ag1 control was Sn96Ag3.2Cu0.8. A no-clean standard SMT assembly process with one reflow was used for board assembly. The ceramic quartz 8.8 mm $\times$ 4.5 mm with two leadless solder joints was a-priori identified to be the critical component with the highest global mismatch from all the components of the assembly.

5.2.1 Finite Element Modelling

Geometric modelling was based upon micrographs taken from cross sectioning of different assemblies. In this way realistic joint geometries could be adopted. No significant difference of the solder geometry was observed for the two different solders and, accordingly, the same model was used for both materials. A half model was meshed with unsymmetrical solder thickness, shown in Figures 43 and 44. Similar models were meshed for comparison analyses, the first one assuming double symmetry and the second one with pre-cracked solder, i.e., solder in the gaps removed. A worst case assumption was made for the asymmetrical solder gap heights, which were taken as left joint 20 µm, right joint 60 µm. After cycling extensive cross-sectioning revealed that this worst case assumption had been unrealistic, because the average joint was almost symmetric with a joint height in the range of 50 µm to 70 µm.

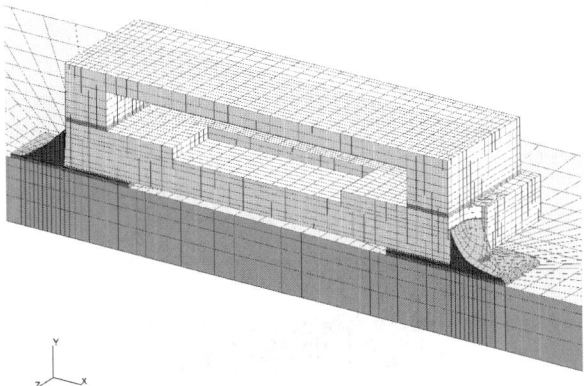

Figure 43. FE-Model with one symmetry plane for SMD quartz 8.8 mm $\times$ 4.5 mm on FR-4. Asymmetric solder gap heights: left joint 20 µm, right joint 60 µm

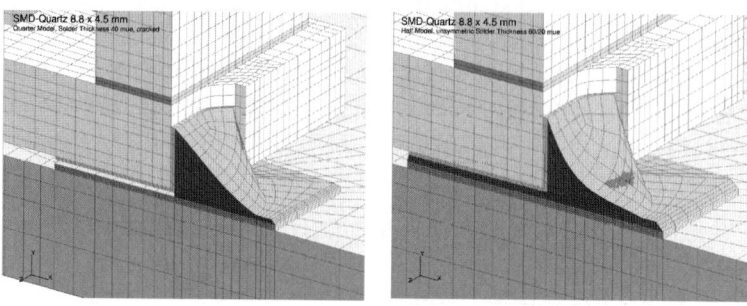

Figure 44. Detail of the vicinity of the right solder joint with and without solder in the gap

For the test cycles at temperatures from –40°C to 125°C and –40°C to 85°C, it was attempted to include the measured profiles. The actual cycles differ slightly from the ideal temperature values. The temperature profiles were defined by decay functions, as can be observed from Figure 45.

Several field cycles were investigated, see Table 12. The field cycles were related to automotive use of type "passenger compartment". A smooth cycle definition, again by decay functions, was applied for the field cycle – 7°C to 25°C, see Figure 46, the other cycles were simplified assuming linear ramps with ramping times between 5 min and 15 min in accordance with Table 12. Opposite to the test cycles the field cycles are not symmetric, i.e., different hold times occur at the temperature extremes.

Table 12. Definition of field cycles and a composed field cycle

Field Cycle	Ramps duration (min)	Occurrence	Contribution to a Composed Cycle
15min25°C/3h–7°C	10	20% of days	13.33%
15min 20°C/3h 45°C	5	70% of days	46.66%
15min20°C/3h 60°C	10	10% of days	6.66%
15min25°C/3h–23°C	15	20% of nights	6.66%
15min25°C/3h 7°C	5	70% of nights	23.33%
22°C/22°C		10% of nights	3.33%

Material modelling was performed analogously to the previous case studies. The CTE values for the component and substrate are of special importance for the theoretical analyses because of the dominant global thermal mismatch induced loads. Accordingly, these CTE-values were measured by

7. Characterization and Modelling of Solder Joint Reliability

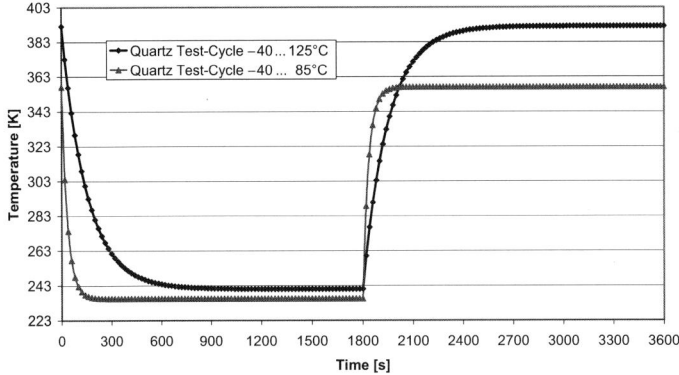

Figure 45. Simulated test cycles at temperatures from –40°C to 125°C and –40°C to 85°C, 1hr

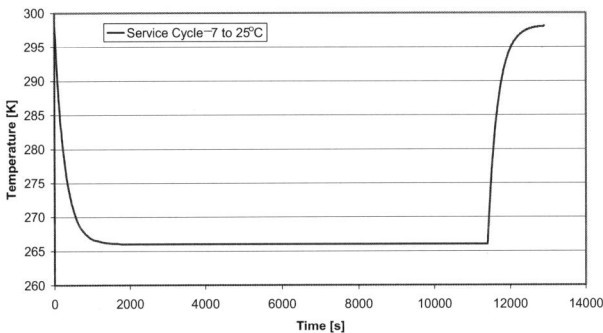

Figure 46. Example for a simulated field cycle at temperatures from –7°C to 25°C

TMA. Temperature and time dependence of the material characteristics as well as their non-linear character was taken into account, see Table 13. The ceramic components and the organic substrates were assumed to behave linearly elastic, where the anisotropic thermal expansion and elastic stiffness, respectively, of the FR-4 board is accounted for. Note that the y-direction is the direction perpendicular to the board. Temperature dependent elastic-plastic behaviour was assumed for the copper metallization as well as the metallization on the components. For SnPb and SnAgCu the secondary creep laws given by Schubert et al., equation (11) and equation (15), were used. For both creep laws the primary creep terms were added as described by equation (21).

Table 13. Selected material data

Material	Constitutive model	Young's modulus [GPa]	CTE. [ppm/K]
Ceramics (Quartz)	Elastic	350	5.4
Glass	Elastic	70	4.1
Termination	elastic-plastic	100	12
Solder PbSn	primary + secondary	36 at 233 K 21 at 398 K	24.0
Solder SnAgCu	primary + secondary	48.5 at 233 K 33 at 523 K	20.0
Pad (Cu)	elastic-plastic	87	16.5
Board FR-4	elastic-orthotropic	22 at 230 K 19.5 at 398 K	$\alpha_x = 12.8$ $\alpha_z = 15$ $\alpha_y = 45.0$

5.2.2 Results

Test results

For the two thermal test cycles failure data was generated by cross sectioning of the quartz components. At certain stages of the cycling process 5 samples were analyzed for each of the solders. The stages were adjusted to the temperature amplitudes, for the 125°C cycle cross sectioning was started at 450 cycles and continued in steps of 50 cycles up to 1050 cycles and for the 85°C cycle the initial cycle number for cross sectioning was selected to be 900 cycles and cross sectioning was continued up to a cycle number of 2000 cycles with steps between 150 and 250 cycles.

For the development of the fatigue failure no fundamental difference was observed between neither the two different solders nor the different cyclic regimes. Cracking always initiated within the solder gap within the bulk of the solders. In exceptional cases also cracking along the interfaces of solder to component or to board interfacial intermetallic layer was seen. The development of multiple cracks in the gap, which link to form one macro-crack starting from the inner gap region was the characteristic failure mode. When the in-depth direction of the joints perpendicular to the component is considered, cracking started earlier at the peripheral parts of the joints. For the evaluation the cross sections at the joint center were used throughout, controlled by length measurements of the pads, which had a circular shape on their outer sides. Because of earlier cracking of the peripheral gap regions these cross sections represent the shortest crack lengths in the in-depth direction. Cracking finalized by propagation of the cracks from the gap

7. Characterization and Modelling of Solder Joint Reliability 433

diagonally through the solder fillet. Cracking at the upper edge of the solder fillet-component interface was also frequently observed, especially for the high cycle numbers of the 85°C cycle. This local mismatch induced cracking, however, never formed the final crack path. In Figure 47 characteristic failure patterns are visualized for both solders on the 800 cycles stage of the 125°C cycle.

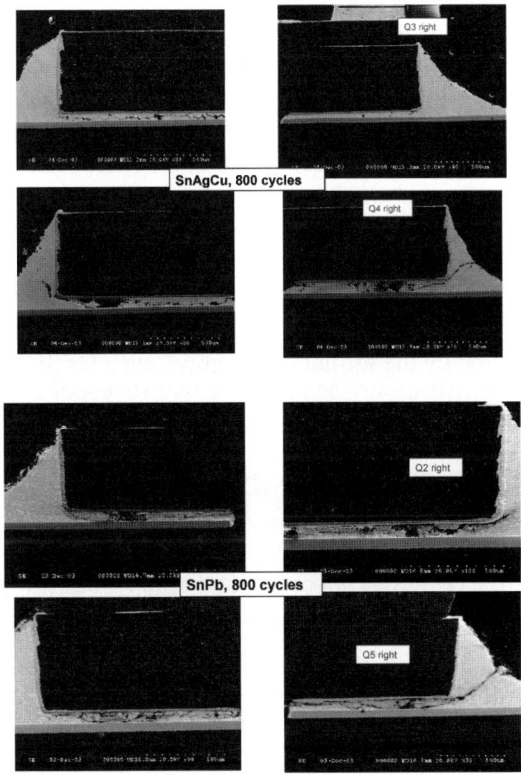

Figure 47. Testing results – cross sections after test cycling –40°C to 125°C

The estimations of mean numbers of cycles to failure were based on Weibull statistics. Cracking was evaluated for the central cross sections of the joints as described above, whereby different crack lengths were evaluated separately. The schematic related to Table 14 illustrates the classified crack lengths of 30%, 70% and 100% of the total joint fracture lengths, while Table 14 gives the data for two different confidence levels. For each quartz the longest crack either on the left hand side or on the right hand side of the component was used to describe the cracking state of the components. For the two test cycles, the Weibull slopes are rather high, i.e., in the range between 4 and 13, for both solders and testing conditions, indicating a narrow scatter band for the tests. It was pointed out by Lau [32] that, for a linear acceleration, the Weibull plots of the two conditions compared should have the same Weibull slope. This holds approximately true for the tin-lead solder, however for the lead-free one the Weibull slope of the 85°C cycle is mostly higher than for the 125°C cycle, indicating a non-linear acceleration between the two test cycles.

If this effect, which might be attributed to the limited test data, is not considered, estimates of the acceleration factors between the two test cycles can simply be calculated from the Weibull characteristic life of both solders. By averaging the numbers for different crack lengths, an acceleration factor of AF = 2.1 is calculated for the SnAgCu solder and AF = 1.8 for SnPb solder for the 125°C cycle compared to the 85°C cycle. The acceleration can also be calculated from formula (5), which results in AF = 2.2. Because of the same cyclic frequency of 1 cycle/hour for both test cycles, the contributions to AF in (1) result only from the different temperature amplitudes (AF_1 = 1.5) and from the different upper dwell temperatures (AF_2 = 1.4). It can be concluded that the differences between the solders are low and the prediction by the formula (5) is good, even for SnAgCu solder.

Table 14. Test results according to Weibull data for lead-free (LF) and leaded solder (PB)

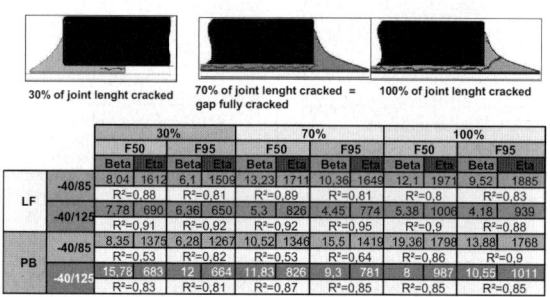

		30%				70%				100%			
		F50		F95		F50		F95		F50		F95	
		Beta	Eta	Beta	Eta	Beta	Eta	Beta	Eta	Beta	Eta	Beta	Eta
LF	-40/85	8,04	1612	6,1	1509	13,23	1711	10,36	1649	12,1	1971	9,52	1885
		R^2=0,88		R^2=0,81		R^2=0,89		R^2=0,81		R^2=0,8		R^2=0,83	
	-40/125	7,78	690	6,36	650	5,3	826	4,45	774	5,38	1006	4,18	939
		R^2=0,91		R^2=0,92		R^2=0,92		R^2=0,95		R^2=0,9		R^2=0,88	
PB	-40/85	8,35	1375	6,28	1267	10,52	1346	15,5	1419	19,36	1798	13,88	1768
		R^2=0,53		R^2=0,82		R^2=0,53		R^2=0,64		R^2=0,86		R^2=0,9	
	-40/125	15,78	683	12	664	11,83	826	9,3	781	8	987	10,55	1011
		R^2=0,83		R^2=0,81		R^2=0,87		R^2=0,85		R^2=0,85		R^2=0,85	

FE-analysis results for test cycles and comparison with test results

For computational efficiency the test cycles are started at the upper dwell temperature and only one cycle is calculated assuming complete relaxation during the upper dwell time. A comparison to the second cycle revealed even for the lower temperature cycle to 85°C a deviation of less than 3%.

7. Characterization and Modelling of Solder Joint Reliability

Characteristic cyclic creep strain distribution patterns within the solder joints are shown in Figure 48 for the cycles –40 to 125°C and –7 to 25°C. The calculations show that the creep strain always concentrates within the solder filled gap underneath the component. It is obvious from the figures that maximum solder straining occurs at the peripheral edge of the solder-filled gap, where cracking is expected to initiate, independent of the solder material.

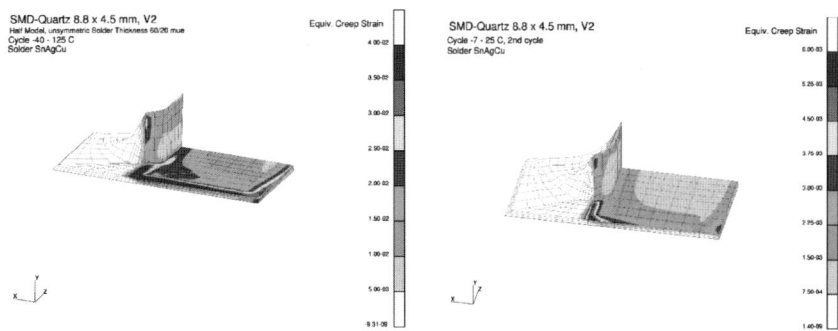

Figure 48. Acc. equivalent creep strain distribution pattern with in a SnAgCu joint due to one thermal cycle of 125 to –40°C and one of 25°C to –7°C

The distribution patterns of deformations and creep strains or energy densities, respectively, are nearly equivalent for the two different solders. They can be considered characteristic for all the different temperature cyclic environments, which only alter the amplitudes, but not the distributions. The calculation of averaged values of creep strain and energy density was done for two crack path, related to the experimental results: Crack path 1 is the area of the solder filled gap underneath the component, crack path 2 is cracking of the fillet in approximately 45° direction. Cracking alongside the latter crack path means total solder joint failure. Both the creep strain and the creep strain energy density criterions were applied and the out coming cycle numbers were again averaged. The results are compared in Figure 49 for macro crack initiation and in Figure 50 for joint failure for the test cycles 125 to –40°C and 85 to –40°C.

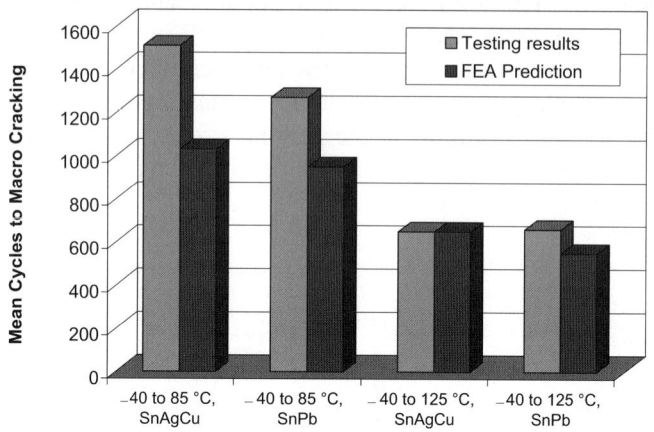

Figure 49. Test and simulation results for macro-cracking (gap 30%) – test cycling

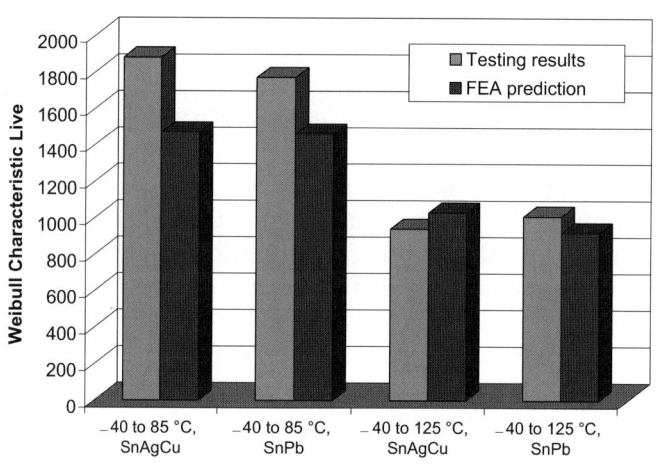

Figure 50. Test and simulation results for 63 % total joint failure – test cycling

It can be seen from the figures that failure prediction by FEA is in the right range on the safe side. The tests and the predictions correlated within a factor of 1.5 for the test cycles.

Gap cracking is predicted by use of the models with complete solder joints. For the asymmetric model, the side with solder gap height of 60 μm was considered, which matches the mean height measured.

7. Characterization and Modelling of Solder Joint Reliability 437

- Fillet cracking is predicted by use of the models with the assumption of fully cracked gaps.
- The difference between the two solders is negligible for both test cycles.
- When the test cycle –40 to 125°C is compared to the –40/85°C cycle, the critical cycle numbers are generally lower, as has to be expected, the averaged acceleration factors for test cycling are AF=1.6 for SnAgCu and AF = 1.5 for SnPb solder, which slightly underestimate the experimental results given above.
- For SnAgCu, slightly higher critical cycle numbers are predicted.

FE-analysis results for field cycles

The investigations on the mission profile cycles aimed at statements concerning the acceleration factors between test cycles and field cycles. Because of the better predictive capabilities of the above-mentioned hypotheses for prediction of gap cracking the investigations on the mission profile cycles were limited to this kind of failure. Accordingly, the calculations were performed only with the unsymmetrical and uncracked model.

Table 15. Estimates of mean cycles to gap cracking numbers for different field cycle conditions and acceleration factors in relation to the test cycle of 85°C to –40°C

Model	Cycle	Solder	Mean number of cycles to gap cracking	Acceleration factor cycle 85 to – 40	Acceleration factor cycle 85 to – 40
			FEA	Analytical estimate	
unsymmetrical, uncracked	25...–7°C	SnAgCu	16,200	16.2	
	20...45°C	SnAgCu	22,700	22.7	
	20...60°C	SnAgCu	8,300	8.3	
	25...+7°C	SnAgCu	56,000	56	
	–23...25°C	SnAgCu	7,100	7.1	
	Composed	SnAgCu	20,500		
unsymmetrical, uncracked	25...–7°C	SnPb	12,700	12.7	19
	20...45°C	SnPb	19,000	19	23
	20...60°C	SnPb	7,200	7.2	7.5
	25...+7°C	SnPb	32,700	32.7	58
	–23...25°C	SnPb	7,200	7.2	8.8
	Composed	SnPb	17,000		

Five different field cycles were investigated as defined above. It turned out after the first runs of field cycles of –7/25°C that it is required to calculate at least two cycles for each field cycle profile, because the stress relaxation in the solders is not complete. For each temperature profile only the results of the second cycle were used for the evaluations.

The critical mean numbers to gap cracking are listed in Table 15. It can be observed from the table that the critical field cycle numbers are generally higher for SnAgCu solder than for SnPb solder. This tendency is the stronger the smaller the cyclic temperature amplitude is. To calculate the acceleration factors, the mean numbers to gap cracking for the field cycles can simply be related to those calculated for the test cycles. This is done for the test cycle of –40/85°C by assuming a critical test cycle number of 1000 cycles for both solders. A similar calculation can easily be done also for the test cycle of –40/125°C.

The results for the composed cycle are established as follows: from the percentage of contribution of each field cycle a percentage of contribution of the failure measures, i.e., creep strain or creep strain energy density, to the fictitious "composed cycle", is calculated. Concerning the "damage potential" of the failure indicators creep strain or creep strain energy density the assumption is analogous to that for the relation of test cycle to field cycle: The cyclic accumulated equiv. creep strain or energy density of a certain amount has the same damaging effect, independent on the cycle duration, mean temperature etc. These effects have already been included in the calculation by use of the solder related temperature dependent creep properties and in the Manson-Coffin coefficients used for the cycles to failure estimation. The duration of the composed cycle is equal to the weighted sum of the field cycles durations divided by their number. In our case of approximately equal durations of the field cycles, it is equal to the duration of the field cycle, i.e., approx. 4 hours. Transitions from one field cycle to another are not considered. It can be seen from Table 15 that the 20/45°C field cycle dominates the composed cycle. This dominance is confirmed also when the acceleration factors are considered, which for the composed cycle do not differ much from the 20/45°C cycle.

In the last column of Table 15, the FE-calculation based acceleration factors are compared to estimates based on the Norris-Landzberg analytical formula (5), described in paragraph 4. Comparisons to the Norris-Landzberg estimates differ again by a factor in the range of 1.5 for the field cycles. FEA results are more conservative compared to the analytical estimates.

5.3 Parametric Study on Chip Size Packages

One of the advanced packaging solutions combining small package sizes, high I/O counts but standard pitches for board mounting are Chip Scale Packages (CSPs), used in diverse geometric realisations. The thermo-mechanical performance of one of the most frequently applied variants, the

Flip-Chip Chip-Scale Packages (FC-CSPs) on organic interposer substrate, was studied [71]. Lead-free solder interconnects were selected. The parametric studies on the CSPs were performed considering a wide variation of geometric and material parameters. Two geometrical versions on organic interposers with different die sizes were investigated theoretically and experimentally by thermal cycling tests of –40°C to 150°C. In the FE-analyses, several additional parameters were examined including BT-interposer thickness, standoff, perimeter vs. full array, and solder-mask defined vs. non-solder-mask defined (NSMD) balls. Underfilling of the CSPs was an additional option.

5.3.1 Chip Size Package with 10 × 10 mm² Die and 230 I/O (CSP 230)

Theoretical studies are presented to characterize parametric effects on the reliability; one of the most important applications of FEA-based virtual prototyping. The parametric studies of the CSPs were performed considering a wide variation of geometric and material parameters. The geometrical versions were based on an area array package with 230 I/O and a 10 mm square die placed on an organic interposer substrate, referred to as CSP 230 throughout the section. The ball pitch was 0.5 mm and the standard standoff 250 µm. For further geometric details compare Figure 51. In the FE-analyses, several parameter variations were examined including BT-interposer substrate thickness, standoff, perimeter vs. full array, and solder-mask defined vs. non-solder-mask defined (NSMD) balls. Second level underfilling was included as an additional option. For the model of the CSP 230, the solder masks were not taken into account.

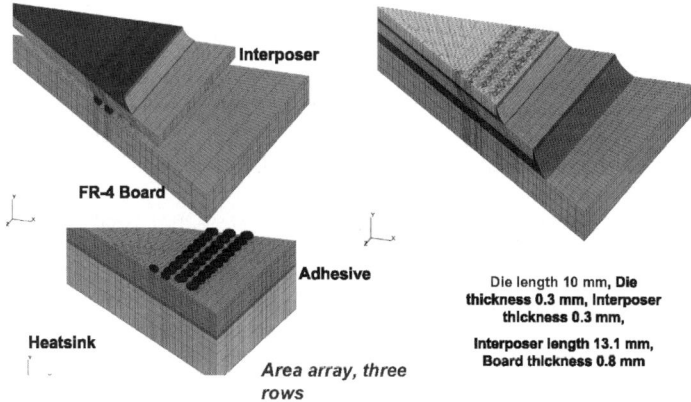

Figure 51. 3D-FE Models for the chip size package with 10 mm × 10 mm die (CSP 230)

In a pre-screening step, a simplified model of the standard CSP 230 configuration was investigated containing a reduced number of first level (FC-bumps) and second level interconnects (CSP-balls). It turned out that for the standard CSP 230 configuration without second level underfill, but with stiff underfill for the first level, the higher loads occur at the balls. Accordingly, the FC-bumps were not taken into account for the parametric studies, but the full ball array was. The double symmetric models shown in the figures contain HEX20 elements only.

A variety of material data is required for modelling. The CSP-PCB assemblies include metals such as Ni-UBM, Cu-traces and metallization, polymeric materials like epoxy mould compounds, solder masks and underfills, as well as BT and FR-4 used as interposer or substrates and the silicon die. Most of these materials have temperature dependent behaviours. FR-4 boards and BT substrates consist of layers of epoxy-glass laminate alternating with thin sheets of copper, and are assumed to be orthotropic. For these analyses, only the overall behaviour of the substrates was of interest. Accordingly, they were modelled as homogeneous materials. The anisotropic thermal expansions of the organic FR-4 and BT substrates as well as their elastic anisotropy are both accounted for in the analyses. Since only high Tg materials were used in the study, one with Tg = 180°C, the other with Tg = 160°C, no CTE changes had to be considered. The silicon die is treated as a linearly elastic body.

Table 16. Selected material data

	Constitutive Model	Young's Modulus (Instantaneous) [GPa]	CTE [10^{-6}/°C]
Chip	elastic	168.0	2.8E-6
Board FR-4	elastic-orthotropic	230 K $E_{x,y}$ 21 E_z 8 423 K $E_{x,y}$ 18 E_z 4	$\alpha_{x,y}$ 15 α_z 45
Interposer Substrate	elastic-orthotropic	230 K $E_{x,y}$ 22 E_z 8.3 423 K $E_{x,y}$ 18 E_z 5.2	$\alpha_{x,y}$ 14 α_z 45
Solder Mask	viscoelastic	230 K 5.5 340 K 3.4 423 K 1	60
Solder SnAg	secondary	210 K 48 473 K 33	20
Copper pad	elastic-plastic	90.0	16.5
Underfill A Tg = 160°C (TMA)	viscoelastic	230 K 12 390 K 7 433 K 4	< Tg 22 > Tg 24

7. Characterization and Modelling of Solder Joint Reliability 441

The behaviour of copper was considered elastic-plastic. The polymeric underfills and solder masks were treated as viscoelastic materials. During the FE-analyses two different underfill materials were used. Underfill A is a highly filled stiff material with Tg = 160°C, which is above the maximum cycle temperature, and underfill B, which is not as stiff, and its Tg is slightly lower than the maximum cyclic temperature. For the purpose of assembly and testing only underfill B was selected because of its better processing properties.

Eutectic Sn96.5Ag3.5 solder was used for the investigations applying the hyperbolic sine law (18) given by Darveaux.

5.3.2   Results for the Parametric Study

The failure indicator used in the study was the equivalent creep strain accumulated over one thermal cycle. Figure 52 exemplifies the characteristic distribution pattern of the accumulated equivalent creep strain at the end of one thermal cycle. The creep strain concentrates at the interface of the balls to the package site, which was also observed for all the other CSP versions investigated. For most of the versions the maximum accumulated equivalent creep strains were calculated for the inner peripheral row. Accordingly, this row is expected to fail first.

For comparison, the maximum equivalent creep strains calculated for each ball are plotted in column diagrams, each column representing one ball (see Figures 53 to 56). Unless stated otherwise, the results refer to the standard configuration with interposer thickness 0.3 mm, board thickness 0.8 mm, and a cycle of 150 to –40°C. Because of the symmetry only one half of the three peripheral rows is shown, i.e., the left hand side of the figures is related to balls in mid-position of the package.

The effect of a peripheral versus a full area array arrangement is shown in Figure 52. Maximum straining reaches similar values, located at an inner row. This well-known effect is caused by the action of the highly mismatched stiff die on the balls. However, the bending situation is very complex, and for a thicker board of 1.5 mm both an increase in maximum straining and a shift of the maximum towards the edge ball are observed. Since the balls are highly strained, actions to reduce ball straining were looked at. One optional action is to apply a second level underfill. Effects of different fictious underfills on ball straining are shown in Figure 54. There, the definition "stiff" is related to underfill A, "middle" describes an underfill with Tg = 135°C, E = 4GPa at –40°C and 2.5GPa at 120°C, the CTEs beeing 50 ppm/K below and 125 ppm/K above Tg, and "soft" describes an underfill with Tg = 80°C, E = 4GPa at –40 C and 1.4GPa at 120°C, the CTEs beeing 70 ppm/K below and 180 ppm/K above Tg. It is obvious from the figure that only the stiff material with the low CTE over the cyclic range really has the potential to improve the balls fatigue life. Note that neither solder masks nor FC-bumps were included.

A surprisingly clear reduction in maximum creep strain is caused by a slightly modified ball geometry, i.e., a ball configuration with on both top and bottom sides NSMD geometries, shown in Figure 55. However, practical realization of this geometry is difficult because of the resulting scatter of solder ball heights and related soldering problems. Another approach with difficulties for practical realization but significant reduction of ball straining is increasing the standoff height, e.g., from 250 µm to 300 µm. The reduction in maximum creep strain can be observed from Figure 56, but difficult processing as well as electrical shorts are likely and prevent pratical use of this design.

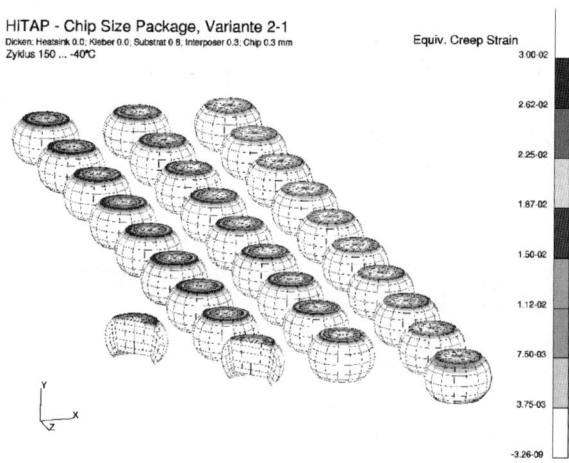

Figure 52. Accumulated equivalent creep strain distribution pattern at the CSP 230 solder balls for one thermal cycle 150...–40°C

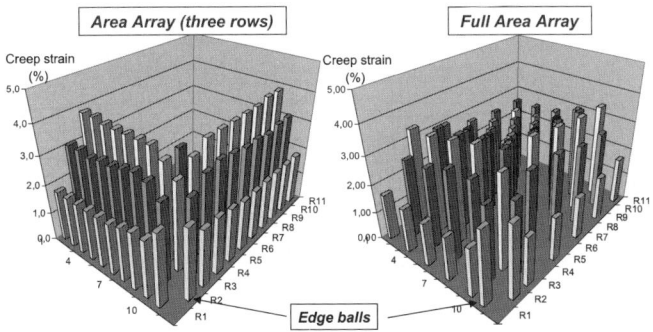

Figure 53. Schematic of the maximum cyclic creep strain at the solder balls of a CSP 230 quarter. Pheripheral (each ball modelled) vs. full area array (every second ball modelled)

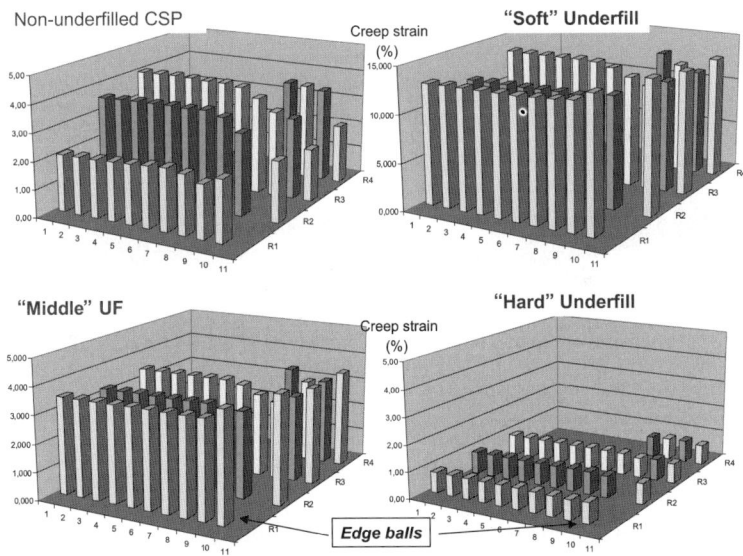

Figure 54. Schematic of the maximum cyclic creep strain at the solder balls of a CSP 230. Effects of different types of second level underfill

444   *Chapter 7*

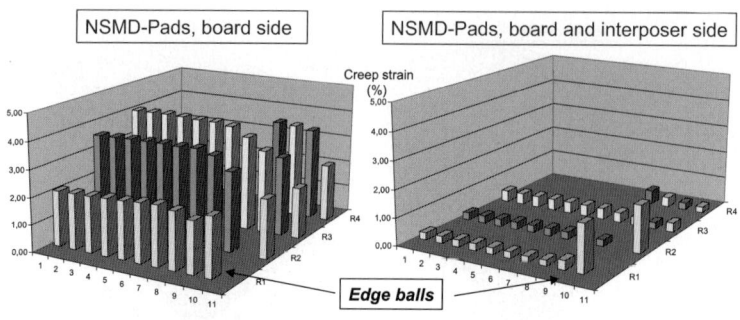

Figure 55. Schematic of the maximum cyclic creep strain at the solder balls of a CSP 230. Effect of solder mask vs. non solder mask defined pads at the interposer

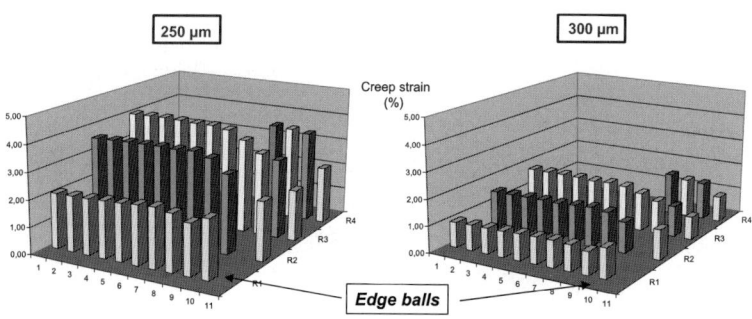

Figure 56. Schematic of the maximum cyclic creep strain at the solder balls of a CSP 230. Effect of ball height

5.3.3 Chip Size Package with 6×6 mm^2 Die and 168 I/O (CSP 168)

In an additional step an analogous CSP containing a 6 mm square die was investigated, which is called CSP 168 and shown in Figure 57. It was also investigated experimentally by air-to-air thermal cycling tests –40°C to 125°C as well as –40°C to 150°C, 1 hour. The assumption of a homogeneous temperature change throughout the assembly was applied in the analyses.

7. Characterization and Modelling of Solder Joint Reliability

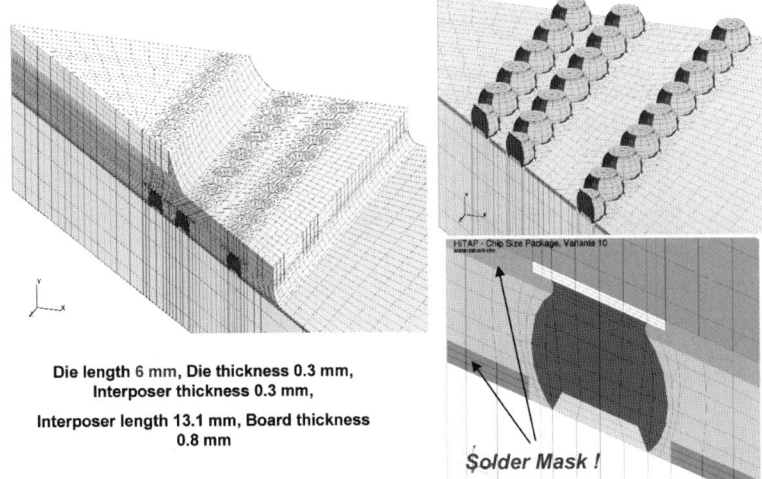

Die length 6 mm, Die thickness 0.3 mm,
Interposer thickness 0.3 mm,
Interposer length 13.1 mm, Board thickness 0.8 mm

Figure 57. 3D-FE model for the chip size package with 6×6 mm^2 die (CSP 168)

For the CSP 168 again a peripheral ball arrangement with 3 rows and 0.5 mm pitch was chosen. Opposite the CSP 230, for the CSP 168 the solder mask layers were considered in the FE model. The solder material was changed to SnAgCu and a softer underfill material B was applied.

Table 17. Additional material data

	Constitutive Model	Young's Modulus (Instantaneous) [GPa]	CTE [10^{-6}/°C]
Solder SnAgCu	Secondary + primary creep	210 K 48 473 K 33	20
Underfill B Tg = 140°C (TMA)	viscoelastic	230 K 9.3 413 K 7.0 443 K 0.1	< Tg 27 > Tg 92

The creep strain distribution pattern for the CSP 168 model, is given in Figure 58 for the underfilled case. Again, maximum straining does not occur in the outer peripheral row, but at the inner rows.

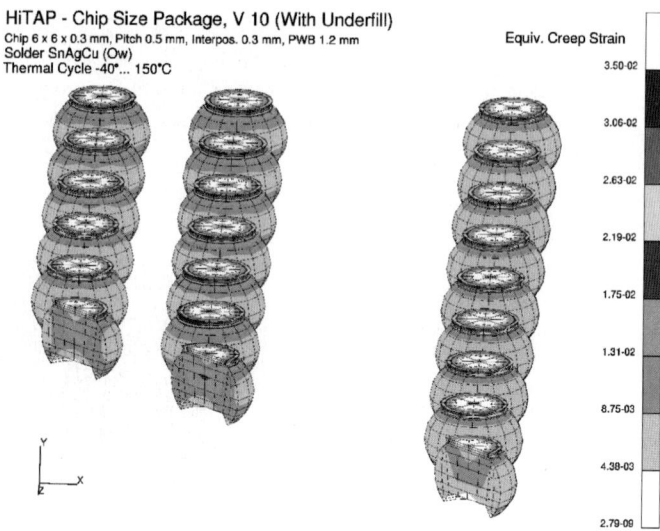

Figure 58. Accumulated equivalent creep strain distribution pattern at the CSP 168 solder balls for one thermal cycle of 150...–40°C. CSP underfilled on 1.2 mm thick board

When the empirical model described in paragraph 6, which assesses the amplitudes of the accumulated creep strain ε_{cr}^{acc} along the damage path for the estimation of the number of mean cycles to failure N_f by application of a Coffin-Manson type relation or an analogous criterion related to the viscoplastic energy density, dissipated per cycle, is applied, the number of mean cycles to macro crack initiation and mean cycles to failure can be estimated. The two values for each failure criterion, i.e., based on creep strain or energy, are calculated by averaging their physical quantities either at a peripheral ring of the mostly strained balls or over the total path of their local maximum. From comparisons between calculation and test results, it became obvious that the failure number calculated for "macro crack initiation" can be identified with a number of "cycles to first failure". The results of this estimation procedure are illustrated in Figure 58. It can be observed from the figure that both failure criteria based on creep strain and energy result in very similar cycles to failure predictions. It is also obvious from the figure that no increase in cyclic reliability is predicted by applying underfill B. These calculation results are somewhat unexpected, since underfill B is relatively stiff, and an improvement of the cyclic reliability was predicted for very stiff underfills. However, in reality the underfill does not form a pure layer but a layer composed of the soldermask layers at board and substrate side and the underfill itself, which is actually included in the CSP 168 model. Due to the low stiffness and high CTE of the composite layer, its stress reduction effect is reduced. If in the model no solder mask

7. Characterization and Modelling of Solder Joint Reliability

layers were included, a remarkably higher ball fatigue life would be predicted, as also can be seen in Figure 59. It is noted here that an additional effect of CSP underfilling was found from thermal cycling tests, the shift of ball failure towards bump failure. This effect depends on the choice of the two underfill materials and the use of solder masks. Because of the modelling simplifycations, the effect was not included in the simulations.

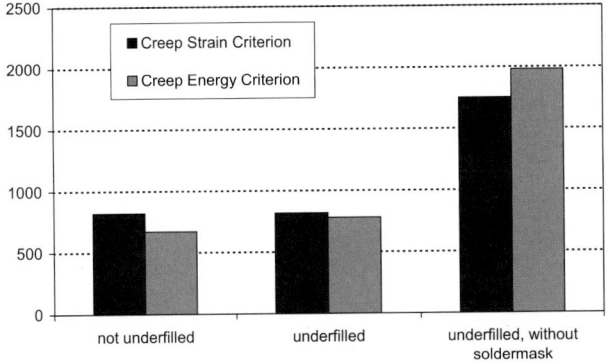

Figure 59. Estimated mean cycles to first failure of solder balls for CSP 168 subjected to temperature cycles 150°C to –40°C

A comparison between failure prediction and test results is given in Table 18. In the tests electrical failure was detected by in-situ resistance measurements. Because of the arrangement of the daisy chains, first level and second level interconnect failures could not be monitored separately. The predicted critical cycles to failure numbers tend to be too high. This difference is likely to be due to deviations of the real packages from the idealized geometrical assumptions.

Table 18. Cycles to failure as estimated from calculation and electrically measured during thermal cycling

	Temperature cycle (°C)	Cycles to first failure		Cycles to 50 % failure	
		Test	Simulation	Test	Simulation
CSP 168	–40 to 150	700	750	900	1600
CSP 168 with UF B	–40 to 150	450	800	1400	1200
CSP 168	–40 to 125	1400		1900	

For the packages selected, failure was analyzed by cross-sectioning. Figure 60 shows characteristic ball cross sections at the inner peripheral ball row. Cracking occurs, as predicted, close to the interposer pad-solder interface starting at the creep strain concentration at the interface edges. The tendency of higher failure probability at the inner ball rows was also confirmed experimentally. However, since the technological focus of the project was the introduction and further development of lead-free soldering technology, soldering errors, badly shaped solder balls and voids could not be fully be avoided during the fabrication of the demonstrator packages. The effects of these process-induced defects on package reliability are likely to decrease package reliability when compared to the predictions. An additional source of failure is seen in the high probability of an inadequate representation of the scattering viscoelastic behaviour of the polymeric materials when they come closer to their Tg temperatures, as it is the case for the cycling range up to 150°C.

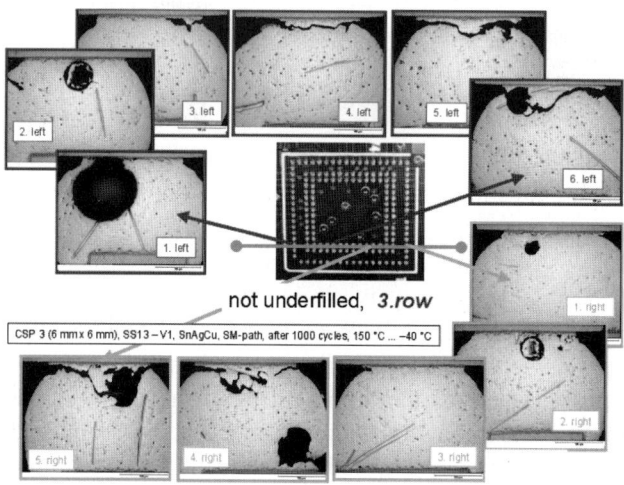

Figure 60. Cross sections of the inner peripheral ball row for a CSP 168 after 1000 cycles at 150 to –40°C

5.3.4 Summary of the Parametric Study

Area array packages differ from other package types mainly by a direct package attach to the board with great distances to neutral point of the solder joints, which can only be reliable if the thermal mismatch between package and PCB is small. However, for the considered FC-CSPs a stiff underfill material is required to cement the chip on the interposer substrate to guarantee a sufficient flip chip bump reliability. Decupling of the die from the CSP balls is limited in this case, what is simultaneously a limiting factor for the ball reliability.

FEA based parametric studies have indicated that both the creep strain and dissipated strain energy density represent suitable indicators to predict cyclic damage trends, also for the lead-free solders. Prediction of trends is the main advantage of the theoretical analyses, while the various deviations of the real package from idealized models, including the process induced defects, require thermal cyclic tests for verification of the trends for selected demonstrators.

Maximum straining was typically calculated at the inner ball row in vicinity of the CSP pads-solder interfaces, were also cracking occured. Major effects on ball fatigue life were theoretically shown to be standoff height, on both sides NSMD defined balls, and a stiff first and second level underfill. The latter trend was also verified experimentally, i.e., the CSP reliability with an underfill layer, softened by composition of the underfill with soldermask layers at board and substrate side, can be worse than that of a non-underfilled CSP. Heatsink attach tends to further decrease ball reliabilty, even in the case of decoupling by a very soft adhesive layer.

In general SnAgCu solder has shown a reliability comparable to or better than SnPb solder, but was in general on a low level for cycling temperatures up to 150°C. However, a higher potential for reliability improvement can be expected not by the choice of solder, but by the choice of improved underfill and solder mask materials.

5.4 Flip Chip on Board Assemblies

Growing demands on performance, cost and the advancement in IC technology have affected packaging and interconnection technology. Among other advanced packaging technologies the flip chip on board technology was developed to handle larger dies, higher I/O counts, lower operating voltages, high power consumption and high clock frequencies. However, their thermo-mechanical reliability, which is discussed in the next paragraph, is one of the factors limiting its use. The major difficulty connected with FCOB solutions originates from the high difference in CTE of the silicon die and the organic substrate material, i.e., the global mismatch. It is obvious that the effects of CTE mismatch become more severe with increasing die size. To circumvent this difficulty, the reliability of FC structures can be enhanced by applying an epoxy-based underfill between the die and the substrate, encapsulating the solder joints. This material, once cured, mechanically couples the IC and substrate together to locally constrain the CTE mismatch. However, the encapsulation step adds additional cost to the manufacturing process. It is therefore interesting to study the effect of underfilling and to see the limits of FC technology without underfilling, which is related to FC on low CTE substrates (e.g., ceramics) or to FCOB assemblies with small die sizes.

The underfilled die mounted on the substrate is a multimaterial structure. The mismatch in coefficients of thermal expansion between the different materials induces stresses and strains if the structure is subjected to temperature changes. The reliability of the solder joints is only one critical issue. Because forming of sharp interface edges cannot be avoided, delamination failure is the second one of the major failure risks. Such sharp edges, causing bi-material notch effects, occur at the underbump metallization-die interfaces, at the bump-metallization interfaces, at the underfill-metallization interfaces, at the solder-mask-underfill interfaces, at the die edges, and at the underfill fillet at the outer periphery of the die.

5.4.1 Flip Chip Assemblies with Small Dies

FE Modelling

Investigations on flip chip assemblies with small dies of size 2.2×2.2 mm^2 and thickness 300 µm mounted on an FR-4 substrate of thickness 400 µm are treated as an example for a typical FC-assembly with small die. The quarter model assuming double symmetry is shown in Figure 61. The area array configuration shown contains 11 eutectic tin-lead bumps with 55 µm bump height. Additional materials used in the model include nickel under bump metallization, copper pads, and solder mask (Table 19). An underfill is optionally added. The metallization behaviour is considered as elastic-plastic. For the tin-lead solder Sn63Pb37 only secondary creep assuming the law of Grivas/Hacke, equation (12), was used. The polymeric materials underfill and solder mask were again treated as viscoelastic materials.

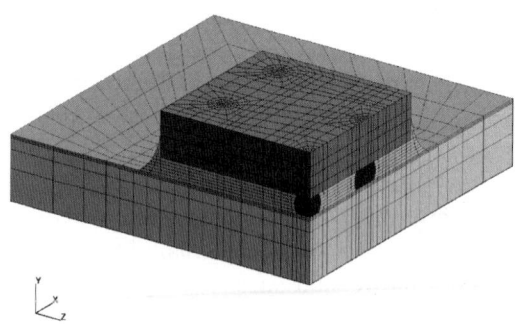

Figure 61. 3D-Finite element model of a FCOB assembly with small die of size of 2.2 mm square: area array bump configuration is exemplified

7. Characterization and Modelling of Solder Joint Reliability

Table 19. Selected material data

	Constitutive Model	Young's Modulus (Instantaneous) [GPa]	CTE [$10^{-6}/°C$]
Chip	elastic	168.0	2.8
Board FR-4	elastic-orthotropic	230 K $E_{x,y}$ 21 E_z 8 423 K $E_{x,y}$ 18 E_z 4	$\alpha_{x,y}$ 15 α_z 45
Solder Mask	viscoelastic	230 K 5.5 340 K 1.4 373 K 0.33	60
Solder PbSn	secondary (equat. (12))	36 at 233 K 21 at 398 K	24.0
Copper pad	elastic-plastic	90.0	16.5
Underfill A Tg = 160°C (TMA)	viscoelastic	233 K 12 393 K 7 433 K 4	< Tg 22 > Tg 24
Underfill D Tg = 130°C (TMA)	viscoelastic	230 K 4 393 K 2.5 413 K 0.85	< Tg 63 > Tg 115

Results – Effects of Underfilling

The shear deformation in the solder bumps of the FCOB without applying an underfill can easily be shown by an FE-analysis, see Figure 62. While for large dies fatigue failure of the bumps occurs immediately, bump straining is not as high in the small die assembly shown in the figure.

The effect of underfilling is, roughly speaking, to shift the overall assembly behaviour from shearing of the soft bumps between two ridged plates to bending of a bi-material plate with adhesive layer, within which the bumps are embedded. This shift of the deformation mode has also been proven experimentally by means of the micro-DAC deformation measurement [5].

The shift from bump shearing to bi-material assembly bending can be observed by comparison of the deformation shown in Figure 62 to that of an underfilled assembly shown in Figure 63. Additionally the bending stress within the die can be observed from the figures, being tensile on the top side and compressive on the bottom side. Note that in the model the boundaries of the board are not allowed to warp freely to account for the action of other dies/components mounted on the board.

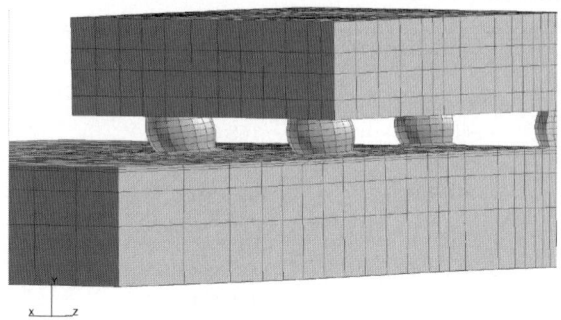

Figure 62. PbSn solder bump deformation for an assembly without underfill on FR-4 board after cooling down from 125°C to –40°C, 20 × magnified

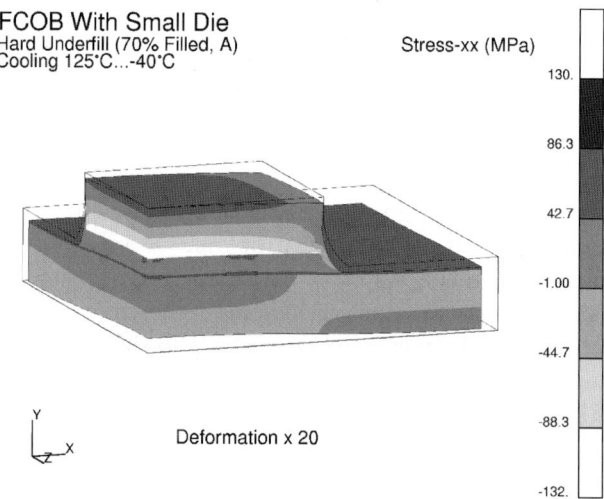

Figure 63. Deformation for an assembly with stiff underfill on FR-4 board after cooling down from 125°C to –40°C, 20x magnified, and bending stress-xx component

Opposite the reduction of shear strain, the presence of underfill might cause remarkable bump creep strain in the transverse board direction that is induced by stretching and compressing the bump during thermal cycling.

7. Characterization and Modelling of Solder Joint Reliability

Note that the creep strain component is compressive after cooling and concentrates in the bump inner region. This load is due to the CTE mismatch between the bumps and the combined solder mask/underfill layer. A high CTE difference between this layer and the solder bumps causes high transverse creep strains, as e.g., shown in Figure 64 for underfill D with a CTE of 63 ppm/K. Even if the underfill is matched to the bumps, the composite layer of underfill/solder mask cannot be matched, because the CTE values of different solder mask materials were measured to be 60-80 ppm/K (below Tg) in most applications. Note that the solder mask can occupy one third or even more of the gap. Accordingly, matching or avoiding a solder-mask layer can often enhance FCOB reliability.

The equivalent creep strain, which is considered to be a measure for the bump fatigue life, is shown in Figure 65 for the same assembly with Underfill D after one thermal cycle. Comparison of both figures yields the equivalent creep strain mainly determined by the transverse normal strain component, and not the shear creep strain component.

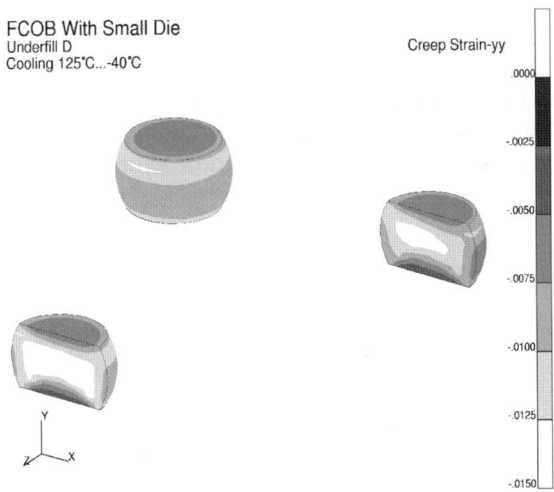

Figure 64. Creep strain component perpendicular to the board direction after cooling down from 125°C to –40°C, Underfill D

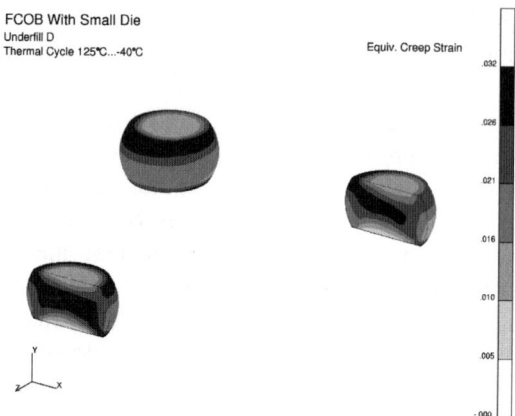

Figure 65. Accumulated equivalent creep strain after one thermal cycle of 125°C to –40°C, Underfill D

The use of a stiff, highly filled material with CTE matched to solder (Underfill A) reduces solder straining, see Figure 66. Additionally, the maximum creep strain is now located at the interface to the die. The high strain inner region of the bump, visible for the assembly with Underfill D vanishes, because the local mismatch-induced normal component of the creep strain reduces drastically. The dissipated creep energy density exhibits the same distribution pattern, see Figure 67.

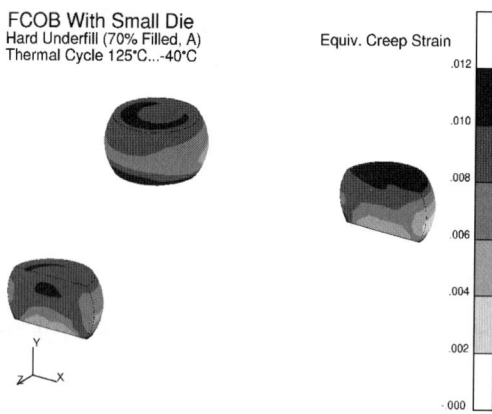

Figure 66. Accumulated equivalent creep strain after one thermal cycle of 125°C to –40°C, Underfill A

7. Characterization and Modelling of Solder Joint Reliability

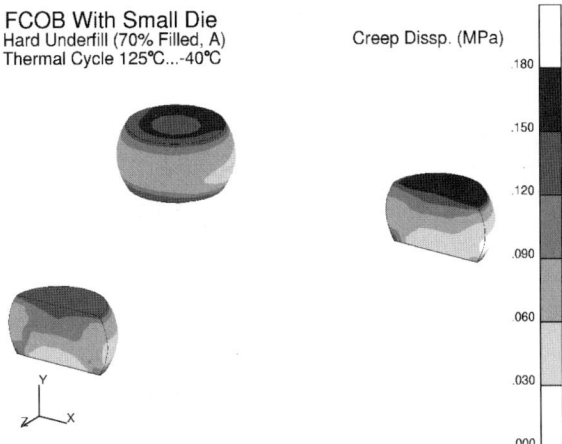

Figure 67. Creep dissipation, energy density after one thermal cycle of 125°C to –40°C, Underfill A

An additional failure risk in underfilled flip chip assemblies is known to be delamination of the polymeric layers from the die or from each other. Unfortunately, no simple criterion can be given for the delamination risk of underfilled flip chips and is not regarded here. If solder creep is assumed to be the limiting factor of the low cycle fatigue life of the flip chip assembly, which is certain only for assemblies without underfill, the Coffin-Manson type methodology results in a much higher fatigue life for the hard underfill (A) with CTE well-matched to solder. For the small chip, even the non-underfilled assembly on organic board has a better fatigue life prediction than those with softly underfilled (underfill D) chips.

5.4.2 Flip Chip Assemblies with Large Dies

Even with the addition of an underfill material, it is assumed that there are limits on the die size used in flip chip applications. It is therefore a challenging question to study the impact of die size on their thermo-mechanical reliability for large ICs attached to organic boards.

Accordingly, the objective was to investigate the reliability of large FCOB assemblies through both thermal cycling tests and numerical simulation [72]. Samples with die sizes from 10 mm to 40 mm and a bump pitch of 500 μm were prepared, see Figure 68. They were subjected to thermal cycling (AATC –55°C/+125°C) to verify the simulation results with the experimental findings.

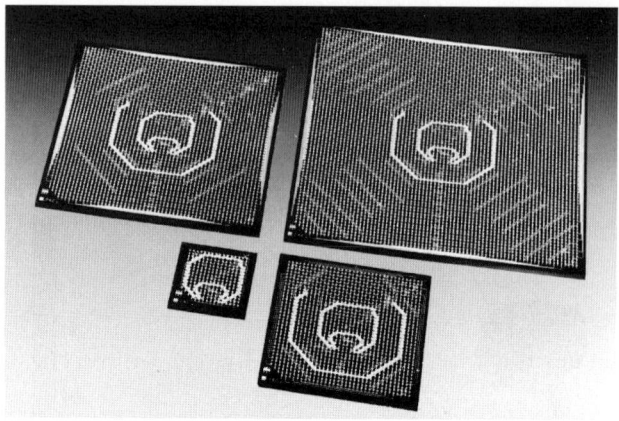

Figure 68. Single chips, IC size 10 mm to 40 mm, I/O counts 250 to 5770

Finite Element Modelling

For purposes of this study, four octant symmetric 3D FE models were created for 10 mm, 20 mm, 30 mm, and 40 mm square silicon dies of 0.625 mm thickness. The dies were mounted onto a 1mm thick FR-4 board in an area-array configuration with a pitch of 500 µm (Figure 69). The interconnects were eutectic SnPb solder joints with bump heights of 90 µm. The remaining gap between the silicon die and the substrate was filled by a solder mask with 20 µm in height and the balance by an underfill material.

Materials used in the structures included FR-4 as board, nickel metallization and copper traces, silicon die, underfill and solder mask. The data different from the other FCOB investigation is given in Table 20. For the solder Sn63Pb37 again only secondary creep was taken into account as stated above. The polymeric materials underfill and solder mask were treated as viscoelastic materials.

Table 20. Materials characteristics additional to Table 19

	Constitutive Model	Young's Modulus (Instantaneous) [GPa]	CTE [$10^{-6}/°C$]
Solder Mask	Viscoelastic	218 K 6 295 K 4.1 373 K 0.33 423 K 0.1	30
Underfill Tg = 130°C (TMA)	Viscoelastic	218 K 9 393 K 6 433 K 4	< Tg 29 > Tg 73

7. Characterization and Modelling of Solder Joint Reliability

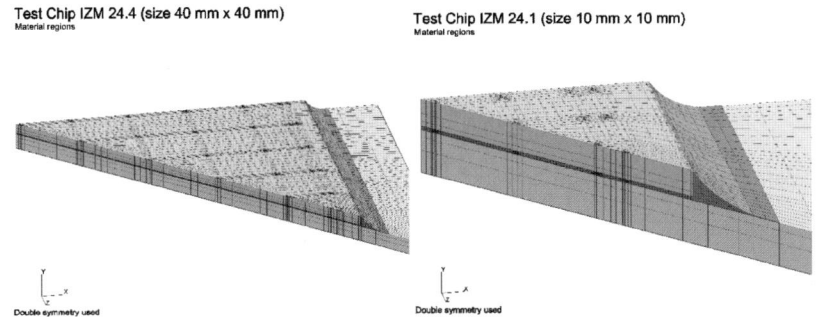

Figure 69. 3D-Finite element models of FCOB assemblies with die sizes of 10 mm and 40 mm

Sometimes particle settling is observed in the underfill material. To quantify effect of this on the solder reliability, the underfill layer was considered two-layered with pure epoxy in one and epoxy filled with silica in the other layer.

The temperature cycle was chosen to range from 125°C to −55°C with 5 min ramp times and 10 min dwell times for a total cycle time of 30 min. The temperature ramps are again defined by decay functions.

The parameters varied included the die size and the effect of particle settling in the underfill. Apart from them, the deformed shapes of the flip chip structures depended on the size and assembly constraints of the printed wiring boards (clamped board or not clamped board).

Results and Discussion

Figure 70 shows plots of the accumulated equivalent creep strain distribution patterns in the outermost solder bumps for flip chip assemblies with a die size of 10 mm and 30 mm. It is evident that creep strain is distributed throughout the solder bump and is not constant. Calculation results for the eutectic solder joints indicate a creep strain concentration near the outer die pad edge. The solder crack is expected to start circularly near the outer die pad edge and to grow underneath the die pad or later along the diagonal direction crossing the solder bump. When comparing the influence of the two die sizes, the accumulated equivalent creep strain distributions in the solder are very similar. Figure 71 illustrates the effects of die size (10 mm to 30 mm) for a not clamped board on the maximum creep strain amplitude across the solder bump. Obviously these creep strains do not depend on the die size.

It was shown above that for non-underfilled flip chips, solder joint deformation is dominated by shear due to global CTE mismatch, and is proportional to the distance to the neutral point (DNP). In case of an underfilled flip chip, there is no major deformation in shearing direction because all parts of die and substrate are tightly adhered and the underfill protects the joints. As a result, the strain caused by the CTE mismatch is converted to the deformation in transverse direction and leaves no major strain in the joint.

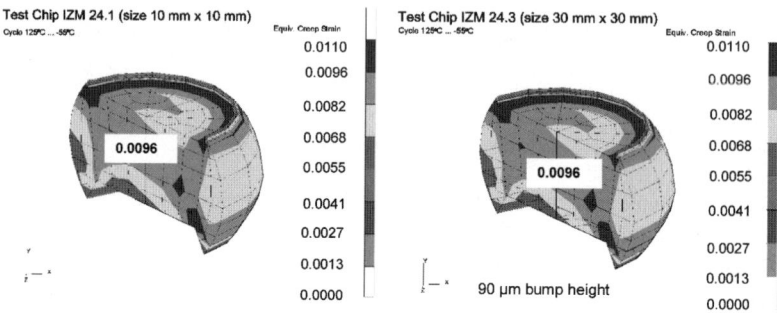

Figure 70. Accumulated equivalent creep strain distribution in the outermost solder bumps (90 µm height) at the end of one temperature cycle

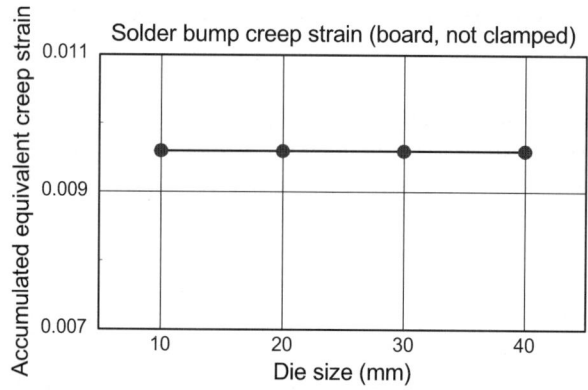

Figure 71. Accumulated equivalent creep strain in the outermost solder bumps (minimum along the damage path) for different die sizes of FCOB assemblies (board, not clamped)

This deformation depends on the CTE and modulus of each material. The solder bump strain is independent of DNP [73].

Sometimes particle settling is present in the underfill material. From the mechanical point of view, this means lowering of the overall attachment stiffness due to an additional "layer" of pure epoxy with low Young's modulus and high CTE. The presence of underfill particle settling might cause remarkable bump creep strain in the transverse direction by stretching and compressing the bumps during thermal cycling. This leads to an increase in the creep strain range (Figure 72) and to a reduction in life time.

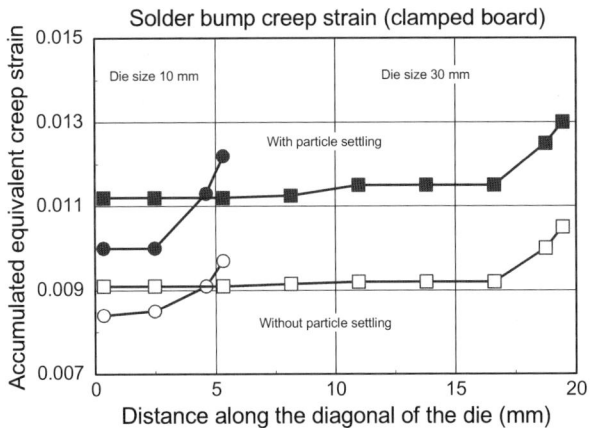

Figure 72. Creep strains in the solder bumps (height 90 µm) in dependence on bump position and particle settling effect for die sizes of 10 mm and 30 mm

The FE analyses have also shown that the fundamental limits on die size for an underfilled flip chip with a chosen clamped board condition are not crucially influenced by the DNP of the outermost solder joint. The major consideration was solder fatigue and cracking in these very large packages where the difference in CTE between the substrate and the die is about one order of magnitude.

The effect of local transverse CTE-mismatch does not only lead to normal straining of the bump, but causes also a ring of tensile peel stress at the die-underfill interface around the under-bump metallization, with the concentration line being the circle at the interface edge (Figure 73). This peel stress, being tensile after cooling, is one major delamination risk [73, 74]. Note that it is local in character and, therefore, almost independent of bump position. Note further that a "high CTE layer", that can be either a pure epoxy layer caused by filler settling or a solder mask layer, can strongly influence the transverse mismatch effects. The influence of such a layer is shown in Figure 74, where the peel stress amplitude rises for an assembly with underfill settling. Although underfilling can reduce the strains in the solder joints, it results in the risk of delamination along the die-underfill interface during temperature cycling. As shown in Figure 73, this peel stress is more influenced through the particle settling effect in the underfill as through the die size.

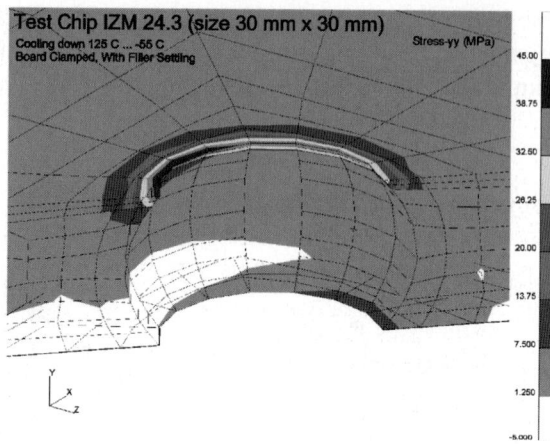

Figure 73. Peel stress σ_{yy} (MPa) in the underfill in bump vicinity (with filler settling, board clamped, thermal load 125°C ... –55°C)

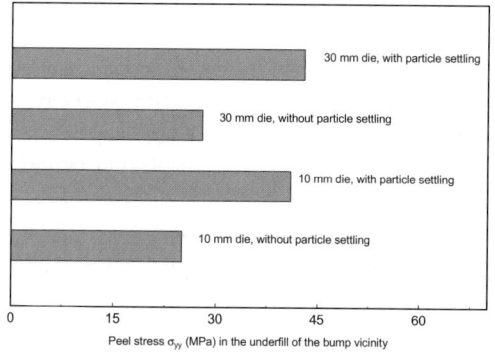

Figure 74. Peel stress σ_{yy} (MPa) in the underfill in bump vicinity (bump height 90 μm) for two die sizes (board clamped, thermal load 125°C ... –55°C)

Local interface delaminations of the underfill from the die passivation can be visualized by scanning acoustic microscopy. Figure 75 shows characteristic delaminations in the bump vicinity for different die sizes. Even at the large 40 mm dies, the sites of failing bumps were independent of location, and depended more on the absolute number of solder bumps on the die. Increasing the number of bumps appears to raise the statistical possibility of failure.

7. Characterization and Modelling of Solder Joint Reliability

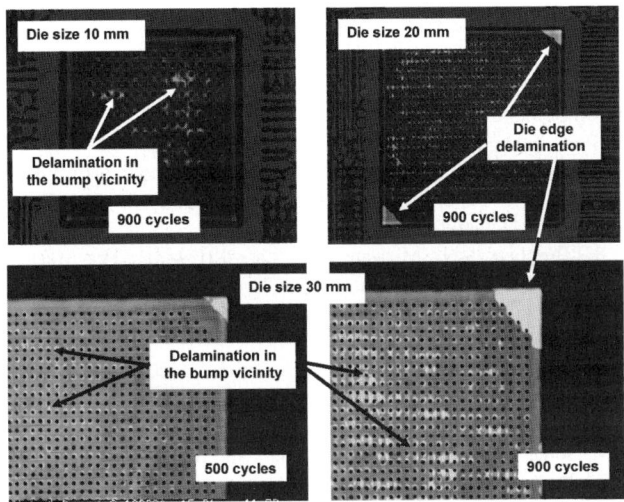

Figure 75. C-SAM images of the FCOBs demonstrating local underfill delaminations close to the solder bumps after thermal cycling 125°C to –55°C

From the simulation point of view, the results can be summarized as follows: The solder bump most at risk of a non-underfilled flip chip is the one at the corner of the die. In a 40 mm die, the bump at the corner is so far from the center that the stresses caused by thermal cycling are predictably large. The effect of underfill – stated rather crudely – is to greatly dampen this lateral motion and to convert some of it into vertical motion. For an ideally underfilled flip chip, the strain fields are relatively independent of die size and DNP. These facts suggest that larger die samples will have increased propensities for fracture not due to increased internal strains and stresses, but due to the fact that they are statistically more likely to contain the critical size flaws under stress conditions for catastrophic crack propagation [75].

The relationship between die size and solder joint failure has a subordinate importance. Manufacturing and process-induced defects are likely to be more important limiting factors (e.g., particle settling). The assembly constraints and the influence of adjacent components may also significantly impact the determination of fundamental limits to die size [76, 77].

6. REFERENCES

[1] Dudek, R., Lead-free Solder Interconnects: Characterization, Testing and Reliability, 3rd Int. Conf. on Benefiting from Thermal and Mechanical Simulation in (Micro)-Electronics (€SIME), Paris, France, April 15-17, 2002, Proc. pp. 62-72.
[2] Lau, H. (ed.), Solder Joint Reliability, Van Nostrand Reinhold, New York, 1991.
[3] Lau, J.H. and Pao, Y.-H., Solder Joint Reliability of BGA, CSP, Flip Chip and Fine Pitch SMT Assemblies, McGraw-Hill, 1997.
[4] Frear, D.R., Burchett, S.N., Morgan, H.S., Lau, J.H. (eds), The Mechanics of Solder Alloy Interconnects, Van Nostrand Reinhold, New York, 1994, pp. 60.
[5] Lau, J.H., Thermal Stress and Strain in Microelectronics Packaging, Van Nostrand Reinhold, 1993.
[6] Manson, S.S., Thermal Stress and Low-Cycle Fatigue, McGraw-Hill, New York, 1966.
[7] Engelmaier, W., Functional Cycles and Surface Mounting Attachment Reliability, Circuit World, Vol. 11, No. 3 (1985), pp. 61-72.
[8] Wild, R.N., Some Fatigue Properties of Solders and Solder Joints IBM Techn. Rep. No. 73Z000421, January 1973.
[9] Engelmaier, W., Turbini L.J., Design for Reliability, A White Paper for SM and Related Technologies for the Surface Mount Council, Mendham/Atlanta, 1995, presented at Workshop Reliability of SMT-Solder Joints, Berlin, October 1996.
[10] Clech, J.-P., Solder Reliability Solutions: a PC-based design-for-reliability tool, Proceedings, Surface Mount International Conference, Sept. 8-12, 1996, San Jose, CA, Vol. I, pp. 136-151.
[11] Clech, J.-P., Assessing the Solder Joint Reliability of Fine Pitch SMT, BGA, Flip-Chip and CSP Assemblies, Course, SMT Nueremberg, June 1998.
[12] de Kluizenaar, E.E., Reliability of Soldered Joints, Tutorial for the 5th International Conference on Interconnection Technology in Electronics, Fellbach, February 1990.
[13] Norris, K.C., Landzberg, A.H., Reliability of Controlled Collapse Interconnections, IBM Journal of Research and Development, May 1969, pp. 266-271.
[14] Clech, J.-P., Handwerker, C., Review and analysis of lead-free solder material properties BGA, Flip-Chip, ESPI Inc., NIST, August 20, 2002.
[15] Biglari, M.H., Oddy, M., Oud, M.A., Davis, P., Pb-Free Solders Based on SnAgCu, SnAgBi, SnCu and SnCu for Wave Soldering of Electronic Assemblies, Proc. Electronics Goes Green 2000+, Berlin, Germany, Sept. 11-13, 2000, pp. 73-82.
[16] Neu, R.W., Scott, D.T., and Woodmansee, M.W., Thermomechanical behaviour of 96Sn-4Ag and Castin Alloy, ASME Transactions, Journal of Electronic Packaging, Vol. 123, No. 3, September 2001, pp. 238-246.
[17] Darveaux, R., Banerji, K., Mawer, A. and Dody, G., Reliability of plastic ball grid array assemblies, Chap. 13, Ball Grid Array Technology, ed. J.H. Lau, McGraw-Hill, 1995, pp. 379-442.
[18] Lau, J.H., Chang, Ch., TMA, DMA, DSC, and TGA of Lead Free Solders, 48th ECTC Electronics Components & Technology Conf., Seattle, Washington, USA, May 25-28, 1998, pp. 1339-1344.
[19] Grivas, D., Murty, K.L., Morris, Jr., J.W., Deformation of Pb-Sn eutectic alloys at relatively high strain rates, Acta Metallurgica, 27, 1979, pp. 731-737.
[20] Weertman, J., J. Appl. Phys. 26, 1213 (1955), 27, 832 (1956), 28, 1185 (1957).
[21] Bird, J.E., Mukherjee, A.K., Dorn, J.E., Israel Univ. Press (1969) 255.
[22] Schubert, A., Walter, H., Dudek, R., Michel, B., Lefranc, G., Otto, J. and Mitic G., Thermomechanical properties and creep deformation of lead-containing and lead-free solders, Proceedings, 2001 International Symposium on Advanced Packaging Materials, pp. 129-134.

[23] Knecht, S., Integrated Matrix Creep: Application to Lifetime Prediction of Eutectic Tin-Lead Solder Joints, Proceedings of Mat. Res. Soc. Symp., Vol. 203, 1991 Materials Research Society.
[24] Hacke, P.L., Sprecher, A.F., Conrad, H., ASTM Symposium Thermomechanical Fatigue Behaviour of Materials, San Diego, 1991, ASTM STP 1186, 1993.
[25] Hacke, P.L., Sprecher, A.F., Conrad, H., in Thermal Stress and Strain in Microelectronics Packaging, 1993, edited by J. H. Lau, pp. 467-499.
[26] Knecht, S., Fox, L., Integrated Matrix Creep: Application to Accelerated Testing and Lifetime Prediction, in Lau, H. (ed.), Solder Joint Reliability, Van Nostrand Reinhold, New York, 1991, pp. 508-544.
[27] Darveaux, R. and Banerji, K., Constitutive relations for tin-based solder joints, IEEE Transactions on Components, Hybrids, and Manufacturing Technology, Vol. 15, No. 6, Dec. 1992, pp. 1013-1024.
[28] Schubert, A., Dudek, R., Döring, R., Walter, H., Auerswald, E., Gollhardt, A., Schuch, B., Sitzmann, H., Michel, B., Lead-free Solder Interconnects: Characterization, Testing and Reliability, 3rd Int. Conf. on Benefiting from Thermal and Mechanical Simulation in (Micro)-Electronics (€SIME), Paris, France, April 15-17, 2002, Proc. pp. 62-72.
[29] Kariya, Y. and Plumbridge, W.J., Mechanical properties of Sn-3.0mass%Ag-0.5mass% Cu alloy, Proceedings, 7th Symposium on Microjoining and Assembly Technology in Electronics, Yokohama, Japan, Feb. 1-2, 2001, pp. 383-388.
[30] Amagai, M., Mechanical reliability in electronic packaging, Microelectronics Reliability 42, 2002, pp. 607-627.
[31] Wiese, S., Meusel, E., Wolter, K.-J., Microstructural dependence of constitutive properties of eutectic SnAg and SnAgCu solders, Proceedings, 53rd Electronic Components & Technology Conference, 2003, pp. 197-206.
[32] Lau, J. H., Dauksher, W., Vianco, P., "Acceleration models, constitutive equations, and reliability of lead-free solders and joints", Proceedings, 54th Electronic Components & Technology Conference, 2003, pp. 229-236.
[33] Vianco, P., Rejent, J., Compression Deformation Response of 95.5Sn-3.9Ag-0.6Cu Solder, UCLA Lead-Free Workshop, October 2002.
[34] Zhang, Q., Dasgupta, A., Haswell, P., Viscoplastic constitutive properties and energy-partitioning model of lead-free Sn3,9Ag0.6Cu solder alloy, Proceedings, 53rd Electronic Components & Technology Conference, 2003, pp. 1862-1868.
[35] Clech, J.-P., Lead-free solder joint reliability, ESPI Inc., Workshop held at IPC/JEDEC 4th Int. Conf. on Lead-Free Electronic Components and Assemblies, Frankfurt, Germany, Oct. 2003.
[36] Pang, J.H.L., Design for Reliability Test and Analysis for Pb-free Solders, in B. Michel, R. Aschenbrenner: The World of Electronic Packaging and System Integration, IZM special edition, ddp goldenbogen, Dresden, 2004, pp. 359-366.
[37] Pang, J.H.L., Xiong, B.S., Neo, C.C., Zhang, X.R. and Low, T.H., Bulk Solder and Solder Joint Properties for Lead Free SnAgCu Solder Alloy, Proceedings, 53rd Electronic Components & Technology Conference, 2003, pp. 673-679.
[38] NIST-Boulder database, Database for Solder Properties with Emphasis on New Lead-free Solders, Properties of Lead-Free Solder, Release 4.0, 2003, by Siewers, T., Liu, S., Smith D.R., Madeni J.C., National Institute of Standards and Technology, Colorado School of Mines, available at http://www.boulder.nist.gov/div835/.
[39] Clech, J.-P., An Obstacle-Controlled Creep Model for Sn-Pb and Sn-Based Lead-Free Solders, Proc. SMTAI Conference, September 2004, Chicago, USA.

[40] Schubert, A., Dudek, R., E. Auerswald, A. Gollhardt, B. Michel, H. Reichl, "Fatigue Life Models for SnAgCu and SnPb Solder Joints Evaluated by Experiments and Simulation", Proceedings, 53rd Electronic Components & Technology Conference, 2003, pp. 603-610.
[41] Deplanque, S., Nuechter, W., Spraul, M. Wunderle, B., Dudek, R., Michel, B., Relevance of Primary Creep in Thermo-Mechanical Cycling for Life-Time Prediction in Sn-Based Solders, Proceedings, 6th EuroSimE Conference, Berlin, Germany, April 2005.
[42] Schubert, A., Dudek, R., Döring, R., Walter, H., Auerswald, E., Gollhardt, A., Schuch, B., Sitzmann, H., Michel, B., Lead-free Solder Interconnects: Characterization, Testing and Reliability, 3rd Int. Conf. on Benefiting from Thermal and Mechanical Simulation in (Micro)-Electronics (€SIME), Paris, France, April 15-17, 2002, Proc. pp. 62-72.
[43] Mayuzumi, M., Onchi, T., Creep Deformation and Rupture Properties of Unirradiated Zircaloy-4 Nuclear Fuel Cladding Tube at Temperatures of 727 to 857 K, Journal of Nuclear Materials, 175 (1990), pp. 135-142.
[44] Clech, J.-P., "An Extension of the Omega Method to Primary and Tertiary Creep of Lead-Free Solders", Proc. 55th ECTC Electronics Components & Technology Conf., Lake Buena Vista, USA, May 2005.
[45] R. Dudek, M. Nylen, A. Schubert, B. Michel, and H. Reichl, An Efficient Approach to Predict Solder Fatigue Life and its Application to SM- and Area Array Components, Proc. of ECTC 47, San Jose, May 1997, pp. 462-471.
[46] Akisanya, A.R., Meng, C.S., Initiation of Fracture at the Interface Corner of Bi-Materials Joints., Joun. of the Mechanics and Physics of Solids 51, 2003, pp. 27-46.
[47] Scherzer, M., Non-Linear Deformed Interface Corner Stress Characterization by Effective Parallel Numerical Methods in: B.L. Karihaloo, Y.W. Mai, M.I. Ripley, R.O. Ritchie (eds.): Advances in Fracture Research, Proceedings of the Ninth International Conference on Fracture, Vol. 4, April 1997, Sydney, Australia, Pergamon Press, Amsterdam, Oxford, New York, 1997, pp. 1959-1970.
[48] Lau, J.H., Shangguan, D., Lau, D., Kung, T., Lee, R., Thermal-Fatigue Life Prediction Equation for Wafer-Level Chip Scale Packages (WLCSP) Lead-Free Solder Joints on Lead-Free Printed Circuit Board, Proceedings, 53rd Electronic Components & Technology Conference, 2004, pp. 1563-1569.
[49] Lau, J.H., Design for Reliability of Lead-Free Solder Joints, in B. Michel, R. Aschenbrenner, The World of Electronic Packaging and System Integration, IZM special edition, ddp goldenbogen, Dresden, 2004, pp. 120-127.
[50] Syed, A.R., Factors Affecting Creep-Fatigue Interaction in Eutectic Sn/Pb Solderr Joints, Advances in Electronic Packaging, InterPack97 (1997), pp. 1535-1542.
[51] Syed, A.R., Predicting Solder Joint Reliability for Thermal, Power, &Bend Cycle within 25% Accuracy, 51st ECTC, 2001, pp. 255-263.
[52] Syed, A.R., Accumulated Creep Strain and Energy Density Based Thermal Fatigue Life Prediction Models for SnAgCu Solder Joints, 54st ECTC, Las Vegas, June 2004, pp. 737-746.
[53] Dasgupta, A., Oyan, C., Barker, D., Pecht, M., Solder Creep-Fatigue Analysis by an Energy-Partitioning Approach, ASME Journ. of Electronic Packaging, Vol. 114, No. 2, (1991), pp. 152-160.
[54] Frear, D.R., Burchett, S.N., Neilsen, M.K., Life Prediction Modeling of Solder Interconnects for Electronic Systems, ASME EEP-vol. 19-2, Advances in electronic packaging, 2(1997) pp. 1515-1522.
[55] Desai, C.S., Review and Evaluation of Approaches for Thermomechanical Analysis for Stress, Failure and Reliability, ASME EEP-vol. 26-2, Advances in electronic packaging, 2(1999), pp. 1713-1740.

[56] Qian, Z., Ren, W., Liu, S., A Damage Coupling Framework of Unified Viscoplasticity for the Fatigue of Solder Alloys, ASME Transactions, Journal of Electronic Packaging, Vol. 121, September 1999, pp. 162-168.
[57] Chandaroy, R., Barasan, C., Damage Mechanics of Surface Mount Technology Solder Joints Under Concurrent Thermal and Dynamic Loading, ASME Journ. of Electronic Packaging, Vol. 121, June 1999, pp. 61-68.
[58] Barasan, C., Chandaroy, R., Thermomechanical Analysis of Solder Joints Under Thermal and Vibrational Loading, ASME Journ. of Electronic Packaging, Vol. 124, March 2002, pp. 60-67.
[59] Zhang, X., Lee, S.-W.R., Pao, Y.-H. A Damage Evolution Model for Thermal Fatigue Analysis of Solder Joints. ASME Journ. of Electronic Packaging, Vol. 122, 2000, pp. 200-206.
[60] Subbarayan, G., Towashiraporn, P., A Hybrid Fracture-Damage Model for Computationally Efficient Fracture Simulations, Proc. 5th EPTC, Signapore, Dec. 2003, pp. 462-469.
[61] Abdul-Baqi, A., Schreurs, P.J.G., Geers, M.G.D., Fatigue Damage Modellling in Solder Interconnections: a Cohesive Zone Approach. Proc. 3rd EuroSimE, Aix-en-Provence, France, April 2003, pp. 235-239.
[62] Sharma, P., Dasgupta, A., Micro-Mechanics of Creep-Fatigue Damage in Pb-Sn Solder Due to Thermal Cycling-Part I: Formulation, ASME Journ. of Electronic Packaging, Vol. 124, Sept. 2002, pp. 292-297.
[63] Sharma, P., Dasgupta, A., Micro-Mechanics of Creep-Fatigue Damage in Pb-Sn Solder Due to Thermal Cycling-Part II: Mechanistic Insights and Cyclic Durability Predictions From Monotonic Data, ASME Journ. of Electronic Packaging, Vol. 124, Sept. 2002, pp. 298-304.
[64] Mueller, W.H.,Understanding Morphology Changes in Solders. Proc. 3rd EuroSimE, Aix-en-Provence, France, April 2003, pp. 479-484.
[65] Ubachs, R.L.J.M., Schreurs, P.J.G., Geers, M.G.D., Microstructural Behaviour of Solder Alloys, Proc. 5th EuroSimE, Brussels, Belgium, May 2004, pp. 543-548.
[66] Ferry, J.D., Viscoelastic Properties of Polymers, John Wiley & Sons, New York, 1980.
[67] Dudek, R., Schubert, A., Michel, B., Nonlinear FE Simulation for Plastic Packaging Applications, Second Workshop Mechanical Reliability of Polymeric Materials and Plastic Packages of IC Devices POLY'99, Paris, France, Dec. 12-15, 1999.
[68] Hall, P.M., Forces, Moments, and Displacements During Thermal Chamber Cycling of Leadless Ceramic Chip Carriers Soldered to Printed Boards, IEEE Transact. on Components, Hybrids, and Manufacturing Technol.,Vol. CHMT-7, No. 4, Dec. 1984, pp. 314-327.
[69] A. Picault, J. F. Autissier, Comparison of Fatigue Life Prediction Based on Numerical Simulation to Testing Results, (in French) Internal report, Jabil Circuit, Meung sur Loire, France, June 2004.
[70] Dudek, R., R. Döring, B. Michel, A. Picault, J. F. Autissier,Thermo-mechanical Reliability Analyses on Solder Joints of Ceramic Components, Proceedings IPC Soldertec, Barcelona, Spain, June 2005.
[71] R. Dudek, R. Doering, B. Michel, G. Petzold, J. Albrecht, C. Wieand, S. Kuhn, "Investigations on the Reliability of Lead-Free CSPs Subjected to Harsh Environments", Proceedings ITherm 2004, Las Vegas, USA, June 2004.
[72] A. Schubert, R. Dudek, R. Leutenbauer, P. Coskina, K.-F. Becker, J. Kloeser, B. Michel, H. Reichl, D. Baldwin, J. Qu, S. Sitaraman*, C.P. Wong, R. Tummala, Numerical and Experimental Investigations of Large IC Flip Chip Attach, 50th Electronic Components & Technology Conf. (ECTC) Las Vegas, USA, May 2000.

[73] Schubert, A., Dudek, R., Vogel, D., Becker, K.-F., Kloeser, J., Michel, B., Reichl, H., Flip Chip Solder Joint Reliability, Proc. Symposium on Advances in Packaging (APACK '99), December 8-10, 1999, Singapore, pp. 234-242.

[74] Dudek, R., Schubert, A., Michel, B., Analysis of Flip Chip Attach Reliability, Internat. Conference Adhesives in Electronics 2000, Espoo, Finland, June 18-21, 2000, Proc. pp 77-85.

7. EXERCISES

1. What kinds of failures are characteristic for solder interconnects?
2. Explain the advantages and drawbacks of theoretical analyses of solder joint reliability when compared to testing methods.
3. Which environmental loadings cause low cycle fatigue loading of interconnects? Explain the terms "test cycle" and "field cycle".
4. What are the major differences between low-cycle fatigue analyses based on approximate analytical models and FE calculation?
5. The introduction of which technologies has enforced the use of theoretical analyses of solder fatigue? Consider the additional functions of the joints besides the electrical connection.
6. What is the main source of thermal fatigue failure for solder joints? Explain the different types of thermal mismatch.
7. Is a solder joint homogeneous? Describe its principal structure.
8. Give some reasons why the thermo-mechanical properties of different solder joints made of the same solder can vary.
9. What are the advantages and limitations of analytical-empirical prognosis of the solder joint reliability?
10. Describe the basic ideas of analytical solder fatigue models. Use a principal sketch to explain the basic equations. What assumptions and simplifications are needed?
11. For what type of component-board assembly can the analytical-empirical prognosis be expected to work best? Give the primary parameters which are taken into consideration in the analytical models.
12. Explain the term "acceleration factor". What rough estimate can be given for this factor with regard to the temperature amplitudes?
13. Within what temperature range do the melting temperatures of most of the soft solders fall?
14. Explain the term "homologous temperature". What are the consequences that follow from the use of metallic materials at homologous temperatures greater than 0.5?
15. Characterize the mechanical behavior of soft solders in the use temperature range. Does an elastic constitutive material model describe the behaviors in an appropriate manner?

7. Characterization and Modelling of Solder Joint Reliability

16. What kinds of non-linear constitutive deformation behaviours can be found within soft solders?
17. Describe the different phases of the creep of metals when subjected to constant stress. Which part is used for most of the constitutive descriptions implemented in the material models of FE-analyses used for low cycle fatigue loading?
18. Explain briefly the deformation mechanisms, which have been assumed to be controlling the secondary creep strain rate of eutectic SnPb solders. What kinds of formulas are used to describe this behaviour mathematically?
19. Within what range are the differences between the secondary creep rates for eutectic SnPb published in different sources? Give some reasons why there seems to exist no unique secondary creep law for SnPb solders.
20. Do the different microstructures of lead-free tin based solder alloys require a creep description fundamentally different from that used for eutectic tin-lead?
21. Describe some general findings concerning the creep resistance of SnAgCu and SnAg solders and compare them to SnPb.
22. Which empirical criteria are used for evaluation of the cyclic equivalent creep strain or dissipated creep strain energy density to link FE calculation results to a critical cycles to failure number? What is the main difference to analytical failure prediction?
23. Describe briefly some proposals given in the literature to improve FEA-based solder fatigue prediction.
24. Which other non-linear or temperature-dependent materials characteristics used in soldered assemblies have to be considered besides that of solder?
25. Make a sketch of a characteristic shear-strain versus shear-stress hysteresis loop at the gap of a solder joint of a ceramic resistor on an organic board, subjected to one thermal cycle. Why does it make sense to start the calculation run at the ramp down period of the upper dwell temperature?
26. Why can parametric studies be useful for the virtual thermo-mechanical prototyping of new package designs, even if the solder creep description contains uncertainties? Is it acceptable to compare failure indicators like creep strain or creep dissipation to other studies, which were based on different creep laws?
27. Make sketches of the characteristic macro crack initiation sites for a lead-free solder joint (e.g., a ceramic component on a PWB) and an area array ball-type solder joint (e.g., a BGA ball). Considering FE-calculation results, what distribution patterns of the creep strain or energy have to be expected for these joints?

28. Describe the reasons for the low fatigue resistance of flip chip on board assemblies and the effects of underfilling. Comment on dependencies related to die size, standoff, and underfill stiffness.
29. Is solder fatigue the only characteristic failure mode for underfilled FCOB assemblies?

Chapter 8

VIRTUAL THERMO-MECHANICAL PROTOTYPING

G.Q. Zhang[1,4], N. Tzannetakis[2], W.D. van Driel[3,4]
[1]*Philips Semiconductors, HTC 60, 5656AG Eindhoven, The Netherlands*
[2]*Noesis Solutions, Interleuvenlaan 68, B-3001 Leuven, Belgium*
[3]*Philips Semiconductors, P.O. Box 30008, 6503HK Nijmegen, The Netherlands*
[4]*Delft University of Technology, Mekelweg 2, 2628CD Delft, The Netherlands*

Abstract: This chapter presents our research and development results for virtual thermo-mechanical prototyping of microelectronics. It starts from a systematic overview on the basic theories and methodologies of advanced simulation based optimisation, consisting of mainly DOE, Response Surface Models and optimisation methods. With detailed description of our Virtual Prototyping (VP) strategy imbedded in a commercial software, focusing on the efficient development of reliable Response Surface Models for the underlying nonlinear responses, several analytical and industrial examples are demonstrated, covering different application objectives of developing design rules and tools, robust design and finding optimal configurations. Our results show that VP is an effective and efficient method to predict, qualify, and optimise the behaviour and/or trends of microelectronics against the actual requirements prior to major physical prototyping, manufacturing investments and reliability qualification tests.

Key words: Virtual Prototyping, design of experiments, response surface model, simulation-based optimisation, design rules, design tool.

1. INTRODUCTION

Along with the major technology development trends characterized by Moore's law and "More than Moore", the business trends of microelectronics are mainly characterized by cost reduction, shorter-time-to-market, and outsourcing. The combination of these technology and business trends leads to increased design complexity, dramatically decreased design margins, increased chances and consequences of failures, decreased product development and qualification times, increased gap between technology advance and development of fundamental knowledge, and increased difficulties to meet quality, robustness and reliability requirements.

Based on the root cause analyses from observed failures of microelectronics during different life cycles, it is found that, among others, thermo-mechanical (thermal, mechanical and thermo-mechanical) related failures account for about 65% of total failures in microelectronics, and they originate mostly from the product/process design phase. Due to the lack of available knowledge on one hand, and non-sufficient R&D effort from both the academia and industry for the new mechanical challenges on the other hand, unfortunately, thermo-mechanical design and qualification of microelectronics are still largely depending on one's experience (in other words, trial and error based). Often, up to 5 cycles (material development/ pre-selection, concept designing, building and testing multiple physical prototypes) are needed, with some qualitative support from numerical simulations. Quality, robustness and reliability are usually dealt with after physical prototyping, wherein reliability qualification testing with duration of 3 months is no exception. Clearly, this experience-based design and qualification method cannot lead to competitive products with shorter-time-to-market, optimised performance, low costs, and guaranteed quality, robustness and reliability. Therefore, there is an urgent need to develop and exploit virtual prototyping methods.

There are more companies today using increasing levels of simulation or optimisation tools to support physical prototyping and testing for designing and qualifying product/process. The engineer starts by defining key design parameters (geometry, materials, process history, etc.) and their design spaces based on application feasibilities and experience. The problem is then modelled and simulated to determine a certain product performance attributor such as stress, damages and failure probability, reliability, durability, and dynamic behaviour. If a problem is detected (crack or delamination for example) via either modelling or experiment, the design is modified by changing the design parameters, the model is modified, and the simulation is run again. With pre-knowledge and proper execution, these simulation and optimisation supports can be considerably faster, less expensive and able to provide more insights than physical prototyping and testing in product/process development.

However, due to the complicated nature of the microelectronic technologies, often associated with multi-physics, strong nonlinear, multi-parameter, multi-process and stochastic responses, it is expensive to achieve accurate and efficient simulation results, in both qualitative and quantitative ways. For many occasions, it is just not possible, due to the bottlenecks of computational time and capabilities. Administrative tasks associated with handling and manipulating data for each simulation run also require large amounts of engineers' time. Conflicting engineering requirements (miniaturized and heterogeneous integrated microelectronic devises with extreme high reliability requirements, for example) demand complex tradeoffs that usually aren't intuitively obvious. Also, deterministic simulations generally fail to account for unavoidable uncertainties and variability in product characteristics, loading, or processing. Engineers supported by some simulations and optimisation in most cases only have enough time to develop a workable and "satisfactory" design that is typically far from the potentially possible optimum.

These issues are being addressed by an emerging class of product development methodology and software known as Virtual Prototyping (VP), see Figure 1. These solutions support engineers in automating their analysis processes, quickly investigating multiple design options, and finding the most optimal design depending on the specific needs. The ultimate aim of virtual prototyping is to predict, qualify, optimise and design the performance and/or trends of microelectronics against the actual requirements prior to major physical prototyping, manufacturing investments and reliability qualification tests. The solutions include the following seamlessly integrated capabilities covering all the major elements of design chain, such as:

- Parametric and predictive reliability models to capture the damage initiation, propagation and the eventual failures under given process and loading history, material systems and geometric design space.
- Process automation captures and automates the simulation process, acting like a "simulation robot" in quickly performing multiple repetitive analyses.
- Design space exploration tools include design of experiments (DOE) for identifying the dominant parameters/or parameter interaction to functional performance and response surface modelling (RSM) to meet other design optimisation needs.
- Advanced probability and statistics software accounts for variability and uncertainties of reliability initiatives such as Design for Six Sigma.
- Numerical optimisation algorithms converge on the best design among all the various alternatives.

Such capabilities have the potential of making a revolutionary impact on the manner of product development by reducing the product development cycle substantially. Whereas conventional CAE arrives at one "satisfactory" design by computing product attributors from design parameters defined by the engineer, VP determines the optimal design parameters necessary to

472   *Chapter 8*

meet the targeted performance attributors, in combination with existing experience. In this way, VP arrives not just at one "satisfactory" design but an optimal design that best satisfies all engineering requirements, enables engineers to reach the ultimate goal of product development: "first time right, first time fast, and first time best".

In this chapter, we start from a systematic overview on the basic theories and methodologies of advanced simulation based optimisation, consisting of mainly DOE, Response Surface and optimisation methods. With detailed description of our VP strategy imbedded in a commercial software, focusing on the efficient development of reliable Response Surface Models for the underlying nonlinear responses. Several industrial examples are demonstrated, covering different application objectives of developing design rule/tool, robust design, and design optimisation. Our results show that VP is an effective and efficient method to predict, qualify and optimise the behaviour and/or trends of microelectronics against the actual requirements prior to major physical prototyping, manufacturing investments and reliability qualification tests.

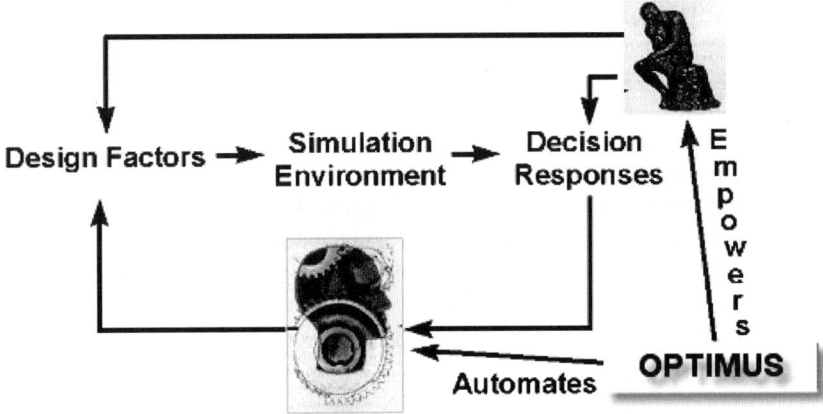

Figure 1. Virtual Prototyping

8. Virtual Thermo-Mechanical Prototyping 473

2. STRATEGY, METHODOLOGY AND PROCEDURES OF VIRTUAL PROTOTYPING

2.1 Strategy and Methodology

Figure 2 shows our strategy and methodology for virtual prototyping, wherein the two core building blocks, 'accurate and efficient prediction models' on the left side and 'advanced simulation-based optimisation methods' on the right side, should be seamlessly integrated, in order to predict, qualify, optimise and design microelectronics against the actual requirements prior to major physical prototyping, manufacturing investments and reliability qualification tests, in an effective and efficient manner.

To achieve the desired benefits, VP solutions utilize a range of competencies and technologies. Figure 3 shows the major factors determining the accuracy and efficiency of predictive reliability simulation models. Figure 4 shows the methodology and procedure to conduct simulation-based optimisation.

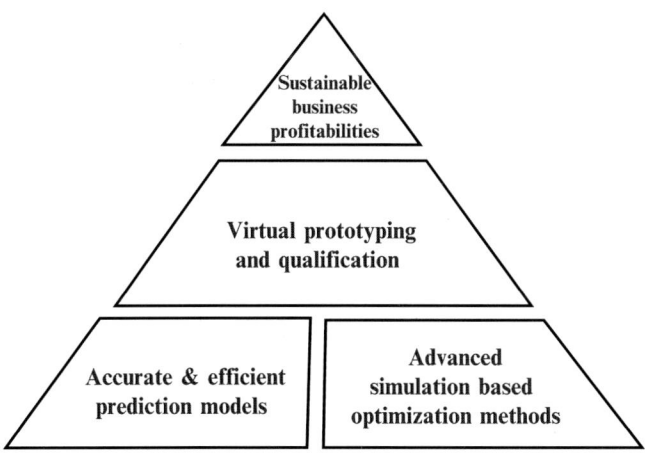

Figure 2. Strategy of VP

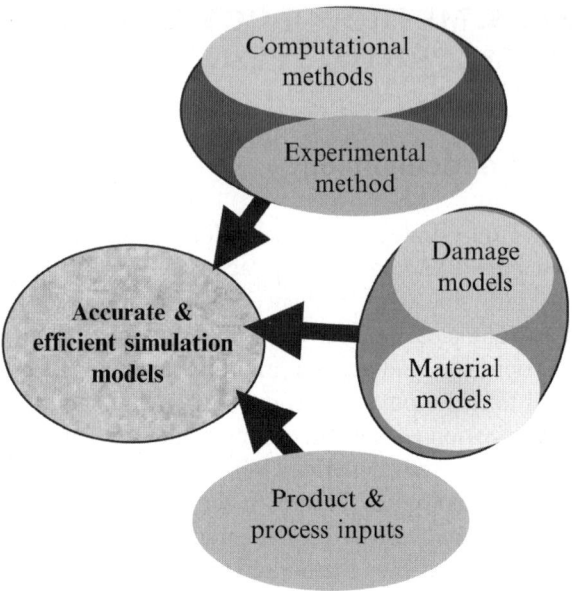

Figure 3. Factors determining the accuracy and efficiency of simulation models

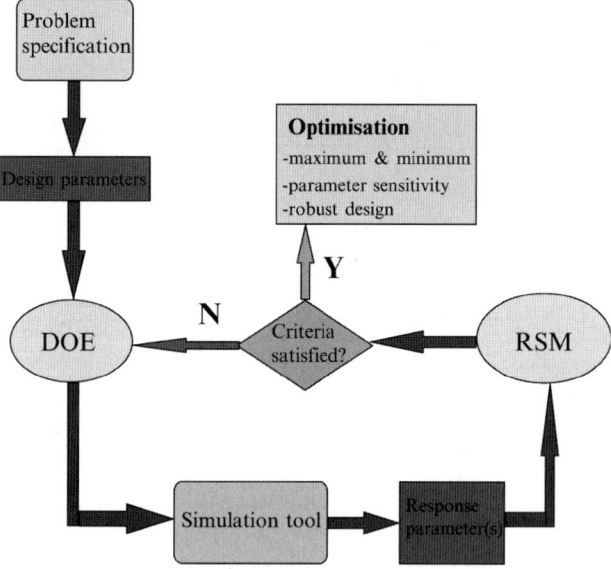

Figure 4. Simulation based optimisation method

8. Virtual Thermo-Mechanical Prototyping

From modelling perspectives, it should be noticed that:

- The key to the success of VP is to develop parametric reliability prediction models covering the complicated nonlinear, multi-physics and stochastic behaviour under given design spaces (material, geometry, process and use conditions), accurately and efficiently. Accuracy is essential, but depending on the application requirements, it can be either qualitative or quantitative. Efficiency is needed due to the fact that it is expensive (if it is not impossible) to simulate the complicated responses of microelectronics covering the whole design space and whole life cycle, and to conduct global design optimisation, such as finding the maximum/minimum, robust design, and parameter sensitivity.

- Another success factor for VP is the co-designing principle. That is to say, the design and qualification of microelectronics should be based on integrated understanding and solutions covering all the major involved disciplines (electric, mechanics, physics, chemistry, metallurgy, etc.) and processes (IC, packaging, assembly, testing, etc.). As one example, it is well known that stress/strain induced from packaging might have significant impact on the design and reliability of board level assembly, and vise verse. As the fast integration of IC, packaging, and assembly technologies, design and qualification based on mono-discipline/process may lead to sub-optimal that can be contra dictionary with the desired global optimal.

- Experiments and tests are still playing and will continue to play an important role in the content of VP. First, they are needed in characterizing material and their interface behaviour (material properties, damage initiation, evolution and failure criteria). Secondly, very often, the correctness and accuracy of the developed reliability prediction models need to be verified via experiments for the whole range of the design spaces, and by covering all the critical processes. One should be aware that "in principle, all the simulation results are wrong, unless one can prove that they are right". The results of VP should also be verified, wherein experiments and tests are the direct and obvious options.

- In industrial application, one usually distinguishes virtual design prototyping with virtual reliability qualification, according to the different phases of product creation process (PCP). Virtual prototyping is aimed at achieving first time right, first time fast and first time best design, according to the integrated functionality and reliability requirements. However, no matter how good the design is, as one of the common industrial practices, reliability qualification tests (usually a expensive process in terms of time and money) are always required by the customers. Therefore, virtual reliability qualification focuses on prediction of the behaviour of microelectronics under reliability qualification tests and real usage conditions, aiming at reducing the number and duration of reliability qualification tests, and better matching and correlating the specified reliability performance with the application conditions.

From the simulation based optimisation perspective, capabilities must include process automation, design space exploration, advanced probability and statistics algorithm, and numerical optimisation algorithms.

- Process automation capabilities allow users to capture the process utilized in implementing their existing analysis and simulation tools. By way of an easy-to-use graphical user interface with which a straightforward workflow diagram is constructed, users interactively define the inputs, outputs, and intermediate processes required for various simulation tools. By pointing and clicking, the user "teaches" the VP system the process, with no programming or writing interface codes required. In this way, users identify key design variables, the design objectives, and the analysis sequence required - with one or multiple simulation codes operating simultaneously or in series. After the process is defined, it can then be readily automated, thus freeing engineers from repetitive manual tasks usually required in going through the multiple iterations for different design alternatives in arriving at an optimal solution. In this respect, VP tool acts as a "simulation robot" in executing the work quickly without user intervention, automating iterative processes where the same procedures are repeated many times, often by many engineers. Work is distributed transparently over available computer resources.

- Design space exploration technologies serve as efficient decision-support tools that enable users to readily identify the best of all possible designs. With advanced design of experiments (DOE) technology, the VP software automatically samples the design space of the large set of all possible combinations of design factors and selects a number of virtual experiments from a set of pre-defined approaches to automatically evaluate the design alternatives and identify which parameters contribute most to the targeted performance. Response Surface Modelling (RSM) is a mathematical model built using the DOE response data. It can predict the design output for any combination of design inputs and is an extremely fast way of calculating new design points. Built from a "cloud of points" in the design space, RSM enables users to quickly understand and visualize how the system response is influenced by any combination of design variables. The method answers questions like: which design factors most affect functional performance? Which performance responses are linked? What design changes are required to achieve performance targets? How robust are the design responses? Which constraints most limit the design?

- Advanced probability and statistics software accounts for variability and uncertainties for reliability initiatives such as Design for Six Sigma. In this way, VP can be used to manage risk and improve product quality, safety and performance by accounting for unavoidable variability of product characteristics, boundary conditions, loading or processing. These tolerances in design variables are thus automatically taken into account to meet robustness and reliability objectives.
- Numerical optimisation algorithms converge on the best design among all the various alternatives. This effectively crosses the chasm between the classical trial and error processes to one of "first time right" design. The result of this sort of optimisation is the best design in meeting the required target objectives. This is Design by Objective, essentially reversing the engineering process, starting from the target result desired and then defining the design parameters needed to achieve that target. A range of algorithms is used to converge on the optimal design including gradient techniques as well as state-of-the-art differential evolution.

Having the two building blocks, namely accurate and efficient reliability prediction models and advanced simulation based optimisation method/tool, is the precondition for doing virtual prototyping. However, the success of VP depends also on many other factors, such as the proper specifications of accuracy covering reliability simulation models, the developed RMS, and the design parameters. These three types of accuracy should be specified properly in an integrated way, to account for their strong interaction effect and to achieve optimal balance between the desired design accuracy on one hand, and the resulting efficiency on the other hand.

As an advanced design tool, VP manages analysis programs from various engineering disciplines; answers the real engineering challenges (such as: Which design meets the design targets the best? What are the design rule?); shifts the paradigm of simulation from a reactive design verification to a proactive and creative design improvement process; stimulates and empowers engineering creativity; enables a systematic approach to design optimisation; adds value to the often-considerable investment in existing simulation technologies; increases the effectiveness and utility of the engineering computer workstations.

2.2 Procedures

The major VP procedures are:
1. *Setting-up an appropriate analysis sequence*

The first task of VP is to guide the user in setting up the Sequence of Analysis functions that describe the complete simulation – much like setting up the test laboratory apparatus. For example, if the important performance characteristics of the design is fatigue life time of solder interconnects, the engineer will use FEM based analysis programs to simulate the fatigue performance, – much like doing fatigue experiments. VP enables the engineer to set up the sequence of analysis tools graphically and specify the flow of information from the Design Factors (which determine the design parameters that the engineer is free to change) to the Decision Responses (the design design attributors which characterize the product performance).

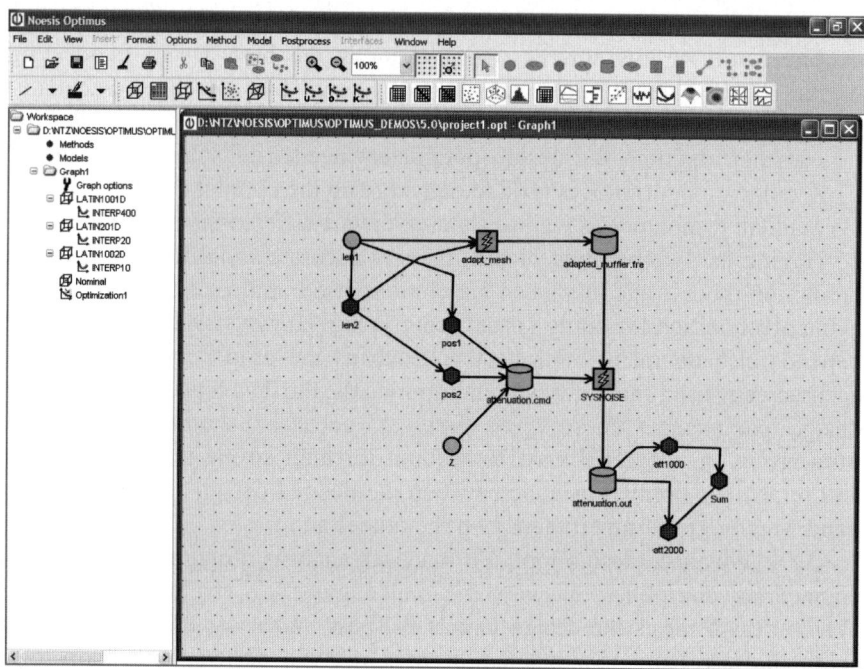

Figure 5. Example OPTIMUS Main Window Depicting the Analysis Sequence [1]

2. *Communicating with simulation programs*

One of the most important aspects to enabling the process of VP is the ability to communicate and interface easily with any existing simulation program. The commercial VP software [1] can achieve this through the introduction of Intelligent Drivers. The engineer, with simple point and click operations, 'teaches' the VP based tools the structure of the Analysis program input file(s) and assigns the Design Factors. The definition of Decision Responses is achieved with similar simple operations guided by Intelligent Parsers.

8. Virtual Thermo-Mechanical Prototyping

3. *Sampling the design space*

The engineer, much like his test counterpart, can then select from a number of pre-defined experimental plans to sample the Design Space – defined by the range of the Design Factors. Design of Experiment (DOE) methodologies automatically generate several alternative designs based on. This is similar to generating many physical prototypes and testing them to see how each of them performs.

4. *Selecting the best design*

The selection of the best design among all the alternatives generated in the previous step is achieved through interactive search, visualization and data analysis techniques, within the framework of Response Surface Modelling and Analysis Module. A complimentary approach uses the advanced numerical optimisation. The latter has no similar procedure in the laboratory environment, where this function can only be fulfilled by experience and trial and error.

This step answers questions such as
- Which of the Design Factors affect the Decision Responses the most?
- Which Decision Responses are inter-linked?
- What amount of change is necessary for each Design Factor to achieve the target values for the Decision Responses?
- How robust are the Design Responses with respect to the Design Factors?
- How does each Design Factor affect the Decision Responses?
- Which Constraints limit the design?

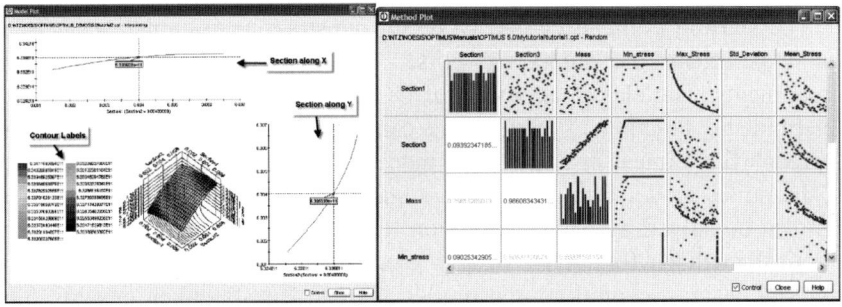

Figure 6. Design Alternatives Analysis – Visual Aids [1]

3. FUNDAMENTS OF SIMULATION-BASED OPTIMISATION

Simulation-based optimisation is optimisation based on and integrated with advanced simulation models that can predict the product/process behaviour accurately and efficiently. Optimisation here refers to different design needs, such as to find the maximum or minimum, to know the parameter sensitivity, to obtain robust design, to develop design rule, to know the probability of failures according to the given derivation of design parameters, etc. In this section, some fundaments of simulation-based optimisation will be reviewed.

3.1 Design of Experiments (DOE)

Design of Experiments technique (DOE) is a systematic approach to get the maximum amount of information out of various types of experiments while minimizing the number of experiments. Many DOE methods have been developed and are available for different kinds of applications. No attempt will be made here to summarize all of those methods. However all of these methods can be classified as two categories:
- Orthogonal designs
- Random designs

3.1.1 Orthogonal Designs

The starting point in classic DOE for constructing response surfaces is that experiments are subject to noise. This typically holds for physical experiments. The classic approach of DOE takes the response surface as a deterministic function of which one can only observe noisy values. The approach fits a linear or a quadratic model to the response surface. Because of the noise, model fitting becomes a statistical parameter estimation issue that can be solved by using regression techniques. This area offers a lot of practical methods for DOE. Widely used methods are fractional and full-factorial designs, central composite designs and Box-Behnken designs. The orthogonality of the designs means that the model parameters are statistically independent; see Figure 7 for a three Level Full Factorial DOE.

For simulation experiments, as long as the simulation models are verified and reliable, the concern for experiment noise can be eliminated. Therefore, the classic DOE, used to account for noise, are less suitable for simulation experiments. Beyond that, a factorial design has some disadvantages:
- Initially it is usually not clear which factor is important and which not. Since the underlying function is deterministic there is a potential hazard

8. Virtual Thermo-Mechanical Prototyping

that some of the initial design points collapse and one or more of the time consuming computer experiments become useless. This is called the collapse problem.
- Most classic DOEs are only applicable for rectangular design regions.
- The number of experiments increases exponentially with the number of levels.

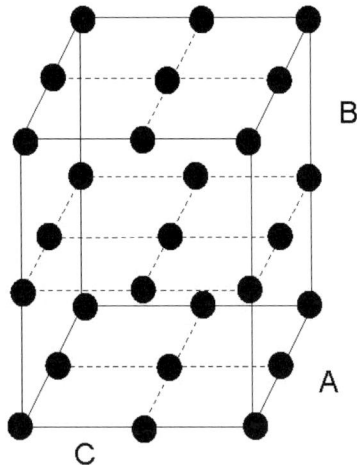

Figure 7. Three level full factorial design of experiments

3.1.2 Random Designs

Another widely used DOE technique is a random DOE. The most common-used random DOE is the so-called Latin Hypercube Designs (LHDs). A LHD can be constructed as follows. Let n be the number of design points that we are willing to simulate. Divide each design parameter dimension in n equidistant levels. The design points are obtained by selecting for each dimension a permutation of the levels and combining these permutations into a design. In this way each level is present in the design and the number of levels is maximized. LHDs do not suffer from the collapse problem, because if in one or more of the factors appear not to be important, every point in the design still gives information about the influence of the other factors on the response. In this way none of the time-consuming computer experiments become useless.

There is still complete freedom in assigning levels to dimensions and therefore numerous LHDs exist. Assigning levels can for instance be done randomly. An efficient way to construct a LHD is to assign the levels in such a way that the resulting design is space filling. Intuitively, a design is space filling if the points are spread out and do not cluster in one portion of the experimental region. As a measure for the space-fillingness of a design we take the minimal distance between two of its design points. The larger this minimal distance the better the design. A design for which the minimal distance is maximal is called a maximin distance design. The required number of experiments for a LHD is determined by the complexity of the underlying model (linear, non-linear, continuous or discontinuous).

3.2 Response Surface Models (RSM)

Response surface modelling (RSM) is a collection of mathematical and statistical techniques that are useful for the modelling and analysis of problems in which a response of interest is influenced by several variables. Using the DOE method combined with the response surface method, the response between the points set up by the design, can be predicted efficiently. The RSM methodology allows for further processing of the DOE results. Examples are the availability of design variables contributions processing, 2D and 3D graphing, and foremost utilization of the RSM for optimisation. As consequences, not only the lengthy and costly simulation runs can be reduced, but also the inherent trend (often non-linear) can be correctly predicted. When utilizing RSMs for optimisation one has to always be cautious and take explicit actions relating to the RSM quality.

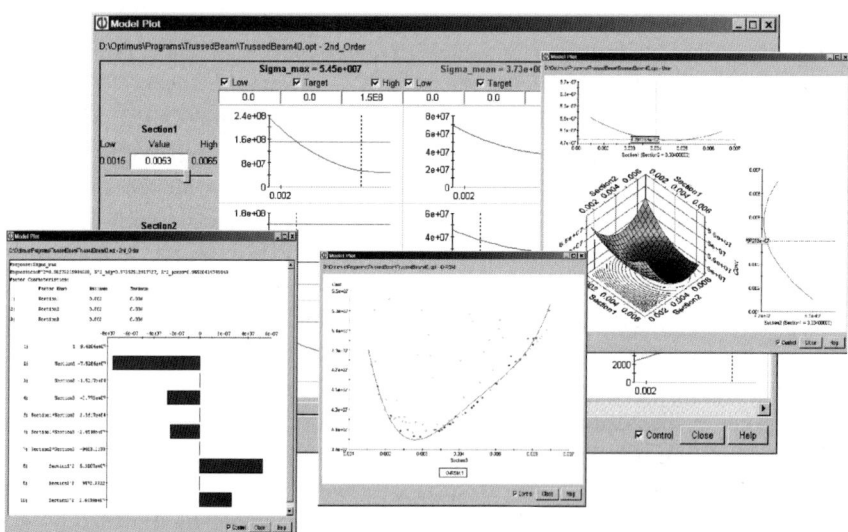

Figure 8. Examples of RSM post-processing plots

8. Virtual Thermo-Mechanical Prototyping

There are three major categories of RSM methods, namely RSM based on Least Squares approximations, RSM based on Interpolating Methods, RSM based on Neural Network methods (including Radial Basis Functions). In the following, the Least Squares based RSM Methodology and the Kriging based method will be reviewed.

3.2.1 Theory of Linear Least Squares Fitting Response Surface Modelling

Given a set of observations, one often wants to condense and summarize the data by fitting it to a model that depends on adjustable parameters. The basic approach of modelling can be described as: You choose or design a Merit function that measures the agreement between the data and the model with a particular choice of parameters. The merit function is conventionally arranged so that small values represent close agreement. The parameters of the model are then adjusted to achieve a minimum in the merit function, yielding best-fit parameters. The general procedure of linear least squares employs a merit function, namely the Sum of squares of the error between the approximated value and the exact observation, which one can show is a maximum likelihood estimator. There are important issues that go beyond the mere finding of best-fit parameters. The most important is what model to choose, thus our intent is to provide a strategy to address this issue in order to limit the total number of observations necessary to a minimum.

In the following we are briefly going to derive the mathematical formulas of the general linear least squares problem. For details the interested reader is referred to [2] and [3]. Suppose that we have data points (experiments):

$$(x_1^i, x_2^i, ..., x_k^i, y^i), \quad i = 1,...,n \tag{1}$$

with κ independent variables $x_1, ... x_K$, and we are fitting them to a model that has m adjustable parameters α_j, $j=1,...,m$. The model predicts a functional relationship between the measured independent and dependent variables:

$$y(x) = y(x; \alpha_1, ..., \alpha_m), \quad x = (x_1, ..., x_k)^t \in \Re^k \tag{2}$$

The general method of Least Squares says, figure out the parameters α_j, $j=1,...,m$ in the model (2), so that the Sum of Squares of the error takes out a minimum, that is:

$$\min_{\alpha_1,...,\alpha_m} \sum_{i=1}^{n} \left[y^i - y(x^i; \alpha_1, ..., \alpha_m) \right]^2 \tag{3}$$

with $x^i = (x_1^i, ..., x_k^i)^t$

The introduced model equation (2) is very general, i.e. it contains any linear and nonlinear relation in both the dependent variables $x_1,...x_K$ and the independent parameters $\alpha_1,...,\alpha_m$. If we restrict ourselves to the case where the model is a linear combination of the form:

$$y(x;\alpha_1,...,\alpha_m) = \sum_{j=1}^{m} \alpha_j X_j(x) \tag{4}$$

We get the general model of a Linear Least Squares Problem, linear with respect to the parameters $\alpha_1,...,\alpha_m$. The functions $X_1(x),...X_m(x)$ are arbitrary fixed functions of $x=(x_1,...x_k) \in \mathfrak{R}^k$. For these linear models we generalize the discussion by defining the optimisation problem we have to solve, namely:

$$\min_{\alpha_1,...,\alpha_m} \sum_{i=1}^{n} \left[y^i - \sum_{j=1}^{m} \alpha_j X_j(x^i) \right]^2 \tag{5}$$

Let $A \in \mathfrak{R}^{n \bullet m}$ be a matrix whose $n \bullet m$ components are constructed from the m basis functions evaluated at the n abscissas $x^i=(x,...x_k)^t$, i.e.:

$$A \equiv [a_{ij}]_{\substack{i=1,...,n \\ j=1,...,m}} = X_j(x^i) \tag{6}$$

Matrix A is called the design matrix of the fitting problem. In general it has more rows than columns, $n \geq m$, since there must be more data points than model parameters to be solved for. Schematically the design matrix looks like:

$$A = \begin{bmatrix} X_1(x^1) & X_2(x^1) & \cdots & X_m(x^1) \\ X_1(x^2) & X_2(x^2) & \cdots & X_m(x^2) \\ \vdots & \vdots & \vdots & \vdots \\ X_1(x^n) & X_2(x^n) & & X_m(x^n) \end{bmatrix} \tag{7}$$

Also let $b \in \mathfrak{R}^n$ be a vector of length n defined by:

$$b_i = y^i, \quad b = (y^1,...,y^n)^t \tag{8}$$

8. Virtual Thermo-Mechanical Prototyping

and denote the m vector whose components are the parameters to be fitted, $\alpha_1,...,\alpha_m$, by α, i.e.:

$$\alpha = (\alpha_1,...,\alpha_m)^t \qquad (9)$$

With these notations we are all set to come up with the so-called Normal Equations.

3.2.2 Solution of the Linear Least Squares Problem by Means of the Normal Equations

We employ the necessary condition of first order of class room optimisation and get: If $\alpha^* = (\alpha_1^*,...,\alpha_m^*)^t$, is a minimizer of problem (5) then:

$$\frac{d}{d\alpha_k}\left(\sum_{i=1}^{n}\left[y^i - \sum_{j=1}^{m}\alpha_j^* X_j(x^i)\right]^2\right) = 0, \quad k=1,...,m \qquad (10)$$

or equivalently:

$$\sum_{i=1}^{n}\left[y^i - \sum_{j=1}^{m}\alpha_j^* X_j(x^i)\right] X_k(x^i) = 0, \quad k=1,...,m \qquad (11)$$

By means of changing the order of summation, we can write (11) as:

$$\sum_{j=1}^{m} q_{kj}\alpha_j^* = \beta_k \qquad (12)$$

where:

$$q_{kj} = \sum_{i=1}^{n} X_j(x^i) X_k(x^i) \text{ and } \beta_k = \sum_{i=1}^{n} y^i X_k(x^i) \qquad (13)$$

or equivalently as:

$$[Q] = [q_{ij}]_{i,j=1,...,m} = A^t \cdot A \in \Re^{m \times m}, \quad [\beta] = A^t \cdot b \qquad (14)$$

and finally as the matrix equation:

$$[Q] \cdot \alpha^* = [\beta] \qquad (15)$$

The equations (15) are called the normal equations of the least squares problem. If *[Q]* is regular they can be solved by standard methods for solving linear systems of equations.

3.2.3 Kriging Based Response Surface Modelling

The Kriging method belongs to the stochastic interpolation methods. The stochastic modelling approach is based on considering the deterministic response y(X) as a realization of a stochastic process, which means that an error is replaced by a random process. One of the most popular methods for such a stochastic model interpolation is Kriging. Kriging is extremely flexible due to the wide range of correlation functions, which may be chosen. Depending on the choice of a correlation function, Kriging can either result in exact interpolation of the data points or smooth interpolation, providing an inexact interpolation. It is worth noticing that Kriging is different than fitting splines and in fact it is believed even better than splines.

For example, computer analysis is deterministic and not subject to a measurement error. Therefore the usual uncertainty derived from least-squares residuals has no meaning. Because of this the response model can be treated as a combination of a polynomial model and an additional factor referring to the deviation from the assumed model.

If matrix x denotes the sampled points and matrix y the output values of the deterministic function of k variables at n points:

$$y^i = f(x_1^i, x_1^i, ..., x_1^i) \quad ; \quad i = 1,...n \qquad (16)$$

where:

$$A = \begin{bmatrix} x_1^1 & x_2^1 & \cdots & x_k^1 \\ x_1^2 & x_2^2 & \cdots & x_k^2 \\ \vdots & \vdots & \vdots & \vdots \\ x_k^n & x_k^n & & x_k^n \end{bmatrix} ; y = \begin{bmatrix} y^1 \\ y^2 \\ \vdots \\ y^n \end{bmatrix} \qquad (17)$$

The most straightforward way to fit a response to a data is by linear regression (approximation of the sampled data):

$$y^i = \sum_{h=1}^{k} \beta_h f_h(x^i) + \varepsilon(x^i) \quad ; \quad 1 = 1,...,n \qquad (18)$$

8. Virtual Thermo-Mechanical Prototyping

where $f_h(xi)$ can be a linear or non-linear function of x_i, β_h are the unknown coefficients to be estimated and finally $\varepsilon(x^i)$ are error terms of the systematic deviation with normal distribution $N(0,\delta^2)$. In this way we can get an interpolation of the sampled data.

As mentioned earlier, the function $\varepsilon(x^i)$ representing the realization of a stochastic process is assumed to have zero mean and covariance V between two inputs u and v given by:

$$V(u,v) = \sigma^2 R(u,v) \tag{19}$$

between $\varepsilon(u)$ and $\varepsilon(v)$ where σ^2 is the process variance and $R(u,v)$ is a correlation. The covariance structure of ε relates to the smoothness of the approximating surface. For a smooth response, a covariance function with some derivates might be adequate, whereas an irregular response might call for a function with no derivates. The fitting procedure can be viewed as a two-stage problem:
- Calculation of the generalized least-squares predictor,
- Interpolation of the residuals at the design points as if there were no regression.

Because computer simulation is deterministic by nature, the error is totally due to modelling error and not to e.g. measurement error, and then it is justified to treat the error εi as a continuous function of x_i:

$$\varepsilon^i = \varepsilon^i(x^i) \tag{20}$$

As the error is the continuous function then the errors could be considered as correlated by the distance function between the points. If the points are close together, then the errors should also be similar, which means high correlation. Then it means finally that in the above approach it can be assumed that the correlation between errors would be related to the distance between the corresponding points. A special weighted distance formula can be used, which in comparison to the Euclidean distance does not weight all the variables equally:

$$d(x^i, x^j) = \sum_{h=1}^{k} \Theta \left| x_h^i - x_h^j \right|^{p_h} \tag{21}$$

where $\Theta \geq 0$ and $p_h \in [1,2]$. Using this distance function, the correlation between the errors can be defined as follows:

$$corr\left[\varepsilon(x^i), \varepsilon(x^j)\right] = e^{-d(x^i, x^j)} \tag{22}$$

The so defined correlation function has obvious properties, which means that in case of small distance the correlation is high while in case of large distance the correlation will approach zero. The values of the correlation function define the correlation matrix R of the order $n \times n$, which has practical meaning in the final response model definition:

$$R = \begin{pmatrix} r_{1,1}, r_{1,2}, \cdots, r_{1,n} \\ \vdots \qquad \vdots \\ r_{1,1}, r_{1,2}, \cdots, r_{1,n} \end{pmatrix} \tag{23}$$

$$r_{i,j} = corr\left[\varepsilon(x^i), \varepsilon(x^j)\right] \tag{24}$$

where the values of matrix R depend on parameters (θ_h, p_h). Thanks to the so defined correlation function and the correlation matrix R it is possible to get a simple linear regression model and avoid a quite complicated functional form of the response. The evaluation of the so defined stochastic model has a very important virtue, which allows replacing the regression terms by the constant value μ, the mean value of the results.

Therefore the stochastic model of the response can be rewritten as follows:

$$y(x^i) = \mu + \varepsilon(x^i); \quad \varepsilon(x^i) \to N(0, \delta^2) \tag{25}$$

It is required to estimate 2k+2 parameters to define the Kriging model: $\mu, \delta_1, ..., \delta_k, \theta_1, ..., \theta_k, p_1, ..., p_k$. This task can be achieved by maximizing the likelihood function F of the sample, which is defined as follows:

$$F = \frac{1}{(2\pi)^{n/2}(\delta^2)^{n/2}|R|^{1/2} e^{\frac{(y-1\mu)' R^{-1}(y-1\mu)}{2\delta^2}}} \tag{26}$$

where 1 denotes the n-vector of ones and y denotes the n-vector of observed function values

The best linear unbiased estimator of the response value y at point x* is defined as:

$$y(x^*) = \hat{\mu} + r' R^{-1}(y - 1\hat{\mu}) \tag{27}$$

where the r is the n-vector matrix given as follows:

$$r_i(x^*) = corr[\varepsilon(x^*), \varepsilon(x^i)] \tag{28}$$

The mean squared error s2(x*) is estimated as follows:

$$s^2(x^*) = E\left[(\hat{y}(x^*) - y(x^*))^2\right] = \hat{\delta}^2\left[1 - r'Rr + \frac{1 - 1'R^{-1}r'}{1'R^{-1}1}\right] \tag{29}$$

3.3 Design Optimisation

Traditionally Design Optimisation deals with the selection of the "best" alternative amongst the many possible designs. It involves the:
- Selection of a set of variables to describe the design alternatives
- Selection of one or more objectives, expressed in terms of the design variables, which we seek to optimise
- Determination of a set of constraints, expressed in terms of the design variables, which must be satisfied by any acceptable design
- And finally the determination of a set of values for the design variables that optimise the objective(s), while satisfying all the constraints.

In mathematical terms one can describe the above as:

$$\min_{x \in \Re^n} \vec{f}(x) = (f_1(x), \ldots, f_k(x))^t \tag{30}$$

$$\begin{aligned}
s.t.: \quad & g(x) \geq 0 \quad where: \quad g: \Re^n \to \Re \\
& h(x) = 0 \quad \quad h: \Re^n \to \Re \\
& x_l \leq x \leq x_u \quad \quad x_l, x_u \in \Re^n
\end{aligned} \tag{31}$$

Note that the above functions (objectives and constraints) can be linear or non-linear, continuous or non-continuous and include continuous and discrete design variables.

There are two major classes of solution methodologies that solve general Non-Linear Programming optimisation problems. Local optimisation methodologies that assume continuity and unique optimal solution. Global optimisation methodologies that can address discontinuities of the design space as well as the possibility of multiple optimal solutions.

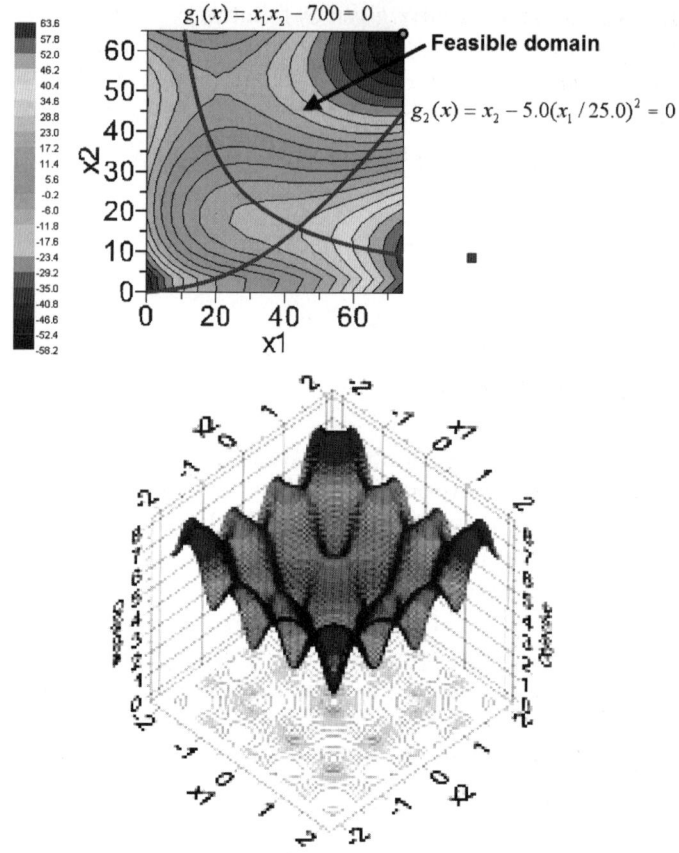

Figure 9. Single and multiple optima examples

Single objective optimisation problems are usually addressed with single objective optimisation methodologies, while multiple objective optimisation problems are addressed with multi-objective optimisation methodologies that produce the so-called Pareto set. Any of the above can handle continuous and discrete parameters. One must acknowledge that difficulty will be added when discrete parameters are involved.

3.3.1 Local Optimisation Methodologies

In solving NLP problems with the aim of finding a local optimum, gradient-based numerical iterative methods can be deployed. A number of methodologies exist, utilizing first order derivatives of the objective and constraint functions in order to iteratively progress towards the optimum. Two of the well-known and established methodologies are the Sequential

8. Virtual Thermo-Mechanical Prototyping

Quadratic Programming (SQP) method and the Generalized Reduced Gradient methodologies. There exist a number of other methods like the Modified Method of Feasible Directions and the Moving Asymptotes method. In the following we present the Sequential Quadratic Programming method [2, 3] to provide the reader with an example of an algorithm so as to expose the various elements of a numerical optimisation method. Note that vectors are written in bold, e.g. x represents the vector of design variables. For a given optimisation problem:

- *Min f(x)*
  Subject to: $h(\mathbf{x}) = 0$ (equality constraints)
  $g(\mathbf{x}) \leq 0$ (inequality constraints)
  Necessary condition for point x* to be a minimizer is the KKT condition (Karush-Kuhn-Tucker condition):

$$\nabla f^* + \lambda^T \nabla h^* + \mu^T \nabla g = 0^T$$
with $\lambda \neq 0$
$$\mu^T g = 0, \mu \geq 0$$

λ and μ are the Lagrange multipliers. An inequality constraint can be inactive, in that case $g_j < 0$, $\mu = 0$. If the inequality constraint is active, $g_j = 0$, $\mu \neq 0$, and it can be handled as an equality constraint. Sufficiency condition for point x* to be a minimizer: Hessian matrix (second derivative) should be positive definite: $H > 0$.

- *SQP* Let us look at a concept of the SQP algorithm.:
  STEP 1. Select starting point x0, initialise λ_0, initialise H, let k (iteration counter) = 0
  STEP 2. First gradient calculation
  STEP 3. Iterate while convergence criteria not met:
    STEP 3a. if $k \neq 0$, update Hessian matrix using the BFGS method
    STEP 3b. Solve quadratic sub-problem with linearized constraints following Newton's method. This results in:
      $\mathbf{s}_k$, the search direction
      λ_{k+1}, the lagrange multipliers
    STEP 3c. Minimize a merit function along $\mathbf{s}_k$, to determine a step length α_k
    STEP 3d. Set $\mathbf{x}_{k+1} = \mathbf{x}_k + \alpha_k \mathbf{s}_k$
    STEP 3e. Calculate gradient for $\mathbf{x}_{k+1}$ and check convergence (KKT condition)
    STEP 3f. Let k = k+1

The merit function is the one proposed by Powell:

$$\Phi(x,\lambda,\mu) = f(x) + \sum w_i h_i + \sum w_j \left|\min\{0,-g_j\}\right|$$

w_i and w_j are the weights that are used to balance the infeasibilities. The values for the weights are taken as follows:
  $w_i = |\lambda_i|$ for $k = 0$ (first iteration)
  $w_{i,k} = \max\{|\lambda_{i,k}|, 0.5(w_{i,k-1}+|\lambda_{i,k}|)\}$ for $k > 0$

Design variables are always normalized to the interval [-1,1] to give equal weight to different dynamic ranges. Constraint values are always normalized by dividing by a range, which is |H-L| with H the high constraint value and L the low constraint value. If only one of both (L or H) is specified, the other is taken such that H = 1.2L if H > 0 and H = L/1.2 if H < 0. The merit function is always a compromise between minimizing f(x) and satisfying the constraints, hence it is possible that a better point is found while constraints are not satisfied, because the total merit function is better. The merit function can be influenced by multiplying the objective function with a factor, which results in a bigger or smaller weight in the merit function.

- *General Algorithmic Procedure*
  Select a starting point
  At that point approximate the original non-linear problem with a Quadratic Programming (QP) sub-problem. This has a quadratic objective function and linear constraints
  Solve the resulting QP sub-problem using special, efficient techniques. The vector connecting the starting point and the solution of the QP sub-problem defines the search direction
  Minimize a special "merit function" along the search direction using a one-dimensional minimization technique. The merit function attempts to balance constraint violation vs. reduction in the objective function
  The minimum of the merit function is the next iteration point. The method should reach feasibility at convergence, but intermediate results can be infeasible

8. Virtual Thermo-Mechanical Prototyping

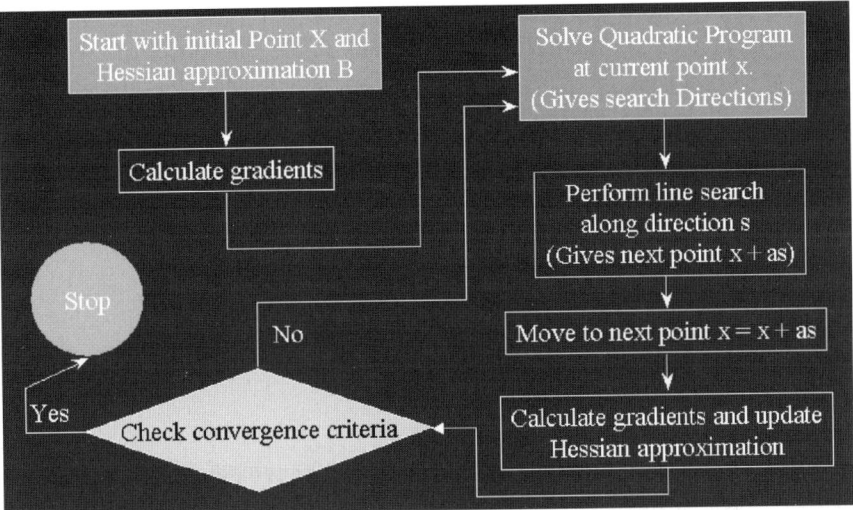

Figure 10. SQP Flowchart

3.3.2 Global Optimisation Methodologies

Evolutionary Computation

Generic Algorithms (GA) were invented by John Holland in the 1960s and were developed by himself and his students and colleagues at the University of Michigan in the 1960s and 1970s. Holland's 1975 book Adaptation in Natural and Artificial Systems presented the genetic algorithm as an abstraction of biological evolution and gave a theoretical framework for adaptation under the GA. Holland's GA is a method for moving from one population of "chromosomes" to a new population by using a kind of natural selection together with genetics inspired operators of crossover, mutation, and inversion. Each chromosome consists of "genes". The selection operator chooses those chromosomes in the population that will be allowed to reproduce, and on average the fitter chromosomes produce more offspring than the less fit ones. Crossover exchanges subparts of two chromosomes, roughly mimicking biological recombination between two single-chromosome organisms; mutation randomly changes the same chromosome; and inversion reverses the order of a contiguous section of the chromosome, thus rearranging the order in which genes are arrayed.

Biological Terminology

All living organisms consist of cells, and each cell contains the same set of one or more chromosomes. A chromosome can be conceptually divided into genes. Very roughly one can think of a gene as encoding a trait, such as eye colour. The different possible settings for a trait are called alleles. Each gene is located at a particular locus (position) on the chromosome.

Many organisms have multiple chromosomes in each cell. The complete collection of genetic material is called the organism's genome. The term genotype refers to the particular set of genes contained in a genome. During sexual reproduction, recombination (or crossover) occurs: in each parent, genes are exchanged between chromosomes to form a new chromosome. Offspring are subject to mutation, in which single nucleotides are changed from parent to offspring, the changes resulting from copying errors. The fitness of an organism is typically defined as the probability that the organism will live to reproduce (viability) or as a function of the number of offspring the organism has (fertility).

Elements of Genetic Algorithms

All GA's have the following elements: populations of chromosomes, selection according to fitness, crossover to produce new offspring, and random mutation of new offspring. The chromosomes in a GA population typically take the form of bit strings. Each locus in the chromosome has two possible alleles: 0 and 1. Each chromosome can be thought of as a point in the search space of candidate solutions. The GA processes populations of chromosomes, successively replacing one such population with another. The GA most often requires a fitness function that assigns a score (fitness) to each chromosome in the current population. The fitness of a chromosome depends on how well that chromosome solves the problem at hand.

GA Operators

The simplest form of genetic algorithm involves three types of operators: selection, crossover, and mutation.

1. Selection

   This operator selects chromosomes in the population for reproduction. The fitter the chromosome, the more times it is likely to be selected to reproduce.

2. Crossover

   This operator randomly chooses a locus and exchanges the subsequences before and after that locus between two chromosomes to create two offspring.

3. Mutation

   This operator randomly flips some of the bits in a chromosome.

A Simple Genetic Algorithm

Start with a randomly generated population of n l-bit chromosomes; Calculate the fitness f(x) of each chromosome x in the population; Repeat the following steps until n offspring have been created:

Select a pair of parent chromosomes from the current population, the probability of selection being an increasing function of fitness. Selection is done with the same chromosome possibly to be selected more than once to become a parent.

With probability p_c (the crossover probability) cross over the pair at a randomly chosen point to form two offspring. If no crossover takes place, form two offspring that are exact copies of their respective parents.

Mutate the two offspring at each locus with probability p_m (the mutation rate) and place the resulting chromosomes in the new population.

If n is odd, one new population member can be discarded at random.

Replace the current population with the new population.

Go to beginning of the loop.

Each iteration of this process is called a generation. A GA is typically iterated for anywhere from 50 to 500 or more generations.

Simulated Annealing [4]

Simulated Annealing algorithms are inspired by the technical process of annealing solids. This technology applied to optimisation formulates a sequence of steps, defining a minimization process by random changes of the design variables and the calculation of a probabilistic acceptance criterion. An evaluated design vector is accepted or rejected following the criterion below. Accept a design if:

$$f(x_{i+1}) - f(x_i) \leq 0 \quad and \quad set \quad x_{i+1} = x_i \tag{32}$$

Otherwise accept the design with the probability:

$$P(f(x_{i+1}) - f(x_i)) = e^{-(f(x_{i+1}) - f(x_i))/k_b T} \tag{33}$$

The factor k_b is the so-called Boltzmann constant and T is the temperature. There exist global optimum convergence theorems for this algorithm, assuming that the reduction of the temperature T per iteration k is taken to be no faster than: $T_k = T_0/\ln(k)$. Applying such a logarithmic decay in the temperature results in long optimisation process. Heuristics are used in order to speed up the temperature decay.

The advantages of simulated annealing are:
- Guarantees high probability of global convergence
- Introduces the possibility to solve mixed integer, discrete or continuous problems
- Insensitive to stochastic disturbances of the objective function

Self Adaptive Evolution Strategy [5]

Evolution strategies imitate in a simplified way biological mutation and selection. The implemented algorithm is a multi-recombinant scheme. It is an evolution scheme based on a population of designs λ. The members of this population are created by recombination and mutation from a set of μ parent designs $x_p^{(k)}$. These parent designs are selected from the total population of designs by selecting the parents with the best fitness from the previous λ designs (offspring at iteration k-1). The multi-recombinant method used here selects multiple parents to generate one offspring. Mutation is independently applied to each design according to the following scheme:

$$x_{m,i}^{(k)} = x_{p,i}^{(k)} + \delta_m^{(k)} s_i^{(k)} \tag{34}$$

with k representing the generation number, p the parent index, m the particular member of the population and $s_i^{(k)}$ the random search direction. Using the intermediate recombination scheme, the step length $\delta_m^{(k)}$ for each population member is constructed on the basis of the μ parents as follows:

$$\delta_m^{(k)} = \frac{1}{\rho} \sum_{j=1}^{\rho=\mu} \delta_{p,z_j(1,\mu)}^{(k)} \tag{35}$$

with $z_j(1,\mu)$ a uniform distributed random integer from the interval $[1,\mu]$.

That is how old generations produce new generation. The new generation fitness is then calculated and new offspring are made. The algorithm has convergence proof, and certain ranges for values of the algorithm parameters have been determined for improved performance.

Differential Evolution [6]

Differential Evolution is a recent approach for the solution of real-valued multi-objective optimisation problems. As is typical for stochastic search algorithms, differential evolution does not require the calculation of the sensitivities.

In the process of the optimisation, DE generates new parameter vectors by adding the weighted difference between a defined number of randomly selected members of the previous population to another member. In its basic strategy, this is the difference of two vectors added to a third:

$$\mathbf{v}_m^{(k+1)} = \mathbf{x}_{r1}^{(k)} + \alpha \cdot (\mathbf{x}_{r2}^{(k)} - \mathbf{x}_{r3}^{(k)}) \tag{36}$$

for m = 1(1)λ with k the generation index, l the population size, $r_1, r_2, r_3 \in [1,\lambda]$, randomly chosen and mutually different. The step length is

8. Virtual Thermo-Mechanical Prototyping

not taken from a fitness selected set of individuals, but rather from a randomly selected individual of the previous population. To increase diversity in the population, crossover is introduced. Here, parts of the previous design vector are inherited by the new design, following a certain probability. This recombination scheme is different than the Self-Adaptive recombination defined above. The selection process is similar to above, if the resulting design has a better fitness value it replaces its parent. This is another distinctive difference between the Differential Evolution and the Self-Adaptive Evolution. Here fitness is tested against the direct predecessor, where as the Self-Adaptive constructs a fitness selected set of parents from the whole population.

3.3.3 Efficient Global Optimisation [7]

The approach is based on fitting response surface models (Kriging Response Surface, for example) to data collected by evaluating the objective and constraint functions at a few points. These response surfaces are then used to visualize input-output relationships, estimate the location of the optimum and suggest points where additional function evaluations may help improve this estimate. The process flow is given in below.

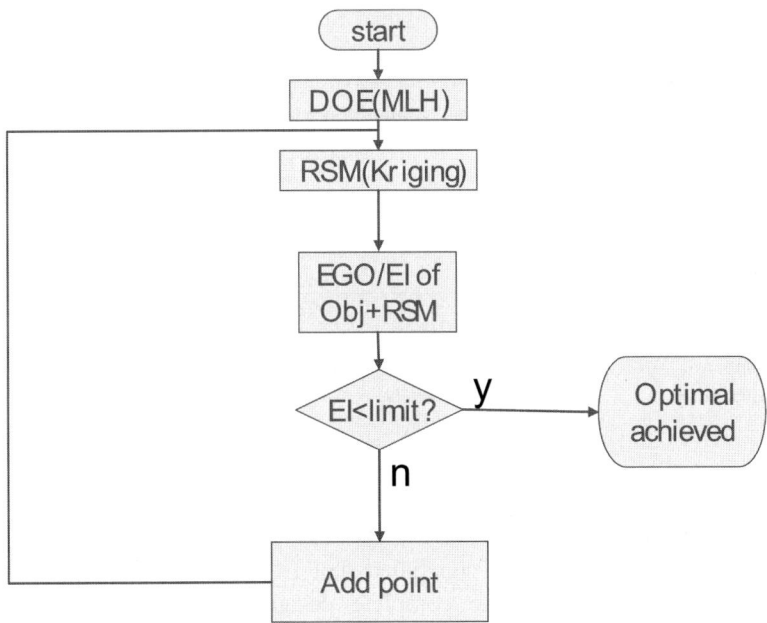

Figure 11. EGO algorithm flowchart

The methodology starts with a set of DOE points determined by a Modified Latin Hypercube DOE. A Kriging model is then fit to the data. A measure of Expected Improvement (EGO/EI) is then calculated.

$$EI(\mathbf{x}) = (f_{min} - \hat{y})\Phi\left(\frac{f_{min} - \hat{y}}{\hat{s}}\right) + \hat{s}\varphi\left(\frac{f_{min} - \hat{y}}{\hat{s}}\right) \qquad (37)$$

A combination of the EI and the global search for the optimum is combined by adding a term in the objective function that introduces the standard deviation of the design variable space

$$f(x) - n \cdot \sigma(x) \qquad (38)$$

A large value of n will guide the EGO algorithm to add points in a wider region within the design space, while a n value closer to zero will make the algorithm behave more as a local optimisation methodology.

Points are added then accordingly. Depending on convergence the algorithm continues by calculating a new Kriging model and calculating the new EGO/EI measure. Termination is reached upon determination of the optimum point, or upon reaching an acceptable value for the EI.

The following table summarizes the properties of the EGO algorithm compared to the classical algorithms like Self Adaptive Evolution (SEVOL) and SQP.

Table 1. Summary of the EGO algorithm, comparison with other algorithms

Criterion	EGO	SEVOL	SQP
Global optimiser behaviour	+	+	– (can arrive in local minimum)
Local optimiser behaviour	+ (uses shape of function)	+/– (convergence speed = low)	+ (uses shape of function)
Convergence to exact result	Normal	normal	Only local
Number of experiments	Low – For same level of accuracy at least a 50% improvement has been demonstrated	high	Low

From the table it can be concluded that EGO is a global optimisation algorithm, since it scans the complete design space, and subsequently adds experiments in the interesting regions. Compared to classical global optimisation algorithms like Self Adaptive Evolution (SEVOL), it needs fewer experiments to reach the same global minimum, because it used the shape of

8. Virtual Thermo-Mechanical Prototyping

the function as information to select new experiments more intelligently. Compared to classical local optimisation algorithms like SQP, it spends a comparable amount of experiments, but has the advantage of being a global optimiser.

Let us see an application case of the EGO algorithm. The minimization of a modified Branin function can be achieved in a fraction of the function calls necessary in comparison to the Self Adaptive Evolution methodology.

$$f(x) = \left(x_2 - \frac{5.1x_1^2}{4\pi^2} + \frac{5x_1}{\pi} - 6\right)^2 + 10\left(1 - \frac{1}{8\pi}\right)\cos(x_1) \quad (39)$$
$$+ 10 + \frac{(x_1 - 3.14)^2}{100} + \frac{(x_2 - 2.28)^2}{100}$$

The function has three minima in the range of $x1 \in [-5;10]$, $x2 \in [0;15]$
(9.41;2.47) => f(x) = 0.792
(3.14;2.28) => f(x) = 0.398
(3.08;12.03) => f(x) = 1.763
The function can be visualized in Figure 12.

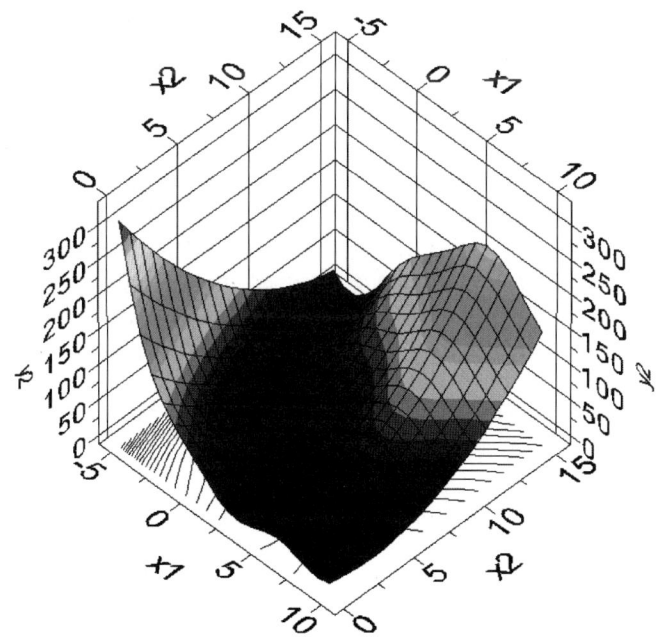

Figure 12. 3D plot of the Modified Branin Function

The EGO algorithm found the optimum as shown in the following figure. The algorithm adds more points in the area of the global optimum.

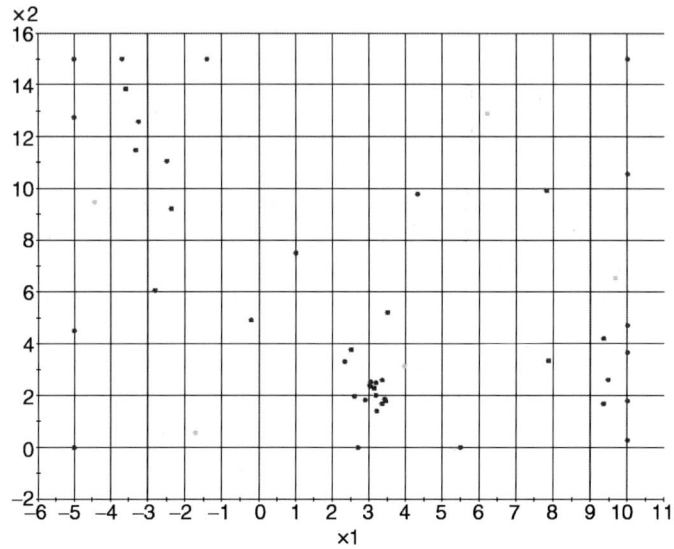

Figure 13. EGO added points in the design space

3.3.4 Multi-Objective Optimisation

Multi-Objective Optimisation algorithms allow to perform optimisations taking into account multiple objectives at the same time. One objective can be a minimization or a maximization of an output. Some commercial optimisation software [1] contain a number of methods that are able to construct the so called Pareto front of the multi-objective optimisation problem, and a number of methods that will find one compromise solution between the different objectives.

The multi-objective optimisation problem is defined as follows:

$$\min_{x \in \Re^n} \vec{f}(x) = (f_1(x),..., f_k(x))^t \qquad (40)$$

8. Virtual Thermo-Mechanical Prototyping

$$s.t.: \quad g(x) \geq 0 \quad where: \quad g: \Re^n \to \Re$$
$$h(x) = 0 \qquad\qquad\quad h: \Re^n \to \Re$$
$$x_l \leq x \leq x_u \qquad\quad x_l, x_u \in \Re^n$$

In other words we wish to determine from among the set of all numbers which satisfy the inequality $g(x) \geq 0$ and the equality constraints $h(x) \geq 0$ that particular set $x^* \in \Re^n$ with $x_l \leq x^* \leq x_u$ which yields the optimum values of all the objective functions $f_1, \ldots, f_k$. Here 'optimise' does not mean simply to find the minimum or maximum of the objective function as it is for a single criterion optimisation problem. It means to find a 'good' solution considering all the objective functions simultaneously. Of course, first we need to know how to designate a particular solution as 'good' or 'bad'. This is the major question, which arises while solving any multi-criteria optimisation problem, and will be discussed here.

In the following let $X = \left\{ x \in \Re^n : \begin{array}{c} g(x) \geq 0 \\ h(x) = 0 \\ x_l \leq x \leq x_u \end{array} \right\}$ denote the set of feasible design vectors.

Methods that generate the Pareto front

In multi-objective optimisation, the Pareto front is defined as the border between the region of feasible points (i.e. design points for which all constraints are satisfied), and the region of infeasible points. If this border is shown as function of the objectives, it is a continuous line (surface in case of more than two objectives) that contains possible optimal combinations of the objectives. Decreasing the value of one objective while keeping the value of the other objective(s) constant would move the design point into the infeasible domain. Increasing the value of one objective while keeping the value of the other objective(s) constant would make the design not optimal. The goal of the different methods that generate the Pareto front is to find a number of points on the Pareto line, by giving different weights to the different objectives.

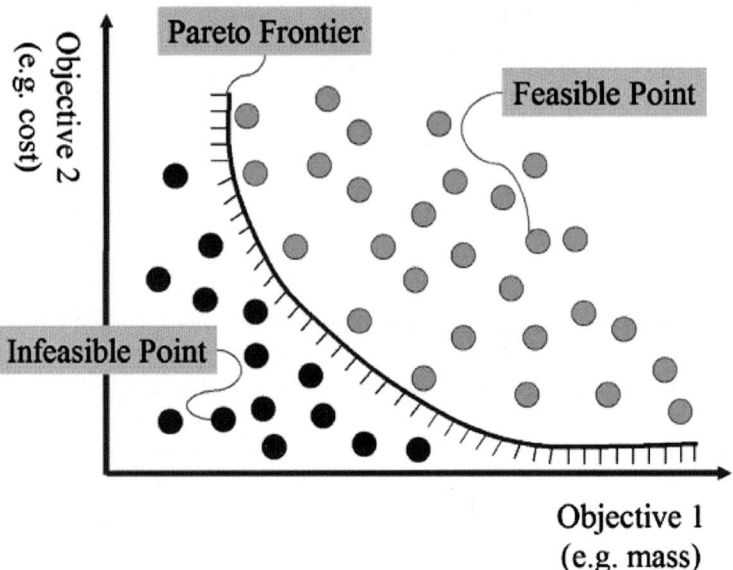

Figure 14. Pareto front

Normal Boundary Intersection method

For this method, different points on the Pareto front are found by performing optimisations that are restricted to lines that are
- Intersecting the chord between the ideal objective values
- And perpendicular to this chord.

In this way the points generated are nicely spread on the pareto front.

Weighted Objective method

Different points on the Pareto front are found by varying the weights ω_i in the following minimization problem:

$$f(x) = \sum_{i=1}^{k} \omega_i f_i(x) \tag{41}$$

8. Virtual Thermo-Mechanical Prototyping

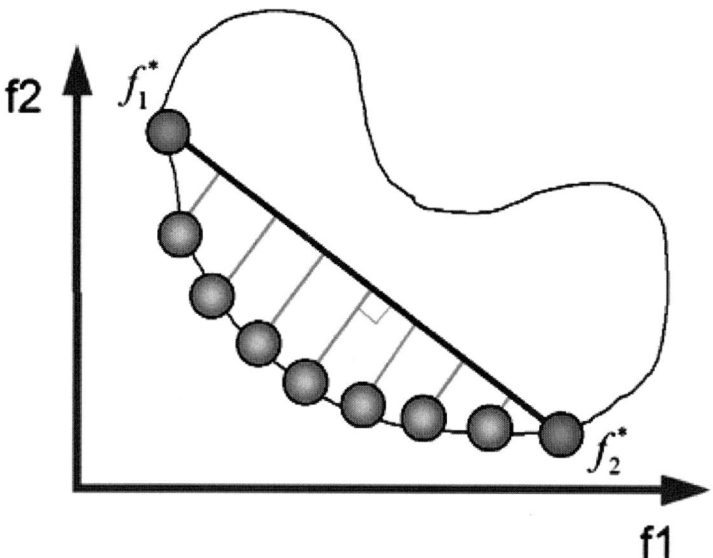

Figure 15. Normal Boundary Intersection method

where $\varpi_i \geq 0$, $\sum_{i=1}^{k}\varpi_i = 1$ are the weighting coefficients, which reflect the preference of the individual objective functions.

Weighted Tchebycheff method

Different points on the Pareto front are found by varying the weights wi in the following minimization problem:

$$\min_{x \in X} \max_{i=1,\ldots,k} \left(\varpi_i \left| f_i(x) - f_i^* \right| \right) \tag{42}$$

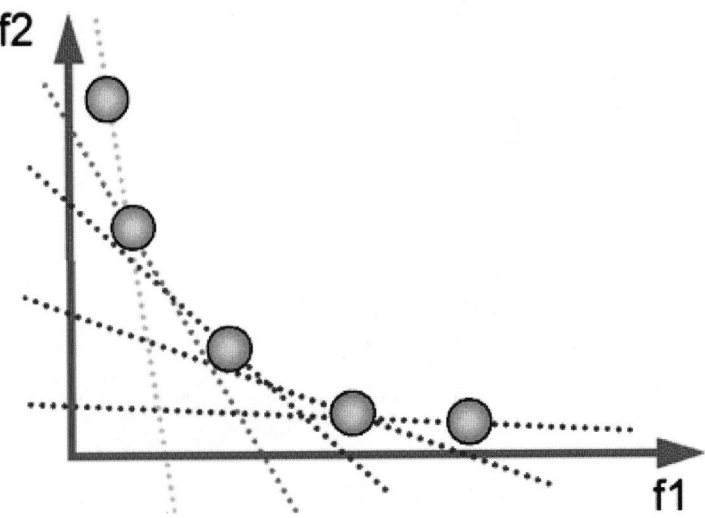

Figure 16. Weighted Objective method

with $f_i^* = \min_{x \in X} f_i(X), i = 1,...,k$, which are the ideal objective values, i.e., the optimal values of each objective function in case of a single-objective optimisation problem.

For generating the Pareto front, the sum of all weights is kept to one.

Min-Max Optimum method

Different points on the Pareto front are found by varying the weights ϖ_i in the following minimization problem:

$$\min_{x \in X} \max_{i=1,...,k} \left(\varpi_i \frac{f_i(x) - f_i^*}{f_i^*} \right), \; f_i^* \geq 0 \qquad (43)$$

8. Virtual Thermo-Mechanical Prototyping

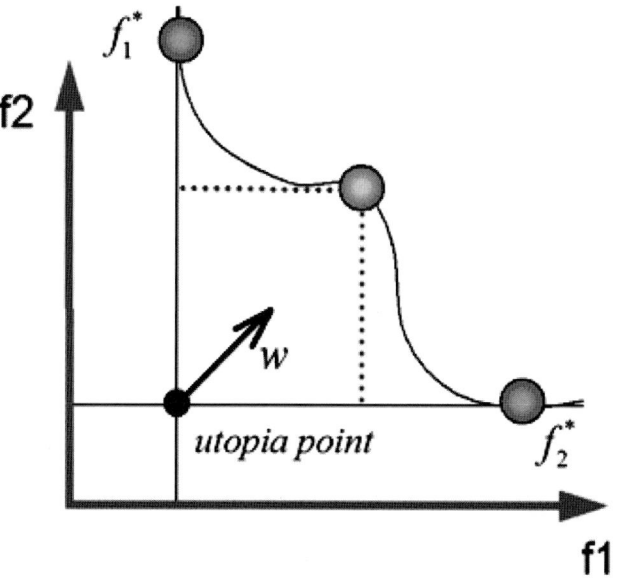

Figure 17. Weighted Tchebycheff method

with $f_i^* = \min_{x \in X} f_i(X), i = 1,...,k,$ which are the ideal objective values, i.e., the optimal values of each objective function in case of a single-objective optimisation problem.

For generating the Pareto front, the sum of all weights is kept to one.

Methods that find one compromise solution
- Trade-Off Method
  For given $l \in \{1,...,k\}$ do:
  Find the minimum of the l'th objective function, i.e., find x^* such that

$$f_l(x^*) = \min_{x \in X} f_l(x) \qquad (44)$$

subject to additional constraints of the form

$$f_i(x) \leq y_i, \text{ for all } i = 1,...,k, i \neq l \qquad (45)$$

where the y_i are assumed values of the objective functions we which not to exceed.

Repeat (1) for different values of y_i. The information derived from a well-chosen set of y_i can be useful in making the decision. The search is stopped when the user finds a satisfactory solution.

It may be necessary to repeat for different indices $l \in \{1,...,k\}$. In order to get a reasonable choice of y_i it is useful to minimize each objective function separately, i.e., let $f_i^*, i = 1,...,k$ be the optimal values of the scalar problems, then take account of the following constraints:

$$f_i(x) \le f_i^* + \Delta f_i, \text{ for all } i = 1,...,k, i \ne l \qquad (46)$$

where Δf_i are given values of function increments.

- **Hierarchical Method**

Without loss of generality let the vector of objectives $(f_1(x),...,f_k(x))'$ be ordered with respect to there importance, i.e., let $f_1(x)$ be the most and $f_k(x)$ the least important objective function, specified by the user. We now minimize each objective separately, adding in, at each step a new constraint, which limits the assumed increase or decrease of the previously considered functions.

Find the minimum of the first objective, i.e., find $x^{(1)} = (x_1^{(1)},...,x_n^{(1)})'$, such that:

$$f_1(x^{(1)}) = \min_{x \in X} f_1(x) \qquad (47)$$

Repeat step (2) for $i = 2,...,k$

Find the minimum of the i'th objective, i.e., find $x^{(i)} = (x_1^{(i)},...,x_n^{(i)})'$, such that:

$$f_i(x^{(i)}) = \min_{x \in X} f_i(x) \qquad (48)$$

with the additional constraints:

$$f_{j-1}(x) \le (1+\frac{\varepsilon_{j-1}}{100}) f_{j-1}(x^{(j-1)}), \text{ for } j = 2,...,i \qquad (49)$$

Note that the solutions of the k scalar problems are all functional-efficient but only the last (k'th) leads to optimality.

- **Method of Distance Function (Euclidean Norm)**
  This method also leads to the scalarization of the problem MMO. Assuming we have bounds $y_i, i = 1,...,k$, usually provided by the user, our goal is to evaluate the functions to approach the y_i-values as close as possible, i.e.,

$$f^p(x) = \left(\sum_{i=1}^{k}|f_i(x) - y_i|^p\right)^{1/p}, 1 \leq p \leq \infty \tag{50}$$

where p is the order of the method:

$$p = 1 \quad f^1(x) = \sum_{i=1}^{k}|f_i(x) - y_i| \quad \text{Goal Programming} \tag{51}$$

$$p = 2 \quad f^2(x) = \left(\sum_{i=1}^{k}|f_i(x) - y_i|^2\right)^{1/2} \quad \text{Euclidean Norm} \tag{52}$$

3.4 Reliability and Robustness Analysis

A reliability analysis aims to estimate the probability that a structure will fail to meet a pre-defined criterion. Figure 18 visualizes a typical reliability problem, with two stochastic parameters X = [x₁, x₂] with a Gaussian distribution. The ovals of increasing size represent lines of equal variance of the parameter steps with respect to the mean value [i₁, i₂]. The monitored output parameter g_S(X) is called the load effect. The limit value gR, which should not be exceeded by the load effect, is known as the resistance effect. The intersection of load and resistance is the Limit State Function (LSF), the subset where g(X) = g_R-g_S(X) = 0. The LSF forms the boundary between the safe area and the failure area in the parameter space X. The nearest failure point in the parameter space is commonly known as the design point X*, at distance â from the mean value. This distance is called the Reliability Index, as it is a direct measure for the reliability: it denotes how many variance steps lie between the parameter mean and the design point X*. A reliability problem is generally not evaluated in X-space, but first transformed to standard normal space U, where parameters are uncorrelated and have a zero-mean, unit variance normal distribution. This transformation prevents numerical problems and simplifies the algebra. In the next sections, some well-established methods are described and visualized in U-space, namely Monte Carlo, FORM, SORM, Importance Sampling and Transformed Importance Latin Hypercube Sampling.

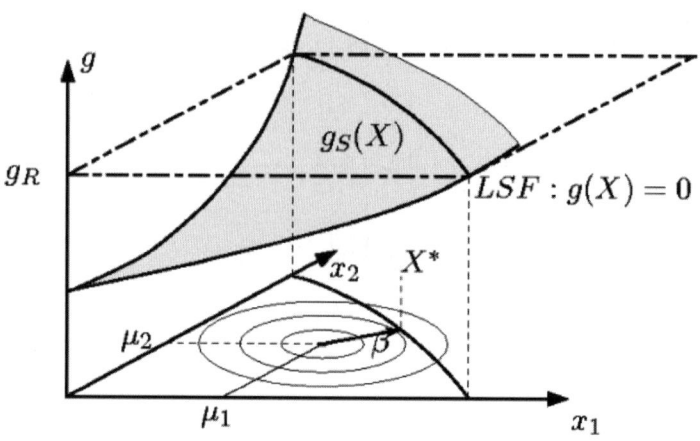

Figure 18. A typical reliability problem

3.4.1 Monte Carlo Simulation

The most straightforward numerical approach to solve a reliability problem is by performing a number of simulations with randomly selected parameter combinations for the given distribution, and to verify for each combination if this results in a failure. One then divides the number of failures by the number of simulations, which yields an approximation of the actual failure probability. This is the standard Monte Carlo (MC) simulation approach. This method is widely used for benchmarking and validation purposes, as it is almost always applicable. A large disadvantage of the method is that a lot of simulations are typically required to accurately estimate the failure probability. Especially when the failure probability is low, the failure density is too low. For example, if the failure probability is 0.003, only three of 1000 samples are expected to produce a failure. As simulations are random, there might as well be 2 or 4 failures, thus influencing the estimate with 33%. A certain number of failures are therefore required, to reduce the influence of a single failure case on the total estimate.

8. Virtual Thermo-Mechanical Prototyping

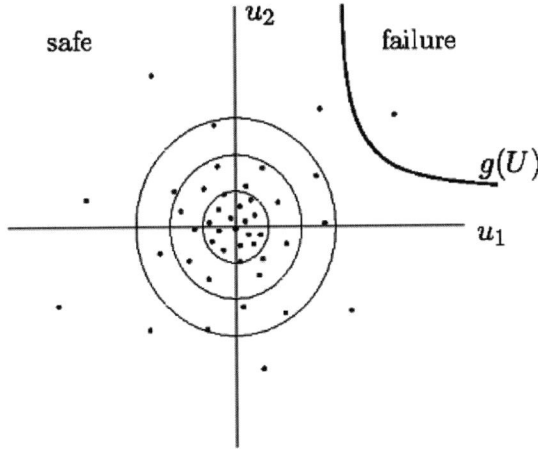

Figure 19. Monte Carlo Simulations

3.4.2 Second Order Reliability Method

The Second Order Reliability Method (SORM) is generally implemented as an extension of FORM algorithm; this is also the case in OPTIMUS. The second-order derivatives of the Limit State Function (LSF) are computed in the FORM design point. They are stored in the Hessian matrix, from which the principal curvatures of the LSF are obtained. The FORM failure probability is then corrected for these curvatures, using the well-established correction algorithm of Hohenbichler and Rackwitz.

When curvatures are as in the FORM failure probability is an overestimation of the actual failure probability: the area in the parameter space between the actual LSF $g(U)$ – that is unknown – and the linear approximation is taken into account as a failure, while it actually belongs to the safe area in the parameter space. Similarly, when the curvatures are in opposite direction, the FORM failure probability is an underestimation of the actual failure probability. The SORM correction algorithm makes a correction for these curvatures, yielding a second-order approximation of the failure probability based on the second-order information obtained in the FORM design point.

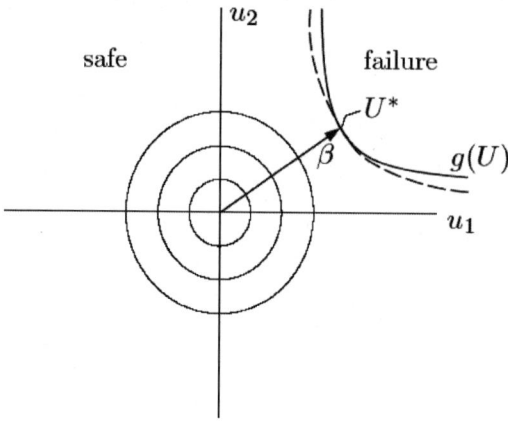

Figure 20. Second Order Reliability Method (SORM)

3.4.3 Importance Sampling Monte Carlo

The general idea behind Importance Sampling (IS) is to place more samples in the failure range in the parameter space, to increase the failure density. The most straightforward IS approach is the Importance Sampling Monte Carlo (ISMC) approach, where the sampling centre of a standard MC sample set is moved from the origin in U-space to the FORM design point U* over a distance â.

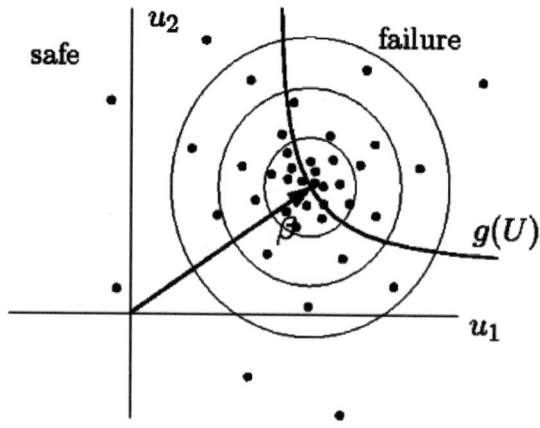

Figure 21. Importance Sampling Monte Carlo

8. Virtual Thermo-Mechanical Prototyping

3.4.4 (Transformed) Importance Latin Hypercube Sampling

Latin Hypercube (LH) sampling is a well-known tool to improve the efficiency of a sampling distribution. To place N samples in a K-dimensional parameter space, the space [0,1] is divided into N parts for every dimension (dimension = number of parameters) thus forming subsets of equal probability. Exactly one sample is then placed in every row, each column, etc., of the design. The figure below shows two LH sampling plans.

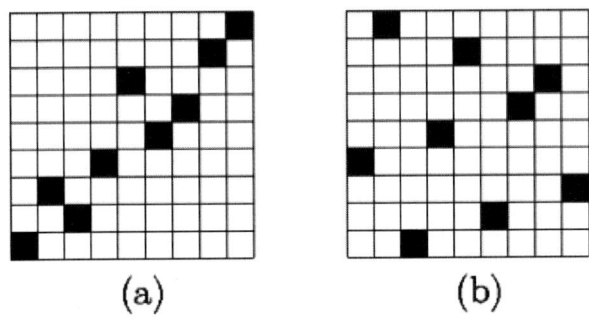

Figure 22. An LH sampling plan (a), with 9 samples in 2 dimensions, and the decorrelated plan (b)

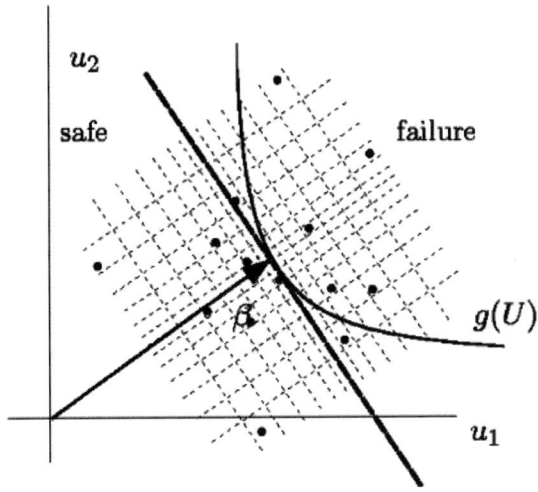

Figure 23. TILHS sampling plan

Transformed Importance Latin Hypercube Sampling (TILHS) is an IS approach based on Latin Hypercube. For a pre-defined number of N samples, an LH plan is created, which is decorrelated and mapped on the cumulative distribution function. The resulting distribution is then placed around the FORM design point and aligned with the linear FORM approximation of the LSF. A TILHS sampling plan guarantees that exactly half the number of samples lies on each side of the FORM approximation of the LSF. TILHS is therefore intrinsically better than ISMC. For both IS approaches, the failure density depends on the shape of the LSF in the vicinity of the design point, not on the linear approximation of the LSF. Importance Sampling can thus be used to improve or validate the FORM estimate of the failure probability. This still requires a reasonably high number of samples, typically much higher than FORM itself needs, as a certain number of failures are needed to sufficiently reduce the c.o.v. However, IS often requires much less samples than the standard MC approach to yield an acceptable result; the advantage of IS increases if the actual failure probability decreases.

4. CASE STUDIES

4.1 Analytical Examples

Problem 1

Objective function	$\text{Min } f(x) = 2x_1^2 + x_2^2 - 48x_1 - 40x_2$
Constraints	$g_1(x) = x_1 + 3x_2 \geq 0$ $g_2(x) = 18 - x_1 - 3x_2 \geq 0$ $g_3(x) = x_1 + x_2 \geq 0$ $g_4(x) = 8 - x_1 - x_2 \geq 0$
Bounds	$0 \leq x_i \leq 6; i = 1, 2$

8. Virtual Thermo-Mechanical Prototyping

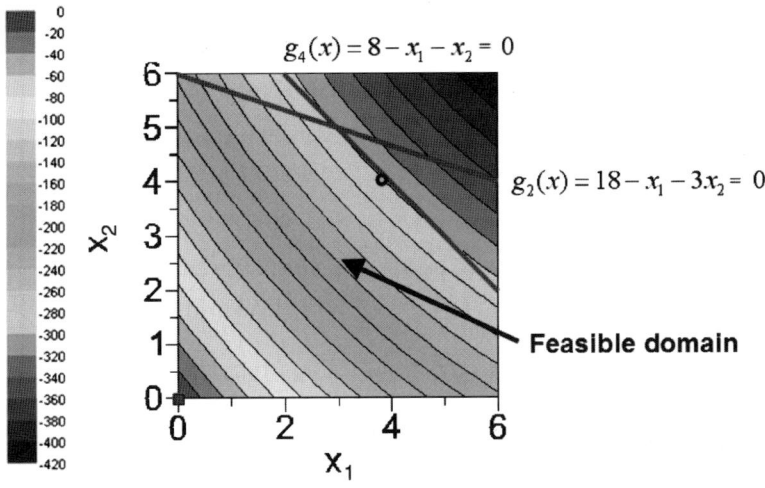

Optimisation results

	Start point	SQP optimum	NLPQL optimum	Exact solution
f(x)	−8.77	−303.9999	−303.9999	−304
x1	0.1	3.9993335	3.9993335	4
x2	0.1	4.0006665	4.0006665	4
Feasible	Yes	Yes	Yes	
# Experiments		12	12	

Problem 2

Objective function	$\min f(x) = 75.1963666677 - 3.8112755343 x_1$ $+ 0.1269366345 x_1^2 - 0.0020567665 x_1^3$ $+ 0.0000103450 x_1^4 - 6.8306567613 x_2$ $+ 0.0302344793 x_1 x_2 - 0.0012813448 x_1^2 x_2$ $+ 0.0000352559 x_1^3 x_2 - 0.0000002266 x_1^4 x_2$ $+ 0.2564581253 x_2^2 - 0.0034604030 x_2^3$ $+ 0.0000135139 x_2^4 + -28.1064434908 \left[\dfrac{1.0}{x_2 + 1.0} \right]$ $- 0.0000052375 x_1^2 x_2^2$ $- 0.0000000063 x_1^3 x_2^2 + 0.0000000007 x_1^3 x_2^3$ $+ 0.0003405462 x_1 x_2^2 - 0.0000016638 x_1 x_2^3$ $- 2.8673112392 e^{(0.0005 x_1 x_2)}$
Constraints	$g_1(x) = x_1 x_2 - 700 \geq 0$ $g_2(x) = x_2 - 5.0 (x_1 / 25.0)^2 \geq 0$
Bounds	$0 \leq x_1 \leq 75$ $0 \leq x_2 \leq 65$

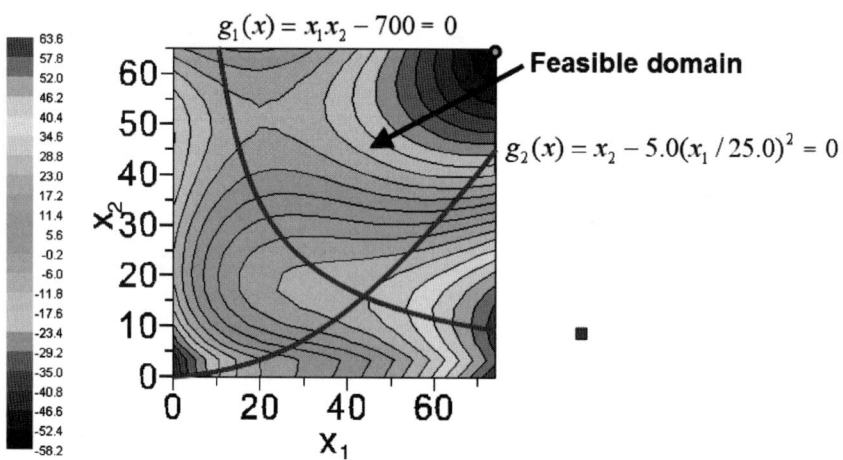

8. Virtual Thermo-Mechanical Prototyping

Optimisation results

	Start point	SQP optimum	NLPQL optimum	Exact solution
f(x)	82.8278	–58.90344	–58.92519	–58.9034
x1	90	75	75.0	75
x2	10	65	65.07	65
Feasible	No	Yes	Yes	
# Experiments		31	24	

Problem 3

Objective function	$\min f(x) = -20 e^{-0.2\sqrt{\frac{1}{2}(x_1^2 + x_2^2)}} - e^{\frac{1}{2}[\cos(2\pi x_1) + \cos(2\pi x_2)]} + 20 + e^1$
Constraints	
Bounds	$-32.768 \leq x_i \leq 32.768; i = 1,2$

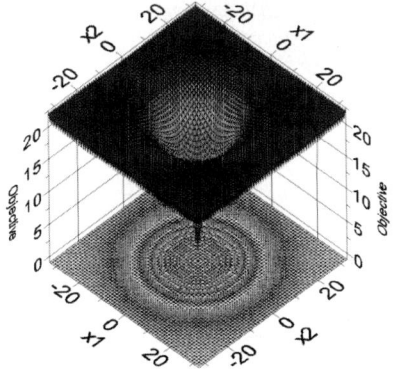 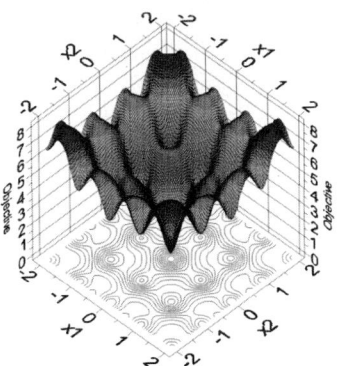

Optimisation results

	Start point	SAE optimum	DE optimum	EGO optimum	Exact solution
f(x)	19.8397	0.28598	0.492972	0.10020	0
x1	29	−0.06391	−0.095119	−0.01238	0
x2	18	8.2985·10−5	0.012390	−0.02517	0
# Experiments		120	140	80	

Problem 4

Objective function	Min $f(x) = -x_1 \sin(\sqrt{abs(x_1)}) \cdot -x_2 \sin(\sqrt{abs(x_2)})$
Constraints	
Bounds	$-500 \leq x_i \leq 500; i = 1,2$

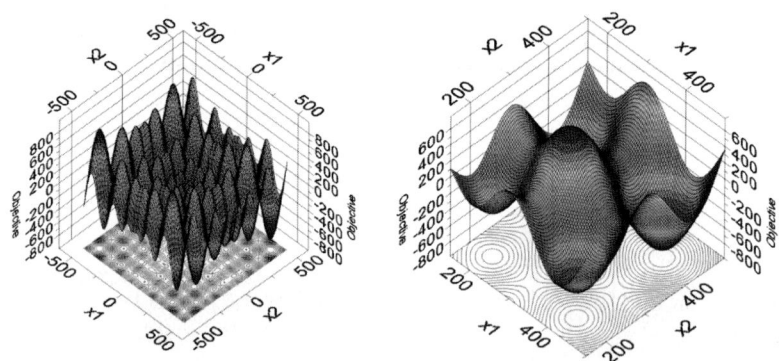

Optimisation results

	Start point	SAE optimum	DE optimum	EGO optimum	Exact solution
f(x)	0	−830.50810	−837.68870	−837.950001	−837.9658

x1	0	425.42606	419.48744	421.301394	420.9687
x2	0	427.22993	420.91138	421.088212	420.9687
# Experiments		90	140	39	

4.2 Industrial Application Cases

4.2.1 HVQFN Map Mould Warpage Optimisation

In Chapter 5 a case is described dealing with the prediction of curing induced warpage in map mould systems of HVQFN (Heatspreader Very thin Quad Flat No lead) package family [28]. HVQFN packages are composed of an Integrated Circuit (IC) attached to an exposed pad and in a later stage encapsulated with moulding compound. For this case, FEM models for the calculation of the HVQFN map warpage after moulding and post mould curing are created. Simulation results show that the effect of curing can be neglected when a dead weight, forcing the map mould to remain flat during PMC, is used. In case of no dead weight used the curing shrinkage will have a large effect on the end warpage. After validation with warpage measurements, the developed FEM model will be used to minimize the warpage, taking into account the time and process dependent visco-elastic behaviour of moulding compound. In this section, simulation-based optimisation is used to optimise the compound behaviour to obtain minimal warpage of map mould systems.

As mentioned before, simulation-based optimisation is based on, and integrated with, advanced simulation models that can predict the product/process behaviour reliably and efficiently. For this HVQFN case, simulation-based optimisation involves finding settings for a number of designs parameters that are optimal with respect to one or more simulated model responses. The following compound properties were chosen as input parameters (design space) together with the geometric design parameter of die thickness:

1. Coefficient of Thermal Expansion (CTE) below Tg (CTE1)
2. Quotient of CTE above Tg and CTE below Tg (CTE2/CTE1)
3. Curing shrinkage
4. 1 Hz storage modulus at room temperature
5. Glass transition temperature (CTE based)
6. Rubbery modulus (T>>Tg)

In the table below, all design variables, their nominal design value and lower and upper design bounds are listed. Figure 24 shows a graphical representation of the compound parameters. For instance, increasing of the E modulus at room temperature causes the left straight line in the figure to shift upward. Automatically a new visco-elastic spectrum of the new "three line curve" is created. Variation of the Tg makes both E modulus and CTE curves to shift horizontally. As stated above, CTE2 is varied indirectly via CTE1 and CTEfact.

Table 2. Design space of the DoE

	Variable	[–]	Nominal	Variation
1	cte1	ppm/°C	8	7 – 9
2	Ctefact	–	4	3 – 5
3	Shrink	vol%/°C	0.25	0 – 0.5
4	Eroom	Mpa	20732	12000 – 30000
5	Tglass	°C	110	100 – 200
6	Einf	Mpa	900	800 – 1300
7	Tdie	Mm	0.25	0.14 – 0.36

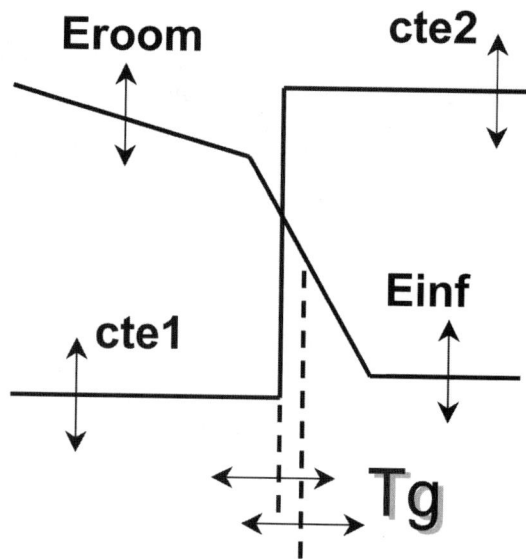

Figure 24. Schematic picture of the compound properties with material variables depicted by arrows

With the following three types of DoE schemes, in total, 249 design parameter combinations were created and subsequently used in parametric FEM model to the determine the warpage output:
1. Nominal design 1 simulation
2. Full factorial $2^7 = 128$ simulations
3. Space filling Latin-hypercube 120 simulations

8. Virtual Thermo-Mechanical Prototyping

The z-displacement output at 9 locations (corners of ¼ map mould and corners of 1½ HVQFN's, see Chapter 5) together with the warpage output are described with RSM developed by using OPTIMUS [1]. A third order Taylor polynomial (120 variables) was used as RSM and this gave a satisfactory regression statistic of ~0.996 for the z-displacements.

However the quality of the associated RSM for warpage, defined as (z max − z min), is not satisfactory. From Figure 25 it can be seen that the warpage is always positive resulting in a sharp point that cannot be accurately described by a Taylor polynomial. This intersecting point is associated with almost flat map moulds. Therefore it was decided to use the 9 RSMs of the 9 z-displacements and put them into Excel for further investigation and optimisation. The apparent spread in the figure near $u_z = 0$ is the result of other z-displacements being maximum or minimum.

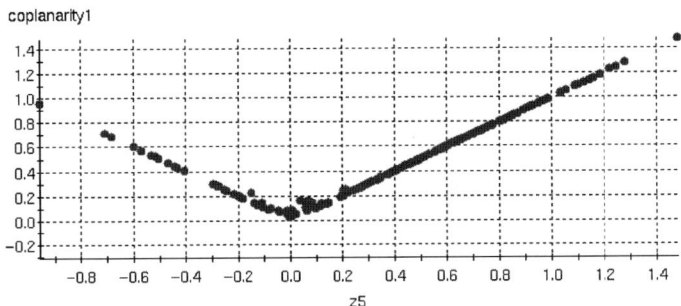

Figure 25. DoE FEM simulation results; warpage (y-axis) versus z-displacement center complete map mould ($u_z = 0$ for 4 corners map mould)

Figures 26 to 28 show some typical Response Surfaces results. All other parameters, which are not plotted on the two input axes, were given nominal value for these figures. As it can be observed in Figure 26 the RSM model is nonlinear. Figures 27 and 28 show that CTE2 and E_{inf} hardly influence the final warpage. This is expected due to the fact that the effect of the curing shrinkage is very small. However in case of no weight these parameters will play a most important role.

Figures 27 to 28 also show that the map warpage can be reduced by decreasing the die thickness; decreasing the Tg; increasing the CTE below Tg and increasing the E modulus at room temperature of the compound. The T_g effect can be explained from the fact that the warpage starts to build up below this temperature. Therefore, the smaller the temperature trajectory below T_g is, the smaller the warpage. The other two effects can be attributed to a larger compound counteracting of the downward bending of the die – lead bimetal.

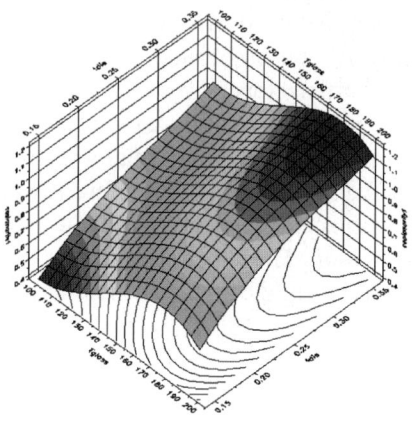

Figure 26. HVQFN map warpage RSM, vertical axis: warpage, left diagonal axis: T_g and right diagonal axis: t_{die}

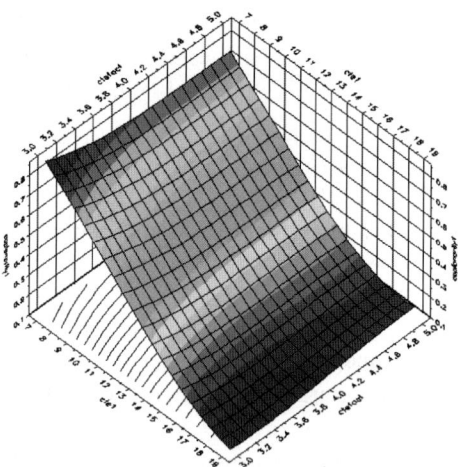

Figure 27. HVQFN map warpage RSM, vertical axis: warpage, left diagonal axis: CTE1 and right diagonal axis: CTE1/CTE2

8. Virtual Thermo-Mechanical Prototyping

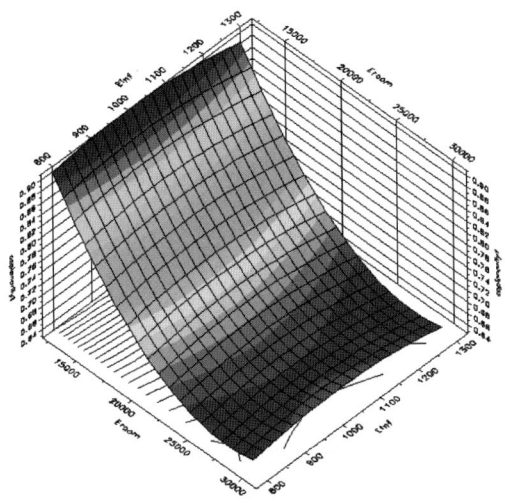

Figure 28. HVQFN map warpage RSM, vertical axis: warpage, left diagonal axis: E_{room} and right diagonal axis: E_{inf}

The 9 RSM's obtained from OPTIMUS were placed in a spreadsheet program. This enables to immediately determine the HVQFN map warpage once supplier data for a compound is known. Figure 29 shows the final warpage determined from the RSMs of 232 randomly chosen compound input parameters (die thickness remains nominal value). It can be seen that the warpage can be significantly reduced with many possible design combinations.

The next step in the warpage minimization process is to select the most suitable moulding compound, by comparing the properties of existing compounds, wherein the warpage associated with different types of compounds is determined as a function of the die thickness. The combination with the lowest warpage can then be tested and verified by experiments. Furthermore the identified "low warpage compounds" will lead to suggestions for the material supplier to create better compounds with lower warpage.

In this case, the design parameters are the die thickness and several specific compound properties. In order to minimize the warpage, accurate RSMs were developed and verified, wherein the nonlinear and strongly

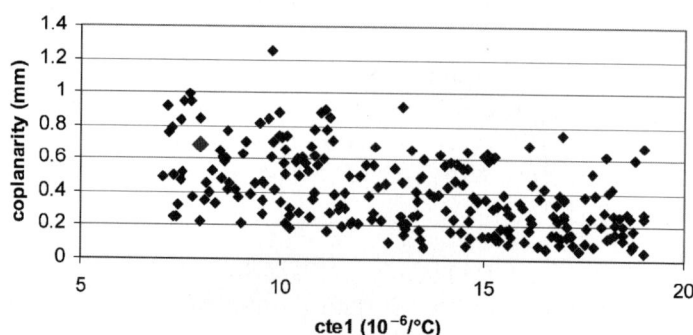

Figure 29. HVQFN map end warpage determined in spread sheet by RSM on 232 random compound property combinations; red dot is nominal

interactive relationships between the design parameters and warpage were established. These RSMs are built into spreadsheet tool, so that the development engineers can easily and efficiently find the influence of design parameters and their variations on the warpage, and to minimize the warpage by choosing proper combinations of compounds and die thickness.

4.2.2 Advanced Structural Similarity Rules for the BGA Package Family

Structure Similarity Rules

Structural similarity determines the extent to which test results from a specific device or family can be considered representative for other similar types. When applied to reliability aspects, structural similarity is indispensable in predicting the reliability performance of types similar to those that have been specifically subjected to reliability testing. The need for having structural similarity rules is not only because reliability testing of every type is time and money consuming, but also because it is often unnecessary to test all due to the inherited similarity of products. Currently structural similarity rules for Integrated Circuit (IC) and package designs, processes and qualifications are used by industry to define efficient Qualification and Reliability Monitoring programs. They imply that the reliability evaluation results from one product may be used for others if they meet the structural similarity criteria. In general, there are three qualification options for IC and package:

8. Virtual Thermo-Mechanical Prototyping

- Option 1: Qualification using existing data.
  Based on existing data the IC and package design and process are qualified.
- Option 2: Bridge Qualification.
  Based on existing data for one or more similar IC/package designs or processes (so called 'bridge' design or process) qualification is granted. This can only be done when certain criteria, so called structural similarity rules are met.
- Option 3: Qualification requiring new/additional testing.
  When both the needed reliability data and structural similarity rules are not available, new or additional testing is required to qualify the design or process.

It is evidential that in the case where a full qualification program is needed, time to market and costs will increase significantly. Thus, any qualification data and existing knowledge of reliability performance that can be used to minimize the number and duration of additional testing will be helpful. For assembly and packaging, a structurally similar package means that it:

- Belongs to the same package type, such as BGA, QFP, or others,
- Has the similar body size,
- Has the similar inner, outer lead or ball spacing,
- Has the similar diepad size,
- Has the similar die size,
- Has the same materials, for leadframe/die attach/moulding compound.

By following the package structural similarity rules, the numbers and duration of reliability qualification tests may be greatly reduced. However, when looking at the existing rules it is clear that they are not all reliably defined. For instance, geometrical parameters such as die-to-pad ratio are not quantitatively included due to the assumption of linear relationships between the design variations and the response attributors. Besides that, these rules are mainly deducted from experience and industrial trial-error results, without full backing and support of reliability physics. Driven by the present development trends of microelectronics, characterized by miniaturization, heterogeneous integration, eco-designing, cost reduction, shorter-time-to-market and outsourcing, the chances and consequences of failures, the design complexity and challenges to meet quality, robustness and reliability requirements are dramatically increased. Therefore, structural similarity rules based on reliability physics (physics of failures) become an urgent need for industry to meet the technology and business development trends.

The Ball Grid Array Package Family

In this case our results to develop advanced structural similarity rules using the state-of-the-art virtual prototyping and qualification techniques are described. As a carrier, the Ball Grid Array (BGA) package family is chosen. In the late 1980s, BGA packages were introduced based on a new multi-layer process. Double-sided flex circuit pairs were stacked and laminated using adhesives to provide vertical connections [1]. During the past years, various new ideas emerged leading to the current BGA packaging concept. The BGA family allows for low profiles and outlines and is currently the standard for high-density IO packages. The concept is based on using an organic laminate, such as FR4, BT, including copper traces that connect to the IC and further encapsulated by using a moulding compound. Many different variations are available on the market, such as TFBGA's, μBGA's, HBGA's, die up of die down, tape-based. But in principle all are using the same concept. In the following study we focus on the BGA version and have parameterised the total design space of the family with six parameters. Figure 30 shows a ¼ schematic of the BGA identifying the parameters. Table 3 lists the ranges of these parameters. The specified design parameters are

- Die thickness,
- Pad-to-body ratio, defined as the ratio between the package body and the copper pad size,
- Die-to-pad ratio, defined as the ratio between the die and the diepad size,
- Body size,
- Body thickness,
- Substrate thickness.

Table 3. Geometric ranges for the design parameters

Parameter	Minimum	Maximum
Die thickness [μm]	280	380
Pad-to-body ratio [%]	15	80
Die-to-pad ratio [%]	15	95
Body size [mm^2]	4×4	44×44
Body thickness [mm]	0.9	1.9
Substrate thickness [mm]	0.36	0.56

Developing accurate reliability prediction models

As the first step towards developing structural similarity rules, accurate, robust and efficient reliability prediction models have to be developed. Herein 3D, non-linear, and multi-physics FEM models are established to calculate the thermo-mechanical and moisture-diffusion responses of BGA for the complete design space as a function of the specified parameters. The developed and verified 3D non-linear FE models include anisotropy for silicon [27], visco-elasticity for moulding compound and die-attach [27],

8. Virtual Thermo-Mechanical Prototyping

elasto-plasticity for copper, and orthotropic visco-elasticity for FR4. They can be used to predict the responoses of BGA, covering the effect of

1. Thermo-mechanical behaviour
- Package warpage, which plays an important role in PCB placement,
- Interface stress levels between moulding compound and substrate, which is related to occurrence and propagation of delamination.
2. Moisture behaviour
- Moisture content, which is related to MSL performance
- Moisture concentration at the interface between moulding compound and substrate, which is related to the occurrence of delamination [6].

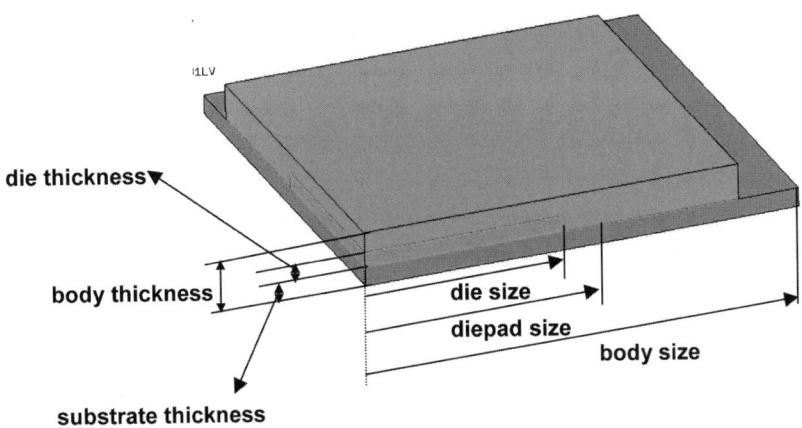

Figure 30. The BGA package style identifying the parameterised design space

Developing similarity rules

By combining the accurate reliability prediction models with advanced simulation-based optimisation methods, such as sequence Design Of Experiments (DOE) and stochastic Response Surface Modelling (RSM) techniques, the 'physics of failure based structural similarity rules' can be developed in the form of Excel tool. By using such a tool, the possible failure mechanisms can be understood and predicted, so the number of reliability qualification tests can be reduced.

For both the thermo-mechanical and the moisture-diffusion responses of the BGA package family, a space-filling Latin-Hypercube-Design consisting of 70 variations is constructed for four output variables. Figure 6 shows the distribution for two design parameters, i.e., body size and pad-to-body ratio. It turned out that for all response parameters quadratic type of RSM models with interactions are sufficiently accurate. Using OPTIMUS [1] automatic running procedure based on cross-validation, the unimportant model terms were deleted. The regression statistics are listed in Table 4, indicating that for all quadratic models the accuracy requirements are satisfied. Figure 32 shows the predicted RSM response versus the FEM results for the DOE points, confirmed the quality of correlation.

Table 4. Compact model regression statistics

Response Parameter	RMSE
Warpage	0.96
Interface stress	0.85
Moisture mass	0.99
Moisture interface concentration	0.98

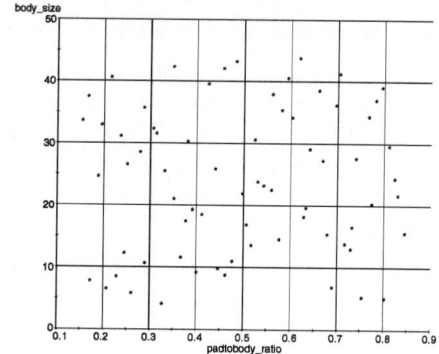

Figure 31. Space filling Latin-Hypercube design of experiments: body size vs. pad-to-body-ratio

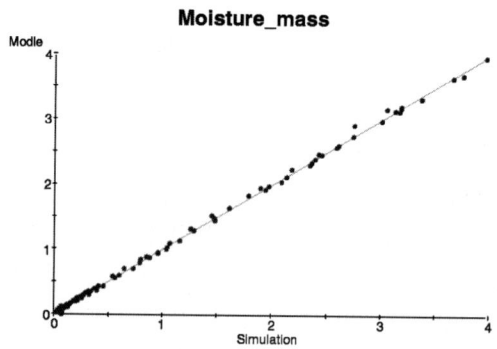

Figure 32. Predicted (RSM) versus simulated (FEM) result

8. Virtual Thermo-Mechanical Prototyping

Table 5 lists the relative importance of the linear polynomial terms describing the RSM functions for i) warpage, ii) interface stress, iii) moisture mass, and iv) moisture concentration. The following can be observed from this table:

1. Thermo-mechanical behaviour
- For the warpage, most relevant terms are the pad-to-body-ratio, body size, and die-to-pad ratio. As a consequence, body thickness, die thickness, and substrate thickness have a marginal effect on the package warpage. Increasing the body size leads to increased warpage and stress levels. When keeping the body size constant, pad-to-body and die-to-pad ratios play a dominant role in the thermo-mechanical response of the package.
- For the interface stress, most relevant terms are the same as those found for warpage. Expect that for interface stress, the substrate and die thickness have some significant effect on the interface stress levels.

2. Moisture induced behaviour
- For the moisture mass in the package, the body size and the body thickness are the dominant parameters. The bigger and thicker the package is, the more moisture ingresses into the package.
- For the moisture concentration at the compound substrate interface, body and substrate thickness are the dominant parameters (which is the path for the moisture to come in). Die thickness, die-to-pad, and pad-to-body ratios seem to have a significant influence as well. All these parameters are proportional to the amount of moisture diffused into the interface.

Table 5. Normalized relative importance of the linear polynomial terms for the four response parameters

Parameter	Warpage	Interface Stress	Moisture Mass	Moisture Concentration
Die thickness	0.08	0.13	−0.01	−0.02
Pad-to-body ratio	1.00	0.59	−0.01	−0.04
Die-to-pad ratio	0.98	1.00	−0.02	−0.08
Body size	0.82	0.48	1.00	−0.01
Body thickness	0.01	0.00	0.12	1.00
Substrate thickness	0.26	−0.18	0.00	0.23

Table 6 and 7 show the 3D responses of the four output parameters as function of the pad-to-body and the die-to-pad ratios (Table 8) and as function of the body thickness and size (Table 9). These 3D responses are the base for the structural similarity rules for the BGA package family. It is observed that when increasing the die-to-pad ratio the risk of interface delamination will be larger. Increasing the die-to-pad ratio also increases the package warpage and more moisture will ingress into the package. Table 9 shows that the body size effect is very dominant for warpage, stress, and moisture mass but not for moisture concentration at interfaces. For the later one, body thickness is a dominant parameter. Many structural similarity rules can be deduced from Table 6 and 7. In this way, the results of non-linear FE models can be made accessible for package designers. Such tools may reduce the design cycles of new package development.

Table 6. 3D responses as function of pad-to-body and die-to-pad ratios

Package Warpage	Interface stress

Moisture mass	Moisture concentration

8. Virtual Thermo-Mechanical Prototyping

Table 7. 3D responses as function of body size and thickness

Package Warpage	Interface stress
Moisture mass	Moisture concentration

4.2.3 Design Tool for IC and Package Interaction

In Chapter 5 two cases are described dealing with the interaction between IC backend and packaging design. It has been demonstrated that for successful development of IC backend structures and processes, it is essential to take into account the influence of packaging and reliability qualification in the earlier phase of IC backend development. The so-called integral design, accounting for all the major loading sources and history of the complete product creation process has to be used for the development of new generation of Semiconductors devices.

In this section, virtual prototyping method is used to generate more accurate and efficient stress design rules for Cu/low IC backend structure and process development, in combination with QFN package. The addressed

failure modes are delamination, cracks and pattern shift. The design parameters are:
1. The thickness of the IC or die, ranging from 0.1 to 0.38 mm
2. The width of the powerline, ranging from 5 nm to 0.10 μm
3. The length of the IC, ranging from 6 to 10 mm
4. The yield strength of the leadframe, ranging from 150 to 600 Mpa
5. The type of passivation material, either TEOS or SiN
6. The type of the moulding compound, either Cel9220 or EME
7. The intrinsic stress in the IC layers, and the critical energy value

Figure 33 shows an example view of the design tool built up using the response surface models. The critical areas of the IC are indicated in red, and the areas with a 75% critical value in yellow. The colours and the area will change automatically with the changes of the input parameters on the right.

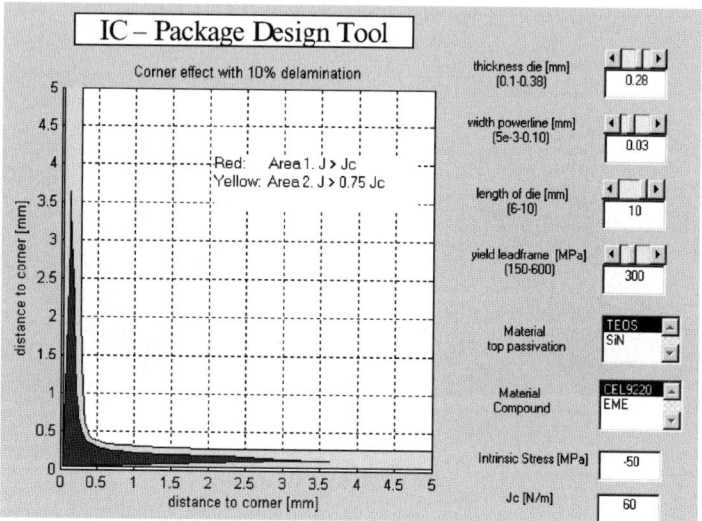

Figure 33. Example view on the IC – package design tool

Figure 34 shows an application example of design tool. In this example, the effect of IC thickness on the failure risk is studied, by comparing IC with thickness of 150 μm and 300 μm. It can be seen clearly that the pattern and the area of the critical areas are different. The thick IC (300 μm) has much less chance of failure. Besides, it is possible to use more area near the edge, which will lead to direct cost saving. This example shows that by using the RMS-based design tool, designers may achieve first time right IC backend structure and processes design prior to any physical testing.

8. Virtual Thermo-Mechanical Prototyping

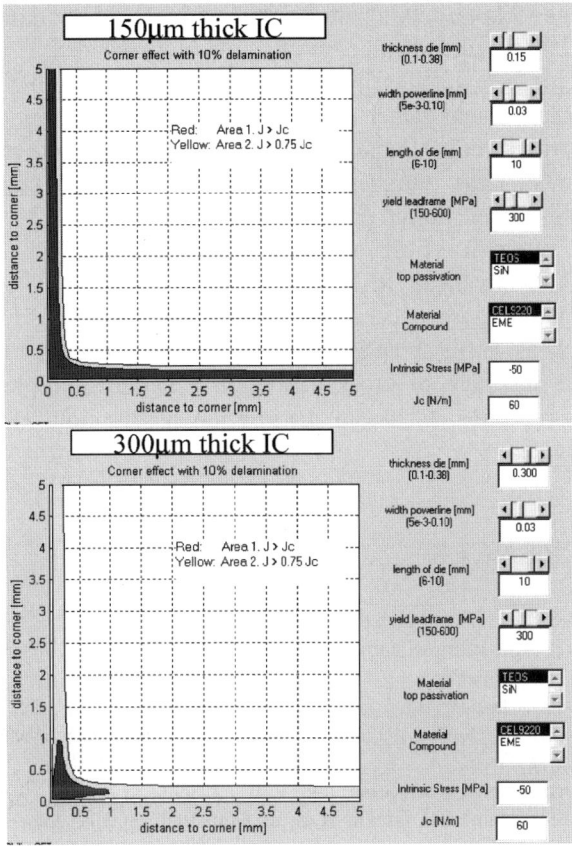

Figure 34. Example tool result: effect of IC thickness

5. CONCLUSIONS

To cope with the development needs of microelectronics, virtual prototyping is an effective and efficient way to predict, qualify, optimise and design microelectronics against the actual requirements prior to major physical prototyping, manufacturing investments and reliability qualification tests. However, the success of VP depends on mainly accurate and efficient prediction models, advanced simulation based optimisation algorithms and methods, and software tool being able to seamlessly and efficiently integrating prediction models with optimisation algorithms. The results of both the strongly nonlinear analytical cases and the real industrial application examples showed in this chapter demonstrate that with sufficient knowledge and proper execution, the added value of VP could be realized.

Nevertheless, the available simulation-based optimisation methods are still not always accurate and rather expensive to deal with design optimisation requirements with strong nonlinear responses; multi-objective targets; multi-level constraints; large numbers of design parameters; combination of continuous with discrete design parameters.

A lot of research effort on developing smart DOE algorithms in conjunction with advanced RSM methods and software is ongoing. The focus is to obtain more accurate RSM with less DOE. However, many questions, such as, infill sampling criteria for multi-objective optimisation, efficiency and accuracy of cross validation scheme, multiple constraints, convergence properties, Gaussian distribution assumption, etc. should be further investigated.

6. REFERENCES

[1] OPTIMUS Rev 5.1, Users Manual, Noesis Solutions, Belgium, 2005.
[2] Y. Panos, Y. Papalambros and Douglas J. Wilde: Principles of Optimal Design, Cambridge University Press.
[3] M.J.D. Powell, A fast algorithm for nonlinearly constrained optimisation calculations. In numerical Analysis, Dundee 1977. G.A. Watso (ed.), Lecture Notes in Mathematics 630, Springer-Verlag, Berlin.
[4] S.S. Rao, Engineering Optimisation, John Wiley & Sons, New York, 1996.
[5] H.-P. Schwefel, Numerical Optimisation of Computer Models, John Wiley & Sons, Chicester, New York, 1981.
[6] R. Storn and K. Price, "Minimizing the real functions of ICEC'96 contest by Differential Evolution", Proceedings of the international Conference on Evolutionary Computation, Nagoya, Japan, 1996.
[7] M. Schonlau, W.J. Welch, and D.R. Jones. Global versus local search in constrained optimisation of computer models. Technical Report RR-97-11, Institute for Improvement in Quality and Productivity, University of Waterloo, Waterloo, Ontario, Canada, 1997.
[8] G. Box, W. Hunter, S. Hunter, Statistics for Experiments, Wiley, New York.
[9] G. Box, N. Draper, H. Smith, Empirical Model-Building and Response Surfaces, Wiley, New York.
[10] N. Draper, H. Smith, Applied Regression Analysis, Wiley, New York.
[11] A. Khuri, J. Cornell, Response Surfaces, Design and Analyses, Marcel Dekker, New York.
[12] P.E. Gill, W. Murray, M.H. Wright, Practical Optimisation, Academic Press, New York.
[13] P.Y. Papalambros, D.G. Wilde, Principles of Optimal Design, Cambridge University Press, Cambridge, UK.
[14] U. Pahner, A General Design Tool for the Numerical Optimisation of Electromagnetic Energy Transducers, Ph.D. Thesis, Katholieke Universiteit Leuven, May 1998.
[15] R.B. Melchers, Structural Reliability Analysis and Prediction, 2nd Edition, 1999.

8. Virtual Thermo-Mechanical Prototyping 533

[16] A.M. Hasofer, N.C. Lind, Exact and Invariant Second-Moment Code Format, Journal of Engineering Mechanics Division, 100 (1), pp. 111-121, 1974.

[17] R. Rackwitz, B. Fiessler, Structural Reliability Under Combined Load Sequences, Computers and Structures, 9, pp. 489-494, 1978.

[18] M. Hohenbichler and R. Rackwitz, Improvement of second-order reliability estimates by importance sampling, Journal of Engineering Mechanics, Vol. 114, pp. 2195-2199 (1988).

[19] A. Olsson, G. Sandberg and O. Dahlblom, On Latin hypercube sampling for structural reliability analysis, Structural Safety 25(1), 2003, pp. 47-68.

[20] G.Q. Zhang and P. Stehouwer, Simulation-based optimisation in virtual prototyping of electronic packaging, Proceedings of 1st Internatioanl Conference of EuroSimE, Eindhoven, 2000, The Netherlands.

[21] G.Q. Zhang, J. Janssen, L.J. Erns, J. Bisschop, Z.N. Liang, F. Kuper, R. Schravendeel, Virtual thermo-mechanical prototyping of electronic packaging using Philips' optimisation strategy, Proc of IMAPS2000, 2000.

[22] G.Q. Zhang, A.A.O. Tay, L.J. Ernst, Virtual thermo-mechanical prototyping of electronic packaging – bottlenecks and solutions of damage modelling, Proceedings of 3rd EPTC conference, 2000.

[23] G.Q. Zhang, A.A.O. Tay, L.J. Ernst, S. Liu, S.F. Qian, H.J.L. Bressers, J. Janssen, Virtual prototyping: Challenges in material characterization and modelling, Proceedings of 51st ECTC, USA, May.

[24] V. Bouwman, G.Q. Zhang, J.W. ter Weeme, Stochastic buckling strength predictions of orthotropic shallow shells Proceeding of 41st AIAA/ASME/ASCE/AHE/ASC conference, USA, 2000.

[25] V. Bouwman, G.Q. Zhang, J.W. ter Weeme, Virtual stability prototyping of doubly curved orthotropic panels, Proceeding of 41st AIAA/ASME/ASCE/AHE/ASC conference, USA, 2000.

[26] G.Q. Zhang, The challenges of virtual prototyping and qualification for future microelectronics, J. Microelectronics Reliability, 43, 1777-1785, (2003).

[27] W.D. van Driel, G.Q. Zhang, J.H.J. Janssen, Response Surface Modelling for Nonlinear Packaging Stresses, ASME Journal of Electronic Packaging, 125(4), 490-497, (2003).

[28] HVQFN™, Amkor Technology Inc., Chandler, USA.

[29] J.G.J. Beijer, J.H.J. Janssen, H.J.L. Bressers, W.D. van Driel, K.M.B. Janse3, D.G. Yang, G.Q. Zhang, Warpage minimization of the HVQFN map mould, Proceedings 6th EuroSimE, 2005, pp. 168-174.

[30] D.G. Yang, K.M.B. Jansen, L.J. Ernst, G.Q. Zhang, W.D. van Driel, H.J.L. Bressers, X.J. Fan, Prediction of Process-Induced Warpage of IC Packages Encapsulated with Thermosetting Polymers, Proc ECTC 2004, pp. 98-105.

[31] D.G. Yang, K.M.B. Jansen, L.J. Ernst, G.Q. Zhang, W.D. van Driel, H.J.L. Bressers, Modelling of Cure-Induced Warpage of Plastic IC Packages, Proc EuroSimE 2004, pp. 33-40.

[32] Jansen, K.M.B., Wang, L., van't Hof, C., Ernst, L.J., Bressers, H.J.L., Zhang, G.Q., Cure, temperature and time dependent constitutive modelling of moulding compounds, Proc EuroSimE 2004, pp. 581-585.

[33] Jansen, K.M.B., Wang, L., Yang, D.G., van't Hof, C., Ernst, L.J., Bressers, H.J.L., Zhang, G.Q., Constitutive modelling of moulding compounds, Proc ECTC 04, pp. 890-894.

[34] van't Hof, C., Wisse, G., Ernst, L.J., Jansen, K.M.B., Yang, D.G., Zhang, G.Q., Bressers, H.J.L., A novel tool for cure dependent viscoelastic characterization of packaging polymers, Proc EuroSimE 2004, pp. 385-390.

[35] D.G. Yang, et al, Parameter sensitivity study of cure-dependent underfill properties on Flip Chip failures, Proc 52th Electronic Components and Technology Conference, San Diego, May 28-31, 2002, pp. 865-872.

7. EXERCISES

1. Consult a design text in a selected discipline and find a worked-out example of an elementary design. Describe the traditional design approach and then formulate an optimisation model statement. Derive any results, if you can, and compare the two approaches. Particularly examine how many design variables are in the problem you derived and how many constraints might be active (satisfied as equalities) at the optimal solution.
2. Derive general expressions for the coefficients of linear, quadratic, and cubic approximations, when the sampling points are equally spaced along the x-axis.
3. Consider an electric motor series cost model with the data given below (Stoecker, 1971).

hp	Cost/$	$/hp
0.50	50	100.00
0.75	60	80.00
1.00	70	70.00
1.50	90	60.00
2.00	110	55.00
3.00	150	40.00
5.00	220	44.00
7.50	305	40.50
15.00	560	37.30

Derive the curve fitting equation (Response Surface Model) $\$/hp = 34.5 + 36(hp)^{-0.865}$

Hint: Draw the curve using the table values and estimate a value for the constant term. For the steep part f the curve, draw its representation on a log-log plot to get values for the coefficient of the second term.

4. Find the minimum of the function $f = (1-x_1)^2 + 100(x_2 - x_1^2)^2$. This is a well-known test function for numerical optimisation (Rosenbrock's "banana" function).
5. Find the global minimum of the function $f = 2x_1^2 + x_1x_2 + x_2^2 + x_2x_3 + x_3^2 - 6x_1 - 7x_2 - 8x_3 + 19$
6. Consider the function $f = -x_2 + 2x_1x_2 + x_1^2 + x_2^2 - 3x_1^2x_2 - 2x_1^3 + 2x_1^4$

(a) Show that the point (1,1) is a possible minimum and that the Hessian matrix is positive-semi definite there
(b) Find a straight line along which the second-order perturbations ∂f is zero
(c) Examine the sign of the third- and fourth-order perturbations along the line found above

7. Sketch graphically the problem
$$\min f(X) = (x_1 + 1)^2 + (x_2 - 2)^2$$
subject to $\quad g_1 = x_1 - 2 \leq 0 \quad g_3 = -x_1 \leq 0$
$\quad\quad\quad\quad\quad g_2 = x_2 - 1 \leq 0 \quad g_4 = -x_2 \leq 0$

Find the optimum graphically. Determine directions of feasible descent at the corner points of the feasible domain. Show the gradient directions of f and g_i's at these points. Verify graphical results analytically using KKT conditions.

8. Consider the problem
$$\min f(x) = x_1^2 + x_2^2 + x_3^2 + 40x_1 + 20x_2 - 3000$$
subject to $\quad g_1 = x_1 - 50 \geq 0$
$\quad\quad\quad\quad\quad g_2 = x_1 + x_2 - 100 \geq 0$
$\quad\quad\quad\quad\quad g_3 = x_1 + x_2 + x_3 - 150 \geq 0$

Solve the problem with one of the methods described in the chapter. Can you verify the solution by checking the KKT conditions?

9. Repeat the above exercise for the problem
$$\max f(x) = 0.020 x_1^4 x_2 x_3^2 / 10^7$$
subject to $\quad g_1 = 675 \geq x_1^2 x_2$
$\quad\quad\quad\quad\quad g_2 = 0.419(10^7) \geq x_1^2 x_3^2$
$\quad\quad\quad\quad\quad 0 \leq x_1 \leq 36 \quad 0 \leq x_2 \quad 0 \leq x_3 \leq 125$

10. A popular purely sectioning procedure is the golden section search. It aims at creating bracket lengths that are in a sense "optimal", that is they minimize the worst possible reduction of the uncertainty interval. Consider a function as in the following figure with three points x_1, x_2, x_3 already placed with the ratio $(x_3 - x_2)/(x_2 - x_1)$ fixed at τ where $\tau > 1$ is a constant. Insert the fourth trial point x_4 so that the potential new brackets $[x_1, x_4]$ and $[x_2, x_3]$ have intervals in the ratio τ. Based on this construction, evaluate the lengths of the intervals $[x_2, x_4]$ and $[x_4, x_3]$. Prove that $\tau \cong 1.618$ (the so-called "golden section ratio" of the ancient Greeks). Write out the steps of an algorithm based on repeating this sectioning.

11. Find the minimum of $f(x) = (10-x)^2 + x$ with the golden section method starting with the interval [5,20].

12. An experiment was conducted to determine the effects of a certain air pollutant on the cardiopulmonary functions of mongrel dogs. The measurement used for this test was pulmonary arterial pressure in pounds per square inch, psi, which should increase under the influence of toxins. A total of 15 dogs were randomly chosen, five were injected with 2 mg sodium chloride (table salt) as a control, five were injected with 1 mg sodium chloride, 1 mg sodium sulfate, and five were injected with 2 mg sodium sulfate. Sodium sulfate is a constituent of air pollution. The data obtained are as follows:

	Pulmonary arterial pressure (psi)				
2 mg sodium chloride	12.3	11.7	12.2	12.1	12.0
1 mg of each	12.8	13.2	11.9	12.9	13.1
2 mg sodium sulfate	13.3	12.7	13.2	12.8	12.0

(a) What type of DOE was used?
(b) Does it make sense to fit orthogonal polynomial to this data?
(c) What additional sample size would it be needed to increasing the confidence in the model?

13. In an attempt to understand why one part is good and another part is bad engineers often take components from the good part and put them in the bad part and vice versa to see if the bad part becomes good, or the good part becomes bad. Assume there are two parts, P1, good and P2, bad, and three components labelled A, B, and C. Design a three factor completely randomised experiment. State any assumptions that you make. Explain what it means for the A component to be significant. What would it mean if there was an AB interaction?

14. In an experiment dealing with the amount of shrink in molten metal used for casting complicated parts, three factors were considered, the amount of aluminium in the metal, 40 and 44 ppm, the pouring spout type, p and m, and the temperature of the molten metal, 750 and 790 degrees. Two different castings for each combination were poured and the percentage shrink recorded. Write the mathematical RSM appropriate for this experiment. Summarize the findings in a report appropriate for metal casting experts who have no statistical background.

Chapter 9

CHALLENGES AND FUTURE PERSPECTIVES

G.Q. Zhang[1,2]
[1]*Philips Semiconductors, HTC 60, 5656AG Eindhoven, The Netherlands*
[2]*Delft University of Technology, Mekelweg 2, 2628CD Delft, The Netherlands*

Abstract: The combination of technology and business trends drives microelectronic technologies into an unknown level of complexity, characterized by ongoing miniaturization down to nano-scale, heterogeneous and multi-functionality, multi-discipline, multi-scale (in both geometry and time), multi-technology, multi-process, multi-material/interface, multi-damage and multi-failure mode. As the consequences, industries are confronted with ever increased design complexity, dramatically decreased design margins, increased chances and consequences of failures, decreased product development and qualification times, increased gap between technology advance and development of fundamental knowledge, and increased difficulties to meet quality, robustness, and reliability requirements. In this chapter, the mechanical related characteristics of microelectronics are summarized and the challenges and future perspectives for mechanics of microelectronics are highlighted, as an attempt to guide the future developments and exploitation of mechanics for microelectronics.

Key words: Mechanics, multi-physics, multi-scale, modelling, simulation, molecular dynamics.

1. INTRODUCTION

Microelectronics has pervaded our lives for the past fifty years, with massive penetration into health, mobility, security, communications, education, entertainment, and virtually every aspect of human lives. In the past decades, as the main stream, these progresses are powered by Moore's law, with two focused development areas, namely, IC miniaturization down to the nano dimension, and system on chip (SoC) based system integration. While the microelectronics community continues to invent new solutions around the world to keep Moore's law alive, there is ever-increasing awareness, R&D effort and business drivers to push the development and application of "More than Moore" (MtM) that are based upon or derived from silicon technologies but do not scale with Moore's law (with typical examples as RF, Power, Sensors and Actuators, SiP, heterogeneous integration, etc.) [1]. This emerging trend is partially trigged by the increasing social needs for high level system integration including non-digital functionalities, the necessity to speed up the innovative product creation and to broaden the product portfolio of existing wafer fabs, and the limiting cost and time factors of advanced SoC development.

Along with the technology development, the business trends of microelectronics are mainly represented by cost reduction, shorter-time-to-market, and outsourcing. The combination of these two trends drives microelectronic technologies into an unknown level of complexity, characterized by ongoing miniaturization down to nano-scale, heterogeneous and multi-functionality, multi-discipline, multi-scale, multi-technology, multi-process, multi-material/interface, multi-damage and multi-failure mode. As the consequences, we are confronted with ever increased design complexity, dramatically decreased design margins, increased chances and consequences of failures, decreased product development and qualification times, increased gap between technology advance and development of fundamental knowledge, and increased difficulties to meet quality, robustness, and reliability requirements.

Mechanics has been playing a prominent role in industrial and technological development for decennia, such as in aerospace and transport industries, and in a very broad application spectrum of mechanical engineering and civil engineering. Computational mechanics, such as Finite Element Methods, have also been impacting various industries and technologies for several decades, especially due to the rapid development of computational capability. However, the available knowledge of mechanics cannot meet the needs of microelectronics development.

In this chapter, based on the technology roadmaps of micro and nano-electronics [2, 3], the associated challenges and future perspectives for mechanics of microelectronics are reviewed, as an attempt to guide the future developments and exploitation of mechanics of microelectronics.

2. MECHANICAL RELATED CHARACTERISTICS OF MICROELECTRONICS

To understand the challenges of microelectronics for mechanics, it is worthwhile to analyse the mechanical related characteristics of microelectronics. To summarize, microelectronic technologies and products are:

9. Challenges and Future Perspectives

- *Multi-scale in both geometric and time domains.* In geometric domain, the typical feature sizes range from nanometers to millimetres. Nanometer configurations are mainly driven by the miniaturization of Integrated Circuits. According to ITRS and Eniac [2, 3], the shrinking of the electronic components on chip is expected to continue for another 15 years on the base of silicon technology. During this period, the critical feature size of the transistors (i.e. the physical gate length) will drop from 32 nm in 2005 (technology node 90 nm) to 6 nm in 2020 (technology node 14 nm), allowing unprecedented leaps in performance and circuit complexity. Other nano scale structures are nanowire with diameters from 5 – 40 nm, nanobelt and nano carbon tube (NCT). Not only IC and nano-devices, the geometric feature sizes of packaging and assembly are also going beyond our visualization capability of bare eyes. For examples, the interconnect wire diameter is smaller than 10 microns, interconnect pitch of nano scale wafer level package smaller than 20 microns, thickness of copper film in PCB smaller than 5 microns, microvia diameter smaller than 20 microns, and wafer thickness smaller than 40 microns. On the other hand, for various end products of microelectronics, it is common to have feature sizes in millimetre scale. Microelectronics has an inherited multi-scale nature in geometry. In time domain, it may vary from nano-second to years (such as the required lifetime of some microelectronic products).
- *Multi-technology, multi-loading, and multi-discipline.* From technology aspect, the major manufacturing chain consists of IC (etching, deposition, electroplating, CMP, singulation, etc.), packaging (die attaching and wafer stacking, wire bonding and solder bumping interconnections, moulding, trimming, storage, transport), assembly (electronic, bio and fluidic) and testing (for both functionality and reliability qualification). It even can go beyond the traditional semiconductors technology. Example is the biosensor development, with the ultimate aim to have a full integrated Microsystems which perform all the steps of a molecular diagnosis from the macro sample to the diagnostic information, including building blocks like preparation, ultra sensitive detection, chemistry process for molecular recognitions and micro fluidics. In order to do so, proven individual technology is necessary but not sufficient. A system approach, that is more than assembling together high performance individual functions, is needed to achieve overall design optimisation. Multi-technology results in multi-loading. Various loading types, such as thermal, mechanical, humidity, chemical, electrical, and combination hereof are commonly presented in different life cycles of various microelectronic products. From discipline

point of view, chemistry, thermal, metallurgy, physics, electrical, mechanics, optics, electromagnetic and bioengineering may all be involved in microelectronic product creation.

- *Multi-material and multi-interface.* Faster than ever various types of new materials are being developed and introduced in microelectronics, such as various functional materials, green materials, and nano-particle enforced materials, in order to overcome the brick wall of microelectronic technologies. Material development is becoming an indispensable and integrated part of new technology development. Joint development effort between material suppliers and technology innovators, starting from the earlier phase of product development, is no exception anymore. To meet the needs of ongoing miniaturization and heterogeneous integration beyond digital functionality, both the number and the type of materials and their interfaces are greatly increased.
- *Multi-failure mechanism, multi-failure mode, and multi-failure location.* Typical failure modes confronted by other industries and applications, such as various cracks, delamination, yield, stress voiding, instability, fatigue, creep, wear, and corrosion are all observed in microelectronics. What are more, "exotic" failure modes, such as pop corning, electro-migration, thermo-migration, stress-migration, stiction, gate oxide breakdown, cratering, sweep, etc. form the new barriers for the microelectronics. Strong interaction among these different failure mechanisms and failure modes in either simultaneous or sequential way, have also been observed. Often these interactions, instead of single failure driving forces, are the direct root causes of failures. Examples are the earlier solder fatigue failures due to underfil delamination, IC metal interconnect extrusion due to dielectric cracking, and stress corrosion cracking due to simultaneous actions of mechanics and corrosion. For a microelectronic device consisting of various building constituents of IC, components and substrate, different failure locations can be expected, depending on the type and history of loadings, the strength of substructures and the associated interface systems.
- *Strongly non-linear behaviour.* Covering geometric nonlinearity (e.g., buckling and postbuckling, large warpage), material nonlinearity (viscoplastic behaviour of solders, creep and pattern shift of metal interconnect, yield of interconnect wires, leadframe, and copper film/trace), and boundary nonlinearity associated with manufacturing and assembly processes (e.g., wire bonding and probing loading, dropping impacts). Due to these nonlinearities, known design insights and rules for certain applications cannot be simply extrapolated in a linear way to new designs. Establishing structural or product similarity becomes less evident.

9. Challenges and Future Perspectives

- *Stochastic in nature.* For microelectronics, it is virtually impossible to design and manufacture product and process with deterministic performance. For design parameters, such as material/interface properties, geometric dimensions, process window, loading intensities, deviations represented by different statistic characteristics and magnitudes are inevitable.
- *Strongly time and temperature dependent.* A lot of materials used in microelectronics have inherited nature of time and temperature dependence. Examples are the visco-elastic behaviour of polymers used as moulding compound, adhesive, dielectric layers, underfill and coating, and the visco-plastic behaviour of solder materials used for interconnecting wherein the homonological temperatures are often higher than 0.5. Applications of these time and temperature dependent materials in microelectronics make both the manufacturing processes and the final products in use also time and temperature dependent.

Note that in the reality, the above-mentioned mechanical characteristics of microelectronics will inevitably interact with each other, resulting in not only a lot *multi's*, but also strong, nonlinear, stochastic, and time/temperature dependent "multi" interactions.

As emphasized in Chapter 8, it is essential to be able to develop accurate and efficient reliability prediction models that can cope with the tough demands and characteristics of microelectronics, wherein mechanics of microelectronics plays a prominent role. Unfortunately, due to the fast development and ever increasing complexity of microelectronic technologies on one hand, and insufficient attention and effort from mechanics community on the other hand, the currently available knowledge of mechanics of microelectronics (including theories, software and hardware tools, methodologies) is insufficient to meet the development needs of microelectronics. Although a lot of relevant knowledge has been successfully incorporated into the development of microelectronics via experience, experiments and modelling, often, it is observed that:

- The prediction results are not always qualitatively reliable.
- It is hard (very time and money consuming), if not impossible, to obtain results with quantitative accuracy.
- Most of the modelling results are empirical and phenomenological based, and case dependent. There is still a long way to go to reach failure prediction without pre-knowledge on failure mechanism and mode.

Given the fact that the development of mechanics of microelectronics is lacking behind of that of microelectronic technologies, industries have to be confronted with two major dilemmas:

- No appropriate mechanical design methods, the existing ones are:
    - Experience and trial-error based,
    - Empirical, phenomenological and case dependent.

- No appropriate mechanical qualification methods, the existing ones are:
    - Time and money consuming,
    - Associated with unclear correlation between application profiles with reliability specifications, and between application profiles with accelerated testing,
    - Not able to be easily extrapolated to outside of the specifications.

It is clear that to meet the needs of development trends of microelectronics, developing more knowledge for mechanics of microelectronics is necessary, in order to enable industries to predict, optimise and design upfront mechanical related performance and reliability of microelectronics.

3. RELIABLE INPUTS

It is essential to have reliable inputs, namely design and failure inputs, in order to establish accurate and efficient mechanics prediction models.

3.1 Design Inputs

There are two types of design inputs. The first type is the design variables or design freedoms. By choosing different values of the selected design variables, design engineers can change and influence the performance of the products. It is the primary task of design engineer to select the proper design variables, in order to meet the product specifications. Examples of design variables are geometric configurations, materials (types and properties), and process parameters. For all these design variables, one needs not only the nominal design value, very often, also the allowed range of variation - the so-called design space. For some microelectronic products, several product types can be created by simply changing the value of some design variables within the feasible design space, to meet the needs of different market sectors. For this case, one would like to achieve not only one "good" design, most importantly, also design rules enabling the realization of a family of "good" designs.

The second type of design inputs is the inherited deviation of some design variables. They are so-called deviation parameters corresponding to some fixed value of design parameters. For microelectronics, there will always be variations about any specification. Thus the properties of a product from any manufacturer will vary about the target (median) value; the further characteristics values are from the median value, the less often they should occur. Compared with design variables, deviation parameters are usually probabilistic in nature. Still one can classify the causes of variation

9. Challenges and Future Perspectives

into two categories: random (chance) causes and assignable causes that can be identified and corrected. Statistical process control (SPC) is to get all the assignable causes identified and eliminated. For examples, in SiP development, the thickness of die attachment layer has been chosen a fixed median value. However, in reality, it is not possible to obtain this targeted adhesive thickness for all the products, due to the presence of unavoidable scatters as functions of materials and processes. For all the materials used in microelectronics, their properties are all probabilistic in nature, depending on not only the material batch itself, but also the process windows. For example, different curing profiles for the moulding compound will lead to different behaviour of encapsulation.

These two types of design inputs determine the product design, actual loading and loading history, boundary conditions, damage initiation and evolution material behaviour, and eventually the life time and performance of the final product. However, acquisition of reliable design variables and deviation parameters is not a trivial task. Especially, it is technically difficult, time and money consuming to extract the probabilistic data via in-situ measurements and observations. Due to business pressure, it is often neglected or over-simplified by industries. Therefore, sufficient effort should be spent on acquiring accurate design inputs, to be able to obtain reliable predictive mechanics models.

3.2 Failure Inputs

For new product development of microelectronics, especially associated with new technologies and product concept, industries would like to know upfront the probability of failures and the associated failure modes and mechanisms, before building and testing multiple physical prototypes, so that the first time right design can be achieved. Unfortunately, it is still not possible, at this moment, to build simulation models being able to upfront predict the mechanical related behaviour of microelectronics, covering all possible failure modes and processes, without pre-knowledge and experience about possible failures. Failure inputs herein refer to the failure modes and their ranking, failure evolution (the history from damage initiation to failure), failure location, failure root causes/mechanisms, their interactions, and failure probabilistic. Knowing failure inputs is of paramount importance for predictive modelling of mechanics of microelectronics.

Due to the complicated nature of microelectronics, it is a challenging task to collect failure inputs. Therefore, failure analysis, aiming to understand the failure causes, the first step towards failure prevention, has become an important engineering field of microelectronics, with reputation of delicate, complex, time-consuming and expansive operation. During the past decades, industry has developed many practices, methods, and tools for failure analysis.

Failure Mode and Effect Analysis (FMEA) is one of the well-established practices for structured analysis of potential failure modes and their effects. Many methods are available to analyse and identify the failure mode and the associated failure mechanism, such as Pareto analysis, Fishbone diagram, Histogram, Scatter diagram, Paynter chart, and Spider graph. In product development, these practices are often applied in a combined way.

Advanced failure analysis tools, both in-situ and off-line, are also becoming more and more important. Examples are [5]: Scanning Electron Microscopy (SEM) – a universal instrument that allows the imaging of surface topography with a large depth of view over a large magnification range (20-200.000 × magnifications). It can be used for the determination of the microstructure and chemical composition of the various structures and layers of microelectronics as well as to characterize the physico chemical characteristics of failures. Transmission Electron Microscopy (TEM) makes use of the interaction between high-energy electrons and a thin specimen, giving rise to a wide variety of secondary signals from the specimen. Through careful separation and detection of these signals, a complete picture of the composition and microstructure of the specimen is possible right down to the atomic scale. Such high spatial resolution, together with high contrast and the availability of crystallographic information make TEM an important tool for multi-purpose failure analysis. Focused Ion Beam Microscopy (FIB) is used for the localized deposition of conducting or insulating material and/or removal of materials using a focused beam of Ga+ ions, with high precision. It is also widely used for the preparation of SEM and TEM cross-section samples. Scanning Acoustic Tomography (SCAT) is used to find internal body cracks and delamination. X-ray analyses is widely used to find package failures like wire-break, fused wires, wiresweep, voids, and missing solder-bumps. Other often-used tools are Optical Microscopy, Nanoprobing, and Atomic Force Microscopy (AFM).

Failure analysis sample preparation is also a challenge issue. It may take the form of sequentially removing layers from the front and/or backside of the IC through etching or polishing thus exposing the suspected layer where the failure occurs. Alternatively, cross sectioning may be performed, permitting the failing part of the device to be viewed in profile.

Special dedicated failure tests have also to be developed to trig and verify specific failure mode and mechanism, such as E-beam testing and infrared Optical Beam Induced Resistance Change (OBIRCH), and many others.

Finding the right failure inputs is demanding professional and multi-disciplinary effort. Often they are underestimated or conducted without systematic and dedicated effort. Without the reliable guidance provided by

9. Challenges and Future Perspectives 545

the knowledge of failure inputs, the reliability and feasibility of predictive modelling cannot be guaranteed.

4. TESTS AND EXPERIMENTS

There are mainly two types of tests used for microelectronics, namely, the functionality test and the reliability qualification test. For functionality test, due to increased design complexity and intensified heterogeneous integration of technology and functionality, mainly driven by "More than Moore", new test strategy, methods, and tools are needed to achieve maximum test coverage with minimum costs. For reliability qualification test, a must for microelectronics, it is essential to establish qualitative and quantitative correlation for failure modes and mechanisms between application profiles with the reliability test specifications aiming to qualify the products in an accelerated manner, based on the profound understanding for the underlying fundaments. Typical packaging level reliability qualification tests are: (un) biased humidity tests (HAST/THB/UHST), High Temperature Operating Life (HTOL), Hot Carrier Lifetime, and Temperature Cycling (TMCL). On the other hand, predictive reliability modelling is playing more and more important role in reliability qualification, aiming at reducing the duration and the number of real qualification tests.

Three types of experiments are widely used in supporting mechanical modelling of microelectronics, namely, experiments for material and interface characterization to provide data for constitutive modelling, experiments for damage and failure criteria extraction to provide quantitative data on the allowable strengths, and experiments for model verification to validate the correctness and accuracy of the developed models.

Similar as the model partitioning of constitutive law and damage model, traditionally, experiments for material/interface characterization and experiments for determining their strength are often separately conducted, for the sake of simplicity. As the consequences of miniaturization and multi-scale, for many microelectronic applications, experiments for constitutive law and damage model cannot be partitioned anymore. A unified law governing both the constitutive and damage behaviour of material and interface needs also innovative and unified experiments. These experimental methods and tools should be able to provide inputs for the integrated models of material and interface behaviour, subjected to multi-scale and multi-disciplinary conditions. Let us take solder fatigue as one example. It is known that the microstructure of the solder alloy is in constant change. Performance and reliability of solder interconnects are directly a function of the alloy microstructure, defined as the combination of grain structure, metallurgical phase, defect distribution, and morphology. However, the currently used solder fatigue prediction models are all empirical based, without considering the dominant link between the histories of constant nano/micro-structural change with the damage evolution. It is partially due to the difficulties in developing suitable experimental techniques to monitor the complete process of microstructure evolution (such as grain coarsening) from solidification until macro failures under actual loading conditions with real product configurations.

It is worthwhile to repeat the well-known motto: "In principle all simulation results are wrong, unless one can prove that they are right" [4]. Conducting verification experiments is one of the direct ways to judge the correctness of any prediction. There are two commonly used methods for verification experiments, in case that the real lifetime verification cannot be obtained directly and efficiently (as it is usually the case). One is to develop special experiments to verify some specific and/or simplified part of the modelling results, focusing on individual influencing factors. Another is to develop extreme case experiments to obtain qualitative conclusion for the total predicted behaviour. Very often, one needs to combine these two methods.

Generally speaking, developing experimental methods and tools with sufficiently accuracy, being able to correlate the experimental conditions with the real loading history and constraints, and preparing samples representing the real product configurations, processing and intrinsic stresses, covering the whole process and product design spaces, can be a formidable challenge. Numerous bottlenecks and problems exist, ranging from sample preparation, experimental set-up, data acquisition and analysis, modelling development, and correlation.

5. MATERIAL AND INTERFACE BEHAVIOUR

Reliable models describing the time, process, and size dependent behaviour (properties, damage initiation, evolution, and failure criteria) of materials and their interfaces are essential for not only predictive modelling, but also material development, material pre-selection and process optimisation. This section will address several related issues deserving more research attention.

5.1 Material Behaviour

Unfortunately, it is still not possible to predict upfront the thermo-mechanical properties and behaviour of materials, according to the given

9. Challenges and Future Perspectives

material composition (such as chemistry for polymer), process conditions, and geometric design. Characterization and modelling are needed.

As mentioned in the earlier section, presently, the characterization and modelling of material behaviour are usually based on the partitioning of constitutive models that describe the material properties with damage models that describe the damage and failures of materials. This practice cannot meet the need of microelectronics with nano-scale and strong multi-disciplinary interaction. Strictly speaking, all material properties are strongly interlinked with damage initiation, evolution and failures. For microelectronics, the conventional constitutive models should be integrated with damage models, to form a law that governs the total behaviour of materials. From computational point of view, the macroscopic equations of physics of failures and the kinetic equations of the micro-structural transformation (including total damage history) should be solved simultaneously in a consistent way.

More quantitatively accurate failure criteria and damage models are needed for accurate failure predictions, even for macro-scale applications. Although commercial FEM tools nowadays can be used to calculate the resulting stress/strain of microelectronics under many complicated loading conditions, failure prediction cannot be realized without the pre-knowledge of failure criteria and dedicated damage modelling. However, developing failure criteria and damage models is not only time consuming, but also a specialist work. Most industries have limited resources and time freedom to do it. Therefore, the added value of predictive modelling is, sometimes, limited to obtain some qualitative design insight. On the other hand, since many materials used in microelectronics are usually time, geometric scale, and processes (stress and strain rate, temperature, constraint, etc.) dependent, it adds extra difficulties to obtain quantitatively accurate failure criteria and damage models. Even for single failure mode prediction, it is tough to describe the size and process dependent damage initiation, evolution and failures, without developing new modelling theories and methods, and experimental techniques. Yet another difficulty is the characterization and modelling of multi-damages, where different failure modes occur simultaneously or consequentially, with possibly interlinked mechanisms. There is no computational framework and algorithm available to deal with the simultaneously occurring failure modes, due to the difficulty in predicting the birth of new failure mode from the parent failure modes.

5.2 Interface Strengths

Interface strengths and interfacial failures are probably the most prevalent and pervasive issues in microelectronic industry. In particular, as more organic materials being used in various types of microelectronic products, and the ongoing trends of miniaturization towards nano-scale, various interfaces, such as polymer and metal, polymer and other adjacent materials, solid and fluid, etc. are becoming more critical. From Integrated Circuits to their packages and to the Microsystems made of both digital and non-digital components, numerous material interfaces exist. Driven by the rapid development of SiP, a dramatically increase in both the number and the type of interfaces is observed. The increasing number of interfaces is mainly pushed by miniaturization, while the increasing type of interfaces is mainly pushed by heterogeneous system integration including non-digital function. As a consequence, not only the chance of interface failure is increased, but also the consequence of interface failure becomes more pronounced. Another well-know example is the backend stacking and bondpad development of Cu/low-k IC technology. Reducing the dielectric constant will naturally lead to lower interface strength between low-k dielectrics and metal interconnect lines, leading to delamination and cracking in backend structures during packaging, assembly, or qualification tests. The quality, robustness, and reliability of microelectronic devices depend, to a large extent, on the adhesion and durability of these interfaces. Debonding or delamination of these interfaces often results in the malfunction or failures of these devices.

Although interfacial adhesion had been studied for decades by numerous researchers, a few studies dealt with the adhesion issue from a multi-disciplinary and multi-scale viewpoint. The vast majority of studies focused on either the chemical, or the physical, or the mechanical aspect alone, and often covered only single geometric scale. Because of such compartmentalized approaches, no effective methodologies, models and tools are available currently for the prediction of interfacial strength in microelectronics, for given product, process, and material characteristics.

The current practice is to first build up the simplified interface to be studied, develop characterization experiment (3 and 4 point bending test, blister test, nanoindentation test, mode mixity test, for examples), conduct the test, and extract data for interfacial damage models. Interfacial strength characterization and failure modelling is still empirical in nature, and interface engineering (design, manufacturing, and qualification) is still trial-error. Thus, it is important to develop a generic framework for prediction of interface strengths incorporating the combined effects of and interactions among physical, chemical, and mechanical bonding, covering the effect of different geometric and time scales.

Macroscopically speaking, the failure of an interface is observed as interfacial fracture. Therefore, interfacial adhesion characterizes the resistance of an interface to interfacial crack initiation and growth. Generally speaking, there are three intrinsic interfacial adhesion mechanisms for polymers on metallic or passivation adherents, namely, physical adsorption, chemical bonding, and mechanical interlocking.

Physical adsorption is responsible for the interaction between permanent dipoles in the molecules, the induction effect, and the dispersion effect. The induction effect is the influence of the dipole moments in polarizing neighbouring molecules, while the dispersion effect is the result of an interaction between the random motions of the electrons of the materials. Once adhesive and substrate are selected, the nature and strength of physical adsorption is theoretically predictable.

Chemical bonding, on the other hand, is based on the primary covalent bond at the interface. One example of chemical bonding is the chemical reactions created by coupling agents at the interface.

Both physical and chemical adhesion mechanisms are on the microscopic scale. At the macroscopic scale, in order to form a strong adhesive joint, mechanical interlocking can be applied, as in the surface treatments applied to metals to provide various topologies. However, strictly speaking, mechanical interlocking is not one of the intrinsic material adhesion mechanisms. It is only a technological means in achieving adhesive bonding as in the case of structural adhesives.

The ultimate aim of characterizing and modelling of material and interface behaviour is to develop material design rules based on multi-disciplinary and multi-scale knowledge, in order to upfront tailor and manipulate material and interface properties and behaviour for specific application needs.

6. MULTI-SCALE MECHANICS

6.1 Introduction

Micro and nanoelectronics is multi-scale in nature. The current available CMOS technology can already manufacture IC with feature size down to several nanometers. To assemble the IC into various packages, Microsystems (such as SiP) and eventually various end products, one has to deal with integration of geometric dimensions from nano to micron to macro-scales. Due to the huge scale difference, size effects will become essential. These size effects are often related with microstructures and their evolution, various gradient effects (such as chemical, electrical, thermal, and mechanical gradients), and surface effect. For example, metals with a grain size of around 10 nanometers can be as much as seven times harder and tougher than their ordinary counterparts with grain sizes in the micrometer range. Tremendous size effects like this are known to play a significant role in miniaturization, and the implementation of these effects in future product and process design is a necessary prerequisite to make optimal use of materials and structures at the nano scale. Presently the required knowledge in this area is substantially under-developed. The product/process behaviour at nano scales cannot be predicted by simply applying the conventional macro-scale based approaches, such as continuum mechanics and thermal management, because they do not include any peculiarity of the small-scale structure of materials, but merely represent an average behaviour. For this reason, they are not directly applicable for current and future development of micro and nanoelectronics. From both application needs and academic challenges, there is an urgent need to understand the multi-scale phenomena and to develop theories, methods and tools to eventually predict the associated responses. The multi-scale method, defined as those, which combine information and capability available from different length and time scales into a coherent or coupled simulation, is becoming more and more important.

To elucidate the concept of the multi-scale method, we consider a traditional example, the deformation behaviour of materials. In the micro-scale, the deformation is actually the combination of dislocations, which appear as extra or displaced planes of atoms inserted into a regular lattice, allow materials to deform without brittle fracture. An important property of dislocations is that they can move through the lattice in response to an applied external loading, thereby allowing slip to propagate. Dislocations and other defects in crystals correspond to specific atomic configurations whose energies can be calculated from quantum mechanics principles.

Besides the physical description of the initialisation/propagation of the crack and dislocation/deformation of the film, multi-scale modelling can also be applied in many product and process development and optimisation of nanotechnologies. Due to the lack of theories, methods, and tools of multi-scale modelling, trial-and-error based design methods are dominating the development of various nanotechnologies and applications, such as nanoelectronics. Due to the two folded challenges (both atomistic and multi scales), it is often not possible to extrapolate existing experience for new applications. Consequently, it leads to high cost, longer time to market, and non-optimised performance.

It is clear that much effort should be spent on developing non-continuum (or enriched continuum) mechanics capable of simulating the behaviour of microelectronics with micron, deep-submicron and nano dimensions, in order to capture the strong and inherited size effects. Secondly, a bridge should be developed to close the gap between non-continuum theories, simulation tools and results of atomistic scale with the continuum theories, simulation tools and results of macro-scale. So that it will be possible to use

9. Challenges and Future Perspectives 551

multi-scale modelling for many applications, such as integrated process modelling starting from wafer, packaging, to systems levels.

In the following, traditional methods for mechanical modelling will be summarized, including the quantum mechanics, molecular dynamics, Monte Carlo method and continuum mechanics. Each method has been developed and optimised for a particular space and time scales. The method of multi-scale modelling and the associated major challenges will also be reviewed.

6.2 Hierarchy of Methods for Mechanical Modelling

6.2.1 Quantum Mechanics Methods

The observable properties of solids are governed by quantum mechanics, as expressed by solutions of a Schrödinger equation for the motion of the electrons and the nuclei. Moreover, the Schrödinger equation describes both the particle nature and the wave nature of the matters. By solving the wave function and energy of the Schrödinger equation of the specific system under boundary and initial conditions, one could obtain energy and the atomic structure of a system of electrons and nuclei. However, because of the inherited difficulty of obtaining even grossly approximate solutions of the full multi-body Schrödinger equation, one typically focuses on reduced descriptions that are believed to capture the essential energetic of the problem of interest.

The density functional theory (DFT), established by Hohenberg, Kohn and Sham [8, 9], is based on the electron density, in term of which the solution of the Schrödinger equation could be given a mathematical basis. The basic idea of the DFT is that the 'real' electrons are replaced by 'effective' electrons with the same total density that move as independent particles in the potential generated by the other electrons and ion cores. But the DFT consist only statements about existence and uniqueness. They do not provide a prescription for determining this functional, which, because of the exchange–correlation energy, is not generally known. The widely adopted local density approximation (LDA) [9] presumes that this functional depends only on the local value of electron density, i.e. the inhomogeneous electron systems can be viewed as locally homogeneous. The LDA has proven to be an accurate approximation, with bond lengths and angles calculated to within a few percent of their experimentally measured values.

The direct use of DFT methods to perform full quantum molecular simulations on real application is limited to hundreds of atoms and a simulated time of a few picoseconds. Thus, the next step of coarse graining the problem is to remove the electronic degrees of freedom by imagining the atoms to be held together by some sort of glue or interatomic potential, thereby allowing large-scale atomistic simulations for millions of atoms and a simulated time of nanoseconds. Such simulations, employing either the embedded-atom method (EAM) [10] type or the Finnis–Sinclair type [11] in metals, and the Stillinger–Weber type [12] or the Tersoff type [13] in covalent materials, have been very useful in investigating generic phenomena in simple systems. Empirical potentials involve the fitting of parameters to a predetermined experimental or ab initio database, which includes physical quantities such as the lattice constant, the elastic constants, the vacancy formation energy, and the surface energy.

6.2.2 Molecular Dynamics

Another theoretical framework for studying the dynamics of solids far from equilibrium is classical molecular dynamics (MD). This is the level at which thermal fluctuations and material decohesions are most naturally described. But computational issues constrain molecular dynamics simulations to about thousands of atoms. In linear dimensions, this corresponds to a constraint of about fifty nanometers.

The computation issues around the ab initio methods consist in the derivation of the electron potential energy and the accuracy of the ab initio method depends upon the choice of the basis sets. The MD method, which is widely used in organic chemistry, is a treatment for the many-particle problems. The MD treats the atoms as solid particles and the wave nature of the matters is neglected. Regarding the atoms as classical particles moving on this surface, the quantum mechanics motion governed by the Schrödinger equation is replaced by Newton's equation of classical mechanics. The movements of the particle are described by the variables like displacement, velocity, and acceleration in MD, rather than the wave function/energy in quantum mechanics method. Moreover, MD assumes no mass changes in the system. Equivalently, the number of atoms in the system remains the same.

There are two primary aspects to the practical implementation of molecular dynamics: (i) the time integration scheme of the equations of motion together with the boundary conditions and constraints on the system and (ii) the choice of the potential functions. A proper numerical integration scheme should be both numerically stable and computationally efficient. The numerical integration of the equations of motion is performed by either explicit or implicit methods. The simple Euler scheme is not appropriate for MD simulations because it lacks numerical stability. The Verlet algorithm is popular in MD simulations because it is stable and memory-efficient, and allows for a reasonably large time-step. Another popular integration method for MD simulations is the predictor-corrector scheme, in particular the Gear algorithm. Similar to the numerical integration scheme, proper use of boundary conditions is crucial to obtaining a physically meaningful MD simulation result. The boundary of a simulation cell is usually within close proximity to each simulated atom, because of limited computational power.

The choice of potential for a molecular dynamics simulation is determined by factors such as the bond type, the desired accuracy, transferability and the available computational resources [14]. Potentials can be categorized broadly as (i) pair potentials, (ii) empirical many-body potentials, and (iii) quantum mechanics potentials. Two-body, or pair, potentials, such as the Lennard-Jones [15] and Morse potentials [16], are used for large-scale simulations where computational efficiency is a significant factor, but where a generic description is sufficient, rather than detailed comparisons with a particular material system. For systems where multibody interactions are important, the Stillinger–Weber, Tersoff, and Brenner potentials are often used for covalent materials, and embedded-atom, effective medium and Finnis–Sinclair potentials are common choices for metals. Such potentials are empirical since they are parameterised by fitting either to a set of experimental measurements or to quantum mechanics calculations of representative atomic configurations. However, large local departures from the coordination or bonding used for the parameterisation can take such potentials outside their domain of validity and lead to unreliable results. This has fostered efforts at deriving interatomic potentials directly from quantum mechanics principles.

The long simulation time for MD is due to the processes such as atomic diffusion that is an inherently multi-scale phenomenon. The integration time step must be small enough to capture the dynamics of the vibration modes of the system, with frequencies in the order of 10^{13} s^{-1}. This requires time steps in the femto-second range. But the residence time between hopes is of the order of microseconds, and the interactions responsible for aggregation phenomena occur over a time scale of milliseconds to minutes. This 'time gap' is evident from the trajectories of atoms, which are complex orbitals localized around their initial sites with only rare excursions to neighbouring sites. In favourable circumstances, molecular dynamics simulations can be extended to microseconds.

6.2.3 Monte Carlo Methods

The basic concept of Monte Carlo methods is that the deterministic equations are replaced by stochastic transitions for the slow processes in the system. Physically speaking, the Monte Carlo methods are the ones to represent the probability nature of matters. In the most general form, Monte Carlo methods [17] are stochastic algorithms for exploring phase space, but their implementation for equilibrium and nonequilibrium calculations is somewhat different. In this approach, if the distribution function is normalized to obtain a probability density function (PDF), we can determine the probability that the random variable x is less than an arbitrary x' by integrating the PDF from the minimum value to random. The integral of the PDF is called the cumulative distribution function (CDF). When the CDF is equated to a uniformly distributed random number, the resulting solution for x gives the desired distribution function.

Kinetic Monte Carlo (KMC) methods [18] are algorithms that solve the master equation by accepting and rejecting transitions with probabilities that yield the correct evolution of a nonequilibrium system. The KMC method represents an additional level of abstraction beyond the MD method. The effect of fast dynamical events is taken into account by using stochastic transition rates for slower events. These transition rates are often represented as the product of an attempt rate and the probability of success per attempt, which is taken as an exponential involving the energy barrier to the process. Every event i is assigned a rate r_i:

$$r_i = v_i \exp(-E_i / k_B T) \tag{1}$$

where v_i is a frequency prefactor, typically of the order of a vibrational frequency (10^{13} s^{-1}) for surface processes, E_i is the free energy barrier for the process and T is the absolute temperature. Although the details of the underlying mechanism for kinetic processes are lost, the explicit calculation of atomic trajectories is avoided, so KMC simulations can be performed over real times, running into seconds, hours or days, as required. In essence, the factor v_i in above equation represents the time scale of the fastest process, which is computed explicitly in the MD method, but the exponential factor increases this time scale in the KMC method to that of the actual transitions. The KMC method thereby offers considerable advantages over the MD method, both in terms of the real time over which the simulation evolves, and the number of atoms included in the simulation, because much of the computational overhead in MD is used to evolve the system between rare events.

The construction of a model for a KMC simulation can often benefit from a related classical or quantum molecular dynamics simulation to identify the important physical process and estimate the prefactors and kinetic barriers. The transition rates are depending on the processes of interest and must be determined either by direct calculation, from a quantum mechanics methods or a molecular dynamics simulation, or data extracted from the experiments. The feasibility of performing detailed simulations over experimental time scale allows various parameterisations to be tested and models of kinetic phenomena to be validated. Such simulations play a key role in several multi-scale modelling strategies.

6.2.4 Continuum Mechanics Method

Continuum mechanics equations, typically in the form of deterministic or stochastic partial differential equations, are at the pinnacle of the coarse-graining hierarchy. In a generic boundary value problem (BVP), the deformation of a body subjected to external forces or prescribed displacements is governed by the (i) equilibrium equations, (ii) constitutive equations, (iii) boundary conditions, and (iv) initial conditions.

The 'weak' form of the boundary value problem is obtained when the equilibrium equations and the boundary conditions are combined into the 'principle of virtual work'. This 'weak form' constitutes the basis for obtaining a numerical solution of the deformation problem via, for example, the finite-element method. Thus, in continuum mechanics Lagrangian formulation of a quasistatic BVP, the principle of virtual work is the vehicle by which the global equilibrium equations are obtained.

There are many benefits of the continuum representation of matter phenomena. Foremost among them is the ability to examine macroscopic regions in space over extended periods of time. This is facilitated by extensive libraries of numerical methods for integrating deterministic and stochastic differential equations. The best known of these is the finite element method. This is a general method for solving differential equations whereby the region of interest is tessellated with a uniform or nonuniform mesh determined by contiguous components called 'elements'. The solution of the differential equation is discretized on the mesh nodes, and interpolated within the elements. A partial (respectively, ordinary) differential equation is thereby replaced by a set of coupled ordinary (respectively, algebraic) equations and solved numerically for the values of the solution at the nodal points. The main advantage of the finite element method is flexibility in geometric complexity, material inhomogeneities, anisotropies and plasticity, all within a computationally efficient framework and reliable results.

Complementing the numerical solution of partial differential equations is the vast analytic methodology for identifying asymptotic scaling regimes and performing stability analyses. Additionally, if a continuum equation can be systematically derived from atomistic principles, there is the possibility of discriminating between inherently atomistic effects and those that find a natural expression in a coarse-grained framework. Continuum equations also provide the opportunity for examining the effect of apparently minor modifications to the description of atomistic processes on the coarse-grained evolution of a system, which, in turn, facilitates the systematic reduction of full models to their essential components.

6.3 Handshaking between Approaches

Each of the methods described in the above sections performs well in a particular level of accuracy, as dictated by the successive elimination of the original degrees of freedom. Density function methods provide a quantum mechanics approach of electrons and nuclei, which is appropriate for processes such as crack formation, chemical reactions, and surface kinetics. Where wave nature is not apparently but atomistic resolution is still necessary, molecular dynamics offers many computational advantages over a full density functional calculation. Monte Carlo methods are especially useful for obtaining statistical information about a system from the transition rates between configurations, whether in equilibrium or driven away from equilibrium. Finally, continuum methods provide a reduced description in terms of continuous fields for the coarse-grained evolution of the system.

The fundamental tenet of multi-scale modelling is that information and capability at each scale is systematically incorporated in a manner that transcends the single-scale description. There are two basic strategies for accomplishing this: global-local method and hybrid method. In the global-local approach, information from a calculation over particular length and time scales is used as input into a more coarse-grained method. This approach presumes that the phenomenon of interest can be separated into processes that operate at distinct length and time scales, as stated in St. Venant's principle. In hybrid multi-scale modelling [19-20], these disparate scales are combined within a single hybrid scheme, typically involving atomistic and continuum calculations. The main theoretical challenge is to merge the two descriptions in a manner that avoids any spurious effects due to this heterogeneity. This approach is well suited to the simulation of fracture, where the complex feedback between the atomic-scale interactions and macroscopic stresses pre-empts a clear-cut separation of scales.

6.3.1 Handshaking between Quantum Mechanics and Molecular Dynamics

The quantum mechanics is based on the Schrodinger's equation and the molecular dynamics is based on the Newtonian mechanics. Therefore, the direct handshaking algorithms between the quantum mechanics and the molecular dynamics are hard to achieve. However, the quantum mechanics could provide a valuable input to the molecular dynamics – the potential functions and its related coefficients. The potential functions describe the energy between the chemical bonds, including the covalent bonds, metallic bonds, Hydrogen bonds, and van der Waals bonds. Besides, these chemical bonds always relate to electron energy levels of the outmost effective orbits. In the MD computation, the potential functions govern the accuracy of the simulation results, and a feasible potential function could reduce the computation cost but maintain the high prediction capability. Nevertheless, using the experiment method to determine the coefficients of the potential function at specific temperatures is still hard to acquire. Accordingly, using the quantum mechanics to describe the binding energy and use it as the input to the molecular dynamics computation would be an efficient solution.

The solution by Abraham [21] to this problem is to employ a trick that allows the algorithm to localize that quantum mechanics value for the energy. The trick is to employ the convention that at the edge of the QM region, each "dangling bond" is "tied off" with an artificial univalent atom. To do this, each atom location that lies at the edge of the QM region is assigned an atom with a hybrid set of electronic properties. In the case of silicon, what is needed is something like a silicon atom with one valence electron. These atoms, called "silogens", have some of the properties of silicon, and some of the properties of hydrogen. They produce a bonding energy with other silicon atoms that is equal to the usual Si-Si bond energy, but they are univalent like a hydrogen atom. This is made possible by the fact that the method is semi-empirical, and so fictitious values for matrix elements can simply be assigned at will. This makes it such that the silogen atoms don't energetically interact with their silogen neighbours, which means that the algorithm can localize their quantum mechanics energetic contributions. Finally, once the problem of localization is solved, the algorithm can assign the energy between atoms that span the threshold between regions that is the average of the Stillinger-Weber potential and the energy from the Hamiltonian in the tight-binding approximation. Again, this creates a seamless expression for energy.

6.3.2 Handshaking between Continuum Mechanics and Molecular Dynamics

To understand the continuum mechanics and MD handshaking algorithm, first envision a plane separating the two regions. Next, recall that in the finite element method, the material to be simulated is covered in meshes. One of the original strengths of the finite element method is that the finite element mesh can be varied in size to suit the simulation's needs, allowing us to vary how fine or coarse the computational grid is in different locations. When the finite element method is being used in a multi-scale model, this feature of the FE mesh becomes especially useful. The first step in defining the handshake region is to ensure that as you approach the plane separating the two domains from the finite element side, the mesh elements of the FE domain are made to coincide with the atoms of the MD domain. (Far away from the plane, the mesh will typically get much coarser).

The next step is to calculate the energy of the "handshake region". This is the region between the last mesh point on one side and the first atom on the other. The technique by Abraham's group [21] used is essentially to calculate this energy twice, once from the perspective of FE, and once from the perspective of MD, and then average the two. Doing the first of these involves pretending that the first row of atoms are actually mesh elements, doing the second involves the opposite—pretending that the last row of mesh element are atoms.

Suppose for example that there is an atom on the MD side of the border. It looks over the border and sees a mesh point. For the purpose of the handshaking algorithm, we treat that mesh point as an atom, calculate the energetic interaction according to the Stillinger-Weber potential, and we divide it by two (remember, we are going to be averaging together the two energetic). We do this for every atom/mesh point pair that spans the border. Since the Stillinger-Weber potential also involves triples, we do the same thing for every triple that spans the border (again dividing by two). This is one half of the "handshaking Hamiltonian". The other half comes from the continuum dynamics' energetic. Whenever a mesh point on the continuum mechanics side of the border looks over and sees an atom, it pretends that atom is a mesh point. Thus, from that imaginary point of view, there are complete elements that span the border (some of whose vertices are mesh points that are "really" atoms). Treating the position of that atom as a mesh point position, the algorithm can calculate the strain in that tetrahedron, and integrate over the energy stored in the tetrahedron. Again, since we are averaging together two Hamiltonians, we divide that energy by two.

We now have a seamless expression for the energy stored in the entire region made up of both the continuous solid and the classical atoms. The gradient of this energy function dictates how both the atoms and the mesh points will move from step to step. In this way, the happenings in the continuum mechanics region are automatically communicated to the molecular dynamics region, and vice versa.

6.4 Summary and Outlook

It is important to note that this area of research – multi-scale mechanics in general, and this multi-scale modelling in particular – is in its relative infancy. And while some researchers have had some success with their models, many important challenges are to be solved. It is probably too early

9. Challenges and Future Perspectives

to say whether or not a particular method of simulation will turn out to be the right way to go about predicting and representing the multi-scale behaviour of micro and nano electronics. The three main challenges facing future development of multi-scale modelling are:

- Simulations at finite temperature: Many of the methods which are described are confined to zero temperature, wherein the basic quantity is a Hamiltonian for the system expressed in terms of the appropriate degrees of freedom. In principle, Hamiltonian methods can be extended to equilibrium at finite temperatures by using the free energy, but inherently nonequilibrium situations are fundamentally different and a general approach is not clear. Theoretically speaking, the coefficients of the potential functions only represent the atomic groups at specific temperature (energy level). The temperature (energy level) change of the atomic system due to applying the external loading (external energy) might influence the potential functions, especial the specific temperatures of which could change the physical state of the molecules.

- Time scales accessible by means of molecular dynamics: The bottleneck of the macroscopic time simulations remains because of the small time step in the classical and quantum molecular dynamics. While this may be sidestepped in certain applications, and acceleration strategies are available for particular situations, a general acceleration methodology would have revolutionary implications that would stretch across many disciplines. Theoretically speaking, two candidate techniques could resolve this issue with lack of the particular accuracy. One is to apply a faster loading speed, which is often 100 to 10,000 times faster than the real loading, is applied onto the system to reduce the total simulation. However, the characteristics of the stress/strain wave propagation would be different from the slower one. Second one is to use a heavier atomic group to replace the independent atoms. The clustered atomic group often contains 10 to 1,000 atoms, and the atomic groups are assumed as the rigid body or the homogenous deformable body. However, the mechanical information inside the clustered atomic groups could not be achieved.

- Wave propagation across atomistic and continuum interfaces. In methods with a specific interface between atomistically resolved and (finite element) continuum regions, high frequency modes emanating from the molecular region must be accommodated by the continuum region. For example, in the hybrid method of the molecular dynamics and the finite element method, the critical region would be modelled by molecular dynamics and the rest by finite element methods. To efficiently reduce the computation time, the mesh density of the finite element area would be much coarser than the one in the molecular dynamics area. However, the finite elements are unable to resolve the small wavelengths, which are generated by the atomistic region. Therefore, the high frequency response from the atomistic region would bounce back to the atomistic region, and this could lead to spurious results in the critical region.

7. MULTI-PHYSICS MODELLING

Multi-physics modelling is another grand challenge for mechanics of microelectronics. Due to the strong size effect, multi-technology, heterogenous function integration, microelectronics becomes strongly multi-disciplinary. In principal, it is unlikely to correctly predict the behaviour of microelectronics based on mono discipline knowledge.

Many important failure modes and mechanisms, such as humidity related failures, electrical overstress, stress corrosion, ion transport induced shortcut, MEMS stiction, void formation, combined diffusion, undesired intermetallic growth, material aging, electro/thermal/stress/chemical migrations, etc. are results of strong multidisciplinary interactions. Taking the CMP process as an example, both mechanical and chemical simulations should be conducted simultaneously. For the characterization and modelling of curing behaviour of polymer materials used in microelectronics, the necessary model ingredients should include cure kinetics describing the evolution of the cure rate, cure dependent heat transfer model, cure dependent thermo-mechanical model and cure dependent initial strain models describing the thermal and chemical driven intrinsic strains. Other well-know multi-physics cases are electromechanical couplings, thermo-mechanical couplings, fluid structural interactions, and microfluidics, coming from sensors, actuators, and micro fluidic components and devices. These models combine electroelastic, thermoelastic, microfluidic, and fluid-structural interactions. Multi-physics simulation and verification capabilities to capture the strong interaction effects between different disciplines are essential for microelectronics.

However, due to the strong interaction between multi-physics and multi-scale, complexity of modelling and data description, the large number and wide range of parameters under investigation, as well as the necessity to control and steer the simulation processes, accurate and efficient simulation of multi-physics and multi-scale systems is still not applicable. Commercial FEM tools are all originated from the needs and knowledge of solving mechanical related problems. Most emerging multi-physics software cannot yet deal with complicated engineering reality with strong nonlinear responses. Robust and easy multi-physics combination is still not there. Therefore, more effort should be spent on the development of sophisticated (multi-physics and multi-scale) models, efficient numerical algorithms, and

9. Challenges and Future Perspectives

user interfaces code integration methods, as well as advanced computational techniques (parallel distributed computing for example).

8. ADVANCED SIMULATION TOOLS

FEM is a well-established technique for predicting thermo-mechanical behaviour of product/process. It has made significant progress especially during the last 20 years due to the rapid development of computer hardware and software. However, the commercially available FEM tools are not specifically developed for applications and needs of microelectronics. More effort is needed for modelling tool development, in the directions of both multi-physics and multi-scale. To do so, the following issues deserve special attentions:

- Developing efficient and robust algorithms and solvers. In many applications, in order to obtain accurate results, the complicated geometric effects with high aspect ratio, nonlinear, time and process dependent material behaviour and the complicated process history should all be considered. Despite the rapid development of computer hardware, often, days are still needed for the results of a single run. Therefore, virtual prototyping and virtual qualification cannot be efficiently conducted without more efficient and robust algorithms and solvers.
- Developing efficient and reliable stochastic simulation methods. Since the design and response parameters are all probabilistic in nature, deterministic modelling alone cannot lead to the optimal solutions. Very often, robust designs are targeted, and failure probability is required. Therefore, efficient and reliable stochastic simulation methods and tools should be further developed.
- The ultimate requirements for the advanced simulation tools being able to deal with multi-physics and multi-scale applications are physically meaningful, computationally efficient, and whenever possible, mathematically consistent.

9. CONCLUSIONS

As mentioned by Einstein, "any intelligent fool can make things bigger, more complex and more violent. It takes a touch of genius and a lot of courage to move in the opposite direction". Despite the increasing awareness and progresses of mechanics of microelectronics achieved in recent years, the currently available knowledge, theories, methodologies, and tools are still lacking behind the rapid development needs of microelectronic technologies. The gap is getting larger. Therefore, it is our hope that the listed challenges in this chapter will draw more attention and effort from worldwide academic communities. Their solutions will definitely enable microelectronic industries to continuously realize faster technology innovation. To be able to do so, structured partnership and strengthened cooperation between industry and academia are vital, in order to speed up new knowledge creation of mechanics of microelectronics, because:

- Industry is leading the technology development. The development of mechanics of microelectronics should match with the existing and anticipated roadmap of microelectronics technologies.
- It is too time and money consuming to take separate actions by industries and academia. By joining the forces, sharing the resource and knowledge, and aligning roadmap and effort, more efficient and effective results can be achieved.

10. REFERENCES

[1] G. Q. Zhang, F. van Roosmalen, M. Graef, The paradigm of More than Moore, Proc. ICEPT 2005.
[2] International Technology Roadmap of Semiconductors (ITRS), 2005.
[3] Strategic Research Agenda of nanoelectronics, Eniac, first edition, 2005.
[4] G. Q. Zhang and W. D. van Driel, Lecture notes: Virtual thermo-mechanical prototyping and qualification of microelectronics and Microsystems, Delft University of Technology, 2006.
[5] pww.qas.sc.philips.com
[6] D. D. Vvedensky, Multi-scale modelling of nanostructures, J. Phys.: Condens. Matter, Vol. 16, pp. R1537–R1576, 2004.
[7] N. M. Ghoniem, E. P. Busso, N. Kioussis and H. Huang, Multi-scale modelling of nanomechanics and micromechanics: an overview, Philosophical Magazine, Vol. 83, pp. 3475–3528, 2003.
[8] P. Hohenberg and W. Kohn, Inhomogeneous electron gas, Phys. Rev. Vol. 136, pp. B864–71, 1964.
[9] W. Kohn and L. J. Sham, Self-consistent equations including exchange and correlation effects, Phys. Rev., Vol. 140, pp. A1133–8, 1965.
[10] M. S. Daw and M. I. Baskes, Semiempirical, quantum mechanical calculation of hydrogen embrittlement in metals, Phys. Rev. Lett., Vol. 50, pp. 1285–8, 1983.
[11] M. W. Finnis and J. E. Sinclair, A simple empirical N-body potential for transition metals, Phil. Mag. A 50, 45–55, 1984.
[12] F. H. Stillinger and T. A. Weber, Computer simulation of local order in condensed phases of silicon, Phys. Rev. B, Vol. 31, pp. 5262–71, 1985.
[13] J. Tersoff, New empirical model for the structural properties of silicon, Phys. Rev. Lett., Vol. 56, pp. 632–5, 1986.
[14] A. E. Carlsson, Beyond pair potentials for transition-metals and semiconductors Solid State Physics: Advances in Research and Applications, Vol. 43, ed. H Ehrenreich and D Turnbull (New York: Academic), pp. 1–91, 1990.

9. Challenges and Future Perspectives

[15] J. E. Jones, On the determination of molecular fields: II. from the equation of state of a gas, Proc. R. Soc. Vol. A 106, pp. 463–77, 1924.
[16] P. M. Morse, Diatomic molecules according to the wave mechanics: II. Vibrational levels, Phys. Rev., Vol. 34, pp. 57–64, 1929.
[17] D. D. McCracken, The Monte Carlo method, Sci. Am., Vol. 192, pp. 90–5, 1955.
[18] N. G. Van Kampen, Stochastic Processes in Physics and Chemistry, North-Holland, Amsterdam, 1981.
[19] S. P. Xiao, T. Belytschko, A bridging domain method for coupling continua with molecular dynamics, Computer Methods in Applied Mechanics and Engineering, Vol. 193, pp. 1645–1669, 2004.
[20] D. Srivastava and S. N. Atluri, Computational nanotechnology: a current perspective, CMES: Computer Modelling in Engineering & Science, Vol. 3, No. 5, pp. 531–538, 2002.
[21] J. Q. Broughton, F. F. Abraham, N. Bernstein and E. Kaxiras, Concurrent coupling of length scales: methodology and application, Phys. Rev. B, Vol. 60, pp. 2391–403, 1999.

Mechanics

SOLID MECHANICS AND ITS APPLICATIONS
Series Editor: G.M.L. Gladwell

Aims and Scope of the Series

The fundamental questions arising in mechanics are: *Why?*, *How?*, and *How much?* The aim of this series is to provide lucid accounts written by authoritative researchers giving vision and insight in answering these questions on the subject of mechanics as it relates to solids. The scope of the series covers the entire spectrum of solid mechanics. Thus it includes the foundation of mechanics; variational formulations; computational mechanics; statics, kinematics and dynamics of rigid and elastic bodies; vibrations of solids and structures; dynamical systems and chaos; the theories of elasticity, plasticity and viscoelasticity; composite materials; rods, beams, shells and membranes; structural control and stability; soils, rocks and geomechanics; fracture; tribology; experimental mechanics; biomechanics and machine design.

1. R.T. Haftka, Z. Gürdal and M.P. Kamat: *Elements of Structural Optimization.* 2nd rev.ed., 1990 ISBN 0-7923-0608-2
2. J.J. Kalker: *Three-Dimensional Elastic Bodies in Rolling Contact.* 1990 ISBN 0-7923-0712-7
3. P. Karasudhi: *Foundations of Solid Mechanics.* 1991 ISBN 0-7923-0772-0
4. Not published
5. Not published.
6. J.F. Doyle: *Static and Dynamic Analysis of Structures.* With an Emphasis on Mechanics and Computer Matrix Methods. 1991 ISBN 0-7923-1124-8; Pb 0-7923-1208-2
7. O.O. Ochoa and J.N. Reddy: *Finite Element Analysis of Composite Laminates.* ISBN 0-7923-1125-6
8. M.H. Aliabadi and D.P. Rooke: *Numerical Fracture Mechanics.* ISBN 0-7923-1175-2
9. J. Angeles and C.S. López-Cajún: *Optimization of Cam Mechanisms.* 1991 ISBN 0-7923-1355-0
10. D.E. Grierson, A. Franchi and P. Riva (eds.): *Progress in Structural Engineering.* 1991 ISBN 0-7923-1396-8
11. R.T. Haftka and Z. Gürdal: *Elements of Structural Optimization.* 3rd rev. and exp. ed. 1992 ISBN 0-7923-1504-9; Pb 0-7923-1505-7
12. J.R. Barber: *Elasticity.* 1992 ISBN 0-7923-1609-6; Pb 0-7923-1610-X
13. H.S. Tzou and G.L. Anderson (eds.): *Intelligent Structural Systems.* 1992 ISBN 0-7923-1920-6
14. E.E. Gdoutos: *Fracture Mechanics.* An Introduction. 1993 ISBN 0-7923-1932-X
15. J.P. Ward: *Solid Mechanics.* An Introduction. 1992 ISBN 0-7923-1949-4
16. M. Farshad: *Design and Analysis of Shell Structures.* 1992 ISBN 0-7923-1950-8
17. H.S. Tzou and T. Fukuda (eds.): *Precision Sensors, Actuators and Systems.* 1992 ISBN 0-7923-2015-8
18. J.R. Vinson: *The Behavior of Shells Composed of Isotropic and Composite Materials.* 1993 ISBN 0-7923-2113-8
19. H.S. Tzou: *Piezoelectric Shells.* Distributed Sensing and Control of Continua. 1993 ISBN 0-7923-2186-3
20. W. Schiehlen (ed.): *Advanced Multibody System Dynamics.* Simulation and Software Tools. 1993 ISBN 0-7923-2192-8
21. C.-W. Lee: *Vibration Analysis of Rotors.* 1993 ISBN 0-7923-2300-9
22. D.R. Smith: *An Introduction to Continuum Mechanics.* 1993 ISBN 0-7923-2454-4
23. G.M.L. Gladwell: *Inverse Problems in Scattering.* An Introduction. 1993 ISBN 0-7923-2478-1

Mechanics

SOLID MECHANICS AND ITS APPLICATIONS
Series Editor: G.M.L. Gladwell

24. G. Prathap: *The Finite Element Method in Structural Mechanics.* 1993 ISBN 0-7923-2492-7
25. J. Herskovits (ed.): *Advances in Structural Optimization.* 1995 ISBN 0-7923-2510-9
26. M.A. González-Palacios and J. Angeles: *Cam Synthesis.* 1993 ISBN 0-7923-2536-2
27. W.S. Hall: *The Boundary Element Method.* 1993 ISBN 0-7923-2580-X
28. J. Angeles, G. Hommel and P. Kovács (eds.): *Computational Kinematics.* 1993
    ISBN 0-7923-2585-0
29. A. Curnier: *Computational Methods in Solid Mechanics.* 1994 ISBN 0-7923-2761-6
30. D.A. Hills and D. Nowell: *Mechanics of Fretting Fatigue.* 1994 ISBN 0-7923-2866-3
31. B. Tabarrok and F.P.J. Rimrott: *Variational Methods and Complementary Formulations in Dynamics.* 1994 ISBN 0-7923-2923-6
32. E.H. Dowell (ed.), E.F. Crawley, H.C. Curtiss Jr., D.A. Peters, R. H. Scanlan and F. Sisto: *A Modern Course in Aeroelasticity.* Third Revised and Enlarged Edition. 1995
    ISBN 0-7923-2788-8; Pb: 0-7923-2789-6
33. A. Preumont: *Random Vibration and Spectral Analysis.* 1994 ISBN 0-7923-3036-6
34. J.N. Reddy (ed.): *Mechanics of Composite Materials.* Selected works of Nicholas J. Pagano. 1994 ISBN 0-7923-3041-2
35. A.P.S. Selvadurai (ed.): *Mechanics of Poroelastic Media.* 1996 ISBN 0-7923-3329-2
36. Z. Mróz, D. Weichert, S. Dorosz (eds.): *Inelastic Behaviour of Structures under Variable Loads.* 1995 ISBN 0-7923-3397-7
37. R. Pyrz (ed.): *IUTAM Symposium on Microstructure-Property Interactions in Composite Materials.* Proceedings of the IUTAM Symposium held in Aalborg, Denmark. 1995
    ISBN 0-7923-3427-2
38. M.I. Friswell and J.E. Mottershead: *Finite Element Model Updating in Structural Dynamics.* 1995 ISBN 0-7923-3431-0
39. D.F. Parker and A.H. England (eds.): *IUTAM Symposium on Anisotropy, Inhomogeneity and Nonlinearity in Solid Mechanics.* Proceedings of the IUTAM Symposium held in Nottingham, U.K. 1995 ISBN 0-7923-3594-5
40. J.-P. Merlet and B. Ravani (eds.): *Computational Kinematics '95.* 1995 ISBN 0-7923-3673-9
41. L.P. Lebedev, I.I. Vorovich and G.M.L. Gladwell: *Functional Analysis.* Applications in Mechanics and Inverse Problems. 1996 ISBN 0-7923-3849-9
42. J. Menčik: *Mechanics of Components with Treated or Coated Surfaces.* 1996
    ISBN 0-7923-3700-X
43. D. Bestle and W. Schiehlen (eds.): *IUTAM Symposium on Optimization of Mechanical Systems.* Proceedings of the IUTAM Symposium held in Stuttgart, Germany. 1996
    ISBN 0-7923-3830-8
44. D.A. Hills, P.A. Kelly, D.N. Dai and A.M. Korsunsky: *Solution of Crack Problems.* The Distributed Dislocation Technique. 1996 ISBN 0-7923-3848-0
45. V.A. Squire, R.J. Hosking, A.D. Kerr and P.J. Langhorne: *Moving Loads on Ice Plates.* 1996
    ISBN 0-7923-3953-3
46. A. Pineau and A. Zaoui (eds.): *IUTAM Symposium on Micromechanics of Plasticity and Damage of Multiphase Materials.* Proceedings of the IUTAM Symposium held in Sèvres, Paris, France. 1996 ISBN 0-7923-4188-0
47. A. Naess and S. Krenk (eds.): *IUTAM Symposium on Advances in Nonlinear Stochastic Mechanics.* Proceedings of the IUTAM Symposium held in Trondheim, Norway. 1996
    ISBN 0-7923-4193-7
48. D. Ieşan and A. Scalia: *Thermoelastic Deformations.* 1996 ISBN 0-7923-4230-5

Mechanics

SOLID MECHANICS AND ITS APPLICATIONS
Series Editor: G.M.L. Gladwell

49. J.R. Willis (ed.): *IUTAM Symposium on Nonlinear Analysis of Fracture.* Proceedings of the IUTAM Symposium held in Cambridge, U.K. 1997   ISBN 0-7923-4378-6
50. A. Preumont: *Vibration Control of Active Structures.* An Introduction. 1997
ISBN 0-7923-4392-1
51. G.P. Cherepanov: *Methods of Fracture Mechanics: Solid Matter Physics.* 1997
ISBN 0-7923-4408-1
52. D.H. van Campen (ed.): *IUTAM Symposium on Interaction between Dynamics and Control in Advanced Mechanical Systems.* Proceedings of the IUTAM Symposium held in Eindhoven, The Netherlands. 1997   ISBN 0-7923-4429-4
53. N.A. Fleck and A.C.F. Cocks (eds.): *IUTAM Symposium on Mechanics of Granular and Porous Materials.* Proceedings of the IUTAM Symposium held in Cambridge, U.K. 1997
ISBN 0-7923-4553-3
54. J. Roorda and N.K. Srivastava (eds.): *Trends in Structural Mechanics.* Theory, Practice, Education. 1997   ISBN 0-7923-4603-3
55. Yu.A. Mitropolskii and N. Van Dao: *Applied Asymptotic Methods in Nonlinear Oscillations.* 1997   ISBN 0-7923-4605-X
56. C. Guedes Soares (ed.): *Probabilistic Methods for Structural Design.* 1997
ISBN 0-7923-4670-X
57. D. François, A. Pineau and A. Zaoui: *Mechanical Behaviour of Materials.* Volume I: Elasticity and Plasticity. 1998   ISBN 0-7923-4894-X
58. D. François, A. Pineau and A. Zaoui: *Mechanical Behaviour of Materials.* Volume II: Viscoplasticity, Damage, Fracture and Contact Mechanics. 1998   ISBN 0-7923-4895-8
59. L.T. Tenek and J. Argyris: *Finite Element Analysis for Composite Structures.* 1998
ISBN 0-7923-4899-0
60. Y.A. Bahei-El-Din and G.J. Dvorak (eds.): *IUTAM Symposium on Transformation Problems in Composite and Active Materials.* Proceedings of the IUTAM Symposium held in Cairo, Egypt. 1998   ISBN 0-7923-5122-3
61. I.G. Goryacheva: *Contact Mechanics in Tribology.* 1998   ISBN 0-7923-5257-2
62. O.T. Bruhns and E. Stein (eds.): *IUTAM Symposium on Micro- and Macrostructural Aspects of Thermoplasticity.* Proceedings of the IUTAM Symposium held in Bochum, Germany. 1999
ISBN 0-7923-5265-3
63. F.C. Moon: *IUTAM Symposium on New Applications of Nonlinear and Chaotic Dynamics in Mechanics.* Proceedings of the IUTAM Symposium held in Ithaca, NY, USA. 1998
ISBN 0-7923-5276-9
64. R. Wang: *IUTAM Symposium on Rheology of Bodies with Defects.* Proceedings of the IUTAM Symposium held in Beijing, China. 1999   ISBN 0-7923-5297-1
65. Yu.I. Dimitrienko: *Thermomechanics of Composites under High Temperatures.* 1999
ISBN 0-7923-4899-0
66. P. Argoul, M. Frémond and Q.S. Nguyen (eds.): *IUTAM Symposium on Variations of Domains and Free-Boundary Problems in Solid Mechanics.* Proceedings of the IUTAM Symposium held in Paris, France. 1999   ISBN 0-7923-5450-8
67. F.J. Fahy and W.G. Price (eds.): *IUTAM Symposium on Statistical Energy Analysis.* Proceedings of the IUTAM Symposium held in Southampton, U.K. 1999   ISBN 0-7923-5457-5
68. H.A. Mang and F.G. Rammerstorfer (eds.): *IUTAM Symposium on Discretization Methods in Structural Mechanics.* Proceedings of the IUTAM Symposium held in Vienna, Austria. 1999
ISBN 0-7923-5591-1

Mechanics

SOLID MECHANICS AND ITS APPLICATIONS
Series Editor: G.M.L. Gladwell

69. P. Pedersen and M.P. Bendsøe (eds.): *IUTAM Symposium on Synthesis in Bio Solid Mechanics.* Proceedings of the IUTAM Symposium held in Copenhagen, Denmark. 1999
ISBN 0-7923-5615-2
70. S.K. Agrawal and B.C. Fabien: *Optimization of Dynamic Systems.* 1999
ISBN 0-7923-5681-0
71. A. Carpinteri: *Nonlinear Crack Models for Nonmetallic Materials.* 1999
ISBN 0-7923-5750-7
72. F. Pfeifer (ed.): *IUTAM Symposium on Unilateral Multibody Contacts.* Proceedings of the IUTAM Symposium held in Munich, Germany. 1999 ISBN 0-7923-6030-3
73. E. Lavendelis and M. Zakrzhevsky (eds.): *IUTAM/IFToMM Symposium on Synthesis of Nonlinear Dynamical Systems.* Proceedings of the IUTAM/IFToMM Symposium held in Riga, Latvia. 2000 ISBN 0-7923-6106-7
74. J.-P. Merlet: *Parallel Robots.* 2000 ISBN 0-7923-6308-6
75. J.T. Pindera: *Techniques of Tomographic Isodyne Stress Analysis.* 2000 ISBN 0-7923-6388-4
76. G.A. Maugin, R. Drouot and F. Sidoroff (eds.): *Continuum Thermomechanics. The Art and Science of Modelling Material Behaviour.* 2000 ISBN 0-7923-6407-4
77. N. Van Dao and E.J. Kreuzer (eds.): *IUTAM Symposium on Recent Developments in Non-linear Oscillations of Mechanical Systems.* 2000 ISBN 0-7923-6470-8
78. S.D. Akbarov and A.N. Guz: *Mechanics of Curved Composites.* 2000 ISBN 0-7923-6477-5
79. M.B. Rubin: *Cosserat Theories: Shells, Rods and Points.* 2000 ISBN 0-7923-6489-9
80. S. Pellegrino and S.D. Guest (eds.): *IUTAM-IASS Symposium on Deployable Structures: Theory and Applications.* Proceedings of the IUTAM-IASS Symposium held in Cambridge, U.K., 6–9 September 1998. 2000 ISBN 0-7923-6516-X
81. A.D. Rosato and D.L. Blackmore (eds.): *IUTAM Symposium on Segregation in Granular Flows.* Proceedings of the IUTAM Symposium held in Cape May, NJ, U.S.A., June 5–10, 1999. 2000 ISBN 0-7923-6547-X
82. A. Lagarde (ed.): *IUTAM Symposium on Advanced Optical Methods and Applications in Solid Mechanics.* Proceedings of the IUTAM Symposium held in Futuroscope, Poitiers, France, August 31–September 4, 1998. 2000 ISBN 0-7923-6604-2
83. D. Weichert and G. Maier (eds.): *Inelastic Analysis of Structures under Variable Loads.* Theory and Engineering Applications. 2000 ISBN 0-7923-6645-X
84. T.-J. Chuang and J.W. Rudnicki (eds.): *Multiscale Deformation and Fracture in Materials and Structures.* The James R. Rice 60th Anniversary Volume. 2001 ISBN 0-7923-6718-9
85. S. Narayanan and R.N. Iyengar (eds.): *IUTAM Symposium on Nonlinearity and Stochastic Structural Dynamics.* Proceedings of the IUTAM Symposium held in Madras, Chennai, India, 4–8 January 1999 ISBN 0-7923-6733-2
86. S. Murakami and N. Ohno (eds.): *IUTAM Symposium on Creep in Structures.* Proceedings of the IUTAM Symposium held in Nagoya, Japan, 3-7 April 2000. 2001 ISBN 0-7923-6737-5
87. W. Ehlers (ed.): *IUTAM Symposium on Theoretical and Numerical Methods in Continuum Mechanics of Porous Materials.* Proceedings of the IUTAM Symposium held at the University of Stuttgart, Germany, September 5-10, 1999. 2001 ISBN 0-7923-6766-9
88. D. Durban, D. Givoli and J.G. Simmonds (eds.): *Advances in the Mechanis of Plates and Shells* The Avinoam Libai Anniversary Volume. 2001 ISBN 0-7923-6785-5
89. U. Gabbert and H.-S. Tzou (eds.): *IUTAM Symposium on Smart Structures and Structonic Systems.* Proceedings of the IUTAM Symposium held in Magdeburg, Germany, 26–29 September 2000. 2001 ISBN 0-7923-6968-8

Mechanics

SOLID MECHANICS AND ITS APPLICATIONS
Series Editor: G.M.L. Gladwell

90. Y. Ivanov, V. Cheshkov and M. Natova: *Polymer Composite Materials – Interface Phenomena & Processes.* 2001
ISBN 0-7923-7008-2
91. R.C. McPhedran, L.C. Botten and N.A. Nicorovici (eds.): *IUTAM Symposium on Mechanical and Electromagnetic Waves in Structured Media.* Proceedings of the IUTAM Symposium held in Sydney, NSW, Australia, 18-22 Januari 1999. 2001
ISBN 0-7923-7038-4
92. D.A. Sotiropoulos (ed.): *IUTAM Symposium on Mechanical Waves for Composite Structures Characterization.* Proceedings of the IUTAM Symposium held in Chania, Crete, Greece, June 14-17, 2000. 2001
ISBN 0-7923-7164-X
93. V.M. Alexandrov and D.A. Pozharskii: *Three-Dimensional Contact Problems.* 2001
ISBN 0-7923-7165-8
94. J.P. Dempsey and H.H. Shen (eds.): *IUTAM Symposium on Scaling Laws in Ice Mechanics and Ice Dynamics.* Proceedings of the IUTAM Symposium held in Fairbanks, Alaska, U.S.A., 13-16 June 2000. 2001
ISBN 1-4020-0171-1
95. U. Kirsch: *Design-Oriented Analysis of Structures.* A Unified Approach. 2002
ISBN 1-4020-0443-5
96. A. Preumont: *Vibration Control of Active Structures.* An Introduction (2nd Edition). 2002
ISBN 1-4020-0496-6
97. B.L. Karihaloo (ed.): *IUTAM Symposium on Analytical and Computational Fracture Mechanics of Non-Homogeneous Materials.* Proceedings of the IUTAM Symposium held in Cardiff, U.K., 18-22 June 2001. 2002
ISBN 1-4020-0510-5
98. S.M. Han and H. Benaroya: *Nonlinear and Stochastic Dynamics of Compliant Offshore Structures.* 2002
ISBN 1-4020-0573-3
99. A.M. Linkov: *Boundary Integral Equations in Elasticity Theory.* 2002
ISBN 1-4020-0574-1
100. L.P. Lebedev, I.I. Vorovich and G.M.L. Gladwell: *Functional Analysis.* Applications in Mechanics and Inverse Problems (2nd Edition). 2002
ISBN 1-4020-0667-5; Pb: 1-4020-0756-6
101. Q.P. Sun (ed.): *IUTAM Symposium on Mechanics of Martensitic Phase Transformation in Solids.* Proceedings of the IUTAM Symposium held in Hong Kong, China, 11-15 June 2001. 2002
ISBN 1-4020-0741-8
102. M.L. Munjal (ed.): *IUTAM Symposium on Designing for Quietness.* Proceedings of the IUTAM Symposium held in Bangkok, India, 12-14 December 2000. 2002
ISBN 1-4020-0765-5
103. J.A.C. Martins and M.D.P. Monteiro Marques (eds.): *Contact Mechanics.* Proceedings of the 3rd Contact Mechanics International Symposium, Praia da Consolação, Peniche, Portugal, 17-21 June 2001. 2002
ISBN 1-4020-0811-2
104. H.R. Drew and S. Pellegrino (eds.): *New Approaches to Structural Mechanics, Shells and Biological Structures.* 2002
ISBN 1-4020-0862-7
105. J.R. Vinson and R.L. Sierakowski: *The Behavior of Structures Composed of Composite Materials.* Second Edition. 2002
ISBN 1-4020-0904-6
106. Not yet published.
107. J.R. Barber: *Elasticity.* Second Edition. 2002   ISBN Hb 1-4020-0964-X; Pb 1-4020-0966-6
108. C. Miehe (ed.): *IUTAM Symposium on Computational Mechanics of Solid Materials at Large Strains.* Proceedings of the IUTAM Symposium held in Stuttgart, Germany, 20-24 August 2001. 2003
ISBN 1-4020-1170-9

Mechanics

SOLID MECHANICS AND ITS APPLICATIONS
Series Editor: G.M.L. Gladwell

109. P. Ståhle and K.G. Sundin (eds.): *IUTAM Symposium on Field Analyses for Determination of Material Parameters – Experimental and Numerical Aspects.* Proceedings of the IUTAM Symposium held in Abisko National Park, Kiruna, Sweden, July 31 – August 4, 2000. 2003
ISBN 1-4020-1283-7
110. N. Sri Namachchivaya and Y.K. Lin (eds.): *IUTAM Symposium on Nonlinear Stochastic Dynamics.* Proceedings of the IUTAM Symposium held in Monticello, IL, USA, 26 – 30 August, 2000. 2003
ISBN 1-4020-1471-6
111. H. Sobieckzky (ed.): *IUTAM Symposium Transsonicum IV.* Proceedings of the IUTAM Symposium held in Göttingen, Germany, 2–6 September 2002, 2003 ISBN 1-4020-1608-5
112. J.-C. Samin and P. Fisette: *Symbolic Modeling of Multibody Systems.* 2003
ISBN 1-4020-1629-8
113. A.B. Movchan (ed.): *IUTAM Symposium on Asymptotics, Singularities and Homogenisation in Problems of Mechanics.* Proceedings of the IUTAM Symposium held in Liverpool, United Kingdom, 8-11 July 2002. 2003
ISBN 1-4020-1780-4
114. S. Ahzi, M. Cherkaoui, M.A. Khaleel, H.M. Zbib, M.A. Zikry and B. LaMatina (eds.): *IUTAM Symposium on Multiscale Modeling and Characterization of Elastic-Inelastic Behavior of Engineering Materials.* Proceedings of the IUTAM Symposium held in Marrakech, Morocco, 20-25 October 2002. 2004
ISBN 1-4020-1861-4
115. H. Kitagawa and Y. Shibutani (eds.): *IUTAM Symposium on Mesoscopic Dynamics of Fracture Process and Materials Strength.* Proceedings of the IUTAM Symposium held in Osaka, Japan, 6-11 July 2003. Volume in celebration of Professor Kitagawa's retirement. 2004
ISBN 1-4020-2037-6
116. E.H. Dowell, R.L. Clark, D. Cox, H.C. Curtiss, Jr., K.C. Hall, D.A. Peters, R.H. Scanlan, E. Simiu, F. Sisto and D. Tang: *A Modern Course in Aeroelasticity.* 4th Edition, 2004
ISBN 1-4020-2039-2
117. T. Burczyński and A. Osyczka (eds.): *IUTAM Symposium on Evolutionary Methods in Mechanics.* Proceedings of the IUTAM Symposium held in Cracow, Poland, 24-27 September 2002. 2004
ISBN 1-4020-2266-2
118. D. Ieşan: *Thermoelastic Models of Continua.* 2004 ISBN 1-4020-2309-X
119. G.M.L. Gladwell: *Inverse Problems in Vibration.* Second Edition. 2004 ISBN 1-4020-2670-6
120. J.R. Vinson: *Plate and Panel Structures of Isotropic, Composite and Piezoelectric Materials, Including Sandwich Construction.* 2005
ISBN 1-4020-3110-6
121. Forthcoming
122. G. Rega and F. Vestroni (eds.): *IUTAM Symposium on Chaotic Dynamics and Control of Systems and Processes in Mechanics.* Proceedings of the IUTAM Symposium held in Rome, Italy, 8–13 June 2003. 2005
ISBN 1-4020-3267-6
123. E.E. Gdoutos: *Fracture Mechanics. An Introduction.* 2nd edition. 2005 ISBN 1-4020-3267-6
124. M.D. Gilchrist (ed.): *IUTAM Symposium on Impact Biomechanics from Fundamental Insights to Applications.* 2005
ISBN 1-4020-3795-3
125. J.M. Huyghe, P.A.C. Raats and S. C. Cowin (eds.): *IUTAM Symposium on Physicochemical and Electromechanical Interactions in Porous Media.* 2005 ISBN 1-4020-3864-X
126. H. Ding, W. Chen and L. Zhang: *Elasticity of Transversely Isotropic Materials.* 2005
ISBN 1-4020-4033-4
127. W. Yang (ed): *IUTAM Symposium on Mechanics and Reliability of Actuating Materials.* Proceedings of the IUTAM Symposium held in Beijing, China, 1–3 September 2004. 2005
ISBN 1-4020-4131-6

Mechanics

SOLID MECHANICS AND ITS APPLICATIONS
Series Editor: G.M.L. Gladwell

128. J.-P. Merlet: *Parallel Robots.* 2006   ISBN 1-4020-4132-2
129. G.E.A. Meier and K.R. Sreenivasan (eds.): *IUTAM Symposium on One Hundred Years of Boundary Layer Research.* Proceedings of the IUTAM Symposium held at DLR-Göttingen, Germany, August 12–14, 2004. 2006   ISBN 1-4020-4149-7
130. H. Ulbrich and W. Günthner (eds.): *IUTAM Symposium on Vibration Control of Nonlinear Mechanisms and Structures.* 2006   ISBN 1-4020-4160-8
131. L. Librescu and O. Song: *Thin-Walled Composite Beams.* Theory and Application. 2006
     ISBN 1-4020-3457-1
132. G. Ben-Dor, A. Dubinsky and T. Elperin: *Applied High-Speed Plate Penetration Dynamics.* 2006   ISBN 1-4020-3452-0
133. X. Markenscoff and A. Gupta (eds.): *Collected Works of J. D. Eshelby.* Mechanics and Defects and Heterogeneities. 2006   ISBN 1-4020-4416-X
134. R.W. Snidle and H.P. Evans (eds.): *IUTAM Symposium on Elastohydrodynamics and Microelastohydrodynamics.* Proceedings of the IUTAM Symposium held in Cardiff, UK, 1–3 September, 2004. 2006   ISBN 1-4020-4532-8
135. T. Sadowski (ed.): *IUTAM Symposium on Multiscale Modelling of Damage and Fracture Processes in Composite Materials.* Proceedings of the IUTAM Symposium held in Kazimierz Dolny, Poland, 23–27 May 2005. 2006   ISBN 1-4020-4565-4
136. A. Preumont: *Mechatronics.* Dynamics of Electromechanical and Piezoelectric Systems. 2006
     ISBN 1-4020-4695-2
137. M.P. Bendsøe, N. Olhoff and O. Sigmund (eds.): *IUTAM Symposium on Topological Design Optimization of Structures, Machines and Materials.* Status and Perspectives. 2006
     ISBN 1-4020-4729-0
138. A. Klarbring: *Models of Mechanics.* 2006   ISBN 1-4020-4834-3
139. H.D. Bui: *Fracture Mechanics.* Inverse Problems and Solutions. 2006 ISBN 1-4020-4836-X
140. M. Pandey, W.-C. Xie and L. Xu (eds.): *Advances in Engineering Structures, Mechanics and Construction.* Proceedings of an International Conference on Advances in Engineering Structures, Mechanics & Construction, held in Waterloo, Ontario, Canada, May 14–17, 2006. 2006   ISBN 1-4020-4890-4
141. G.Q. Zhang, W.D. van Driel and X.J. Fan: *Mechanics of Microelectronics.* 2006
     ISBN 1-4020-4934-X
142. Q.P. Sun and P. Tong (eds.): *IUTAM Symposium on Size Effects on Material and Structural Behavior at Micron- and Nano-Scales.* Proceedings of the IUTAM Symposium held in Hong Kong, China, 31 May–4 June, 2004. 2006   ISBN 1-4020-4945-5

springer.com